Operator Theory: Advances and
Applications
Vol. 150

Limit Operators and Their Applications in Operator Theory

Vladimir Rabinovich
Steffen Roch
Bernd Silbermann

Springer Basel AG

Authors:

Vladimir Rabinovich
Instituto Politécnico Nacional
ESIME Zacatenco
Avenida IPN
Mexico, D.F. 07738
Mexico
vladimir_rabinovich@hotmail.com

Steffen Roch
Department of Mathematics
Technical University of Darmstadt
Schlossgartenstrasse 7
64289 Darmstadt
Germany
roch@mathematik.tu-darmstadt.de

Bernd Silbermann
Department of Mathematics
Technical University of Chemnitz
09107 Chemnitz
Germany
silbermn@mathematik.tu-chemnitz.de

2000 Mathematics Subject Classification 47L80; 35S05, 47A53, 47G30, 65R20

ISBN 978-3-0348-9619-1 **ISBN 978-3-0348-7911-8 (eBook)**
DOI 10.1007/978-3-0348-7911-8

A CIP catalogue record for this book is available from the
Library of Congress, Washington D.C., USA

Bibliographic information published by Die Deutsche Bibliothek
Die Deutsche Bibliothek lists this publication in the Deutsche Nationalbibliografie; detailed bibliographic data is available in the Internet at <http://dnb.ddb.de>.

Originally published by Birkhäuser Verlag AG 2004
Softcover reprint of the hardcover 1st edition 2004

Part of Springer Science+Business Media
Printed on acid-free paper produced from chlorine-free pulp. TCF ∞
Cover design: Heinz Hiltbrunner, Basel

ISBN 3-7643-7081-5

9 8 7 6 5 4 3 2 1

www.birkhäuser-science.com

Contents

Preface

This text has two goals. It describes a topic: band and band-dominated operators and their Fredholm theory, and it introduces a method to study this topic: limit operators.

Band-dominated operators. Let $H = l^2(\mathbb{Z})$ be the Hilbert space of all squared summable functions $x : \mathbb{Z} \to \mathbb{C}$, $i \mapsto x_i$ provided with the norm

$$\|x\|^2 := \sum_{i \in \mathbb{Z}} |x_i|^2.$$

It is often convenient to think of the elements x of $l^2(\mathbb{Z})$ as two-sided infinite sequences $(x_i)_{i \in \mathbb{Z}}$.

The standard basis of $l^2(\mathbb{Z})$ is the family of sequences

$$(e_i)_{i \in \mathbb{Z}} \quad \text{where} \quad e_i = (\ldots, 0, 0, 1, 0, 0, \ldots)$$

with the 1 standing at the ith place. Every bounded linear operator A on H can be described by a two-sided infinite matrix $(a_{ij})_{i,j \in \mathbb{Z}}$ with respect to this basis, where $a_{ij} = \langle Ae_j, e_i \rangle$. The band operators on H are just the operators with a matrix representation of finite band-width, i.e., the operators for which $a_{ij} = 0$ whenever $|i-j| > k$ for some k. Operators which are in the norm closure of the algebra of all band operators are called band-dominated. Needless to say that band and band-dominated operators appear in numerous branches of mathematics. Archetypal examples come from discretizations of partial differential operators.

It is easy to check that every band operator can be uniquely written as a finite sum $\sum d_k V_k$ where the d_k are multiplication operators (i.e., they are given by a diagonal matrix with respect to the standard basis), and where the V_k are the shift operators $e_j \mapsto e_{j+k}$. Conversely, every finite sum of this form is a band operator. This equivalence allows us to think of band operators as being composed of two kinds of generators – multiplication operators and shift operators.

Fredholmness. We will be mainly concerned with the Fredholm properties of band-dominated operators. A bounded linear operator A on H is called a Fredholm operator if both its kernel $\{x \in H : Ax = 0\}$ and its cokernel $H/(AH)$ are finite-dimensional linear spaces. Equivalently, an operator A is Fredholm if its coset $A + K(H)$ is invertible in the Calkin algebra $L(H)/K(H)$ where $L(H)$ stands for the algebra of all bounded linear operators on H and $K(H)$ for the ideal of $L(H)$ consisting of the compact operators. In particular, the property of being Fredholm is invariant with respect to compact perturbations. Thus, no finite part of the matrix representation $(a_{ij})_{i,j\in\mathbb{Z}}$ of $A \in L(H)$ is responsible for the Fredholmness of A, and the whole information on the Fredholm properties of A is hidden at infinity, i.e., in the asymptotic behavior of the entries a_{ij} .

Limit operators. How can one draw information from infinity? A convenient way is to fix a basis vector e_k, to shift the operator $A \mapsto V_{-n}AV_n$, and to observe the evolution of the vectors $V_{-n}AV_ne_k$ as n tends to $\pm\infty$. This has to be done for each basis vector e_k, which amounts to considering the behavior of the sequence $(V_{-n}AV_n)$ for large n with respect to the strong convergence of operators.

Assume for a moment that we are in the lucky case where the entries of each diagonal of A stabilize at infinity (i.e., where the limits $\lim_{i\to\pm\infty} a_{i+k,i}$ exist for every integer k). Then the strong limits of the sequence $(V_{-n}AV_n)$ as $n \to +\infty$ and $n \to -\infty$ exist, and these limits tell us exactly how the operator looks at infinity. What happens in the general situation where the entries of each diagonal are allowed to form an arbitrary bounded sequence? Then we cannot expect that the sequence $(V_{-n}AV_n)$ converges strongly but, hopefully, certain subsequences will still do. Indeed, using a Cantor diagonal argument, we will even get that if A is band-dominated, then *every sequence $h : \mathbb{N} \to \mathbb{Z}$ which tends to (plus or minus) infinity possesses a subsequence $g : \mathbb{N} \to \mathbb{Z}$ such that the sequence of the operators*

$$V_{-g(n)}AV_{g(n)}$$

converges strongly as $n \to \infty$. The strong limit of this sequence is called the limit operator of A with respect to the sequence g, and the set of all limit operators of a given operator A is called the operator spectrum of A. The crucial and surprising point is that the operator spectrum of a band-dominated operator contains exactly the information from infinity which is needed to decide whether the operator is Fredholm or not. The precise statement is: *a band-dominated operator is Fredholm if, and only if, its limit operators are invertible and if the norms of their inverses are uniformly bounded.*

Contents of the book. We will not restrict our attention to band-dominated operators on $l^2(\mathbb{Z})$; rather we consider band-dominated operators on $l^p(\mathbb{Z}^N, X)$ where N is a positive integer, $1 \le p \le \infty$, and where X is a complex Banach space. The main reason for this is that, after a suitable discretization, functions in $L^p(\mathbb{R}^N)$ become sequences in $l^p(\mathbb{Z}^N, X)$ with $X = L^p([0, 1]^N)$, and that a related discretization, applied to (wide classes of) convolution and pseudodifferential operators on $L^p(\mathbb{R}^N)$, indeed produces band-dominated operators on the discrete space $l^p(\mathbb{Z}^N, X)$. Thus, the theory of band-dominated operators which will be developed in the first two chapters can immediately be applied to convolution and pseudodifferential operators to reproduce some known facts and to uncover some new properties of these and other operators.

The inclusion of the 'exotic' case $p = \infty$ and the consideration of sequences with values in infinite-dimensional Banach spaces involve some subtleties. For example, we are limited when working with strong convergence of operators, because the projections $P_n : l^p(\mathbb{Z}) \to l^p(\mathbb{Z})$ which replace all entries of a sequence (x_k) with $|k| > n$ by zero converge strongly to the identity operator if and only if $p \neq \infty$. Another point is that the notions of Fredholmness and of invertibility at infinity (which are synonym for operators on spaces of sequences which take values in a finite-dimensional Banach space) become basically different for operators on spaces of sequences with values in infinite-dimensional Banach spaces. The applications we have in mind suggest to give preference to the aspect of invertibility at infinity over Fredholmness.

So we start in the first chapter with modifying the standard concepts of strong convergence, compactness and Fredholmness by introducing the notions of $\mathcal{P}$-strong convergence, $\mathcal{P}$-compactness and $\mathcal{P}$-Fredholmness. Here, $\mathcal{P}$ is a given approximate identity, for instance, the sequence (P_n) we encountered in the preceding paragraph. Based on these $\mathcal{P}$-notions, we introduce the general concept of a limit operator in Section 1.2.

In case $1 < p < \infty$ and $\dim X < \infty$, all $\mathcal{P}$-notions reduce to their usual meanings. Thus, readers who are exclusively interested in band-dominated operators on scalar-valued sequences can skip the first part of this chapter. It is perhaps also a good advice for a first reading of Chapters 1 and 2 to skip Section 1.1, to ignore the $\mathcal{P}$, and to set $X = \mathbb{C}$ in what follows.

The second chapter is the heart of the book. Here we introduce band-dominated operators on the spaces $l^p(\mathbb{Z}^N, X)$ and prove that they possess sufficiently many limit operators in the sense that a band-dominated operator is $\mathcal{P}$-Fredholm if and only if all of its limit operators are invertible and if the norms of their inverses are uniformly bounded.

The status of the uniform invertibility hypothesis is not really evident at this moment. On the one hand, it might be quite hard to check the uniform boundedness of the norms of the inverses of a family of operators explicitly; so the condition is very unpleasant from this practical point of view. On the other hand, we do not know any example of a band-dominated operator, all limit operators of which are invertible, but which fails to be $\mathcal{P}$-Fredholm. Moreover, it turns out that there are large and important classes of band-dominated operators for which we can prove that the uniformity of the invertibility is indeed redundant. In particular, this happens for band-dominated operators in the Wiener algebra (which includes all band operators) as well as for band-dominated operators with slowly oscillating coefficients. These results will also be presented in Chapter 2. Partially, these results are based on the compatibility of the limit operators method with another local principle, which is due to Allan and Douglas. Thus, a large part of the second chapter is devoted to the study of the relations between these local theories.

The last part of Chapter 2 deals with the problem of calculating the index of a Fredholm band-dominated operator A in terms of its limit operators. Here we restrict ourselves to band-dominated operators on $l^2(\mathbb{Z})$ with scalar-valued coefficients. Under these assumptions, we get an index formula of astonishing simplicity. For, choose an arbitrary limit operator B_+ of A with respect to a sequence tending to $+\infty$ as well as an arbitrary limit operator B_- of A with respect to a sequence tending to $-\infty$. Then the index of A is the sum of the local indices of $B_\pm$ at $\pm\infty$,

$$\operatorname{ind} A = \operatorname{ind}_+ B_+ + \operatorname{ind}_- B_-.$$

In Chapters 3 and 4, we are going to specify the results on Fredholmness obtained for general band-dominated operators to convolution operators and pseudodifferential operators on $L^p(\mathbb{R}^N)$, respectively. The key steps are to embed these operators into suitable operator algebras of Wiener type and to discretize in an appropriate manner these operators to get band-dominated operators on a discrete l^p-space of vector-valued sequences. A similar approach is chosen in Chapter 5 in order to illustrate the applicability of the limit operators method to study the Fredholmness of pseudodifference operators on $l^p(\mathbb{Z}^N)$-spaces with weight. Particular attention is paid to Phragmen-Lindelöf type theorems on the exponential decay at infinity of solutions to pseudodifference equations, to the description of the essential spectrum of discrete Schrödinger operators, and to the decay of their eigenfunctions at infinity.

Chapter 6 shifts the attention from analysis to numerical analysis. We consider the finite section method for the approximate solution of equations with band-dominated system matrices. The basic observation here is that the sequence of the finite sections of a band-dominated operator on $\mathbb{Z}^N$ can be interpreted as a band-dominated operator on $\mathbb{Z}^{N+1}$. Moreover, the sequence is stable if and only if the corresponding operator is $\mathcal{P}$-Fredholm. Thus, the results from Chapter 2 apply immediately to yield stability results for the finite section method. In the Hilbert space case $p = 2$, these stability results will be further used to derive characterizations of the asymptotic behavior of the norms, condition numbers, eigenvalues, pseudo-eigenvalues, and Rayleigh quotients of the finite section matrices.

So far we have only discussed the application of the limit operators method to the Fredholm theory of band-dominated (and related) operators. The goal of the final Chapter 7 is to indicate that the range of the applicability of the limit operators method is much larger. So we will develop an axiomatic scheme which covers most applications of the limit operators method. As concrete examples we consider the Fredholmness of convolution operators as well as of convolutions combined with (non-Carleman) shifts on the Heisenberg group.

Preliminaries. We assume that the reader has basic knowledge in linear Functional Analysis. For Chapters 4 and 7, a first acquaintance with pseudodifferential operators and with non-commutative harmonic analysis would be helpful.

Acknowledgements. The authors are grateful to their friends and colleagues Albrecht Böttcher, Marko Lindner and Alexander Rogozhin who read large parts of the manuscript, suggested many improvements and also pointed out a few mistakes to us. We also thank the Publishers for their kind and stimulating co-operation. Finally, we would like to appreciate our sincere thanks to the German Research Foundation (DFG) and to the CONACYT which supported the research on this subject as well as the writing of this book by several travel and research grants (DFG grants 436 RUS 17/67/98 and 436 RUS 17/24/01, and CONACYT project 32424-E). This book would not exist without the generous support by these institutions.

Chapter 1

Limit Operators

1.1 Generalized compactness, generalized convergence

We start with recalling how the class of the compact operators on a Banach space determines both the strong operator topology and the set of the Fredholm operators. This characterization is then taken as a starting point to introduce generalized notions of strong convergence and Fredholmness.

1.1.1 Compactness, strong convergence, Fredholmness

Let E be a Banach space with norm $\|.\|_E$ and with Banach dual E^*. All Banach spaces occurring in this book are supposed to be linear spaces over the field $\mathbb{C}$ of the complex numbers. By $L(E)$ we denote the Banach algebra of the bounded linear operators on E with norm

$$\|A\|_{L(E)} := \sup_{x \in E \setminus \{0\}} \|Ax\|_E / \|x\|_E.$$

We further let $K(E)$ stand for the closed ideal of the compact operators in $L(E)$. The identity operator on E will be denoted by I. Finally, we will write $\operatorname{Ker} A$ and $\operatorname{Im} A$ for the kernel $\{x \in E : Ax = 0\}$ and the range $\{Ax : x \in E\}$ of A, respectively.

Strong convergence. A sequence (A_n) of operators $A_n \in L(E)$ is said to *converge strongly* if the sequence $(A_n x)$ converges in the norm of E for each $x \in E$. Then $Ax := \lim A_n x$ defines an operator A on E which is called the *strong limit* of the sequence (A_n) and which we denote by $\text{s-lim}\, A_n$. One also says that the A_n converge strongly to A and writes $A_n \to A$ *strongly.*

In general, the adjoint sequence (A_n^*) of a strongly convergent sequence (A_n) fails to be strongly convergent. Thus, the sequence (A_n) is said to *converge *-strongly* if both sequences (A_n) and (A_n^*) converge strongly on E and E^*, respectively. In this case,

$$\text{s-lim}\, A_n^* = (\text{s-lim}\, A_n)^*.$$

Theorem 1.1.1 (Banach-Steinhaus) *Let the sequence (A_n) of bounded linear operators on E be strongly convergent to an operator A. Then A is a bounded and linear operator on E, the sequence (A_n) is uniformly bounded, and*

$$\|A\|_{L(E)} \leq \liminf \|A_n\|_{L(E)}.$$

We agree upon calling a non-empty set $\mathcal{M} \subset L(E)$ of operators *uniformly invertible* if the operators in $\mathcal{M}$ are invertible and if the norms of their inverses are uniformly bounded.

Proposition 1.1.2

(a) *If $A_n \to A$ and $B_n \to B$ strongly, then $A_n + B_n \to A + B$ and $A_n B_n \to AB$ strongly.*

(b) *If $A_n \to A$ strongly, and if the operators A_n are uniformly invertible, then A has a trivial kernel and a closed range, and $A_n^{-1}A \to I$ strongly.*

(c) *If $A_n \to A$ *-strongly, and if the operators A_n are uniformly invertible, then A is invertible, and $A_n^{-1} \to A^{-1}$ *-strongly.*

Proof. The proof of (a) is straightforward. It makes use of the uniform boundedness of the sequence (A_n) due to the Banach-Steinhaus theorem. The last assertion of (b) is a consequence of the estimate

$$\|A_n^{-1}Ax - x\| \leq \sup \|A_n^{-1}\| \, \|Ax - A_n x\|$$

and of the strong convergence of A_n to A. Thus, letting n go to infinity in $\|A_n^{-1}Ax\| \leq C\|Ax\|$, we obtain $\|x\| \leq C\|Ax\|$ for all $x \in E$, i.e., the operator A is bounded below. This boundedness implies that A has a trivial kernel and a closed range. If, in addition, $A_n^* \to A^*$ strongly, then the same arguments yield that A^* has a trivial kernel, too. Since

$$\text{clos Im}\, B =^{\perp} (\text{Ker}\, B^*) := \{x \in E : f(x) = 0 \text{ for all } f \in \text{Ker}\, B^*\}$$

for every bounded linear operator B, the range of A is all of E. Thus, A is invertible. The strong convergence of A_n^{-1} to A^{-1} follows from

$$\|A_n^{-1}x - A^{-1}x\| \leq \sup \|A_n^{-1}\| \, \|A^{-1}\| \, \|Ax - A_n x\|,$$

and the strong convergence of adjoint sequence follows similarly. □

Compactness and strong convergence. The notions of compactness and of strong convergence are intimately related by the following theorem.

Theorem 1.1.3 *Let $A_n, A \in L(E)$. Then $\|A_n x - Ax\|_E \to 0$ for every $x \in E$ if and only if $\|A_n K - AK\|_{L(E)} \to 0$ for every $K \in K(E)$.*

Proof. Let (A_n) be a strongly convergent sequence with strong limit A, and let K be a compact operator, i.e., suppose the set $M := \{Kx : \|x\| \leq 1\}$ is relatively compact. Assume that

$$\|A_nK - AK\| = \sup_{\|x\|\le 1} \|A_nKx - AKx\| = \sup_{y\in M} \|A_ny - Ay\|$$

does not converge to zero as $n \to \infty$. Then there are an $\varepsilon > 0$ as well as an infinite sequence $(y_n) \subseteq M$ such that $\|A_ny_n - Ay_n\| > \varepsilon$. Since M is relatively compact, one can extract a subsequence (z_n) of (y_n) which converges to a $z \in E$. For every n, this choice implies

$$\begin{aligned}\varepsilon < \|A_nz_n - Az_n\| &\le \|(A_n - A)(z_n - z)\| + \|(A_n - A)z\| \\ &\le \sup_k \|A_k - A\|\|z_n - z\| + \|(A_n - A)z\|.\end{aligned}$$

The right-hand side of this inequality becomes smaller than any prescribed $\varepsilon > 0$ if only n is large enough. This contradiction verifies the 'only if' part of the assertion.

For the 'if'-part, given $x \in E$, choose a functional $f \in E^*$ such that $\|f\| = 1$ and $f(x) = \|x\|$, which is possible by the Hahn-Banach theorem, and consider the operator

$$K_xy := f(y)x, \quad y \in E. \tag{1.1}$$

This operator is of rank one, hence compact, and $\|K_x\|_{L(E)} = \|x\|_E$. Since $\|A_nK - AK\| \to 0$ for every compact operator K by hypothesis, one has

$$\|A_nx - Ax\|_E = \|K_{A_nx} - K_{Ax}\|_{L(E)} = \|A_nK_x - AK_x\|_{L(E)} \to 0$$

for every $x \in E$. Hence, $A_n \to A$ strongly. □

Consequently, the strong convergence of A_n to A, considered as operators on E, is equivalent to the strong convergence of A_n to A, considered as operators of left multiplication on $K(E)$. In Section 1.1.4 we are going to employ this equivalence in order to introduce other kinds of strong convergence.

Actually, the algebra $L(E)$ is isometrically and isomorphically embedded into $L(K(E))$ via its *left regular representation*, which associates with every operator $A \in L(E)$ an operator A_l on $K(E)$ acting as multiplication from the left,

$$A_l : K(E) \to K(E), \quad K \mapsto AK.$$

Proposition 1.1.4 *If $A \in L(E)$, then $A_l \in L(K(E))$, and the mapping $A \mapsto A_l$ is an isometry.*

Proof. Evidently, the operator A_l acts linearly on $K(E)$, and the estimate $\|AK\| \le \|A\|\,\|K\|$ shows that A_l is bounded and that $\|A_l\|_{L(K(E))} \le \|A\|_{L(E)}$. For the reverse estimate, let K_x be as in (1.1). Then

$$\begin{aligned}\|A_l\|_{L(K(E))} &= \sup_{K\in K(E)\setminus\{0\}} \frac{\|AK\|_{L(E)}}{\|K\|_{L(E)}} \ge \sup_{x\in E\setminus\{0\}} \frac{\|AK_x\|_{L(E)}}{\|K_x\|_{L(E)}} \\ &= \sup_{x\in E\setminus\{0\}} \frac{\|K_{Ax}\|_{L(E)}}{\|K_x\|_{L(E)}} = \sup_{x\in E\setminus\{0\}} \frac{\|Ax\|_E}{\|x\|_E} = \|A\|_{L(E)},\end{aligned}$$

which proves the assertion. □

Fredholm operators. An operator $A \in L(E)$ is called *Fredholm* if both its kernel $\operatorname{Ker} A$ and its cokernel $\operatorname{Coker} A := E/\operatorname{Im} A$ have finite dimension. In this case, the range of A is closed, and the integer

$$\operatorname{ind} A := \dim \operatorname{Ker} A - \dim \operatorname{Coker} A$$

is called the *index* of A. Let $\operatorname{Fred}(E)$ stand for the set of the Fredholm operators on E.

Theorem 1.1.5

(a) *If A is Fredholm and K is compact, then $A+K$ is Fredholm, and* $\operatorname{ind}(A+K) = \operatorname{ind} A$.

(b) $\operatorname{Fred}(E)$ *is open in $L(E)$, and the function* $\operatorname{ind} : \operatorname{Fred}(E) \to \mathbb{Z}$ *is continuous.*

(c) $\operatorname{Fred}(E)$ *is a semi-group under multiplication, and* ind *is a homomorphism from* $\operatorname{Fred}(E)$ *into the additive group $\mathbb{Z}$.*

(d) *If A is Fredholm, then A^* is Fredholm, and* $\operatorname{ind} A^* = -\operatorname{ind} A$.

Compactness and Fredholmness. The ideal of the compact operators determines the set of the Fredholm operators in the following sense.

Theorem 1.1.6 *An operator $A \in L(E)$ is Fredholm if and only if the coset $A+K(E)$ is invertible in the quotient algebra $L(E)/K(E)$, the Calkin algebra of E.*

The norm and the spectrum of the coset $A + K(E)$ in $L(E)/K(E)$ are referred to as the *essential norm* $\|A\|_{ess}$ and the *essential spectrum* $\sigma_{ess}(A)$ of A, respectively.

The proofs of the preceding theorems can be found in standard textbooks on Functional Analysis. Immediate consequences of Theorem 1.1.6 are the invariance of the Fredholm property under small and under compact perturbations as well as the semi-group property of $\operatorname{Fred}(E)$.

1.1.2 $\mathcal{P}$-compactness

Let E be a Banach space, and let $\mathcal{P} = (P_n)_{n=0}^\infty$ be a bounded sequence of operators in $L(E)$. We call $\mathcal{P}$ an *increasing approximate projection* if, for every $m \in \mathbb{N}$, there is an $N(m) \in \mathbb{N}$ such that

$$P_n P_m = P_m P_n = P_m \quad \text{for all } n \geq N(m), \tag{1.2}$$

and $\mathcal{P}$ is a *decreasing approximate projection* if, for every $m \in \mathbb{N}$, there is an $N(m) \in \mathbb{N}$ such that

$$P_n P_m = P_m P_n = P_n \quad \text{for all } n \geq N(m). \tag{1.3}$$

If $\mathcal{P} = (P_n)$ is an increasing approximate projection, then the sequence (Q_n) with $Q_n := I - P_n$ forms a decreasing approximate projection which we call *associated* to $\mathcal{P}$.

An (increasing or decreasing) approximate projection (P_n) is said to be *proper* if $P_n \neq 0$ and $P_n \neq I$ for all n. If (P_n) is a proper increasing approximate projection, then $\|P_n\| \geq 1$ for all sufficiently large n.

It is also clear that $\mathcal{P}^* := (P_n^*)$ is an increasing (decreasing) approximate projection in $L(E^*)$ whenever $\mathcal{P} = (P_n)$ is an increasing (decreasing) approximate projection in $L(E)$. Further, every infinite subsequence of an increasing (decreasing) approximate projection is an increasing (decreasing) approximate projection again.

In what follows, we will mainly be concerned with proper and increasing approximate projections, which therefore will be simply called *approximate projections.*

$\mathcal{P}$-compactness. Every approximate projection $\mathcal{P} = (P_n)$ gives rise to a substitute of the ideal of the compact operators.

Definition 1.1.7 *An operator $K \in L(E)$ is* $\mathcal{P}$-compact *if*

$$\|KP_n - K\| \to 0 \quad \textit{and} \quad \|P_nK - K\| \to 0 \quad \textit{as } n \to \infty.$$

By $K(E, \mathcal{P})$ we denote the set of all $\mathcal{P}$-compact operators on E, and by $L(E, \mathcal{P})$ the set of all operators $A \in L(E)$ for which both AK and KA are $\mathcal{P}$-compact whenever K is $\mathcal{P}$-compact.

It is immediate from this definition that all operators in $\mathcal{P}$ are $\mathcal{P}$-compact and that

$$K(E^*, \mathcal{P}^*) = \{K^* : K \in K(E, \mathcal{P})\}, \quad L(E^*, \mathcal{P}^*) = \{A^* : A \in L(E, \mathcal{P})\}. \tag{1.4}$$

Proposition 1.1.8 *Let $\mathcal{P} = (P_n)$ be an approximate projection and $Q_n := I - P_n$.*

(a) *An operator $A \in L(E)$ belongs to $L(E, \mathcal{P})$ if and only if, for every $k \in \mathbb{N}$,*

$$\|P_kAQ_n\| \to 0 \quad \textit{and} \quad \|Q_nAP_k\| \to 0 \quad \textit{as } n \to \infty. \tag{1.5}$$

(b) *$L(E, \mathcal{P})$ is a closed subalgebra of $L(E)$ which contains the identity operator, and $K(E, \mathcal{P})$ is a closed ideal of $L(E, \mathcal{P})$.*

Proof. (a) The conditions (1.5) are clearly necessary for $A \in L(E, \mathcal{P})$. Let, conversely, A satisfy (1.5), and let $K \in K(E, \mathcal{P})$. Given $\varepsilon > 0$, choose r such that $\|K - P_rK\| < \varepsilon$, and choose N such that $\|Q_nAP_r\| < \varepsilon$ for all $n \geq N$. Then

$$\|Q_nAK\| \leq \|Q_nA\| \, \|K - P_rK\| + \|Q_nAP_r\| \, \|K\| < \varepsilon \, (\|Q_nA\| + \|K\|)$$

for all $n \geq N$. The other conditions can be checked similarly.

(b) It is immediate from the definitions that $L(E,\mathcal{P})$ is a subalgebra of $L(E)$ and that $K(E,\mathcal{P})$ is contained in $L(E,\mathcal{P})$ and forms an ideal of this algebra. To get the closedness of $K(E,\mathcal{P})$, let K_m be $\mathcal{P}$-compact and $\|K_m - K\| \to 0$. Choose r such that $\|K - K_r\| \sup \|Q_n\| < \varepsilon/2$, and N such that $\|K_r Q_n\| < \varepsilon/2$ for all $n \geq N$. Then

$$\|KQ_n\| \leq \|KQ_n - K_r Q_n\| + \|K_r Q_n\| < \varepsilon$$

for all $n \geq N$. The 'dual' assertion $\|Q_n K\| \to 0$ can be checked analogously.

Let now (A_m) be a sequence in $L(E,\mathcal{P})$ which converges in the norm to $A \in L(E)$. If K is $\mathcal{P}$-compact, then $(A_m K)$ and (KA_m) are sequences in $K(E,\mathcal{P})$ which converge in the norm to AK and KA, respectively. Since $K(E,\mathcal{P})$ is closed, one has AK, $KA \in K(E,\mathcal{P})$ and, hence, $A \in L(E,\mathcal{P})$. □

It is also easy to see that $K(E, \mathcal{P})$ is the smallest closed ideal of $L(E, \mathcal{P})$ which contains the operators P_m constituting $\mathcal{P}$.

Invertibility in $L(E, \mathcal{P})$. A delicate question is that of the inverse closedness of $L(E, \mathcal{P})$ in $L(E)$. Recall that a (not necessarily closed) subalgebra $\mathcal{B}$ of a Banach algebra $\mathcal{A}$ with identity is called *inverse closed in $\mathcal{A}$* if whenever an element $b \in \mathcal{B}$ is invertible in $\mathcal{A}$, its inverse b^{-1} already belongs to $\mathcal{B}$. In case $\mathcal{A}$ is a C^*-algebra with identity, every symmetric and closed subalgebra of $\mathcal{A}$ which contains the identity is inverse closed.

For the inverse closedness of $L(E,\mathcal{P})$ in $L(E)$ we need some stronger properties of $\mathcal{P}$. Given an approximate projection $\mathcal{P} = (P_n)_{n=0}^{\infty}$ in $L(E)$, we set $S_0 := P_0$ and $S_n := P_n - P_{n-1}$ for $n \geq 1$. Further, for every bounded subset U of $\mathbb{R}$, we define $P_U := \sum_{k \in \mathbb{N} \cap U} S_k$ and $Q_U := I - P_U$. (Thus, P_k has still the same meaning as above, whereas $P_{\{k\}} = S_k$.) The approximate projection $\mathcal{P}$ is called *uniform* if

$$\sup \|P_U\| < \infty, \quad \text{the supremum over all bounded } U \subseteq \mathbb{R}. \tag{1.6}$$

Theorem 1.1.9 *If $\mathcal{P}$ is a uniform approximate projection, then $L(E,\mathcal{P})$ is an inverse closed subalgebra of $L(E)$.*

Proof. Let again (Q_n) stand for the approximate projection which is associated to $\mathcal{P}$. Further we will write $m \ll n$ if $P_k Q_n = Q_n P_k = 0$ for all $k \leq m$, and we denote the supremum in (1.6) by C.

Let the operator $A \in L(E,\mathcal{P})$ be invertible in $L(E)$. By (1.5), what we have to show is that

$$\|P_k A^{-1} Q_n\| \to 0 \quad \text{and} \quad \|Q_n A^{-1} P_k\| \to 0 \quad \text{as } n \to \infty.$$

Given $\varepsilon > 0$, choose and fix a positive integer m with $\|A^{-1}\|^2 \|A\|/m < \varepsilon$, and choose integers

$$0 = r_1^{(1)} < r_2^{(1)} < r_3^{(1)} < r_4^{(1)} < r_1^{(2)} < \cdots < r_4^{(m-1)} < r_1^{(m)} < r_2^{(m)} < r_3^{(m)} < r_4^{(m)}$$

such that

$$k + r_l^{(i)} \ll k + r_{l+1}^{(i)} \quad \text{and} \quad k + r_4^{(j)} \ll k + r_1^{(j+1)}$$

for all $1 \le l \le 3$, $1 \le i \le m$ and $1 \le j \le m-1$ and such that

$$\|P_{k+r_1^{(i)}} A Q_{k+r_2^{(i)}}\| < \varepsilon / \|A^{-1}\|^2, \tag{1.7}$$

$$\|Q_{k+r_3^{(i)}} A P_{k+r_2^{(i)}}\| < \varepsilon / \|A^{-1}\|^2, \tag{1.8}$$

$$\|P_{k+r_3^{(i)}} A Q_{k+r_4^{(i)}}\| < \varepsilon / \|A^{-1}\|^2, \tag{1.9}$$

$$\|Q_{k+r_1^{(i+1)}} A P_{k+r_4^{(i)}}\| < \varepsilon / \|A^{-1}\|^2. \tag{1.10}$$

That is, given $r_1^{(i)}$ we choose $r_2^{(i)} > r_1^{(i)}$ such that $k + r_2^{(i)} \gg k + r_1^{(i)}$ and that (1.7) holds, then $r_3^{(i)} > r_2^{(i)}$ such that $k + r_3^{(i)} \gg k + r_2^{(i)}$ and that (1.8) is satisfied, then $r_4^{(i)} > r_3^{(i)}$ which fulfills $k + r_4^{(i)} \gg k + r_3^{(i)}$ and (1.9), and finally $r_1^{(i+1)} > r_4^{(i)}$ such that $k + r_1^{(i+1)} \gg k + r_4^{(i)}$ and that (1.10) is valid.

Let $n \gg k + r_4^{(m)}$. We set

$$U_i := (k + r_1^{(i)},\, k + r_3^{(i)}], \quad V_i := (k + r_2^{(i)},\, k + r_4^{(i)}],$$

$$V_i' := [0,\, k + r_2^{(i)}], \quad U_i' := [0,\, k + r_3^{(i)}].$$

Then, since $n \gg k + r_4^{(m)} \gg k + r_2^{(i)}$ for all i,

$$\begin{aligned} P_k A^{-1} Q_n &= P_k P_{V_i'} A^{-1} Q_n \\ &= P_k A^{-1} P_{U_i'} A P_{V_i'} A^{-1} Q_n + P_k A^{-1} Q_{U_i'} A P_{V_i'} A^{-1} Q_n \end{aligned} \tag{1.11}$$

with

$$\begin{aligned} \|P_k A^{-1} Q_{U_i'} A P_{V_i'} A^{-1} Q_n\| &\le C(C+1)\, \|A^{-1}\|^2 \|Q_{U_i'} A P_{V_i'}\| \\ &= C(C+1)\, \|A^{-1}\|^2 \|Q_{k+r_3^{(i)}} A P_{k+r_2^{(i)}}\| \\ &\le C(C+1)\varepsilon \end{aligned} \tag{1.12}$$

due to (1.8) (observe that $\|Q_U\| = \|I - P_U\| \le 1 + C$). Further, since $n \gg k + r_3^{(i)}$,

$$\begin{aligned} & P_k A^{-1} P_{U_i'} A P_{V_i'} A^{-1} Q_n \\ &= -\, P_k A^{-1} P_{U_i'} A Q_{V_i'} A^{-1} Q_n \\ &= -\, P_k A^{-1} P_{U_i} A Q_{V_i'} A^{-1} Q_n - P_k A^{-1} P_{U_i' \setminus U_i} A Q_{V_i'} A^{-1} Q_n \\ &= -\, P_k A^{-1} P_{U_i} A P_{V_i} A^{-1} Q_n - P_k A^{-1} P_{U_i} A Q_{k+r_4^{(i)}} A^{-1} Q_n \\ &\quad -\, P_k A^{-1} P_{U_i' \setminus U_i} A P_{V_i} A^{-1} Q_n - P_k A^{-1} P_{U_i' \setminus U_i} A Q_{k+r_4^{(i)}} A^{-1} Q_n \\ &= -\, P_k A^{-1} P_{U_i} A P_{V_i} A^{-1} Q_n - P_k A^{-1} P_{U_i'} A Q_{k+r_4^{(i)}} A^{-1} Q_n \\ &\quad -\, P_k A^{-1} P_{U_i' \setminus U_i} A P_{V_i} A^{-1} Q_n. \end{aligned} \tag{1.13}$$

For the middle term on the right-hand side of (1.13), we have by (1.9),

$$\begin{aligned}&\|P_k A^{-1} P_{U_i'} A Q_{k+r_4^{(i)}} A^{-1} Q_n\| \\ &\quad \leq C(C+1)\,\|A^{-1}\|^2 \|P_{k+r_3^{(i)}} A Q_{k+r_4^{(i)}}\| \leq C(C+1)\varepsilon \end{aligned} \tag{1.14}$$

and, for the last term, due to (1.7),

$$\begin{aligned}&\|P_k A^{-1} P_{U_i'\setminus U_i} A P_{V_i} A^{-1} Q_n\| \\ &\quad \leq \|P_k A^{-1} P_{k+r_1^{(i)}} A Q_{k+r_2^{(i)}} P_{k+r_4^{(i)}} A^{-1} Q_n\| \\ &\quad \leq C^2(C+1)\,\|A^{-1}\|^2 \|P_{k+r_1^{(i)}} A Q_{k+r_2^{(i)}}\| \\ &\quad \leq C^2(C+1)\varepsilon \end{aligned} \tag{1.15}$$

where we used that

$$\begin{aligned} P_{V_i} &= P_{(k+r_2^{(i)},\, k+r_4^{(i)}]} = Q_{k+r_2^{(i)}} - Q_{k+r_4^{(i)}} \\ &= Q_{k+r_2^{(i)}} - Q_{k+r_2^{(i)}} Q_{k+r_4^{(i)}} = Q_{k+r_2^{(i)}} P_{k+r_4^{(i)}}. \end{aligned}$$

From (1.11)–(1.15) we conclude that

$$P_k A^{-1} Q_n = -P_k A^{-1} P_{U_i} A P_{V_i} A^{-1} Q_n + D_i \tag{1.16}$$

where D_i is an operator with norm less than $c\varepsilon$ with c being a constant independent of i and ε. Summarizing the identities (1.16) we get

$$\begin{aligned} mP_k A^{-1} Q_n &= -\sum_{i=1}^m P_k A^{-1} P_{U_i} A P_{V_i} A^{-1} Q_n + \sum_{i=1}^m D_i \\ &= -P_k A^{-1} P_{U_1\cup\cdots\cup U_m} A P_{V_1\cup\cdots\cup V_m} A^{-1} Q_n + \sum_{i=1}^m D_i \\ &\quad - \sum_{i=1}^m P_k A^{-1} P_{U_i} A P_{(V_1\cup\cdots\cup V_m)\setminus V_i} A^{-1} Q_n \end{aligned} \tag{1.17}$$

since $U_i \cap U_j = V_i \cap V_j = \emptyset$ for $i \neq j$. For the first item in (1.17) we find

$$\|P_k A^{-1} P_{U_1\cup\cdots\cup U_m} A P_{V_1\cup\cdots\cup V_m} A^{-1} Q_n\| \leq C^3(C+1)\,\|A^{-1}\|^2 \|A\|,$$

and for every term in the last sum in (1.17) we obtain

$$\begin{aligned}&\|P_k A^{-1} P_{U_i} A P_{(V_1\cup\cdots\cup V_m)\setminus V_i} A^{-1} Q_n\| \\ &\quad \leq C(C+1)\,\|A^{-1}\|^2 \|P_{U_i} A P_{(V_1\cup\cdots\cup V_m)\setminus V_i}\|. \end{aligned} \tag{1.18}$$

Since

$$P_{U_i} = P_{k+r_3^{(i)}} - P_{k+r_1^{(i)}} = P_{k+r_3^{(i)}} - P_{k+r_1^{(i)}} P_{k+r_3^{(i)}} = Q_{k+r_1^{(i)}} P_{k+r_3^{(i)}}$$

(recall that $k + r_1^i \ll k + r_3^{(i)}$) and, similarly,

$$P_{U_i} = Q_{k+r_1^{(i)}} - Q_{k+r_3^{(i)}} = P_{k+r_1^{(i)}} Q_{k+r_3^{(i)}}$$

as well as

$$\begin{aligned} P_{(V_1\cup\cdots\cup V_m)\setminus V_i} &= P_{(V_1\cup\cdots\cup V_{i-2})} + P_{(k+r_2^{(i-1)},\, k+r_4^{(i-1)}]} \\ &\quad + P_{(k+r_2^{(i+1)},\, k+r_4^{(i+1)}]} + P_{(V_{i+2}\cup\cdots\cup V_m)} \\ &= P_{k+r_4^{(i-1)}} P_{(V_1\cup\cdots\cup V_{i-2})} + P_{k+r_4^{(i-1)}} Q_{k+r_2^{(i-1)}} \\ &\quad + Q_{k+r_2^{(i+1)}} P_{k+r_4^{(i+1)}} + Q_{k+r_2^{(i+1)}} P_{(V_{i+2}\cup\cdots\cup V_m)}, \end{aligned}$$

we can further estimate the right-hand side of (1.18) by

$$\begin{aligned} &\|P_{U_i} A P_{(V_1\cup\cdots\cup V_m)\setminus V_i}\| \\ &\quad\le \|P_{U_i} A P_{k+r_4^{(i-1)}}\| \, \|P_{(V_1\cup\cdots\cup V_{i-2})} + Q_{k+r_2^{(i-1)}}\| \\ &\qquad + \|P_{U_i} A Q_{k+r_2^{(i+1)}}\| \, \|P_{k+r_4^{(i+1)}} + P_{(V_{i+2}\cup\cdots\cup V_m)}\| \\ &\quad\le C(1+2C)\, \|Q_{k+r_3^{(i)}} A P_{k+r_4^{(i-1)}}\| + 2C(C+1)\, \|P_{k+r_3^{(i)}} A Q_{k+r_2^{(i+1)}}\| \\ &\quad\le C(1+2C)\, \|Q_{k+r_3^{(i)}} Q_{k+r_1^{(i)}} A P_{k+r_4^{(i-1)}}\| \\ &\qquad + 2C(C+1)\, \|P_{k+r_3^{(i)}} A Q_{k+r_4^i} Q_{k+r_2^{(i+1)}}\| \;\le\; d\varepsilon \end{aligned}$$

with a constant d independent of i and ε due to (1.9) and (1.10). Inserting these estimates into (1.17) and dividing by m, we arrive at

$$\begin{aligned} \|P_k A^{-1} Q_n\| &\le \frac{1}{m}(C^3(C+1)\, \|A^{-1}\|^2 \|A\| + mc\varepsilon + md\varepsilon) \\ &\le (C^3(C+1) + c + d)\varepsilon \end{aligned}$$

for all $n \gg k + r_4^{(m)}$. Thus, $\|P_k A^{-1} Q_n\| \to 0$, and the dual assertion $\|Q_n A^{-1} P_k\| \to 0$ can be checked analogously. □

Equivalent approximate projections. We call the approximate projections $\mathcal{P} = (P_n)$ and $\mathcal{P}' = (P'_n)$ *equivalent* if

$$\lim_{n\to\infty} P_m P'_n = \lim_{n\to\infty} P'_n P_m = P_m \quad \text{and} \quad \lim_{n\to\infty} P_n P'_m = \lim_{n\to\infty} P'_m P_n = P'_m$$

for all $m \ge 0$. For example, every infinite subsequence of an approximate projection $\mathcal{P}$ is equivalent to $\mathcal{P}$.

Lemma 1.1.10 *The approximate projections $\mathcal{P}$ and $\mathcal{P}'$ are equivalent if and only if $K(E, \mathcal{P}) = K(E, \mathcal{P}')$.*

Proof. Let $\mathcal{P} = (P_n)$, $\mathcal{P}' = (P_n')$, and set $Q_n := I - P_n$ and $Q_n' := I - P_n'$. Further let $K \in K(E, \mathcal{P})$. Then, for all m, $n \geq 0$,

$$KQ_n' = (K - KP_m)Q_n' + KP_mQ_n'.$$

Given $\varepsilon > 0$, choose m such that $\|K - KP_m\| < \varepsilon$ and $N(m)$ such that

$$\|P_mQ_n'\| = \|P_m - P_mP_n'\| < \varepsilon \quad \text{whenever } n \geq N(m).$$

Then

$$\|KQ_n'\| \leq \varepsilon \sup_n \|Q_n'\| + \varepsilon\|K\| \quad \text{for all } n \geq N(m),$$

whence $\|KQ_n'\| \to 0$. Similarly one gets $\|Q_n'K\| \to 0$, i.e., $K \in K(E, \mathcal{P}')$.

Conversely, the equality $K(E, \mathcal{P}) = K(E, \mathcal{P}')$ implies that $P_m \in K(E, \mathcal{P}')$ for all m, i.e.,

$$\lim_{n\to\infty} P_mP_n' = P_m = \lim_{n\to\infty} P_n'P_m.$$

Changing the roles of $\mathcal{P}$ and $\mathcal{P}'$, we get the second condition for the equivalence of approximate projections. □

Consequently, if $\mathcal{P}$ and $\mathcal{P}'$ are equivalent, then $L(E, \mathcal{P}) = L(E, \mathcal{P}')$.

1.1.3 $\mathcal{P}$-Fredholmness

Let again $\mathcal{P} = (P_n)$ be an approximate projection on the Banach space E, and set $Q_n = I - P_n$. The following definition is motivated by Theorem 1.1.6.

Definition 1.1.11 *An operator $A \in L(E, \mathcal{P})$ is called* $\mathcal{P}$-Fredholm *if the coset $A + K(E, \mathcal{P})$ is invertible in the quotient algebra $L(E, \mathcal{P})/K(E, \mathcal{P})$.*

This definition implies that the $\mathcal{P}$-Fredholmness of an operator is invariant both under sufficiently small and under $\mathcal{P}$-compact perturbations, and that the product of $\mathcal{P}$-Fredholm operators is $\mathcal{P}$-Fredholm again. It is also clear that if $\mathcal{P}$ and $\mathcal{P}'$ are equivalent approximate projections, then an operator is $\mathcal{P}$-Fredholm if and only if it is $\mathcal{P}'$-Fredholm. We call the norm and the spectrum of the coset $A + K(E, \mathcal{P})$ in $L(E, \mathcal{P})/K(E, \mathcal{P})$ the *$\mathcal{P}$-essential norm* and the *$\mathcal{P}$-essential spectrum* of A.

Proposition 1.1.12 *An operator $A \in L(E, \mathcal{P})$ is $\mathcal{P}$-Fredholm if and only if there exist operators C, $D \in L(E, \mathcal{P})$ and an $m \in \mathbb{N}$ such that*

$$Q_mAC = Q_m \quad \text{and} \quad DAQ_m = Q_m. \tag{1.19}$$

Operators $A \in L(E, \mathcal{P})$ which satisfy (1.19) are also called *invertible at infinity.*

Proof. If $A \in L(E, \mathcal{P})$ is $\mathcal{P}$-Fredholm, then there exist operators $C \in L(E, \mathcal{P})$ and $K \in K(E, \mathcal{P})$ such that $AC = I + K$. Multiplying this equality by Q_r from the left-hand side and adding P_r to both sides yield $Q_rAC + P_r = I + Q_rK$. Choose

r such that $\|Q_r K\| < 1$ and m such that $Q_m P_r = 0$ and, hence, $Q_m Q_r = Q_m$ (which is possible due to (1.2)). Then $I + Q_r K$ is invertible, and we get

$$Q_r AC(I + Q_r K)^{-1} + P_r(I + Q_r K)^{-1} = I.$$

Multiplying this equality by Q_m from the left-hand side yields the invertibility at infinity of A from the right-hand side. (Observe that the operator $C(I + Q_r K)^{-1}$ lies in $L(E, \mathcal{P})$ by Neumann series.) The invertibility at infinity from the left-hand side follows analogously.

Conversely, (1.19) implies $AC = I - P_m + P_m AC =: I + K_1$ and $DA = I - P_m + DAP_m =: I + K_2$ with K_1, K_2 $\mathcal{P}$-compact. Hence, A is $\mathcal{P}$-Fredholm. □

1.1.4 $\mathcal{P}$-strong convergence

Let $\mathcal{P} = (P_n)_{n=0}^{\infty} \subset L(E)$ be an approximate projection. Theorem 1.1.3 suggests the following definition.

Definition 1.1.13 *Let $A_n \in L(E, \mathcal{P})$. The sequence (A_n)* converges $\mathcal{P}$-strongly *to $A \in L(E)$ if, for all $K \in K(E, \mathcal{P})$, both*

$$\|(A_n - A)K\|_{L(E)} \to 0 \quad \textit{and} \quad \|K(A_n - A)\|_{L(E)} \to 0.$$

In this case we write $A_n \to A$ $\mathcal{P}$-strongly or $A = \mathcal{P}\text{-}\lim A_n$.

The $\mathcal{P}$-strong convergence of bounded sequences can be characterized as follows.

Proposition 1.1.14 *If (A_n) is a bounded sequence in $L(E, \mathcal{P})$, then (A_n) converges $\mathcal{P}$-strongly to $A \in L(E)$ if and only if*

$$\|(A_n - A)P_m\| \to 0 \quad \textit{and} \quad \|P_m(A_n - A)\| \to 0 \quad \textit{for every fixed } P_m \in \mathcal{P}. \quad (1.20)$$

Proof. Since $\mathcal{P} \subseteq K(E, \mathcal{P})$, the $\mathcal{P}$-strong convergence of A_n to A implies (1.20). Conversely, let (1.20) be satisfied for operators A_n and A which are uniformly bounded, and let K be $\mathcal{P}$-compact. Then, for every P_m,

$$\begin{aligned} \|(A_n - A)K\| &\leq \|(A_n - A)P_m K\| + \|(A_n - A)(I - P_m)K\| \\ &\leq \|(A_n - A)P_m\| \, \|K\| + \|A_n - A\| \, \|(I - P_m)K\|. \end{aligned}$$

The right-hand side of this estimate becomes as small as desired if m is large enough. Thus, $\|(A_n - A)K\| \to 0$ for every K, and the dual condition follows analogously. □

For example, it is immediate from (1.2) that $P_n \to I$ $\mathcal{P}$-strongly. On the other hand, (1.20) indicates that the notion of $\mathcal{P}$-strong convergence has some serious defects. For example, the $\mathcal{P}$-strong limit in not unique in general (choose $\mathcal{P}$ as the constant sequence (P) with a non-trivial projection P). So we will have to impose further conditions on $\mathcal{P}$ which guarantee, for example, the uniqueness of the $\mathcal{P}$-strong limit.

Approximate identities. The approximate projection $\mathcal{P} = (P_n)$ is called an *approximate identity* if

$$\sup_n \|P_n x\| \geq \|x\| \quad \text{for each } x \in E, \tag{1.21}$$

and it is called a *symmetric approximate identity* if, besides (1.21),

$$\sup_n \|P_n^* f\| \geq \|f\| \quad \text{for each } f \in E^*.$$

For $A \in L(E, \mathcal{P})$, consider the operators of left and right multiplication

$$A_l : K \mapsto AK \quad \text{and} \quad A_r : K \mapsto KA$$

on the Banach space $K(E, \mathcal{P})$. The following proposition is the $\mathcal{P}$-analogue of Proposition 1.1.4. It shows that $L(E, \mathcal{P})$ is topologically embedded into the Banach algebra $L(K(E, \mathcal{P}))$ under each of the mappings $A \mapsto A_l$ (= left regular representation) and $A \mapsto A_r$ (= right regular representation). As a consequence we get that the $\mathcal{P}$-strong convergence of A_n to A is essentially equivalent to the (usual) strong convergence of the operators $(A_n)_l$ and $(A_n)_r$ to A_l and A_r on $K(E, \mathcal{P})$, respectively.

Proposition 1.1.15 *Let $\mathcal{P} = (P_n) \subseteq L(E)$ be an approximate identity, $A \in L(E, \mathcal{P})$, and set $C_{\mathcal{P}} := \sup \|P_n\|$. Then*

$$\|A_r\|_{L(K(E,\mathcal{P}))} \leq \|A\|_{L(E)} \leq C_{\mathcal{P}} \|A_r\|_{L(K(E,\mathcal{P}))} \tag{1.22}$$

and

$$\|A_l\|_{L(K(E,\mathcal{P}))} \leq \|A\|_{L(E)} \leq C_{\mathcal{P}}^3 \|A_l\|_{L(K(E,\mathcal{P}))}. \tag{1.23}$$

Proof. The first inequality in (1.22) is obvious. For the second one, let $A \in L(E, \mathcal{P})$ and $\varepsilon > 0$. Choose an $x_0 \in E$ with $\|x_0\| = 1$ such that $\|Ax_0\| \geq \|A\| - \varepsilon$, and let $P_n \in \mathcal{P}$ be such that $\|P_n A x_0\| \geq \|Ax_0\| - \varepsilon$, which is possible due to (1.21). Then, since $\mathcal{P} \subset K(E, \mathcal{P})$,

$$\sup_{K \in K(E,\mathcal{P}) \setminus \{0\}} \frac{\|KA\|}{\|K\|} \geq \frac{\|P_n A\|}{\|P_n\|} \geq \frac{\|P_n A x_0\|}{C_{\mathcal{P}}} \geq \frac{1}{C_{\mathcal{P}}} (\|A\| - 2\varepsilon).$$

Thus, $\|A\| \leq C_{\mathcal{P}} \|A_r\|$.

The first inequality in (1.23) is again obvious. Let $\varepsilon > 0$. As we have just seen, there is an m such that

$$\|P_m A\| \geq \|P_m A\| / \|P_m\| \geq \frac{1}{C_{\mathcal{P}}} \|A\| - \varepsilon.$$

Choose n such that $\|P_m A Q_n\| = \|P_m A - P_m A P_n\| < \varepsilon$ (see Proposition 1.1.8 (a)). Then

$$C_{\mathcal{P}}^2 \frac{\|AP_n\|}{\|P_n\|} \geq C_{\mathcal{P}} \|AP_n\| \geq \|P_m A P_n\| \geq \|P_m A\| - \|P_m A Q_n\| \geq \frac{1}{C_{\mathcal{P}}} \|A\| - 2\varepsilon.$$

Hence, $\|A\| \leq C_{\mathcal{P}}^3 \|A_l\|_{L(K(E,\mathcal{P}))}$. □

Corollary 1.1.16 *Let $\mathcal{P}$ be an approximate identity. Then no sequence in $L(E,\mathcal{P})$ possesses more than one $\mathcal{P}$-strong limit.*

Indeed, if $\|K(A_n - A)\| \to 0$ and $\|K(A_n - B)\| \to 0$ for all $\mathcal{P}$-compact operators K, then $\|K(A - B)\| = 0$ for all $K \in K(E,\mathcal{P})$. Hence, $(A - B)_r = 0$, whence $A - B = 0$ by the preceding proposition. □

Here are some properties of $\mathcal{P}$-convergent sequences which are more or less immediate consequences of Propositions 1.1.2 and 1.1.15. In particular the hypotheses in assertion (c) are chosen such that the invertibility of the $\mathcal{P}$-strong limit follows without effort from Proposition 1.1.2. We will discuss this invertibility problem under weaker assumptions for $\mathcal{P}$ in Section 1.1.5.

Proposition 1.1.17 *Let $\mathcal{P}$ be an approximate identity, and let (A_n) and (B_n) be sequences of operators in $L(E,\mathcal{P})$ which converge $\mathcal{P}$-strongly to operators A, $B \in L(E)$, respectively.*

(a) *The operator A belongs to $L(E,\mathcal{P})$, the sequence (A_n) is uniformly bounded, and*
$$\|A\|_{L(E)} \leq C_{\mathcal{P}} \liminf \|A_n\|_{L(E)}.$$
In particular, $L(E,\mathcal{P})$ is a closed subspace of $L(E)$ with respect to $\mathcal{P}$-strong convergence.

(b) *$A_n + B_n \to A + B$ and $A_n B_n \to AB$ $\mathcal{P}$-strongly.*

(c) *If $\mathcal{P}$ is uniform, if the P_n converge *-strongly to the identity operator, and if the operators A_n are invertible for all sufficiently large n and the norms of their inverses are uniformly bounded, then A is invertible, and $A_n^{-1} \to A^{-1}$ $\mathcal{P}$-strongly.*

(d) *If $\mathcal{P}$ is symmetric, then $A_n^* \to A^*$ $\mathcal{P}^*$-strongly.*

Proof. Let $A_n \to A$ $\mathcal{P}$-strongly and $K \in K(E,\mathcal{P})$. Then we have $A_n K \in K(E,\mathcal{P})$ and $KA_n \in K(E,\mathcal{P})$ since, by assumption, $A_n \in L(E,\mathcal{P})$. Further, $\|A_n K - AK\| \to 0$ and $\|KA_n - KA\| \to 0$. Since $K(E,\mathcal{P})$ is closed, this implies that AK and KA belong to $K(E,\mathcal{P})$, i.e., A is in $L(E,\mathcal{P})$.

Consequently, the convergence $\|(A_n - A)K\| \to 0$ for all $K \in K(E,\mathcal{P})$ is equivalent to the strong convergence of the operators A_n to A on $K(E,\mathcal{P})$ if we identify operators on E with their action on $K(E,\mathcal{P})$ as left multiplication operators. Similarly, the condition $\|K(A_n - A)\| \to 0$ for all $K \in K(E,\mathcal{P})$ is equivalent to the strong convergence of the operators A_n to A on $K(E,\mathcal{P})$, but now considered as acting as right multiplication operators on this Banach space. This shows that $\mathcal{P}$-convergence is essentially strong convergence on a specified Banach space. Thus, by the Banach-Steinhaus theorem, the sequence $((A_n)_r)$ of right multiplication operators is uniformly bounded, and $\|A_r\| \leq \liminf \|(A_n)_r\|$. Taking into account (1.22) we arrive at assertion (a). Similarly, (b) follows from Proposition 1.1.2.

To prove (c), notice that the invertibility of A follows from Proposition 1.1.2 (c). By Theorem 1.1.9, the inverse of A belongs to $L(E, \mathcal{P})$. Thus, for every $K \in K(E, \mathcal{P})$, we have $A^{-1}K \in K(E, \mathcal{P})$ and consequently,

$$\|(A^{-1} - A_n^{-1})K\| \leq \|A_n^{-1}\| \, \|(A_n - A)A^{-1}K\| \leq C\|(A_n - A)A^{-1}K\| \to 0.$$

The dual condition follows similarly, and assertion (d) is obvious. □

Approximate identities with specific properties. In what follows we will often meet approximate projections $\mathcal{P} = (P_n)$ with

$$\sup_n \|P_n x\| = \|x\| \quad \text{for every } x \in E. \tag{1.24}$$

Evidently, every approximate projection which satisfies this condition is an approximate identity. Condition (1.24) implies that $\|P_n\| \leq 1$ for all n. Together with (1.2), this shows that $\|P_n\| = 1$ for all sufficiently large n and that $C_{\mathcal{P}} = 1$. Thus, under the assumption (1.24), the inequalities in (1.22) and (1.23) become equalities. Moreover, the limit $\lim_{n\to\infty} \|P_n x\|$ exists and

$$\lim_{n\to\infty} \|P_n x\| = \|x\| \quad \text{for every } x \in E. \tag{1.25}$$

Indeed, let $x \in E$ and $\varepsilon > 0$. Choose m such that $\|x\| - \varepsilon \leq \|P_m x\|$ and N such that

$$\|P_m x\| = \|P_m P_n x\| \leq \|P_n x\| \quad \text{for all } n \geq N.$$

Thus, $\|x\| - \varepsilon \leq \|P_n x\| \leq \|x\|$ for all sufficiently large n.

Moreover, an obvious modification of the proof of Proposition 1.1.15 under the additional assumption (1.24) also yields that

$$\lim \|P_n A\| = \|A\| \quad \text{for all } A \in L(E) \tag{1.26}$$

and

$$\lim \|A P_n\| = \|A\| \quad \text{for all } A \in L(E, \mathcal{P}). \tag{1.27}$$

Another class of approximate identities is provided by approximate projections which converge strongly to the identity operator. (It is again obvious that the strong convergence $P_n \to I$ forces condition (1.21).) We will call an approximate identity $\mathcal{P} = (P_n)$ *perfect* if both $P_n \to I$ and $P_n^* \to I^*$ strongly. Perfect approximate identities are symmetric.

If $\mathcal{P}$ is a perfect approximate identity, then Theorem 1.1.3 implies that $K(E) \subseteq K(E, \mathcal{P})$. This shows that Fredholm operators are $\mathcal{P}$-Fredholm, and that $\mathcal{P}$-strong convergence implies common strong convergence. For the latter, notice that $\mathcal{P}$-strong convergence of (A_n) to A implies $\|A_n K_x - AK_x\| \to 0$ where K_x is as in (1.1), whence $\|A_n x - Ax\| \to 0$ for every $x \in E$.

Conversely, if all operators P_n are compact in the common sense, then every operator $K \in K(E, \mathcal{P})$ is the uniform limit of the compact operators $P_n K$, whence $K(E, \mathcal{P}) \subseteq K(E)$. Thus, if $\mathcal{P}$ is perfect and $\mathcal{P} \subset K(E)$, then $K(E, \mathcal{P}) = K(E)$ and $L(E, \mathcal{P}) = L(E)$, and all '$\mathcal{P}$-notions' reduce to their common meaning.

1.1.5 Invertibility of $\mathcal{P}$-strong limits

In order to get information about the invertibility of the limit of a $\mathcal{P}$-strongly convergent sequence without assuming the symmetry or the strong convergence of the approximate identity $\mathcal{P}$, we have to restrict the class of sequences under consideration. To have a frame in which we can work, let $\mathcal{F}(E)$ stand for the set of all bounded sequences $(A_n)_{n\geq 0}$ of operators in $L(E)$. Provided with the operations

$$(A_n) + (B_n) := (A_n + B_n), \quad (A_n)(B_n) := (A_n B_n), \quad \alpha(A_n) := (\alpha A_n)$$

and with the norm $\|(A_n)\| := \sup \|A_n\|$, this set becomes a Banach algebra with identity element (I). If E is a Hilbert space, then the involution $(A_n)^* := (A_n^*)$ makes $\mathcal{F}(E)$ even to a C^*-algebra.

To motivate the following definition, recall that an operator $A \in L(E)$ belongs to $L(E, \mathcal{P})$ if, for every $m \geq 0$,

$$\|P_m A Q_n\| \to 0 \quad \text{and} \quad \|Q_n A P_m\| \to 0 \quad \text{as } n \to \infty. \tag{1.28}$$

Definition 1.1.18 *A sequence $(A_n) \in \mathcal{F}(E)$ belongs to the class $\mathcal{F}(E, \mathcal{P})$ if, for every $m \geq 0$,*

$$\sup_{k\geq 0} \|P_m A_k Q_n\| \to 0 \quad \textit{and} \quad \sup_{k\geq 0} \|Q_n A_k P_m\| \to 0 \quad \textit{as } n \to \infty.$$

It follows immediately from (1.28) that if $(A_n) \in \mathcal{F}(E, \mathcal{P})$, then every operator A_n belongs to $L(E, \mathcal{P})$ and that, conversely, for every operator $A \in L(E, \mathcal{P})$, the constant sequence (A) belongs to $\mathcal{F}(E, \mathcal{P})$. It is also evident that the sequence (P_n) is in $\mathcal{F}(E, \mathcal{P})$.

Theorem 1.1.19 *$\mathcal{F}(E, \mathcal{P})$ is a closed subalgebra of $\mathcal{F}(E)$. If the approximate identity $\mathcal{P}$ is uniform, then $\mathcal{F}(E, \mathcal{P})$ is inverse closed in $\mathcal{F}(E)$.*

Proof. It is obvious that $\mathcal{F}(E, \mathcal{P})$ is a linear space. We show that the product of two sequences (A_n), $(B_n) \in \mathcal{F}(E, \mathcal{P})$ also belongs to $\mathcal{F}(E, \mathcal{P})$. For m, $r \geq 0$, we have

$$\begin{aligned} \sup_k \|P_m A_k B_k Q_n\| &\leq \\ &\leq \sup_k \|P_m A_k P_r B_k Q_n\| + \sup_k \|P_m A_k Q_r B_k Q_n\| \\ &\leq C \sup_k \|P_r B_k Q_n\| + C \sup_k \|P_m A_k Q_r\|. \end{aligned} \tag{1.29}$$

We choose and fix an r such that the second term in (1.29) becomes as small as desired, and then an n_0 such that the first term in (1.29) becomes smaller than a given constant for all $n \geq n_0$. This shows that $\mathcal{F}(E, \mathcal{P})$ is an algebra, and the proof of its closedness is also fairly standard.

The inverse closedness of $\mathcal{F}(E, \mathcal{P})$ in $\mathcal{F}(E)$ can be verified in exactly the same way as the inverse closedness of $L(E, \mathcal{P})$ in $L(E)$ in Theorem 1.1.9. □

Let E_0 stand for the closure in E of $\cup_m \mathrm{Im}\, P_m$.

Lemma 1.1.20

(a) *E_0 consists of all $x \in E$ with $Q_n x \to 0$.*

(b) *E_0 is a closed linear subspace of E.*

(c) *The subspace E_0 is invariant for operators in $L(E, \mathcal{P})$.*

Proof. Let $x \in E_0$ and $\varepsilon > 0$. Choose $m \in \mathbb{N}$ and $y \in E$ such that $\|x - P_m y\| < \varepsilon$, and choose N such that $Q_n P_m = P_m Q_n = 0$ for all $n \geq N$. Thus, for all $n \geq N$,

$$\|Q_n x\| \leq \|Q_n(x - P_m y)\| + \|Q_n P_m y\| \leq \|Q_n\| \, \|x - P_m y\| \leq C\varepsilon,$$

yielding $Q_n x \to 0$. Conversely, $Q_n x \to 0$ for some $x \in E$ implies $x = \lim P_n x$, i.e., $x \in E_0$. Assertion (b) follows immediately from (a). For (c), let $A \in L(E, \mathcal{P})$ and $x \in E$ be such that $Q_n x \to 0$. Then, for all m,

$$\|Q_n A x\| \leq \|Q_n A P_m x\| + \|Q_n A Q_m x\| \leq \|Q_n A P_m\| \, \|x\| + C\|Q_m x\|,$$

where the right-hand side becomes less than any $\varepsilon > 0$ if m is chosen such that $\|Q_m x\| < \varepsilon/2$ and if n is sufficiently large. Thus, $Q_n A x \to 0$ and $Ax \in E_0$. □

Here is the desired generalization of Proposition 1.1.17 (c).

Proposition 1.1.21 *Let $\mathcal{P}$ be a uniform approximate identity, and let (A_n) be a sequence in $\mathcal{F}(E, \mathcal{P})$ with $\mathcal{P}$-strong limit A. If all operators A_n are invertible, and if the norms of their inverses are uniformly bounded, then $A|_{E_0}$ is invertible, and $A_n^{-1}|_{E_0} \to (A|_{E_0})^{-1}$ strongly.*

Proof. From $P_m x = A_n^{-1} A_n P_m x$ we conclude that $\|P_m x\| \leq C\|A_n P_m x\|$ for all $x \in E$ and $m \geq 0$. Passage to the limit as $n \to \infty$ yields $\|P_m x\| \leq C\|A P_m x\|$, and letting m go to infinity in case $x \in E_0$ shows that

$$\|x\| \leq C\|Ax\| \quad \text{for all } x \in E_0. \tag{1.30}$$

From Proposition 1.1.17 we know that $A \in L(E, \mathcal{P})$. Thus, by Lemma 1.1.20, E_0 is an invariant subspace for A, and from (1.30) we conclude that the operator $A|_{E_0}$ has a trivial kernel and a closed range. We claim that the range of $A|_{E_0}$ is all of E_0. For every k, $m \geq 0$, we have

$$\begin{aligned} \|(I - AA_n^{-1})P_k\| &= \|(A_n - A)A_n^{-1}P_k\| \\ &= \|(A_n - A)(P_m + Q_m)A_n^{-1}P_k\| \\ &\leq C\|(A_n - A)P_m\| + C\|Q_m A_n^{-1} P_k\|. \end{aligned}$$

Given $\varepsilon > 0$, choose and fix m such that $\|Q_m A_n^{-1} P_k\| < \varepsilon$ uniformly with respect to n which can be done since the sequence (A_n^{-1}) belongs to $\mathcal{F}(E)$ and, hence, to $\mathcal{F}(E, \mathcal{P})$ by Theorem 1.1.19. Further, choose n_0 such that $\|(A_n - A)P_m\| < \varepsilon$ for all $n \geq n_0$. These choices guarantee that $\|(I - AA_n^{-1})P_k\| \leq 2C\varepsilon$ for all $n \geq n_0$, whence

$$AA_n^{-1} P_k x \to P_k x \quad \text{for all } x \in E, \; k \geq 0 \tag{1.31}$$

as $n \to \infty$. Since $\mathcal{P}$ is uniform, the operators A_n^{-1} belong to $L(E, \mathcal{P})$, hence, they leave the space E_0 invariant. Thus, (1.31) shows that $P_m x$ is the norm limit of vectors in $\operatorname{Im} A|_{E_0}$. Since we already know that this range is closed, we obtain $P_m x \in \operatorname{Im} A|_{E_0}$ for every $x \in E$ and $m \geq 0$. Employing the closedness of AE_0 once more we get $\operatorname{Im} A|_{E_0} = E_0$ whence the invertibility of $A|_{E_0}$.

Finally, it follows from $\|(I - AA_n^{-1})P_k\| \to 0$ that

$$\|A_n^{-1}P_k - (A|_{E_0})^{-1}P_k\| \leq \|(A|_{E_0})^{-1}\| \, \|(I - AA_n^{-1})P_k\| \to 0,$$

whence the strong convergence of $A_n^{-1}|_{E_0}$ to $(A|_{E_0})^{-1}$. □

If, in particular, $P_n \to I$ strongly, then the preceding proposition guarantees the invertibility of $A = \mathcal{P}\text{-lim}A_n$ on $E = E_0$ without further symmetry assumptions on $\mathcal{P}$ (but only for operators (A_n) in $\mathcal{F}(E, \mathcal{P})$, whereas Proposition 1.1.17 (c) gives this result for sequences in $\mathcal{F}(E)$). Let us also mention that in Chapters 2 and 6 we will meet situations where E_0 is a proper subspace of E, but where E can be identified with the second dual of E_0, which also can be used to prove the invertibility of the $\mathcal{P}$-strong limits on E, not only on E_0.

1.2 Limit operators

The theory of limit operators which we are going to introduce in this section will provide us with an adequate tool to investigate band and band-dominated operators.

1.2.1 Limit operators and the operator spectrum

Let N be a positive integer. We suppose that the additive group $\mathbb{Z}^N$ acts continuously on the Banach space E, i.e., that there is a bounded family $\mathcal{V} = \{V_k\}_{k \in \mathbb{Z}^N}$ of operators $V_k \in L(E)$ such that $V_k V_l = V_{k+l}$ for all k, $l \in \mathbb{Z}^N$ and $V_0 = I$. It is possible to formulate and prove many of the following notions and results also for more general (in particular, non-commutative) locally compact topological groups in place of $\mathbb{Z}^N$. We renounce to consider these more general settings here since they would not only blow up the size of the book essentially, but would also be redundant for most of the applications we have in mind. Only in Chapters 4 and 7, we will have to deal with actions of discrete Heisenberg groups in place of $\mathbb{Z}^N$. In these cases, the analogues of the following notions and results are evident.

Let $\mathcal{H}$ stand for the set of all sequences $h : \mathbb{N} \to \mathbb{Z}^N$ which tend to infinity in the sense that, for every $R > 0$, there is an m_0 such that $|h(m)| > R$ for all $m \geq m_0$. Further we assume that there is an approximate identity $\mathcal{P} = (P_m)$ on E which is related with the group action $\mathcal{V}$ as follows. For every m, $n \in \mathbb{N}$,

$$\text{there is an } R > 0 \text{ such that } P_m V_k P_n = 0 \text{ for all } |k| > R, \tag{1.32}$$

and for every $m \in \mathbb{N}$ and $k \in \mathbb{Z}^N$,

$$\text{there is an } n_0 \in \mathbb{N} \text{ such that } P_m V_k Q_n = Q_n V_k P_m = 0 \text{ for all } n \geq n_0. \tag{1.33}$$

The latter condition ensures that $\mathcal{V} \subseteq L(E, \mathcal{P})$ due to Proposition 1.1.8 (a).

Definition 1.2.1 *Let $A \in L(E, \mathcal{P})$, and let $h \in \mathcal{H}$. The operator $A_h \in L(E)$ is called* limit operator of A with respect to h *if*

$$A_h = \mathcal{P}\text{-lim}_{m\to\infty} V_{-h(m)} A V_{h(m)}. \tag{1.34}$$

The set $\sigma_{op}(A)$ of all limit operators of A is called the operator spectrum*of A.*

An operator can possess only one limit operator with respect to a given sequence (Corollary 1.1.16) which justifies the notation A_h for the limit operator. Observe also that if g is a subsequence of $h \in \mathcal{H}$, then g is also in $\mathcal{H}$, and if the limit operator A_h exists for an operator $A \in L(E, \mathcal{P})$, then A_g also exists, and $A_g = A_h$. Let us further recall from Proposition 1.1.14 that A_h is the limit operator of A with respect to the sequence h if and only if

$$\lim_{n\to\infty} \|(V_{-h(n)} A V_{h(n)} - A_h) P_m\|_{L(E)} = 0$$

and

$$\lim_{n\to\infty} \|P_m (V_{-h(n)} A V_{h(n)} - A_h)\|_{L(E)} = 0$$

for every $P_m \in \mathcal{P}$.

Here are some elementary properties of limit operators which are immediate consequences of Proposition 1.1.17.

Proposition 1.2.2 *Let $\mathcal{P}$ be an approximate identity, let $h \in \mathcal{H}$, and let A, B be operators in $L(E, \mathcal{P})$ for which the limit operators A_h and B_h exist. Then*

(a) *$\|A_h\| \leq C\|A\|$ with a constant C independent of A and h.*

(b) *the limit operators $(A+B)_h$ and $(AB)_h$ exist and $(A+B)_h = A_h + B_h$ and $(AB)_h = A_h B_h$.*

(c) *if $\mathcal{P}$ is perfect and A is invertible, then A_h is invertible, the limit operator $(A^{-1})_h$ exists, and $(A^{-1})_h = (A_h)^{-1}$.*

(d) *if $\mathcal{P}$ is symmetric, then the limit operator $(A^*)_h$ (taken with respect to $\mathcal{P}^*$) exists and $(A^*)_h = (A_h)^*$.*

(e) *if C, $C^{(m)} \in L(E, \mathcal{P})$ are operators with $\|C - C^{(m)}\| \to 0$, and if the limit operators$(C^{(m)})_h$ exist for all sufficiently large m, then the limit operator C_h exists, and $\|C_h - (C^{(m)})_h\| \to 0$.*

Proposition 1.2.3 *The operator spectrum of an operator in $L(E, \mathcal{P})$ is bounded and closed with respect to $\mathcal{P}$-strong convergence.*

Proof. The boundedness of the operator spectrum follows from Proposition 1.2.2 (a). For the proof of its closedness, let $\tilde{A} \in L(E)$ be the $\mathcal{P}$-strong limit of a sequence $(A^{(k)})_{k=1}^{\infty}$ of limit operators of an operator $A \in L(E, \mathcal{P})$.

Due to the definition of a limit operator, there exists an increasing sequence of points $h(1), h(2), \ldots \in \mathbb{Z}^N$ such that

$$\|(V_{-h(k)}AV_{h(k)} - A^{(k)})P_l\| < 1/k \quad \text{for all} \quad l = 1, \ldots, k.$$

and

$$\|P_l(V_{-h(k)}AV_{h(k)} - A^{(k)})\| < 1/k \quad \text{for all} \quad l = 1, \ldots, k.$$

We claim that $\tilde{A}$ is the limit operator of A with respect to the sequence $(h(k))_{k=1}^{\infty}$.

Let $P_m \in \mathcal{P}$ and $\varepsilon > 0$. Choose k_1 such that

$$\|(A^{(k)} - \tilde{A})P_m\| < \varepsilon/2 \quad \text{for all } k \geq k_1.$$

and k_2 such that $1/k_2 < \varepsilon/2$. Then, for all $k \geq \max\{m, k_1, k_2\}$,

$$\|(V_{-h(k)}AV_{h(k)} - \tilde{A})P_m\| \leq \|(V_{-h(k)}AV_{h(k)} - A^{(k)})P_m\| + \|(A^{(k)} - \tilde{A})P_m\| < \varepsilon.$$

The 'dual' condition follows analogously. This proves our claim. □

If B is a limit operator of an operator A then $V_{-k}BV_k$ is also a limit operator of A for every $k \in \mathbb{Z}^N$. Indeed, let h be a sequence in $\mathcal{H}$ such that $B = A_h$. Then

$$V_{-k-h(n)}AV_{h(n)+k} = V_{-k}V_{-h(n)}AV_{h(n)}V_k \to V_{-k}BV_k$$

$\mathcal{P}$-strongly as $n \to \infty$. Thus, operator spectra are *shift invariant.* In combination with the preceding proposition, this has the following nice consequence.

Corollary 1.2.4 *Every limit operator of a limit operator of A is a limit operator of A.*

1.2.2 Operators with rich operator spectrum

Operators $A \in L(E, \mathcal{P})$ which possess limit operators A_h for sufficiently many sequences h are of particular interest.

Definition 1.2.5 *An operator $A \in L(E, \mathcal{P})$ is called an* operator with rich operator spectrum *or simply a* rich operator *if every sequence $h \in \mathcal{H}$ contains an infinite subsequence g such that the limit operator of A with respect to g exists. The set of all operators with rich spectrum will be denoted by $L^{\$}(E, \mathcal{P})$.*

For example, the operators V_k are rich, and $\sigma_{op}(V_k) = \{V_k\}$ for all $k \in \mathbb{Z}^N$. The richness of $\mathcal{P}$-compact operators is part of the following proposition. Moreover, assertion (c) of that proposition states that richness is essentially a sequential compactness condition with respect to $\mathcal{P}$-convergence.

Proposition 1.2.6 *Let $\mathcal{P}$ be an approximate identity.*

(a) *$L^\$(E, \mathcal{P})$ is a closed subalgebra of $L(E, \mathcal{P})$.*

(b) *$K(E, \mathcal{P})$ is a closed ideal of $L^\$(E, \mathcal{P})$. In particular, $\sigma_{op}(K) = \{0\}$ for every $\mathcal{P}$-compact operator K.*

(c) *An operator $A \in L(E, \mathcal{P})$ is rich if and only if the set $\{V_{-k}AV_k\}_{k\in\mathbb{Z}^N}$ of its shifts is relatively sequentially compact with respect to the $\mathcal{P}$-strong convergence.*

Proof. (a) Let A, $B \in L^\$(E, \mathcal{P})$ and $h \in \mathcal{H}$. By hypothesis, there are subsequences f of h and g of f such that A_f and B_g exist. From Proposition 1.1.12 (b) we know that then $(A+B)_g$ and $(AB)_g$ exist, hence, $L^\$(E,\mathcal{P})$ is an algebra.

Now let $(A^{(k)})_{k=1}^\infty$ be a sequence in $L^\$(E, \mathcal{P})$ which tends uniformly to an operator $A \in L(E, \mathcal{P})$, and let $h \in \mathcal{H}$. By hypothesis, we can find a subsequence g_1 of h such that $A^{(1)}_{g_1}$ exists, then a subsequence g_2 of g_1 such that $A^{(2)}_{g_2}$ exists, and so on. Proceeding in this way, we obtain for each $k \geq 2$ a subsequence g_k of g_{k-1} such that $A^{(k)}_{g_k}$ exists. Define a new sequence g by $g(k) := g_k(k)$. Evidently, g is a subsequence of h, and the limit operators $A^{(m)}_g$ exist for all m. Then, by Proposition 1.2.2 (e), the limit operator A_g exists, too. Thus, $A \in L^\$(E, \mathcal{P})$, showing the closedness of $L^\$(E, \mathcal{P})$ in $L(E, \mathcal{P})$.

(b) Our next objective is the inclusion $K(E, \mathcal{P}) \subset L^\$(E, \mathcal{P})$. Let $h \in \mathcal{H}$ and $P_m \in \mathcal{P}$. Then, for every $K \in K(E, \mathcal{P})$ and $P_n \in \mathcal{P}$,

$$\begin{aligned} \|V_{-h(k)}KV_{h(k)}P_m\| &\leq \|V_{-h(k)}KP_nV_{h(k)}P_m\| + \|V_{-h(k)}KQ_nV_{h(k)}P_m\| \\ &\leq C\|P_nV_{h(k)}P_m\| + C\|KQ_n\| \end{aligned}$$

with a constant C (recall the uniform boundedness of the families $\mathcal{P}$ and $\mathcal{V}$). Given $\varepsilon > 0$, choose n such that $C\|KQ_n\| < \varepsilon$ and then k_0 such that $C\|P_nV_{h(k)}P_m\| < \varepsilon$ for all $k \geq k_0$ (compare (1.32)). Then $\|V_{-h(k)}KV_{h(k)}P_m\| \leq 2\varepsilon C$ for all $k \geq k_0$. The convergence $\|P_mV_{-h(k)}KV_{h(k)}\| \to 0$ follows analogously. Hence, $K(E, \mathcal{P}) \subset L^\$(E, \mathcal{P})$, and the operator spectrum of a $\mathcal{P}$-compact operator is the singleton $\{0\}$.

(c) Let $(V_{-h(n)}AV_{h(n)})_{n\in\mathbb{N}}$ be a sequence in $\{V_{-k}AV_k\}_{k\in\mathbb{Z}^N}$. If the sequence h is bounded, then it possesses a constant subsequence g, and $(V_{-g(n)}AV_{g(n)})_{n\in\mathbb{N}}$ is evidently a convergent sequence. If h is unbounded, then it has a subsequence l which tends to infinity. Then, by definition of richness, there is a subsequence g of l such that $(V_{-g(n)}AV_{g(n)})_{n\in\mathbb{N}}$ converges $\mathcal{P}$-strongly. The reverse implication is obvious. □

Corollary 1.2.7 *If $A \in L^\$(E, \mathcal{P})$, then $\sigma_{op}(A)$ is sequentially compact with respect to the $\mathcal{P}$-strong convergence.*

Indeed, the set $\{V_{-k}AV_k\}_{k\in\mathbb{Z}^N}$ is relatively sequentially compact with respect to the $\mathcal{P}$-strong convergence by Proposition 1.2.6 (c). By definition, the operator spectrum of A is contained in the $\mathcal{P}$-strong closure of this set, and the operator spectrum is $\mathcal{P}$-strong closed by Proposition 1.2.3. Being a closed subset of a sequentially compact set, $\sigma_{op}(A)$ is sequentially compact with respect to the $\mathcal{P}$-strong convergence.

The perfect case. In what follows we will mainly be concerned with the $\mathcal{P}$-Fredholm properties of operators in $L^{\$}(E, \mathcal{P})$ and with the invertibility of their limit operators. The most complete information on invertibility of limit operators, which we have at our disposal is Proposition 1.2.2 (c), where a *perfect* approximate identity is assumed. So we start with focusing our attention on the perfect case.

Proposition 1.2.8 *Let $\mathcal{P}$ be a perfect approximate identity. Then $L^{\$}(E, \mathcal{P})$ is inverse closed in $L(E, \mathcal{P})$. If $\mathcal{P}$ is moreover uniform, then $L^{\$}(E, \mathcal{P})$ is inverse closed in $L(E)$.*

Proof. Let $A \in L^{\$}(E, \mathcal{P})$ be invertible in $L(E, \mathcal{P})$, and let $h \in \mathcal{H}$. Then there is a subsequence g of h such that A_g exists. By Proposition 1.2.2 (c), the limit operator $(A^{-1})_g$ also exists. Thus, A^{-1} has rich spectrum, which reveals the inverse closedness in $L(E, \mathcal{P})$. The inverse closedness in $L(E)$ follows from Theorem 1.1.9. □

Proposition 1.2.9 *Let $\mathcal{P}$ be a perfect approximate identity. If $A \in L(E, \mathcal{P})$ is a $\mathcal{P}$-Fredholm operator, then all limit operators of A are invertible, and the norms of their inverses are uniformly bounded.*

Proof. Let $A \in L(E, \mathcal{P})$ be a $\mathcal{P}$-Fredholm operator, i.e., there exist operators $D \in L(E, \mathcal{P})$ and $T_1, T_2 \in K(E,\mathcal{P})$ such that $DA = I + T_1$ and $AD = I + T_2$. If $h \in \mathcal{H}$ is a sequence such that the limit operator A_h exists, then, for every $\mathcal{P}$-compact operator K,

$$K = V_{-h(n)} I V_{h(n)} K = V_{-h(n)} D V_{h(n)} V_{-h(n)} A V_{h(n)} K - V_{-h(n)} T_1 V_{h(n)} K$$

and, consequently,

$$\|K\| \leq C \|V_{-h(n)} A V_{h(n)} K\| + \|V_{-h(n)} T_1 V_{h(n)} K\|.$$

Passage to the limit $n \to \infty$ yields

$$\|K\| \leq C\|A_h K\| \quad \text{for all } K \in K(E, \mathcal{P})$$

(recall from Proposition 1.2.6 that $(T_1)_h$ exists and is the zero operator). Analogously,

$$\|K\| \leq C\|A_h^* K\| \quad \text{for all } K \in K(E^*, \mathcal{P}^*).$$

Since $\mathcal{P}$ is perfect, we have $K(E) \subseteq K(E, \mathcal{P})$ as well as $K(E^*) \subseteq K(E^*, \mathcal{P}^*)$. Thus, we can replace K by a rank one operator as in (1.1) in the latter two estimates to get

$$\|x\| \leq C\|A_h x\| \quad \text{and} \quad \|f\| \leq C\|A_h^* f\|$$

for all $x \in E$ and $f \in E^*$. Hence, A_h is invertible, and $\|(A_h)^{-1}\| \leq C$. □

Proposition 1.2.10 *Let $\mathcal{P}$ be a perfect and uniform approximate identity. Then $L^{\$}(E, \mathcal{P})/K(E, \mathcal{P})$ is an inverse closed subalgebra of $L(E, \mathcal{P})/K(E, \mathcal{P})$.*

Proof. Let $A \in L^{\$}(E, \mathcal{P})$ be a $\mathcal{P}$-Fredholm operator and let $D \in L(E, \mathcal{P})$ be a regularizer of A (i.e., the coset $D + K(E, \mathcal{P})$ is the inverse of $A + K(E, \mathcal{P})$). Since any two regularizers of A differ by a $\mathcal{P}$-compact operator, and since every $\mathcal{P}$-compact operator belongs to $L^{\$}(E, \mathcal{P})$, it is sufficient to prove that $D \in L^{\$}(E, \mathcal{P})$ for at least one regularizer of A. We choose D such that $DAQ_m = Q_m$ for some $m \in \mathbb{N}$ (compare Proposition 1.1.12).

Now let $h \in \mathcal{H}$. Since A is rich, there is a subsequence g of h for which A_g exists. For every $\mathcal{P}$-compact operator K we then have

$$
\begin{aligned}
&V_{-g(n)}DV_{g(n)}A_gK \\
&\quad = V_{-g(n)}DV_{g(n)}(A_g - V_{-g(n)}AQ_mV_{g(n)})K + V_{-g(n)}DAQ_mV_{g(n)}K \\
&\quad = V_{-g(n)}DV_{g(n)}(A_g - V_{-g(n)}AV_{g(n)}V_{-g(n)}Q_mV_{g(n)})K + V_{-g(n)}Q_mV_{g(n)}K.
\end{aligned}
$$

The right-hand side of this equality tends to K in the operator norm (observe that $\|V_{-g(n)}Q_mV_{g(n)}K - K\| = \|V_{-g(n)}P_mV_{g(n)}K\| \to 0$ by Proposition 1.2.6). Hence,

$$V_{-g(n)}DV_{g(n)}A_g \to I \quad \mathcal{P}\text{-strongly as } n \to \infty.$$

Since A_g is invertible (Proposition 1.2.9), and since $L(E, \mathcal{P})$ is inverse closed in $L(E)$ (Theorem 1.1.9), this yields the $\mathcal{P}$-strong convergence of $V_{-g(n)}DV_{g(n)}$ to A_g^{-1}. □

The preceding results suggest that limit operators are closely related with Fredholmness. One of our main objectives in this book is to single out classes of operators for which the converse of Proposition 1.2.9 is true.

The non-perfect case. As in Section 1.1.5, to derive the invertibility of the limit operators of an invertible operator under weak conditions for $\mathcal{P}$, we have to restrict the class of operators under consideration.

Definition 1.2.11 *An operator $A \in L(E)$ belongs to the class $L(E, \mathcal{P}, \mathcal{V})$ if, for every $m \geq 0$,*

$$\sup_{k \in \mathbb{Z}^N} \|P_mV_{-k}AV_kQ_n\| \to 0 \quad \text{and} \quad \sup_{k \in \mathbb{Z}^N} \|Q_nV_{-k}AV_kP_m\| \to 0 \quad \text{as } n \to \infty.$$

Theorem 1.2.12 *Let $\mathcal{P}$ be an approximate identity. Then $L(E, \mathcal{P}, \mathcal{V})$ is a closed subalgebra of $L(E)$ which contains $K(E, \mathcal{P})$ as its closed ideal. If $\mathcal{P}$ is uniform, then $L(E, \mathcal{P}, \mathcal{V})$ is inverse closed in $L(E)$.*

Proof. The closedness and inverse closedness of $L(E, \mathcal{P}, \mathcal{V})$ can be checked as in Theorem 1.1.19 (one can also refer directly to this theorem because, after rearranging, the sequence $(V_{-k}AV_k)$ can be viewed of as an element of $\mathcal{F}(E, \mathcal{P})$).

We are left with the implication $K(E, \mathcal{P}) \subset L(E, \mathcal{P}, \mathcal{V})$. Let $K \in K(E, \mathcal{P})$. Since $\|P_sKP_s - K\| \to 0$ as $s \to \infty$, and since $L(E, \mathcal{P}, \mathcal{V})$ is closed, it is sufficient to show that $P_sKP_s \in L(E, \mathcal{P}, \mathcal{V})$ for every s. Let us check, for example, that

$$\lim_{n\to\infty} \sup_{k\in\mathbb{Z}^N} \|P_m V_{-k} P_s K P_s V_k Q_n\| = 0$$

for every $m \in \mathbb{N}$. By (1.32), there is an $R > 0$ such that $P_m V_{-k} P_s = 0$ for all $|k| \geq R$. So it remains to show that

$$\lim_{n\to\infty} \|P_m V_{-k} P_s K P_s V_k Q_n\| = 0 \quad \text{for every } k \in \mathbb{Z}^N \text{ with } |k| < R.$$

This assertion follows immediately from (1.33). □

Specifying Proposition 1.1.21 to the present context essentially yields the following.

Proposition 1.2.13 *Let $\mathcal{P}$ be a uniform approximate identity. If $A \in L(E, \mathcal{P}, \mathcal{V})$ is invertible and if the limit operator A_h exists for a sequence $h \in \mathcal{H}$, then $A_h|_{E_0}$ is invertible, $(A^{-1}|_{E_0})_h$ exists, and $(A_h|_{E_0})^{-1} = (A^{-1}|_{E_0})_h$. Moreover, the norms of the inverses of the operators $A_h|_{E_0}$ are uniformly bounded.*

Proof. Let $A \in L(E, \mathcal{P}, \mathcal{V})$ be invertible, and let $h \in \mathcal{H}$ be a sequence for which the limit operator A_h exists. Then the sequence $(V_{-h(n)}AV_{h(n)})$ belongs to $\mathcal{F}(E, \mathcal{P})$ and is invertible in $\mathcal{F}(E)$. Thus, $A_h|_{E_0}$ is an invertible operator by Proposition 1.1.21. The $\mathcal{P}$-strong convergence of $V_{-h(n)}A^{-1}|_{E_0}V_{h(n)}$ to $(A_h|_{E_0})^{-1}$ on E_0 follows in the standard way:

$$\begin{aligned} &\|V_{-h(n)}A^{-1}V_{h(n)}P_m - (A_h|_{E_0})^{-1}P_m\| \\ &\quad \leq \sup_k \|V_k\|^2 \, \|A^{-1}\| \, \|(A_h - V_{-h(n)}AV_{h(n)})(A_h|_{E_0})^{-1}P_m\| \to 0 \end{aligned}$$

and the dual assertion can be checked analogously. Finally, the uniform boundedness of the inverses of the limit operators is a consequence of

$$\|x\| < C \, \|A_h x\| \quad \text{for all } x \in E_0$$

with a constant C independent of the sequence h which can be seen as in (1.30). □

If, in particular, $P_n \to I$ strongly, then Proposition 1.2.13 yields the invertibility of A_h on $E = E_0$ without symmetry assumptions for $\mathcal{P}$ (but only for operators A in $L(E, \mathcal{P}, \mathcal{V})$, whereas Proposition 1.2.9 gives this result for operators in $L(E, \mathcal{P})$).

1.3 Algebraization

The algebraic properties of the mapping $A \mapsto A_h$ are summarized in Proposition 1.2.2. It says essentially that if $h \in \mathcal{H}$ and if $\mathcal{B}$ is a subalgebra of $L(E, \mathcal{P})$ such that the limit operator A_h exists for every $A \in \mathcal{B}$, then the mapping $A \mapsto A_h$ is an algebra homomorphism from $\mathcal{B}$ into $L(E)$. But, of course, this mapping is not an algebra homomorphism on all of $L^\$(E, \mathcal{P})$; it is not even defined on $L^\$(E, \mathcal{P})$. In this section we will briefly discuss two ways to algebraize the concept of limit operators in order to get algebra homomorphisms.

1.3.1 Algebraization by restriction

Here we will single out subalgebras of $L^\$(E, \mathcal{P})$ on which the mapping $A \mapsto A_h$ acts homomorphically.

Proposition 1.3.1 *Let $\mathcal{B}$ be a separable subset of $L^\$(E, \mathcal{P})$. Then every sequence $h \in \mathcal{H}$ possesses a subsequence g such that the limit operator A_g exists for every $A \in \mathcal{B}$.*

Proof. Let $\{B_n\}_{n\geq 1}$ be a dense subset of $\mathcal{B}$, and let $h_0 := h \in \mathcal{H}$. Since the operators B_n are rich, we find, for every $n \geq 1$, a subsequence h_n of h_{n-1} such that the limit operator of B_n with respect to h_n exists. Set $g(n) := h_n(n)$. Then $g \in \mathcal{H}$, and the limit operators $(B_n)_g$ exist for every n.

If now $A \in \mathcal{B}$ and $h \in \mathcal{H}$, then there is a subsequence (A_n) of $\{B_n\}$ which tends to A in the norm, and there is a subsequence g of h such that all limit operators $(A_n)_g$ exist. By Proposition 1.2.2 (e), the limit operator A_g exists, too. □

For each subset $\mathcal{B}$ of $L(E, \mathcal{P})$, we let $\mathcal{H}_\mathcal{B}$ stand for the set of all sequences $h \in \mathcal{H}$ for which the limit operator A_h exists for each $A \in \mathcal{B}$.

Proposition 1.3.2 *Let $\mathcal{B}$ be a separable subalgebra of $L^\$(E, \mathcal{P})$. Then the mapping $A \mapsto A_h$ is a continuous algebra homomorphism on $\mathcal{B}$ for every $h \in \mathcal{H}_\mathcal{B}$, and*

$$\sigma_{op}(A) = \{A_h : h \in \mathcal{H}_\mathcal{B}\} \quad \textit{for every } A \in \mathcal{B}.$$

Proof. The first assertion is a consequence of Proposition 1.2.2 (and holds for any subalgebra of $L(E, \mathcal{P})$). For the second one, let A_h be a limit operator of $A \in \mathcal{B}$. By Proposition 1.3.1, there is a subsequence g of h which belongs to $\mathcal{H}_\mathcal{B}$. Thus, $A_h = A_g \in \{A_h : h \in \mathcal{H}_\mathcal{B}\}$. □

One can push this result a little bit further by including $\mathcal{P}$-compact operators.

Proposition 1.3.3 *Let $\mathcal{B}$ be a separable subalgebra of $L^\$(E, \mathcal{P})$, and let $\mathcal{A}$ be the smallest closed subalgebra of $L(E)$ which contains $\mathcal{B}$ and the ideal $K(E, \mathcal{P})$. Then every sequence $h \in \mathcal{H}$ possesses a subsequence g such that the limit operator A_g exists for every $A \in \mathcal{B}$, the mapping $A \mapsto A_h$ is a continuous algebra homomorphism on $\mathcal{A}$ for every $h \in \mathcal{H}_\mathcal{A}$, and*

$$\sigma_{op}(A) = \{A_h : h \in \mathcal{H}_\mathcal{A}\} \quad \textit{for every } A \in \mathcal{A}.$$

Proof. Let $A \in \mathcal{A}$, and let $\{B_n\}_{n\geq 1}$ be a dense subset of $\mathcal{B}$. There are a subsequence (A_n) of $\{B_n\}$ and a sequence (K_n) of $\mathcal{P}$-compact operators such that $A_n + K_n$ tends to A in the norm. The operators $A_n + K_n$ are rich by assumption and by Proposition 1.2.6. Thus, the assertion follows as in the preceding proposition. □

From Proposition 1.2.6 we further know that $K_h = 0$ for every $\mathcal{P}$-compact operator K and every sequence $h \in \mathcal{H}$. Thus, if $\mathcal{A}$ is as in Proposition 1.3.3 and $h \in \mathcal{H}_{\mathcal{A}}$, then the mapping

$$W_h : \mathcal{A}/K(E, \mathcal{P}) \to L(E), \quad A + K(E, \mathcal{P}) \mapsto A_h \tag{1.35}$$

is well defined, and it is a continuous algebra homomorphism. This homomorphism is unital if $\mathcal{A}$ contains the identity operator.

1.3.2 Symbol calculus

In this section we will see how to associate an operator-valued function smb A with every operator $A \in L^{\$}(E, \mathcal{P})$ such that the set of the values of smb A coincides with the operator spectrum $\sigma_{op}(A)$ and such that the mapping $A \mapsto \text{smb}\, A$ is a continuous algebra homomorphism. In this case we call smb a *symbol mapping* and smb A the *symbol* of A.

Given $A \in L(E, \mathcal{P})$, let $\mathcal{H}_A$ denote the set of all sequences $h \in \mathcal{H}$ such that the limit operator A_h exists. A most natural candidate for the symbol of A is the function

$$\text{smb}^0 A : \mathcal{H}_A \to L(E), \quad h \mapsto A_h. \tag{1.36}$$

There arise obvious difficulties if one wants to add or to multiply such 'symbols'. Indeed, $\text{smb}^0 A$ and $\text{smb}^0 B$ are defined on $\mathcal{H}_A$ and $\mathcal{H}_B$, respectively. Thus, their 'sum' $\text{smb}^0 A + \text{smb}^0 B$ is naturally defined on $\mathcal{H}_A \cap \mathcal{H}_B$ only. On the other hand, the 'symbol' $\text{smb}^0(A+B)$ of $A+B$ is defined on $\mathcal{H}_{A+B}$ which can be a much larger set than $\mathcal{H}_A \cap \mathcal{H}_B$. For example, $\mathcal{H}_A \cap \mathcal{H}_{-A} = \mathcal{H}_A$, whereas $\mathcal{H}_{A-A} = \mathcal{H}_0 = \mathcal{H}$. Thus, the mapping smb^0 defined by (1.36) is not a symbol mapping in our sense. To avoid these difficulties, we are going to introduce an equivalence relation which will allow us to identify $\text{smb}^0 A + \text{smb}^0(-A)$ with $\text{smb}^0 0$ for all $A \in L^{\$}(E, \mathcal{P})$. For, we first have to define an equivalence relation on $\mathcal{H}$ as follows. Two sequences $g, h \in \mathcal{H}$ are *equivalent* if there exist $k, l \in \mathbb{N}$ such that

$$g(k+n) = h(l+n) \quad \text{for all } n \in \mathbb{N}. \tag{1.37}$$

Definition 1.3.4 *A subset D of $\mathcal{H}$ is an* admissible domain *if the following conditions are satisfied:*

(a) *Every sequence $h \in \mathcal{H}$ possesses a subsequence which belongs to D.*

(b) *If $h \in D$, and if $g \in \mathcal{H}$ is equivalent to a subsequence of h, then $g \in D$.*

Evidently, $\mathcal{H}$ itself is an admissible domain, and so are all sets $\mathcal{H}_A$ with $A \in L^{\$}(E, \mathcal{P})$.

Proposition 1.3.5 *The intersection of an at most countable set of admissible domains is an admissible domain again.*

Proof. Let D_1, D_2, ... be admissible domains and $D^* := \cap_k D_k$. It is evident that D^* satisfies property (b) of an admissible domain. We claim that D^* is subject to condition (a).

Let $h \in \mathcal{H}$. Since D_1 is admissible, there is a subsequence h_1 of h which belongs to D_1. Analogously, there is a subsequence h_2 of h_1 which belongs to D_2, and so on. We form a new sequence h^* by $h^*(k) := h_k(k)$. This sequence belongs to D^*. Indeed, the sequence h^* is equivalent to a subsequence of the subsequence $(h^*(k+1), h^*(k+2), \ldots)$ of h_k. Hence, h^* belongs to D_k for every k due to property (b) of the admissible domain D_k. □

Let $\mathcal{F}_E^0$ denote the set of all bounded functions X which are defined on an admissible domain $D(X) \subseteq \mathcal{H}$ and take values in $L(E)$, and which own the following property: If $h \in D(X)$ and if $g \in \mathcal{H}$ is equivalent to a subsequence of h then $X(h) = X(g)$. For example, all functions $\text{smb}^0 A$ defined by (1.36) belong to $\mathcal{F}_E^0$.

We provide $\mathcal{F}_E^0$ with operations as follows: Given X, $Y \in \mathcal{F}_E^0$, the *sum* $X+Y$ (resp. the *product* XY) is the function which is defined on $D(X) \cap D(Y)$ and takes the value $X(h) + Y(h)$ (resp. $X(h)Y(h)$) at $h \in D(X) \cap D(Y)$. Further, if $\alpha \in \mathbb{C}$, then αX is the function which is defined on $D(X)$ and takes the value $\alpha X(h)$ at $h \in D(X)$. Finally, we set

$$\|X\| := \sup\{\|X(h)\|, \; h \in D(X)\}.$$

Evidently, both operations satisfy the associativity and distributivity laws, and the addition is also commutative. The (additive) zero and (multiplicative) identity element are the functions

$$0 : \mathcal{H} \to L(E), \quad h \mapsto 0 \quad \text{and} \quad I : \mathcal{H} \to L(E), \quad h \mapsto I$$

with $D(0) = D(I) = \mathcal{H}$ respectively. But observe that, in general, $X - X \neq 0$ because of the different domains of definition of the functions $X - X$ and 0. Also, $\|.\|$ satisfies the usual properties of a norm with the only exception that $\|X\| = 0$ does not necessarily imply $X = 0$.

The announced equivalence relation on $\mathcal{F}_E^0$ which is aimed to solve these problems is defined by

$$X \sim Y \quad \text{if and only if} \quad X|_{D(X)\cap D(Y)} = Y|_{D(X)\cap D(Y)}.$$

Proposition 1.3.6 *The relation $\sim$ is an equivalence relation on $\mathcal{F}_E^0$ which is compatible with the operations and with the norm.*

Proof. The reflexivity and symmetry of $\sim$ are evident. For the transitivity of $\sim$, let $X \sim Y$ and $Y \sim Z$, and let $h \in D(X) \cap D(Z)$. By property (a) of admissible domains, there is a subsequence g of h which lies in $D(Y)$. Further, by property (b),

g belongs to every of the sets $D(X)$, $D(Y)$, $D(Z)$. Now we have $X(h) = X(g) = Y(g)$ because of $X \sim Y$, and we have $Z(h) = Z(g) = Y(g)$ because of $Y \sim Z$. Hence, $X(h) = Y(g) = Z(h)$ for all $h \in D(X) \cap D(Z)$.

Let now $X_1 \sim Y_1$ and $X_2 \sim Y_2$. We will show that this implies $X_1 + X_2 \sim Y_1 + Y_2$. The compatibility of $\sim$ with the other operations follows analogously.

What we have to check is the equality

$$X_1 + X_2|_{D(X_1+X_2)\cap D(Y_1+Y_2)} = Y_1 + Y_2|_{D(X_1+X_2)\cap D(Y_1+Y_2)}.$$

By definition, $D(X_1 + X_2) = D(X_1) \cap D(X_2)$ and $D(Y_1 + Y_2) = D(Y_1) \cap D(Y_2)$. So, our claim is equivalent to

$$X_1 + X_2|_{D(X_1)\cap D(X_2)\cap D(Y_1)\cap D(Y_2)} = Y_1 + Y_2|_{D(X_1)\cap D(X_2)\cap D(Y_1)\cap D(Y_2)}.$$

Let $h \in D(X_1) \cap D(X_2) \cap D(Y_1) \cap D(Y_2)$. Then

$$X_1(h) = Y_1(h) \quad \text{since } h \in D(X_1) \cap D(Y_1) \text{ and } X_1 \sim Y_1,$$

$$X_2(h) = Y_2(h) \quad \text{since } h \in D(X_2) \cap D(Y_2) \text{ and } X_2 \sim Y_2.$$

Hence, $X_1(h) + X_2(h) = Y_1(h) + Y_2(h)$.

Finally, if $W(X) \subseteq L(E)$ denotes the set of the values of the function X, then $X \sim Y$ implies that $W(X) = W(Y)$. Indeed, if $A \in W(X)$, then there is an $h \in D(X)$ with $X(h) = A$, and we can find a subsequence g of h which is in both $D(X)$ and $D(Y)$. Hence, $A = X(h) = X(g) = Y(g)$, i.e., $A \in W(Y)$. This shows that $\|X\| = \|Y\|$ whenever $X \sim Y$. □

Let $\mathcal{F}_E$ denote the set of all equivalence classes $X^\sim$ of elements of $\mathcal{F}_E^0$ with respect to the equivalence relation $\sim$. By the preceding proposition, it is correct to define operations on $\mathcal{F}_E$ by

$$X^\sim + Y^\sim := (X+Y)^\sim, \quad X^\sim Y^\sim := (XY)^\sim, \quad \alpha X^\sim := (\alpha X)^\sim$$

and a norm on $\mathcal{F}_E$ by

$$\|X^\sim\| := \|X\|.$$

Proposition 1.3.7 *$\mathcal{F}_E$ is a Banach algebra.*

Proof. The proof is straightforward. We will only check that $\|X^\sim\| = 0$ implies $X \sim 0$ and that the normed space $\mathcal{F}_E$ is complete.

Let $\|X^\sim\| = 0$ and let X be a representative of the coset $X^\sim$. Then $X = 0$ on $D(X)$ and

$$X|_{D(X)\cap D(0)} = X|_{D(X)} = 0|_{D(X)} = 0|_{D(X)\cap D(0)},$$

hence, $X \sim 0$. Together with the above remarks this shows that $\mathcal{F}_E$ is a normed algebra.

Let now $(X_n^\sim)$ be a Cauchy sequence in $\mathcal{F}_E$, choose representatives $X_n \in X_n^\sim$, and set $D^* := \cap_n D(X_n)$.

The Cauchy property of $(X_n^\sim)$ implies that, given $\varepsilon > 0$ there is an n_0 such that

$$\|X_n|_{D^*} - X_m|_{D^*}\| \le \varepsilon \quad \text{for all } m,\, n \ge n_0. \tag{1.38}$$

In particular, $(X_n(h))$ is a Cauchy sequence in $L(E)$ for every $h \in D^*$. Hence, $(X_n(h))$ is convergent, and we can define a function X on D^* by

$$X(h) := \lim_{n\to\infty} X_n(h) \quad \text{for every } h \in D^*.$$

The function belongs to $\mathcal{F}_E^0$. Indeed, $D(X) = D^*$ is an admissible domain due to Proposition 1.3.5, and if g is equivalent to a subsequence of $h \in D^*$, then $X_n(g) = X_n(h)$ for all n and, thus, $X(g) = X(h)$.

Our next goal is the boundedness of X. From (1.38) we conclude

$$\sup_{h\in D^*} \|X_n(h) - X_m(h)\| \le \varepsilon \quad \text{for all } m,\, n \ge n_0.$$

Letting m go to infinity yields

$$\sup_{h\in D^*} \|X_n(h) - X(h)\| \le \varepsilon \quad \text{for every } n \ge n_0 \tag{1.39}$$

and, consequently, $\|X\| \le \|X_n|_{D^*}\| + \varepsilon \le \|X_n\| + \varepsilon$.

So we have $X \in \mathcal{F}_E^0$, and the convergence of $X_n^\sim$ to $X^\sim$ in $\mathcal{F}_E$ follows from

$$\|X_n^\sim - X^\sim\| = \sup_{h\in D^*} \|X_n(h) - X(h)\|$$

and (1.39). □

In case E is a Hilbert space, one can define an involution on $\mathcal{F}_E$ in an obvious way which makes $\mathcal{F}_E$ to a C^*-algebra.

Definition 1.3.8 *The* symbol *of the operator* $A \in L^\$(E, \mathcal{P})$ *is the coset* smb $A := (\text{smb}^0 A)^\sim$ *with* $\text{smb}^0 A$ *defined by* (1.36).

This definition is correct because $\mathcal{H}_A$ is an admissible domain and $\text{smb}^0 A \in \mathcal{F}_E^0$ for all $A \in L^\$(E, \mathcal{P})$.

Corollary 1.3.9 *The mapping* smb : $L^\$(E, \mathcal{P}) \to \mathcal{F}_E$ *is a continuous algebra homomorphism.*

Proof. Let us check, for example, that $\text{smb}\, A + \text{smb}\, B = \text{smb}\,(A+B)$. By definition, $\text{smb}^0 A + \text{smb}^0 B$ is the function, which is defined on $\mathcal{H}_A \cap \mathcal{H}_B$ by $h \mapsto A_h + B_h$, and $\text{smb}^0(A+B)$ is the function defined on $\mathcal{H}_{A+B}$ by $h \mapsto (A+B)_h$. Both functions coincide on $\mathcal{H}_A \cap \mathcal{H}_B \cap \mathcal{H}_{A+B}$. □

Corollary 1.3.10 *If* $A \in L^\$(E,\mathcal{P})$ *is* $\mathcal{P}$*-Fredholm, then* smb A *is invertible in* $\mathcal{F}_E$.

1.4 Comments and references

The first appearance of limit operators is in Favard's paper [53], where they are used to verify the existence of almost-periodic solutions of ordinary differential equations with almost-periodic coefficients. Later, Muhamadiev [109, 110] applied limit operators to the question of solvability of elliptic partial differential equations in $\mathbb{R}^n$. The limit operators method has been developed further in the papers [93, 95, 94, 118, 126, 137] for the study of the Fredholm property of wide classes of pseudodifferential operators and convolution operators on $\mathbb{R}^n$ and $\mathbb{Z}^n$. Note also the paper [27], where the limit operators method has been applied to the computation of the essential spectrum of singular integral operators on Carleson curves acting in general weighted L^2-spaces. See also the monograph [72] for lots of applications in numerical analysis. We will give impressions of these results in Chapters 3, 4 and 6. Observe that in all of these papers, the method of limit operators is applied to a concrete class of operators acting on a concrete Banach space.

Generalized notions of compactness have been introduced in [31, 52, 149], for example. Theorem 1.1.9 and its proof are (with minor modifications) taken from Kozak and Simonenko [87]. The original formulation of their result says that the inverse of an operator of local type is of local type again. It has been already employed in [149] to study the stability of approximation methods on spaces with supremum norm.

There are other algebras besides $L(E, \mathcal{P})$ which can be associated with an approximate projection $\mathcal{P}$ in a natural way. A good candidate would be $OLT(E, \mathcal{P})$, the algebra of all operators $A \in L(E)$ with $\|P_nA - AP_n\| \to 0$ as $n \to \infty$. One easily checks that $OLT(E, \mathcal{P})$ is a closed subalgebra of $L(E)$ and that $K(E, \mathcal{P}) \subseteq OLT(E, \mathcal{P}) \subseteq L(E, \mathcal{P})$. Moreover, it is *evident* that the algebra $OLT(E, \mathcal{P})$ is inverse closed in $L(E)$. So, working with operators in $OLT(E, \mathcal{P})$ seems to be much easier than working with operators from $L(E, \mathcal{P})$. We give $L(E, \mathcal{P})$ preference over $OLT(E, \mathcal{P})$ for at least two reasons: $L(E, \mathcal{P})$ is not only the largest algebra for which the $\mathcal{P}$-compact operators form an ideal; this algebra is also (in contrast to $OLT(E, \mathcal{P})$) invariant with respect to the substitution of $\mathcal{P}$ by an equivalent approximate projection. Thus, when dealing with operators in $L(E, \mathcal{P})$, we can switch over between equivalent approximate projections and choose that one which fits to our actual purposes.

1.4 Comments and references

Chapter 2

Fredholmness of Band-dominated Operators

In this chapter we introduce a class of operators for which the converse of Corollary 1.3.10 can be proved: if the symbol of the operator is invertible, then the operator is $\mathcal{P}$-Fredholm. The class under consideration consists of band and band-dominated operators which act on l^p-spaces over $\mathbb{Z}^N$. In the forthcoming chapters we will point out that this class is large enough to include, e.g., discretizations of convolution operators and of pseudodifferential operators.

2.1 Band-dominated operators

We start with fixing the Banach spaces E, the approximate identities $\mathcal{P}$, and the group actions $\mathcal{V}$, which will be used throughout this chapter. Then we continue with introducing the basic objects of this chapter: band and band-dominated operators with operator-valued entries. The main result of the section is Theorem 2.1.6, which provides us with different characterizations of band-dominated operators.

2.1.1 Function spaces on $\mathbb{Z}^N$

For N a positive integer, let $\mathbb{Z}^N$ denote the set of all N-tuples $x = (x_1, \ldots, x_N)$ of integers, provided with the norm $|x| = |x|_\infty := \max\{|x_1|, \ldots, |x_N|\}$, and let X stand for a fixed complex Banach space. For $p \geq 1$, let $l^p(\mathbb{Z}^N, X)$ and $l^\infty(\mathbb{Z}^N, X)$, respectively, stand for the Banach space of all functions f on $\mathbb{Z}^N$ with values in X such that

$$\|f\|_p^p := \sum_{x \in \mathbb{Z}^N} \|f(x)\|_X^p < \infty \quad \text{and} \quad \|f\|_\infty := \sup_{x \in \mathbb{Z}^N} \|f(x)\|_X < \infty.$$

Let further $c_0(\mathbb{Z}^N, X)$ refer to the closed subspace of $l^\infty(\mathbb{Z}^N, X)$ which consists of all functions f with

$$\lim_{x \to \infty} \|f(x)\|_X = 0.$$

In case $X = \mathbb{C}$, we will simply write $l^p(\mathbb{Z}^N)$ and $c_0(\mathbb{Z}^N)$ in place of $l^p(\mathbb{Z}^N, X)$ and $c_0(\mathbb{Z}^N, X)$.

For $1 \le p < \infty$, the dual space of $l^p(\mathbb{Z}^N, X)$ can be identified with $l^q(\mathbb{Z}^N, X^*)$ where $1/p + 1/q = 1$, and the dual of $c_0(\mathbb{Z}^N, X)$ is isomorphic to $l^1(\mathbb{Z}^N, X^*)$. If, in particular, X is a reflexive Banach space, then the spaces $l^p(\mathbb{Z}^N, X)$ are reflexive for all $1 < p < \infty$. If $X = H$ is a Hilbert space, then $l^2(\mathbb{Z}^N, H)$ is a Hilbert space with respect to the inner product

$$\langle f, g \rangle := \sum_{x \in \mathbb{Z}^N} \langle f(x), g(x) \rangle_H .$$

In what follows, we agree upon using the notation E^∞ to refer to one of the spaces $l^p(\mathbb{Z}^N, X)$ with $1 \le p \le \infty$ or to $c_0(\mathbb{Z}^N, X)$, whereas the symbol E will be used if only the spaces $l^p(\mathbb{Z}^N, X)$ with $1 < p < \infty$ or $c_0(\mathbb{Z}^N, X)$ are taken into consideration. Our main emphasis will be on the spaces E, whereas the spaces $l^1(\mathbb{Z}^N, X)$ and $l^\infty(\mathbb{Z}^N, X)$ will play a major role only in Section 2.5, which is devoted to operators in the Wiener algebra.

Given $m \in \mathbb{Z}^N$, let s_m stand for the function on $\mathbb{Z}^N$ which is $I \in L(X)$ at m and 0 at all other points. The corresponding operator on $L(E^\infty)$ of multiplication by s_m will be abbreviated to S_m. For $n \ge 0$, we define P_n as the sum $\sum_{|m| \le n} S_m$ and we set $Q_n := I - P_n$. The operators P_n and Q_n are projections, and

$$Q_n P_m = P_m Q_n = 0 \quad \text{whenever } n > m.$$

Hence, the family $\mathcal{P} := (P_n)$ constitutes a uniform approximate identity on each of the spaces E^∞, and $\mathcal{P}$ is a perfect approximate identity for all spaces E. It is non-perfect but still symmetric for the space $l^1(\mathbb{Z}^N, X)$, whereas it is neither perfect nor symmetric for $l^\infty(\mathbb{Z}^N, X)$. To see the latter, let $f \in l^\infty(\mathbb{Z})^*$ be a functional of norm one which vanishes on $c_0(\mathbb{Z})$. The existence of such functionals is guaranteed by the Hahn-Banach theorem. Then $(P_n^* f)(x) = f(P_n x) = 0$ for all $x \in l^\infty(\mathbb{Z})$ and all n, whence the non-symmetry of $\mathcal{P}$. Moreover, the same functional also yields a compact operator which does not belong to $L(l^\infty(\mathbb{Z}), \mathcal{P})$. Indeed, choose a non-zero constant function $g \in l^\infty(\mathbb{Z})$ and consider the rank one operator $Kx := f(x)g$. Then $P_m K Q_n x = f(Q_n x) P_m g = f(x) P_m g$ is independent of n and non-zero if x is accordingly chosen.

Finally, for $k \in \mathbb{Z}^N$, let V_k refer to the shift operator

$$(V_k f)(x) = f(x - k), \quad x \in \mathbb{Z}^N. \tag{2.1}$$

Clearly, $V_k \in L(E^\infty)$ and $\|V_k\|_{L(E^\infty)} = 1$, and (1.32) as well as (1.33) are satisfied. Thus, the family $\mathcal{V} := \{V_k : k \in \mathbb{Z}^N\}$ constitutes a group action on E^∞ in the sense of Section 1.2.1.

2.1.2 Matrix representation

It is often convenient to think of the operators in $L(E^\infty)$ as matrices with entries in $L(X)$. For $n \in \mathbb{Z}^N$, consider the restriction and extension operators

$$R_n : \operatorname{Im} S_n \to X, \quad (\ldots, 0, x_n, 0, \ldots) \mapsto x_n$$

and

$$E_n : X \to \operatorname{Im} S_n, \quad x_n \mapsto (\ldots, 0, x_n, 0, \ldots),$$

with the x_n standing at the nth place in the sequence. With every operator $A \in L(E^\infty)$, we associate the matrix $(A_{ij})_{i,j\in\mathbb{Z}^N}$ where $A_{ij} := R_i S_i A S_j E_j$. Then, if $u := (u_j)_{j\in\mathbb{Z}^N} \in E^\infty$ is a sequence with finite support, the vector $Au =: v = (v_i)_{i\in\mathbb{Z}^N}$ is given by

$$v_i = R_i S_i A u = R_i S_i A \sum_j S_j E_j u_j = \sum_j A_{ij} u_j,$$

thus, A acts as a usual matrix. The more interesting question is: Given two operators A, $B \in L(E^\infty)$ with the same matrix representation, is then necessarily $A = B$? Of course, the answer is *no* in general: The matrix representation of the operator $K \in L(l^\infty)$ considered at the end of Section 2.1.1 is the zero matrix, but $K \neq 0$.

Proposition 2.1.1 *Both the operators in $L(E)$ and the operators in $L(E^\infty, \mathcal{P})$ are uniquely determined by their matrix representation.*

Proof. First observe that, for every operator $A \in L(E^\infty)$ and for each n, the matrix representation (A_{ij}) of A determines the operators $P_n A P_n$ completely. This follows from $P_n = \sum_{|m|\le n} S_m$ and

$$P_n A P_n = \sum_{|i|,|j|\le n} S_i A S_j = \sum_{|i|,|j|\le n} E_i A_{ij} R_j S_j.$$

Now, if $p < \infty$, then the projections P_n converge strongly to the identity operator. Hence, $P_n A P_n \to A$ strongly for every $A \in L(E)$, showing that A is uniquely determined by its matrix representation. If $p = \infty$, the P_n converge $\mathcal{P}$-strongly to I. Thus, $P_n A P_n \to A$ $\mathcal{P}$-strongly for every operator $A \in L(E^\infty, \mathcal{P})$, what again yields the assertion. □

2.1.3 Operators of multiplication

Let $l^\infty(\mathbb{Z}^N, L(X))$ stand for the Banach algebra of all functions a on $\mathbb{Z}^N$ with values in $L(X)$ and

$$\|a\|_\infty := \sup_{x\in\mathbb{Z}^N} \|a(x)\|_{L(X)} < \infty,$$

and denote by $c_0(\mathbb{Z}^N, L(X))$ the closed ideal of $l^\infty(\mathbb{Z}^N, L(X))$ which consists of all functions a with

$$\lim_{x\to\infty} \|a(x)\|_{L(X)} = 0.$$

We abbreviate $l^\infty(\mathbb{Z}^N, L(\mathbb{C}))$ to $l^\infty(\mathbb{Z}^N)$, and we will often identify a function $a \in l^\infty(\mathbb{Z}^N)$ with the function $x \mapsto a(x)I \in L(X)$ which belongs to $l^\infty(\mathbb{Z}^N, L(X))$.

Every function $a \in l^\infty(\mathbb{Z}^N, L(X))$ gives rise to an operator on E^∞ via

$$(af)(x) = a(x)f(x), \quad x \in \mathbb{Z}^N.$$

We call this operator the *operator of multiplication by* a and denote it by aI. Evidently, $aI \in L(E^\infty)$ and $\|aI\|_{L(E^\infty)} = \|a\|_\infty$.

Proposition 2.1.2 *For $a \in l^\infty(\mathbb{Z}^N, L(X))$, the operator aI belongs to $L(E^\infty, \mathcal{P})$. This operator belongs to $K(E^\infty, \mathcal{P})$ if and only if $a \in c_0(\mathbb{Z}^N, L(X))$.*

Proof. Let $K \in K(E^\infty, \mathcal{P})$. Since aI commutes with every operator Q_n, we have

$$\|aKQ_n\| \le \|aI\|\,\|KQ_n\| \to 0 \quad \text{and} \quad \|KaQ_n\| = \|KQ_n aI\| \le \|KQ_n\|\,\|aI\| \to 0$$

as $n \to \infty$. Similarly, $\|Q_n KaI\| \to 0$ and $\|Q_n aK\| \to 0$, whence $aI \in L(E^\infty, \mathcal{P})$.

For the second assertion observe that $aI \in K(E^\infty, \mathcal{P})$ if and only if, given $\varepsilon > 0$, there is an $n_0 \in \mathbb{N}$ such that $\|aQ_n\| \le \varepsilon$ for all $n \ge n_0$ or, equivalently, such that $\|a(x)\| \le \varepsilon$ for all x with $|x| > n_0$. The latter is equivalent to $a \in c_0(\mathbb{Z}^N, L(X))$. □

We proceed with two equivalent characterizations of multiplication operators.

Proposition 2.1.3 *An operator $A \in L(E^\infty)$ is an operator of multiplication by a function in $l^\infty(\mathbb{Z}^N, L(X))$ if and only if*

$$AS_m = S_m A \quad \textit{for all } m \in \mathbb{Z}^N. \tag{2.2}$$

Proof. It is evident that (2.2) holds if A is a multiplication operator. For the reverse implication, let A be subject to (2.2), and let E_n and R_n be defined as in Section 2.1.2. By M_A we denote the operator of multiplication by the function

$$a : \mathbb{Z}^N \to L(X), \quad m \mapsto R_m S_m A S_m E_m.$$

Evidently, $a \in l^\infty(\mathbb{Z}^N, L(X))$, and for all $m \in \mathbb{Z}^N$ and all $u = (u_m)_{m\in\mathbb{Z}^N} \in E^\infty$, we have

$$S_m M_A u = E_m a(m) u_m = E_m R_m S_m A S_m E_m u_m = S_m A S_m u = S_m A u$$

due to (2.2). This necessarily implies that $M_A u = Au$ for all $u \in E^\infty$, i.e., A is a multiplication operator. □

Thus, an operator is an operator of multiplication if and only if its matrix representation is a diagonal matrix. For another criterion, let $t = (t_1, \ldots, t_N) \in \mathbb{R}^N$, consider the function e_t which is defined at $x = (x_1, \ldots, x_N) \in \mathbb{R}^N$ by

$$e_t(x) := \exp\left(i(t_1x_1 + t_2x_2 + \cdots + t_Nx_N)\right), \tag{2.3}$$

and denote the restriction of a function a on $\mathbb{R}^N$ onto $\mathbb{Z}^N$ by $\hat{a}$.

Theorem 2.1.4

(a) *An operator $A \in L(E)$ is the operator of multiplication by a function in $l^\infty(\mathbb{Z}^N, L(X))$ if and only if*

$$\hat{e}_t A = A\hat{e}_t I \quad \textit{for all } t \in \mathbb{R}^N. \tag{2.4}$$

(b) *An operator $A \in L(E^\infty, \mathcal{P})$ is the operator of multiplication by a function in $l^\infty(\mathbb{Z}^N, L(X))$ if and only if (2.4) holds.*

Proof. Clearly, every multiplication operator satisfies (2.4). For the reverse direction, we claim that if an operator $A \in L(E^\infty)$ is subject to condition (2.4), then

$$A\hat{b}P_n = \hat{b}AP_n \quad \text{for every bounded continuous function } b \text{ on } \mathbb{R}^N. \tag{2.5}$$

Once this is verified, and if $A \in L(E^\infty, \mathcal{P})$, we take the $\mathcal{P}$-strong limit of both sides of (2.5) which yields $A\hat{b}I = \hat{b}A$ for all b and, hence, $AS_m = S_mA$ for all m. Then assertion (b) follows from Proposition 2.1.3. In case $A \in L(E)$, we can take the usual strong limit in place of the $\mathcal{P}$-strong one which yields the conclusion without further restriction on A.

To verify (2.5), let $\mathcal{L}$ denote the closure with respect to the supremum norm of the linear hull of all functions e_t, and let $M_n := [-n, n]^N$. Then, due to (2.4),

$$\hat{a}A = A\hat{a}I \quad \text{for all } a \in \mathcal{L}. \tag{2.6}$$

The Weierstraß' approximation theorem implies that the restriction of $\mathcal{L}$ onto M_n is all of $C(M_n)$. Hence, if b is a bounded continuous function on $\mathbb{R}^N$ then, for every fixed n, there is a function $a \in \mathcal{L}$ such that $a|_{M_n} = b|_{M_n}$. Consequently,

$$A\hat{b}P_nf = A\hat{a}P_nf = \hat{a}AP_nf \quad \text{for all } f \in E^\infty \tag{2.7}$$

due to (2.6). Further, the function a can be chosen such that it takes at $x \in \mathbb{Z}^N \setminus M_n$ an arbitrarily prescribed value. Since the left-hand side of (2.7) is independent of a, this shows that the function AP_nf vanishes outside $\mathbb{Z}^N \cap M_n$. Hence,

$$A\hat{b}P_nf = \hat{a}AP_nf = \hat{b}AP_nf \quad \text{for all } f \in E^\infty$$

which proves our claim (2.5). □

There are operators on $L(l^\infty(\mathbb{Z}))$ which satisfy (2.4), but which fail to be multiplication operators (see [91], Remark 2.1.9).

2.1.4 Band and band-dominated operators

The basic objects of our interest are band and band-dominated operators. They are constituted by operators of multiplication and by the operators of shift introduced in (2.1).

Definition 2.1.5 *A* band operator *is a finite sum of the form* $\sum_\alpha a_\alpha V_\alpha$ *where* $\alpha \in \mathbb{Z}^N$ *and* $a_\alpha \in l^\infty(\mathbb{Z}^N, L(X))$. *A* band-dominated operator *is the uniform limit of a sequence of band operators.*

To justify this notation note that, in case $X = \mathbb{C}$ and $N = 1$ and with respect to the standard basis of E^∞, band operators are given by matrices with finite band width. Observe also that the class of band operators is independent of the concrete space E^∞, whereas the class of band-dominated operators depends heavily on E^∞. We denote this class by $\mathcal{A}_{E^\infty}$.

Every multiplication operator and every shift operator belong to the algebra $L(E^\infty, \mathcal{P})$. Moreover, if A is a multiplication operator and $k \in \mathbb{Z}^N$, then $V_{-k}AV_k$ is a multiplication operator again. Thus, the band operators form an algebra and $\mathcal{A}_{E^\infty}$ is a closed subalgebra both of $L(E^\infty)$ and $L(E^\infty, \mathcal{P})$.

For an equivalent characterization of band-dominated operators, we will have recourse to the class $BUC(\mathbb{R}^N)$ of the *b*ounded and *u*niformly *c*ontinuous complex-valued functions on $\mathbb{R}^N$. This class forms a C^*-algebra with respect to pointwisely defined operations and involution and to the supremum norm. In particular, the functions e_t defined in (2.3) are in $BUC(\mathbb{R}^N)$.

For arbitrary vectors $t = (t_1, \ldots, t_N)$ and $x = (x_1, \ldots, x_N)$ in $\mathbb{R}^N$ we define their product by $tx := (t_1x_1, \ldots, t_Nx_N)$. Given a function φ on $\mathbb{R}^N$ and vectors $t, r \in \mathbb{R}^N$, we define $\varphi_t(x) := \varphi(tx)$ and set $\varphi_{t,r}(x) := \varphi_t(x - r)$. Recall further that, for any bounded complex-valued function φ on $\mathbb{R}^N$, $\hat{\varphi}$ stands for the restriction of φ onto $\mathbb{Z}^N$. The function $\hat{\varphi}$ belongs to $l^\infty(\mathbb{Z}^N)$ and, conversely, every function in $l^\infty(\mathbb{Z}^N)$ is of this form. Instead of $\widehat{\varphi_{t,r}}$ we will write $\hat{\varphi}_{t,r}$.

Finally, for every subset F of $\mathbb{R}^N$ or of $\mathbb{Z}^N$, we let χ_F stand for the characteristic function of this set, i.e., for the function which is 1 on F and 0 outside F and, given subsets F and G of $\mathbb{R}^N$, we let

$$\operatorname{dist}(F, G) := \inf\{|s - t|, \, s \in F, \, t \in G\}$$

denote their distance.

Theorem 2.1.6 *Let $A \in L(E)$ or $A \in L(E^\infty, \mathcal{P})$. Then the following conditions are equivalent:*

(a) *The operator A is band-dominated.*

(b) *For every $\varepsilon > 0$, there is an M such that whenever F and G are subsets of $\mathbb{Z}^N$ with* $\operatorname{dist}(F, G) > M$, *then*

$$\|\chi_F A \chi_G I\|_{L(E)} < \varepsilon. \tag{2.8}$$

(c) *For every $\varphi \in BUC(\mathbb{R}^N)$,*

$$\lim_{t \to 0} \sup_{r \in \mathbb{R}^N} \|\hat{\varphi}_{t,r} A - A\hat{\varphi}_{t,r} I\|_{L(E)} = 0. \tag{2.9}$$

(d) *For every $\varphi \in BUC(\mathbb{R}^N)$,*

$$\lim_{t \to 0} \|\hat{\varphi}_t A - A\hat{\varphi}_t I\|_{L(E)} = 0. \tag{2.10}$$

(e) *Condition* (2.10) *holds for the function* $\varphi(x_1, \ldots, x_N) := \exp i(x_1 + \cdots + x_N)$.

Proof. We will prove the assertion under the assumption $A \in L(E)$. For $A \in L(E^\infty, \mathcal{P})$, the proof is completely analogous (the only difference being that then assertion (b) of Theorem 2.1.4 is invoked). Observe also that the implications (a) $\Rightarrow$ (b) $\Rightarrow$ (c) $\Rightarrow$ (d) $\Rightarrow$ (e) are valid for arbitrary operators $A \in L(E^\infty)$.

(a) ⇒ (b): If A is a band operator then, evidently, $\chi_F A \chi_G I = 0$ whenever dist(F, G) is large enough. Thus, (2.8) holds for band operators A. The validity of (2.8) for arbitrary band-dominated operators $A \in \mathcal{A}_E$ follows by a simple limit argument. Indeed, let (A_n) be a sequence of band operators which converges to A in the norm. Given $\varepsilon > 0$, choose an n such that $\|A - A_n\| < \varepsilon/2$, and choose M such that $\|\chi_F A_n \chi_G I\| < \varepsilon/2$ whenever dist$(F, G) > M$. Then

$$\|\chi_F A \chi_G I\| \le \|\chi_F A_n \chi_G I\| + \|\chi_F (A - A_n) \chi_G I\| < \varepsilon/2 + \|A - A_n\| < \varepsilon.$$

(b) ⇒ (c): It is sufficient to verify (2.9) in the case of a real-valued and non-negative function $\varphi \in BUC(\mathbb{R}^N)$. First let $r = 0$. For every $\varepsilon > 0$ and $k \in \mathbb{N}$, consider the sets

$$U^1_{k,t} := \{x \in \mathbb{R}^N : \varphi_t(x) \ge k\varepsilon\}, \qquad U^2_{k,t} := \{x \in \mathbb{R}^N : \varphi_t(x) \ge (k - 1/2)\varepsilon\}.$$

Since φ is a bounded function, there is an $m \in \mathbb{N}$ such that $U^1_{k,t}$ and $U^2_{k,t}$ are empty sets for $k > m$. Set

$$\varphi_t^{(\varepsilon,1)} := \varepsilon \sum_{k=1}^m \chi_{U^1_{k,t}} \quad \text{and} \quad \varphi_t^{(\varepsilon,2)} := \varepsilon \sum_{k=1}^m \chi_{U^2_{k,t}}$$

which is in accordance with our earlier definition of φ_t for a function φ. Clearly,

$$\|\hat{\varphi}_t - \hat{\varphi}_t^{(\varepsilon,1)}\|_{l^\infty_{L(X)}} < \varepsilon \quad \text{and} \quad \|\hat{\varphi}_t - \hat{\varphi}_t^{(\varepsilon,2)}\|_{l^\infty_{L(X)}} < \varepsilon$$

for every $t \in \mathbb{R}^N$. Let now $A \in L(E)$ be subject to condition (2.8). Then

$$\begin{aligned}
&\|A\hat{\varphi}_t I - \hat{\varphi}_t A\| \\
&\quad \le \|A(\hat{\varphi}_t I - \hat{\varphi}_t^{(\varepsilon,2)} I)\| + \|A\hat{\varphi}_t^{(\varepsilon,2)} I - \hat{\varphi}_t^{(\varepsilon,1)} A\| + \|(\hat{\varphi}_t - \hat{\varphi}_t^{(\varepsilon,1)})A\| \\
&\quad \le 2\varepsilon \|A\| + \|A\hat{\varphi}_t^{(\varepsilon,2)} I - \hat{\varphi}_t^{(\varepsilon,1)} A\|.
\end{aligned}$$

The latter term is not greater than

$$\varepsilon \,\|\sum_{k=1}^m \hat{\chi}_{\mathbb{R}^N \setminus U^1_{k,t}} A \hat{\chi}_{U^2_{k,t}} I\| + \varepsilon \|\sum_{k=1}^m \hat{\chi}_{U^1_{k,t}} A \hat{\chi}_{\mathbb{R}^N \setminus U^2_{k,t}} I\|. \tag{2.11}$$

Since $\lim_{t\to 0} \text{dist}\,(U^1_{k,t}, \mathbb{R}^N \setminus U^2_{k,t}) = \infty$, there exists an $\delta > 0$ such that, for all t with $|t| < \delta$,

$$\|\sum_{k=1}^m \hat{\chi}_{U^1_{k,t}} A \hat{\chi}_{\mathbb{R}^N \setminus U^2_{k,t}} I\| < 1.$$

To estimate the first term in (2.11), consider the sets

$$\begin{aligned}
U^3_{k,t} &:= \{x \in \mathbb{R}^N : (k - 5/8)\varepsilon < \varphi_t(x) < k\varepsilon\}, \\
U^4_{k,t} &:= \{x \in \mathbb{R}^N : (k - 1/2)\varepsilon \le \varphi_t(x) < (k + 1/2)\varepsilon\}.
\end{aligned}$$

Then

$$\lim_{t\to 0} \text{dist}\,((\mathbb{R}^N \setminus U^1_{k,t}) \setminus U^3_{k,t},\, U^2_{k,t}) = \infty, \qquad \lim_{t\to 0} \text{dist}\,(U^2_{k,t} \setminus U^4_{k,t},\, U^3_{k,t}) = \infty$$

and, consequently,

$$\varepsilon \| \sum_{k=1}^{m} \hat{\chi}_{\mathbb{R}^N \setminus U^1_{k,t}} A \hat{\chi}_{U^2_{k,t}} I \| \le \varepsilon \| \sum_{k=1}^{m} \hat{\chi}_{U^3_{k,t}} A \hat{\chi}_{U^4_{k,t}} I \| + \varepsilon.$$

Since $\chi_{U^3_{j,t}} \cap \chi_{U^3_{k,t}} = \emptyset$ and $\chi_{U^4_{j,t}} \cap \chi_{U^4_{k,t}} = \emptyset$ whenever $j \neq k$, we can further estimate

$$\| \sum_{k=1}^{m} \hat{\chi}_{U^3_{k,t}} A \hat{\chi}_{U^4_{k,t}} I \| \le \|A\|.$$

Combining the derived estimates we conclude for every $\varepsilon > 0$ the existence of a δ such that for all $|t| < \delta$,

$$\|A\hat{\varphi}_t I - \hat{\varphi}_t A\| \le 3\varepsilon + \varepsilon\|A\|.$$

This verifies (2.9) in case $r = 0$, and in the very same way one can check that the convergence in (2.9) is uniform with respect to $r \in \mathbb{R}^N$.
The implications (c) $\Rightarrow$ (d) $\Rightarrow$ (e) are trivial. To check the implication (e) $\Rightarrow$ (a), let the function e be defined by $e(x_1, \ldots, x_N) := \exp i(x_1 + \cdots + x_N)$. Note that the functions e_t are just given by (2.3). Thus, the operator $A \in L(E)$ is subject to condition (e) if and only if

$$\lim_{t\to 0} \|\hat{e}_t A - A\hat{e}_t I\| = 0. \tag{2.12}$$

The identity $e_s e_t = e_{s+t}$ implies

$$\|\hat{e}_t A \hat{e}_{-t} I - \hat{e}_s A \hat{e}_{-s} I\| \le \|\hat{e}_{t-s} A - A\hat{e}_{t-s} I\|,$$

which together with (2.12) yields the continuity of the function

$$f_A : \mathbb{R}^N \to L(E), \quad t \mapsto \hat{e}_t A \hat{e}_{-t} I.$$

Moreover, the functions $t \mapsto \hat{e}_t I$ and $t \mapsto \hat{e}_{-t} I$ are 2π-periodic in each variable t_j, and so is the function f_A. We fix $\lambda > (N-1)/2$ and consider the Riesz polynomials

$$R^\lambda_n(f_A;\, t) := \sum_{|k| \le n} \left(1 - \frac{|k|}{n+1}\right)^\lambda A_k e^{i\langle t,\, k\rangle}$$

of f where

$$A_k := (2\pi)^{-N} \int_{[0,2\pi]^N} (\hat{e}_t A \hat{e}_{-t} I) e^{-i\langle t,\, k\rangle}\, dt \quad \in L(E).$$

Then

$$\|R_n^\lambda(f_A;\,0) - f_A(0)\| = \|R_n^\lambda(f_A;\,0) - A\|_{L(E)} \to 0 \quad \text{as } n \to \infty \tag{2.13}$$

([83], Chapter I, Section 2.2). In case $N = 1$ one can choose $\lambda = 1$ which yields the standard Fejer polynomials.

In view of (2.13), it remains to show that $R_n^\lambda(f_A;\,0)$ is a band operator for each n. This can be seen as follows. The operators $A_k V_{-k}$ commute with each of the multiplication operators $\hat{e}_s I$:

$$\hat{e}_s A_k V_{-k} \hat{e}_{-s} I = (2\pi)^{-N} \int_{[0,2\pi]^N} (\hat{e}_{s+t} A \hat{e}_{-s-t} I) e^{-i\langle s+t,\,k\rangle}\, dt\, V_{-k} = A_k V_{-k}.$$

Hence, by assertion (a) of Theorem 2.1.4, $A_k V_{-k}$ is the operator of multiplication by a certain function $a_k \in l^\infty(\mathbb{Z}^N, L(X))$. Thus, $A_k = a_k V_k$, and

$$R_n^\lambda(f_A;\,0) = \sum_{|k| \le n} \left(1 - \frac{|k|}{n+1}\right)^\lambda a_k V_k$$

is a band operator. □

Here are a few simple but useful consequences of this theorem.

Proposition 2.1.7 *The algebra $\mathcal{A}_{E^\infty}$ contains the ideal $K(E^\infty, \mathcal{P})$.*

Proof. Let T be $\mathcal{P}$-compact and $\varphi \in BUC(\mathbb{R}^N)$. It is easy to see that the operators $\hat{\varphi}_t I$ converge $\mathcal{P}$-strongly to $\varphi(0)I$ as $t \to 0$. Hence, by the definition of $\mathcal{P}$-strong convergence, both $T\hat{\varphi}_t I$ and $\hat{\varphi}_t T$ tend to $\varphi(0)T$ in the operator norm. This implies that $\|\hat{\varphi}_t T - T\hat{\varphi}_t I\| \to 0$ as $t \to 0$. Since φ is an arbitrary element of $BUC(\mathbb{R}^N)$, the operator T belongs to $\mathcal{A}_{E^\infty}$ due to Theorem 2.1.6 (d). □

Proposition 2.1.8 *The algebra $\mathcal{A}_{E^\infty}$ is inverse closed in $L(E^\infty)$.*

Proof. Let $A \in \mathcal{A}_{E^\infty}$ be invertible in $L(E^\infty)$. Then A is invertible in $L(E^\infty, \mathcal{P})$ by Theorem 1.1.9, and the estimate

$$\|\hat{\varphi}_t A^{-1} - A^{-1}\hat{\varphi}_t I\| \le \|A^{-1}\|^2\, \|\hat{\varphi}_t A - A\hat{\varphi}_t I\| \to 0,$$

which holds for every function $\varphi \in BUC(\mathbb{R}^N)$, shows via Theorem 2.1.6 (d) that A^{-1} belongs to $\mathcal{A}_{E^\infty}$. □

Proposition 2.1.9 *The quotient algebra $\mathcal{A}_{E^\infty}/K(E^\infty, \mathcal{P})$ (which can be formed by Proposition* 2.1.7*) is inverse closed in the Calkin algebra $L(E^\infty, \mathcal{P})/K(E^\infty, \mathcal{P})$.*

Proof. Let A be a $\mathcal{P}$-Fredholm band-dominated operator, let $B \in L(E^\infty, \mathcal{P})$ be a regularizer of A, and let T_1 and T_2 stand for the $\mathcal{P}$-compact operators $BA - I$

and $AB - I$, respectively. Further we abbreviate the commutator $CD - DC$ of two operators by $[C, D]$. One easily checks that, for every function $\varphi \in BUC(\mathbb{R}^N)$,

$$[B, \hat{\varphi}_t I] = [T_1, \hat{\varphi}_t I]\, B - B\, [A, \hat{\varphi}_t I]\, B - [B, \hat{\varphi}_t I]\, T_2.$$

The proof of Proposition 2.1.7 shows that $\|[T_1, \hat{\varphi}_t I]\, B\| \to 0$ as $t \to 0$, and Theorem 2.1.6 implies the convergence $\|B\,[A, \hat{\varphi}_t I]\, B\| \to 0$ because of $\mathcal{A}_{E^\infty} \subseteq L(E^\infty, \mathcal{P})$. Finally, it is clear that $[B, \hat{\varphi}_t I] \to 0$ $\mathcal{P}$-strongly whence, due to the $\mathcal{P}$-compactness of T_2, $\|[B, \hat{\varphi}_t I]\, T_2\| \to 0$. Hence, $\|[B, \hat{\varphi}_t I]\| \to 0$, whence $B \in \mathcal{A}_{E^\infty}$ by Theorem 2.1.6 (d) again. □

2.1.5 Limit operators of band-dominated operators

Our next goal is limit operators of band-dominated operators with respect to the group action $\mathcal{V} = (V_k)_{k \in \mathbb{Z}^N}$ specified in Section 2.1.1.

Proposition 2.1.10 *Limit operators of multiplication, band and band-dominated operators on E^∞ are operators of the same kind.*

Proof. Let A be a multiplication operator and let A_h be the limit operator of A with respect to a sequence $h \in \mathcal{H}$. Thus, $\mathcal{P}\text{-lim } V_{-h(n)} A V_{h(n)} = A_h$. From Proposition 1.1.17 we infer that

$$\mathcal{P}\text{-lim } S_m V_{-h(n)} A V_{h(n)} = S_m A_h \quad \text{and} \quad \mathcal{P}\text{-lim } V_{-h(n)} A V_{h(n)} S_m = A_h S_m$$

for every $m \in \mathbb{Z}^N$. Since the operators $V_{-h(n)} A V_{h(n)}$ are multiplication operators again, we get $S_m A_h = A_h S_m$. Thus, A_h is a multiplication operator by Proposition 2.1.3. Since the only limit operator of the shift operator V_k is V_k itself, the other assertions follow easily by an approximation argument. □

It turns out that, conversely, every multiplication (band, band-dominated) operator is the limit operator of a certain multiplication (band, band-dominated) operator. This can be seen by means of the following construction called the *inflation* of an operator A by a sequence $h \in \mathcal{H}$.

Let $h : \mathbb{N} \to \mathbb{Z}^N$ be a sequence with

$$|h(k) - h(n)|_\infty \geq 3k \quad \text{for all } k > n. \tag{2.14}$$

It is easy to define such sequences recursively: One starts with an arbitrary $h(0)$ and chooses $h(1)$ such that

$$|h(1) - h(0)|_\infty \geq 3.$$

Then $h(2)$ is chosen such that

$$|h(2) - h(0)|_\infty \geq 6 \quad \text{and} \quad |h(2) - h(1)|_\infty \geq 6,$$

etc. From (2.14) one concludes

$$|h(k)|_\infty \geq |h(k) - h(0)|_\infty - |h(0)|_\infty \geq 3k - |h(0)|_\infty, \tag{2.15}$$

which shows that each sequence with property (2.14) tends to infinity. Conversely, every sequence tending to infinity possesses a subsequence with property (2.14).

Let $k \in \mathbb{N}$. To each operator $A \in L(E^\infty, \mathcal{P})$, we associate the operator

$$A_k := \sum_{n=0}^{k} V_{h(n)} P_n A P_n V_{-h(n)},$$

and we claim that the sequence $(A_k)_{k\in\mathbb{N}}$ converges $\mathcal{P}$-strongly. Let $i \in \mathbb{Z}^N$. Then, by (2.15),

$$|i - h(n)|_\infty \geq |h(n) - h(0)|_\infty - |h(0) - i|_\infty \geq 3n - |h(0) - i|_\infty > n$$

whenever $2n > |h(0) - i|_\infty$. Thus, $P_n S_{i-h(n)} = 0$ for these n, and one gets for $k > |h(0) - i|_\infty/2$

$$\begin{aligned} A_k S_i &= \sum_{n=0}^{k} V_{h(n)} P_n A P_n V_{-h(n)} S_i = \sum_{n=0}^{k} V_{h(n)} P_n A P_n S_{i-h(n)} V_{-h(n)} \\ &= \sum_{n=0}^{[|h(0)-i|_\infty/2]} V_{h(n)} P_n A P_n S_{i-h(n)} V_{-h(n)} = A_{[|h(0)-i|_\infty/2]} S_i \end{aligned}$$

where $[y]$ refers to the integer part of $y \in \mathbb{R}$. Thus, the sequence (A_k) converges strongly on a dense subset of $K(E^\infty, \mathcal{P})$. Moreover, for every $i \in \mathbb{Z}^N$, there is at most one $n \in \mathbb{N}$ such that

$$V_{h(n)} P_n A P_n V_{-h(n)} S_i = V_{h(n)} P_n A P_n S_{i-h(n)} V_{-h(n)} \neq 0.$$

Indeed, if this condition is satisfied for some n, then $|i - h(n)|_\infty \leq n$, and for $m \neq n$ one gets

$$\begin{aligned} |i - h(m)|_\infty &\geq |h(m) - h(n)|_\infty - |h(n) - i|_\infty \\ &\geq \left\{ \begin{array}{lll} 3m - n & \text{if} & m > n \\ 3n - n & \text{if} & n > m \end{array} \right. > m, \end{aligned}$$

whence $P_m S_{i-h(m)} = 0$. Thus,

$$\|A_k\| = \max_{0\leq n\leq k} \|P_n A P_n\| \leq \|A\|, \tag{2.16}$$

i.e., (A_k) is a uniformly bounded sequence. The Banach-Steinhaus theorem implies the $\mathcal{P}$-strong convergence of that sequence.

Definition 2.1.11 *Let $A \in L(E^\infty, \mathcal{P})$ and let $h \in \mathcal{H}$ be a sequence which satisfies* (2.14). *Then the h-*inflation *of A is the operator*

$$I_h(A) := \sum_{n=0}^{\infty} V_{h(n)} P_n A P_n V_{-h(n)}$$

where the series is understood in the sense of the $\mathcal{P}$-strong convergence.

Lemma 2.1.12

(a) *I_h maps $L(E^\infty, \mathcal{P})$ into $L(E^\infty, \mathcal{P})$ isometrically.*

(b) *I_h maps multiplication, band, and band-dominated operators to operators of the same kind.*

Proof. (a) From Proposition 1.1.17 we infer that I_h maps $L(E^\infty, \mathcal{P})$ into itself. To verify the isometry of I_h, we first observe that, in analogy to (2.16), $\|I_h(A)\| = \sup_n \|P_n A P_n\|$. Thus, by the Banach-Steinhaus theorem and by (2.16),

$$\|I_h(A)\| \leq \liminf \|A_k\| \leq \|A\|,$$

whereas Banach-Steinhaus and the identity $\mathcal{P}$-$\lim P_n A P_n = A$ imply

$$\|A\| \leq \liminf \|P_n A P_n\| \leq \sup \|P_n A P_n\| = \|I_h(A)\|.$$

(b) Evidently, I_h maps band operators of a certain band width to band operators with the same band width. This shows the first two assertions. Let now A be band-dominated and (A_n) be a sequence of band operators which converges to A in the norm. Then $(I_h(A_n))$ is a sequence of band operators which converges to $I_h(A)$ due to the continuity of I_h. Hence, $I_h(A)$ is band-dominated. □

Lemma 2.1.13 *Let $A \in L(E^\infty, \mathcal{P})$ and let h be a sequence which satisfies* (2.14). *Then the limit operator of $I_h(A)$ with respect to the sequence h exists, and it is equal to A.*

Indeed, this is a consequence of the identity $V_{-h(n)} I_h(A) V_{h(n)} P_n = P_n A P_n$. □

Theorem 2.1.14 *Let $A \in L(E^\infty, \mathcal{P})$ and $h \in \mathcal{H}$. Then there exist an operator $B \in L(E^\infty)$ as well as a subsequence g of h such that the limit operator B_g of B with respect to g exists and that $B_g = A$.*

To see this, choose a subsequence g of h with property (2.14), and apply the preceding proposition to the operator $B := I_g(A)$. □

The following corollary follows immediately from Lemma 2.1.12 (b) and Theorem 2.1.14.

Corollary 2.1.15 *Every multiplication (band, band-dominated) operator on E^∞ is the limit operator of a certain multiplication (band, band-dominated) operator on E^∞.*

There is some feeling that the limit operator A_h tends to be a simpler object than the original operator A. For example, the $\mathcal{P}$-strong limit of the sequence $(V_{-n}AV_n)_{n\in\mathbb{Z}^N}$ as $n \to \infty$ is, if it exists, a shift invariant operator. But in general, limit operators of A are $\mathcal{P}$-strong limits of certain *subsequences* of $(V_{-n}AV_n)$, and the preceding corollary shows that these limit operators need not to possess any further special structure. Thus, the only sense in which A_h is simpler than A is that the spectrum of A_h is contained in the essential spectrum of A.

2.1.6 Rich band-dominated operators

In what follows we will have to deal with operators which are both band-dominated and possess a rich operator spectrum. We will refer to these operators as *rich band-dominated operators*, and we will write $\mathcal{A}^{\$}_{E^\infty}$ for the set of all of these operators. Thus, $\mathcal{A}^{\$}_{E^\infty} = \mathcal{A}_{E^\infty} \cap L^{\$}(E^\infty, \mathcal{P})$. Analogously, we will use the terms *rich multiplication operator* and *rich band operator*.

The algebra $\mathcal{A}^{\$}_{E^\infty}$ is a closed subalgebra of $L(E^\infty)$, and $K(E^\infty, \mathcal{P})$ is a closed ideal of $\mathcal{A}^{\$}_{E^\infty}$ (Proposition 1.2.6 and Proposition 2.1.7). Further, $\mathcal{A}^{\$}_{E}$ is an inverse closed subalgebra of $L(E)$ (Propositions 1.2.8 and 2.1.8), and $\mathcal{A}^{\$}_{E}/K(E, \mathcal{P})$ is a closed and inverse closed subalgebra of $L(E, \mathcal{P})/K(E, \mathcal{P})$ by Propositions 1.2.10 and 2.1.9.

The rich multiplication operators can be characterized as follows.

Theorem 2.1.16 *The operator aI of multiplication by $a \in l^\infty(\mathbb{Z}^N, L(X))$ is rich if and only if the set $\{a(x) : x \in \mathbb{Z}^N\}$ of values of a is relatively compact with respect to the norm topology on $L(X)$.*

Proof. Let $\{a(x) : x \in \mathbb{Z}^N\}$ be relatively compact, let $h \in \mathcal{H}$, and let $x_1, x_2, \ldots$ be any enumeration of the points of $\mathbb{Z}^N$. Then, for every $f \in E^\infty$ and $k \in \mathbb{N}$,

$$(V_{-h(m)}aV_{h(m)}f)(x_k) = a(x_k + h(m))f(x_k).$$

The set $\{a(x_1 + h(m))\}_{m=1}^\infty$ is relatively compact in $L(X)$ by assumption. Hence, there exists a subsequence g_1 of h such that the sequence $(a(x_1 + g_1(m)))_{m=1}^\infty$ is norm convergent. The same reasoning yields a subsequence g_2 of g_1 such that the sequence $(a(x_2 + g_2(m)))_{m=1}^\infty$ converges in the norm. We proceed in this manner and define a new sequence g by $g(m) := g_m(m)$. Evidently, g is a subsequence of h, and the sequences $(a(x_k + g(m)))_{m=1}^\infty$ are norm convergent for every k. We denote the limit $\lim_{m\to\infty} a(x + g(m))$ by $\tilde{a}(x)$. It is easy to see that $\|\tilde{a}I\|_\infty \le \|aI\|_\infty$.

We are going to show that $\tilde{a}I$ is just the limit operator of aI with respect to the sequence g. Indeed, for every $P_n \in \mathcal{P}$,

$$\begin{aligned}&\|(V_{-g(m)}aV_{g(m)} - \tilde{a}I)P_n\|_{L(E^\infty)}\\ &= \|(V_{-g(m)}aV_{g(m)} - \tilde{a}I)P_n\|_\infty \le \max_{|x|\le n} \|a(x + g(m)) - \tilde{a}(x)\|_{L(X)}.\end{aligned}$$

The maximum is taken over a finite set. Consequently,

$$\lim_{m\to\infty} \|(V_{-g(m)}aV_{g(m)} - \tilde{a}I)P_n\|_{L(E^\infty)} = 0$$

for every $P_n \in \mathcal{P}$. The 'dual' assertion follows analogously. Hence, $(aI)_g = \tilde{a}I$ as desired, which proves the 'if'-part of the assertion.

For the 'only if'-part, let $(a(h(m)))_{m=1}^{\infty}$ be an arbitrary sequence of values of a which is ordered in such a way that the sequence h belongs to $\mathcal{H}$. Since $aI \in L^{\$}(E^{\infty}, \mathcal{P})$, there exists a subsequence g of h such that the limit operator A_g exists. This limit operator is an operator of multiplication by a function $a_g \in l^{\infty}(\mathbb{Z}^N, L(X))$ again. Because

$$\|(V_{-g(m)}aV_{g(m)} - A_g)S_0\|_{\infty} = \|a(g(m)) - a_g(0)\|_{L(X)} \to 0,$$

the sequence $(a(h(m)))$ possesses a convergent subsequence. □

For example, if X is a finite-dimensional space, then every multiplication operator aI with $a \in l^{\infty}(\mathbb{Z}^N, L(X))$ is rich (since balls in $L(X)$ are norm compact). Similarly, if $a \in l^{\infty}(\mathbb{Z}^N)$ and $B \in L(X)$, then the operator of multiplication by the function $m \mapsto a(m)B$ belongs to $L^{\$}(E^{\infty}, \mathcal{P})$, whereas the operator of multiplication by the function $m \mapsto S_m$ fails to be rich.

Corollary 2.1.17 *If X has finite dimension, then $\mathcal{A}^{\$}_{E^{\infty}} = \mathcal{A}_{E^{\infty}}$.*

The following approximation property often simplifies the treatment of band-dominated operators with rich operator spectrum.

Theorem 2.1.18 *Every rich band-dominated operator on E^{∞} is the norm limit of a sequence of rich band operators.*

Proof. Let $A \in L(E^{\infty})$ be a band-dominated operator with rich operator spectrum. For $k \in \mathbb{Z}^N$, we define the operators $A^{(k)}$ as in the proof of the implication (e) ⇒ (a) of Theorem 2.1.6, i.e., we let

$$A^{(k)} := c^{(k)} \int_{[0,2\pi]^N} (\hat{e}_t A \hat{e}_{-t} I) e^{-i(t,k)}\, dt.$$

Here we integrate a continuous function. Thus, the integral exists in the Riemann sense, and $A^{(k)}$ is the norm limit of the corresponding Riemann sums

$$\sum_{r,\,m} e^{-i(t_{r,\,m},k)} \hat{e}_{t_{r,\,m}} A \hat{e}_{-t_{r,\,m}} |I_{r,\,m}|$$

where we subdivide $[0,\, 2\pi]^N$ into cubes $I_{r,\,m}$, $m = 1, \ldots,\, r^N$ with side length $2\pi/r$, and where the points $t_{r,\,m} \in I_{r,\,m}$ are arbitrarily chosen. Since $\mathcal{A}_{E^{\infty}}$ is an algebra, all Riemann sums belong to $\mathcal{A}_{E^{\infty}}$, and since this algebra is closed, we also have $A^{(k)} \in \mathcal{A}_{E^{\infty}}$. Hence, the sequence of band operators, which is the outcome of the last step of the proof of Theorem 2.1.6 and which converges to A in the operator norm, consists of operators with rich operator spectrum. □

Proposition 2.1.19 *Limit operators of rich multiplication (band, band-dominated) operators on E^{∞} are rich multiplication (band, band-dominated) operators again.*

Proof. Let aI be a multiplication operator with rich spectrum, and let $\tilde{a}I$ be a limit operator of aI. (From Proposition 2.1.10 we know that every limit operator of aI is a multiplication operator.) It is evident from the proof of Theorem 2.1.16 that all values $\tilde{a}(x)$ belong to the closure of the set $\{a(x) : x \in \mathbb{Z}^N\}$ of the values of a. The compactness of this closure implies that $\{\tilde{a}(x) : x \in \mathbb{Z}^N\}$ is a relatively compact set. Hence, $\tilde{a}I$ has a rich operator spectrum by Theorem 2.1.16 again.

The assertion for band and band-dominated operators follows by simple algebraic and continuity arguments. □

2.2 $\mathcal{P}$-Fredholmness of rich band-dominated operators

We have seen in Proposition 1.2.9 that all limit operators of a $\mathcal{P}$-Fredholm operator with rich spectrum are invertible and that the norms of their inverses are uniformly bounded. The main result of this section (Theorem 2.2.1 below) states that, at least for rich band dominated operators, the converse is also true: If all limit operators of $A \in \mathcal{A}_E^{\$}$ are invertible, and if the norms of their inverses are uniformly bounded, then A is $\mathcal{P}$-Fredholm.

Another goal of this section is to establish symbol calculi for the $\mathcal{P}$-Fredholmness of rich band-dominated operators.

In this section, we will exclusively deal with operators on the spaces E.

2.2.1 The main theorem on $\mathcal{P}$-Fredholmness

Here is the announced basic $\mathcal{P}$-Fredholm criterion.

Theorem 2.2.1 *A rich band-dominated operator $A \in L(E)$ is $\mathcal{P}$-Fredholm if and only if all of its limit operators are invertible and if*

$$\sup\{\|(A_h)^{-1}\| : A_h \in \sigma_{op}(A)\} < \infty.$$

As already mentioned, the necessity part of the theorem follows immediately from Proposition 1.2.9. The remainder of this subsection is devoted to the proof of the sufficiency part.

Let $\varphi : \mathbb{R} \to [0, 1]$ be a continuous function with

$$\varphi(x) \begin{cases} = 1 & \text{for } |x| \le 1/3 \\ > 0 & \text{for } |x| < 2/3 \\ = 0 & \text{for } |x| \ge 2/3. \end{cases} \tag{2.17}$$

We further suppose that the family $\{\varphi_\alpha^2\}_{\alpha\in\mathbb{Z}}$ with $\varphi_\alpha(x) := \varphi(x - \alpha)$ forms a partition of unity on $\mathbb{R}$ in the sense that

$$\sum_{\alpha\in\mathbb{Z}} \varphi_\alpha(x)^2 = 1 \quad \text{for all } x \in \mathbb{R}.$$

This condition can be forced as follows: If $f : \mathbb{R} \to [0, 1]$ is a continuous function satisfying (2.17) in place of φ, then the non-negative function φ defined by

$$\varphi(x)^2 := \frac{f(x)}{\sum_{\alpha\in\mathbb{Z}} f(x-\alpha)}, \quad x \in \mathbb{R},$$

has the desired property. (Note that this definition makes sense since the series in the denominator is strictly positive and has only finitely many non-vanishing terms for each fixed x.)

Given $x = (x_1, \ldots, x_N) \in \mathbb{R}^N$, $\alpha \in \mathbb{Z}^N$, and $R > 0$, we define $\varphi^{(N)}(x) := \varphi(x_1)\cdots\varphi(x_N)$, $\varphi_\alpha^{(N)}(x) := \varphi^{(N)}(x-\alpha)$ and $\varphi_{\alpha,R}^{(N)}(x) := \varphi_\alpha^{(N)}(x/R)$. (This notation differs slightly from that one used in the preceding section.) Further, let $\psi : \mathbb{R} \to [0, 1]$ be a continuous function which also satisfies (2.17) in place of φ, but with the constants 1/3 and 2/3 replaced by 3/4 and 4/5, respectively. For this function, we define $\psi_{\alpha,R}^{(N)}$ analogously. Clearly, $\varphi_{\alpha,R}^{(N)}\psi_{\alpha,R}^{(N)} = \varphi_{\alpha,R}^{(N)}$ for all α and R, and the family $\{(\varphi_{\alpha,R}^{(N)})^2\}_{\alpha\in\mathbb{Z}^N}$ forms a partition of unity on $\mathbb{R}^N$ for every fixed R.

The following construction goes back to Simonenko [168].

Proposition 2.2.2 *Define $\varphi_{\alpha,R}^{(N)}$ and $\psi_{\alpha,R}^{(N)}$ as above. If $\{A_\alpha\}_{\alpha\in\mathbb{Z}^N}$ is a bounded family of operators in $L(E, \mathcal{P})$, then the series $\sum_{\alpha\in\mathbb{Z}^N} \hat{\varphi}_{\alpha,R}^{(N)} A_\alpha \hat{\psi}_{\alpha,R}^{(N)} I$ converges in the $\mathcal{P}$-strong operator topology of E for every fixed R, and*

$$\|\sum_{\alpha\in\mathbb{Z}^N} \hat{\varphi}_{\alpha,R}^{(N)} A_\alpha \hat{\psi}_{\alpha,R}^{(N)} I\|_{L(E)} \le 2^N \sup_{\alpha\in\mathbb{Z}^N} \|A_\alpha\|_{L(E)}.$$

Proof. We first show that the series converges strongly. To start with, let $N = 1$ (in which case we drop the superscript $^{(1)}$) and $E = l^p(\mathbb{Z}, L(X))$ with $1 < p < \infty$ (the case $p = 1$ does also cause no difficulties). Since

$$\operatorname{supp}\varphi_{\alpha,R} \cap \operatorname{supp}\varphi_{\beta,R} = \operatorname{supp}\psi_{\alpha,R} \cap \operatorname{supp}\psi_{\beta,R} = \emptyset$$

whenever $\alpha \ne \beta$ and both α and β are even, we get

$$\begin{aligned}\|\sum_{\alpha\in2\mathbb{Z}} \hat{\varphi}_{\alpha,R} A_\alpha \hat{\psi}_{\alpha,R} u\|^p &= \sum_{\alpha\in2\mathbb{Z}} \|\hat{\varphi}_{\alpha,R} A_\alpha \hat{\psi}_{\alpha,R} u\|^p \\ &\le \sup_\alpha \|A_\alpha\| \sum_{\alpha\in2\mathbb{Z}} \|\hat{\psi}_{\alpha,R} u\|^p \le \sup_\alpha \|A_\alpha\|^p \, \|u\|^p\end{aligned}$$

for every $u \in E$. Analogously,

$$\|\sum_{\alpha\in2\mathbb{Z}+1} \hat{\varphi}_{\alpha,R} A_\alpha \hat{\psi}_{\alpha,R} u\|^p \le \sup_\alpha \|A_\alpha\|^p \, \|u\|^p.$$

These estimates show that the series $\sum_{\alpha\in\mathbb{Z}} \hat{\varphi}_{\alpha,R} A_\alpha \hat{\psi}_{\alpha,R}$ converges strongly on E, and from

$$\|\sum_{\alpha\in\mathbb{Z}} \hat{\varphi}_{\alpha,R} A_\alpha \hat{\psi}_{\alpha,R} u\| \le \|\sum_{\alpha\in2\mathbb{Z}} \hat{\varphi}_{\alpha,R} A_\alpha \hat{\psi}_{\alpha,R} u\| + \|\sum_{\alpha\in2\mathbb{Z}+1} \hat{\varphi}_{\alpha,R} A_\alpha \hat{\psi}_{\alpha,R} u\|$$

we conclude that

$$\| \sum_{\alpha \in \mathbb{Z}} \hat{\varphi}_{\alpha,R} A_\alpha \hat{\psi}_{\alpha,R} \| \le 2 \sup_\alpha \|A_\alpha\|. \tag{2.18}$$

In case $N = 1$ and $E = c_0(\mathbb{Z}, L(X))$, we have

$$\| \sum_{\alpha \in 2\mathbb{Z}} \hat{\varphi}_{\alpha,R} A_\alpha \hat{\psi}_{\alpha,R} u\|_\infty = \max_{\alpha \in 2\mathbb{Z}} \|\hat{\varphi}_{\alpha,R} A_\alpha \hat{\psi}_{\alpha,R} u\|_\infty \le \sup_\alpha \|A_\alpha\| \, \|u\|$$

which, together with an analogous estimate for odd α, also implies (2.18).

For $N > 1$, we proceed by induction. Writing points $x \in \mathbb{R}^N$ as $(x', x_N) \in \mathbb{R}^{N-1} \times \mathbb{R}$, we have

$$\varphi^{(N)}_{\alpha,R}(x) = \varphi^{(N-1)}_{\alpha',R}(x')\varphi^{(1)}_{\alpha_N,R}(x_N) \quad \text{and} \quad \psi^{(N)}_{\alpha,R}(x) = \psi^{(N-1)}_{\alpha',R}(x')\psi^{(1)}_{\alpha_N,R}(x_N),$$

whence

$$\| \sum_{\alpha \in \mathbb{Z}^N} \hat{\varphi}^{(N)}_{\alpha,R} A_\alpha \hat{\psi}^{(N)}_{\alpha,R} I\| = \| \sum_{\alpha_N \in \mathbb{Z}} \hat{\varphi}^{(1)}_{\alpha_N,R} B_{\alpha_N} \hat{\psi}^{(1)}_{\alpha_N,R} I\|$$

with

$$B_{\alpha_N} := \sum_{\alpha' \in \mathbb{Z}^{N-1}} \hat{\varphi}^{(N-1)}_{\alpha',R} A_{(\alpha',\alpha_N)} \hat{\psi}^{(N-1)}_{\alpha',R} I.$$

The assumption of the induction guarantees that

$$\|B_{\alpha_N}\| \le 2^{N-1} \sup_{\alpha' \in \mathbb{Z}^{N-1}} \|A_{(\alpha',\alpha_N)}\| \le 2^{N-1} \sup_{\alpha \in \mathbb{Z}^N} \|A_\alpha\|,$$

hence,

$$\| \sum_{\alpha \in \mathbb{Z}^N} \hat{\varphi}^{(N)}_{\alpha,R} A_\alpha \hat{\psi}^{(N)}_{\alpha,R} I\| \le 2 \sup_{\alpha_N \in \mathbb{Z}} \|B_{\alpha_N}\| \le 2^N \sup_{\alpha \in \mathbb{Z}^N} \|A_\alpha\|$$

by what has been already shown. Thus, the series $\sum_{\alpha \in \mathbb{Z}^N} \hat{\varphi}^{(N)}_{\alpha,R} A_\alpha \hat{\psi}^{(N)}_{\alpha,R} I$ defines a bounded linear operator on E. Since

$$P_m \hat{\varphi}^{(N)}_{\alpha,R} = \hat{\psi}^{(N)}_{\alpha,R} P_m = 0$$

for every fixed m whenever α is large enough, this sequence also converges in the sense of the $\mathcal{P}$-strong convergence. □

We will use this result to construct regularizers of $\mathcal{P}$-Fredholm band operators. Henceforth we will omit the superscript $^{(N)}$ at the functions φ and ψ.

Proposition 2.2.3 *Let $A \in \mathcal{A}_E$, and let $\psi_{\alpha,R}$ be as above. Suppose there is a constant $M > 0$ such that, for all positive integers R, there is a $\rho(R) > 0$ such that, for all $\alpha \in \mathbb{Z}^N$ with $|\alpha| \ge \rho(R)$, there are operators $B_{\alpha,R} \in L(E, \mathcal{P})$ and $C_{\alpha,R} \in L(E, \mathcal{P})$ with $\|B_{\alpha,R}\|_{L(E)} \le M$, $\|C_{\alpha,R}\|_{L(E)} \le M$ and*

$$B_{\alpha,R} A \hat{\psi}_{\alpha,R} I = \hat{\psi}_{\alpha,R} A C_{\alpha,R} = \hat{\psi}_{\alpha,R} I.$$

Then the operator A is $\mathcal{P}$-Fredholm, and the $\mathcal{P}$-essential norm of any regularizer of A is not greater than $2^{N+1} M$.

Proof. To start with, let A be a band operator. Let φ be a partition of unity satisfying (2.17). By the preceding proposition, the series

$$\sum_{|\alpha|\geq\rho(R)} \hat{\varphi}_{\alpha,R}B_{\alpha,R}\hat{\varphi}_{\alpha,R}I = \sum_{|\alpha|\geq\rho(R)} \hat{\varphi}_{\alpha,R}B_{\alpha,R}\hat{\varphi}_{\alpha,R}\hat{\psi}_{\alpha,R}I$$

converges $\mathcal{P}$-strongly to a certain operator B_R with $\|B_R\| \leq 2^N M$, and from Proposition 1.1.17 (a) we conclude that $B_R \in L(E, \mathcal{P})$. Because of

$$\lim_{R\to\infty} \text{dist}\,(\text{supp}\,\varphi_{\alpha,R},\ \text{supp}\,(1-\psi_{\alpha,R})) = \infty,$$

we further have

$$\hat{\varphi}_{\alpha,R}A = \hat{\varphi}_{\alpha,R}A\hat{\psi}_{\alpha,R}I \tag{2.19}$$

for all α and all sufficiently large R. Thus, if R is large enough, then

$$\begin{aligned} B_R A &= \sum_{|\alpha|\geq\rho(R)} \hat{\varphi}_{\alpha,R}B_{\alpha,R}\hat{\varphi}_{\alpha,R}A\hat{\psi}_{\alpha,R}I \\ &= \sum_{|\alpha|\geq\rho(R)} \hat{\varphi}_{\alpha,R}B_{\alpha,R}A\hat{\psi}_{\alpha,R}\hat{\varphi}_{\alpha,R}I + T_R \\ &= \sum_{|\alpha|\geq\rho(R)} \hat{\varphi}^2_{\alpha,R}I + T_R \ = I - \sum_{|\alpha|<\rho(R)} \hat{\varphi}^2_{\alpha,R}I + T_R \end{aligned} \tag{2.20}$$

where

$$T_R = \sum_{|\alpha|\geq\rho(R)} \hat{\varphi}_{\alpha,R}B_{\alpha,R}[\hat{\varphi}_{\alpha,R}I,\ A]\hat{\psi}_{\alpha,R}I.$$

The $\mathcal{P}$-strong convergence of the latter series is a consequence of Proposition 2.2.2 again, and $T_R \in L(E, \mathcal{P})$ by Proposition 1.1.17 (a). Further, Theorem 2.1.6 (c) implies that

$$\lim_{R\to\infty} \|[\hat{\varphi}_{\alpha,R}I,\ A]\|_{L(E)} = 0$$

uniformly with respect to $\alpha \in \mathbb{Z}^N$. Thus, $\|T_R\| \to 0$ as $R \to \infty$.

We fix a sufficiently large R such that $\|T_R\| < 1/2$. Then $I + T_R$ is invertible and $\|(I+T_R)^{-1}\| \leq 2$. Multiplying (2.20) by $(I+T_R)^{-1}$ from the left-hand side yields

$$(I+T_R)^{-1}B_R A = I - (I+T_R)^{-1} \sum_{|\alpha|<\rho(R)} \hat{\varphi}_{\alpha,R}I. \tag{2.21}$$

Since $\sum_{|\alpha|<\rho(R)} \hat{\varphi}_{\alpha,R}I$ is $\mathcal{P}$-compact, and since $(I+T_R)^{-1} \in L(E, \mathcal{P})$, this equality shows that $(I+T_R)^{-1}B_R$ is a left-sided regularizer of A with $\|(I+T_R)^{-1}B_R\| \leq 2^{N+1}M$. A right-sided regularizer can be constructed analogously. This settles the assertion for band operators.

To deal with the general case, we first verify that if the hypotheses of the proposition are satisfied for an operator $A \in \mathcal{A}_E$ with respect to a constant M, and if $A' \in \mathcal{A}_E$ is an operator with $\|A - A'\| < \varepsilon$ and $\varepsilon M < 1$, then these hypotheses are also satisfied for A' with respect to $M' := M/(1 - \varepsilon M)$.

Indeed, let $B_{\alpha,R}$ be such that $\|B_{\alpha,R}\| \le M$ and $B_{\alpha,R} A \hat{\psi}_{\alpha,R} I = \hat{\psi}_{\alpha,R} I$. Then

$$\begin{aligned} B_{\alpha,R} A' \hat{\psi}_{\alpha,R} I &= B_{\alpha,R} A \hat{\psi}_{\alpha,R} I - B_{\alpha,R}(A - A')\hat{\psi}_{\alpha,R} I \\ &= \hat{\psi}_{\alpha,R} I - B_{\alpha,R}(A - A')\hat{\psi}_{\alpha,R} I = (I - B_{\alpha,R}(A - A'))\hat{\psi}_{\alpha,R} I. \end{aligned}$$

Since $\|B_{\alpha,R}(A - A')\| \le \varepsilon M < 1$, the operator $I - B_{\alpha,R}(A - A')$ is invertible, and for $B'_{\alpha,R} := (I - B_{\alpha,R}(A - A'))^{-1} B_{\alpha,R}$ one has $\|B'_{\alpha,R}\| \le M/(1 - \varepsilon M) = M'$ and $B'_{\alpha,R} A' \hat{\psi}_{\alpha,R} I = \hat{\psi}_{\alpha,R} I$.

Let now $A \in L(E)$ be a band-dominated operator which satisfies the assumptions of the proposition. We choose a sequence (A_n) of band operators with $\|A - A_n\| \le 1/n$. As we have just seen, for n sufficiently large, the operators A_n also satisfy these assumptions with $M_n := M/(1 - M/n)$ in place of M. Hence, by what has already been proved, the operators A_n are $\mathcal{P}$-Fredholm, and the $\mathcal{P}$-essential norm of any regularizer B_n of A_n is not greater than $2^{N+1} M_n$. It remains to recall a standard fact from Banach algebra theory saying that if a_n and a are elements of a unital Banach algebra with $\|a_n - a\| \to 0$, and if the a_n are invertible and $\sup \|a_n^{-1}\| \le M < \infty$, then a is invertible and $a^{-1} \le M$. We apply this result in the algebra $\mathcal{A}_E / K(E, \mathcal{P})$ with $a := A + K(E, \mathcal{P})$ and $a_n := A_n + K(E, \mathcal{P})$ to finish the proof. □

Proposition 2.2.4 *Let $A \in \mathcal{A}_E$ and suppose that the limit operator A_h with respect to the sequence $h \in \mathcal{H}$ exists and is invertible. Then, for each function $\varphi \in l^\infty(\mathbb{Z}^N, L(X))$ with finite support, there is a number m_0 such that, for all $m \ge m_0$, there are operators B_m and C_m in $\mathcal{A}_E$ with*

$$\|B_m\| \le 2\,\|(A_h)^{-1}\|, \qquad \|C_m\| \le 2\,\|(A_h)^{-1}\|$$

and

$$B_m A V_{h(m)} \varphi V_{-h(m)} = V_{h(m)} \varphi V_{-h(m)} A C_m = V_{h(m)} \varphi V_{-h(m)}.$$

Proof. Given $\varphi \in l^\infty(\mathbb{Z}^N, L(X))$, choose a function $\chi \in l^\infty(\mathbb{Z}^N, L(X))$ with finite support such that $\chi\varphi = \varphi$. It is immediate from the definition of a limit operator that

$$V_{-h(m)} A V_{h(m)} \chi I = A_h \chi I + T_m \tag{2.22}$$

where the T_m are operators in $\mathcal{A}_E$ with $\|T_m\| \to 0$ as $m \to \infty$. Multiplying (2.22) by A_h^{-1} from the left-hand side and by $\varphi V_{-h(m)}$ from the right-hand side yields

$$A_h^{-1} V_{-h(m)} A V_{h(m)} \varphi V_{-h(m)} = (I + A_h^{-1} T_m) \varphi V_{-h(m)}. \tag{2.23}$$

Choose m_0 such that $\|A_h^{-1} T_m\| < 1/2$ for all $m \ge m_0$. Then $I + A_h^{-1} T_m$ is invertible and $\|(I + A_h^{-1} T_m)^{-1}\| < 2$, and multiplication of (2.23) by $V_{h(m)}(I + A_h^{-1} T_m)^{-1}$ from the left-hand side gives

$$V_{h(m)}(I + A_h^{-1} T_m)^{-1} A_h^{-1} V_{-h(m)} A V_{h(m)} \varphi V_{-h(m)} = V_{h(m)} \varphi V_{-h(m)}.$$

Thus, the 'left-sided' assertion of the proposition holds with the operators B_m being specified as $V_{h(m)}(I + A_h^{-1}T_m)^{-1}A_h^{-1}V_{-h(m)}$, and its 'right-sided' analogue can be verified in the same manner. □

Now we are in position to prove the sufficiency part of Theorem 2.2.1. Let $A \in \mathcal{A}_E^\$$, let all limit operators of A be invertible, and let

$$M_A := \sup\{\|A_h^{-1}\| : A_h \in \sigma_{op}(A)\} < \infty,$$

but assume contrary to what we want that A is not $\mathcal{P}$-Fredholm. Proposition 2.2.3 implies that, for $M := M_A$, there is an $R > 0$ such that, for all $\rho > 0$, there is an $\alpha_1 \in \mathbb{Z}^N$ with $|\alpha_1| > \rho$ such that $BA\hat{\psi}_{\alpha_1,R}I \neq \hat{\psi}_{\alpha_1,R}I$ for all operators $B \in L(E, \mathcal{P})$ with $\|B\| \le M_A$. A repeated use of this construction (where, in the second step, $|\alpha_1|$ plays the role of the ρ and α_2 is accordingly chosen) yields a sequence $(\alpha_k) \subseteq \mathbb{Z}^N$ with $\alpha_k \to \infty$ such that

$$BA\hat{\psi}_{\alpha_k,R}I \neq \hat{\psi}_{\alpha_k,R}I \tag{2.24}$$

for all k and all $B \in L(E, \mathcal{P})$ with $\|B\| \le M_A$.

Since A is rich, there is a subsequence $h := (\alpha_{k_m}R)_{m=0}^\infty$ of $(\alpha_k R)_{k=0}^\infty$ (which also tends to infinity) such that the limit operator A_h exists. By hypothesis, A_h is invertible, and $\|A_h^{-1}\| \le M_A$. From Proposition 2.2.4 we know that, for every function $\xi \in l^\infty(\mathbb{Z}^N, L(X))$ with finite support, there are operators $B_m \in \mathcal{A}_E \subset L(E, \mathcal{P})$ such that

$$\|B_m\| \le 2\,\|A_h^{-1}\| \le 2\,M_A$$

and

$$B_m AV_{-h(m)}\hat{\xi}V_{h(m)} = V_{-h(m)}\hat{\xi}V_{h(m)}.$$

Choosing $\xi := \psi_{0,R}$, we have

$$V_{-h(m)}\hat{\xi}V_{h(m)} = V_{-\alpha_{k_m}R}\hat{\xi}V_{\alpha_{k_m}R} = \hat{\psi}_{\alpha_{k_m},R}I$$

and, consequently,

$$B_m A\hat{\psi}_{\alpha_{k_m},R}I = \hat{\psi}_{\alpha_{k_m},R}I$$

with certain operators $B_m \in L(E, \mathcal{P})$ with $\|B_m\| \le M_A$. This contradicts (2.24). Hence, our assumption was wrong, and A is $\mathcal{P}$-Fredholm. □

It will be a main goal of Sections 2.3–2.5 to single out classes of band-dominated operators for which the *uniform boundedness* of the norms of the inverse limit operators is redundant, i.e., for which the *invertibility* of all limit operators already guarantees the $\mathcal{P}$-Fredholmness. A simple criterion is the following.

Proposition 2.2.5 *Let $A \in \mathcal{A}_E^\$$, and let all limit operators of A be invertible. If $\sigma_{op}(A)$ lies in a finite-dimensional subspace of $L(E)$, then*

$$\sup\{\|(A_h)^{-1}\| : A_h \in \sigma_{op}(A)\} < \infty.$$

Proof. By Proposition 1.2.3 and by hypothesis, the operator spectrum $\sigma_{op}(A)$ is a bounded and closed subset of a finite-dimensional space. Hence it is compact, and the continuous function $A_h \mapsto \|A_h^{-1}\|$ on $\sigma_{op}(A)$ attains its maximum. □

2.2.2 Weakly sufficient families of homomorphisms

As mentioned in Section 1.3, it is often advantageous to consider suitable subalgebras of the algebra of the rich band-dominated operators on which the mappings $A \mapsto A_h$ act as algebra homomorphisms. In this section, we will reformulate Theorem 2.2.1 in an appropriate way. For this goal, we introduce the notions of weakly sufficient and sufficient families of homomorphisms. These notions as well as the results of this section will be needed in Sections 4.3.2 and 6.3 only.

In this subsection we will work on Hilbert spaces only, i.e., we let $p = 2$ and X be a Hilbert space.

Definition 2.2.6 *Let $\mathcal{B}$ be a C^*-algebra with identity element e, and let $\mathcal{B}'$ be a symmetric, unital and dense subalgebra of $\mathcal{B}$. For every element t of a set T, let $\mathcal{B}_t$ be a C^*-algebra with identity element e_t, and let $W_t : \mathcal{B} \to \mathcal{B}_t$ be a unital *-homomorphism. Then we call $\{W_t\}_{t\in T}$ a* weakly sufficient family of homomorphisms for $\mathcal{B}'$ *if the following assertions are equivalent for every $b \in \mathcal{B}'$:*

(a) *b is invertible in $\mathcal{B}$.*

(b) *$W_t(b)$ is invertible in $\mathcal{B}_t$ for every $t \in T$, and*

$$\sup_{t\in T} \|(W_t(b))^{-1}\| < \infty.$$

We call $\{W_t\}_{t\in T}$ a sufficient family of homomorphisms for $\mathcal{B}'$ *if, for every $b \in \mathcal{B}'$, the assertion* (a) *is equivalent to*

(c) *$W_t(b)$ is invertible in $\mathcal{B}_t$ for every $t \in T$.*

Theorem 2.2.7 *Let $\mathcal{B}$, $\mathcal{B}'$, $\mathcal{B}_t$ and W_t be as in Definition* 2.2.6. *If $\{W_t\}_{t\in T}$ is a weakly sufficient family for $\mathcal{B}'$, then*

$$\|b\| = \sup_{t\in T} \|W_t(b)\|_t \quad \text{for every } b \in \mathcal{B}. \tag{2.25}$$

If the family $\{W_t\}_{t\in T}$ is sufficient for $\mathcal{B}'$, then the supremum in (2.25) *is a maximum.*

Proof. *-Homomorphisms are contractions. Hence,

$$\|b\| \geq \sup_{t\in T} \|W_t(b)\|_t \quad \text{for every } b \in \mathcal{B}.$$

Suppose there is a $b \in \mathcal{B}$ such that $\|b\| > \sup_{t\in T} \|W_t(b)\|_t$. Then there is also an element $b \in \mathcal{B}'$ with that property. By the C^*-axiom,

$$\|b^*b\| > \sup_{t\in T} \|W_t(b^*b)\|_t.$$

Set $d := \|b^*b\| - \sup_{t\in T} \|W_t(b^*b)\|_t$. Then $\|W_t(b^*b)\| \leq \|b^*b\| - d < \|b^*b\|$ for every $t \in T$. Hence, all operators $W_t(b^*b - \|b^*b\|e) = W_t(b^*b) - \|b^*b\|e_t$ are invertible,

and

$$\|(W_t(b^*b - \|b^*b\|e))^{-1}\| = \sup_{x\in\sigma(W_t(b^*b))} |(x - \|b^*b\|)^{-1}|$$
$$\geq \sup_{x\in[0,\rho(W_t(b^*b))]} |(x - \|b^*b\|)^{-1}| = (\|b^*b\| - \rho(W_t(b^*b)))^{-1} < 1/d$$

where $\rho(b)$ denotes the spectral radius of $b \in \mathcal{B}$. Thus,

$$\sup_{t\in T} \|(W_t(b^*b - \|b^*b\|e))^{-1}\| \leq 1/d.$$

Since $\{W_t\}$ is a weakly sufficient family, the latter estimate implies the invertibility of $b^*b - \|b^*b\|e$ in $\mathcal{B}$ which is impossible. This shows (2.25).

Let now $\{W_t\}$ be a sufficient family of homomorphisms, and suppose there is a $b \in \mathcal{B}'$ such that

$$\|W_t(b)\| < \sup_{t\in T} \|W_t(b)\| \quad \text{for all } t \in T. \tag{2.26}$$

Again by the C^*-axiom we get that (2.26) also holds with b replaced by b^*b. Since the norm of a self-adjoint element coincides with its spectral radius, (2.26) can be rewritten as

$$\rho(W_t(b^*b)) < \sup_{t\in T} \rho(W_t(b^*b)) \quad \text{for all } t \in T. \tag{2.27}$$

Denote the supremum on the right-hand side of (2.27) by M. The elements $W_t(b^*b - Me) = W_t(b^*b) - Me_t$ are invertible for all $t \in T$ since $\rho(W_t(b^*b)) < M$. Thus, the sufficiency of $\{W_t\}$ yields the invertibility of $b^*b - Me$ in $\mathcal{B}$. Then, clearly, $b^*b - me$ is invertible for all m belonging to a certain neighborhood U of M. On the other hand, since $\sup_{t\in T} \rho(W_t(b^*b)) = M$, for every neighborhood U of M there is a $t_U \in T$ such that $m_U := \rho(W_{t_U}(b^*b)) \in U$. The element $W_{t_U}(b^*b) - m_U e_{t_U}$ is not invertible, because the spectral radius of a non-negative element belongs to the spectrum of this element. Hence, $b^*b - m_U e$ is not invertible. This contradiction proves the assertion. □

Proposition 2.2.8 *Let $\mathcal{B}$, $\mathcal{B}'$, $\mathcal{B}_t$ and W_t be as in Definition* 2.2.6. *If the family $\{W_t\}$ is weakly sufficient for $\mathcal{B}'$, then it is also weakly sufficient for $\mathcal{B}$.*

Proof. The implication (a) ⇒ (b) in Definition 2.2.6 holds for every $b \in \mathcal{B}$. Thus, let $b \in \mathcal{B}$ be an element which satisfies condition (b), and set

$$C := \sup_{t\in T} \|(W_t(b))^{-1}\|.$$

Choose a sequence (b_n) of elements in $\mathcal{B}'$ which converges to b in the norm. There is an n_0 such that $\|b - b_n\| < 1/(2C)$ for all $n \geq n_0$. Then, since the W_t are contractions,

$$\|W_t(b) - W_t(b_n)\| < 1/(2C) \leq 1/(2\|(W_t(b))^{-1}\|)$$

for every $n \geq n_0$ and $t \in T$. Thus, by Neumann series, $W_t(b_n)$ is invertible and

$$\|(W_t(b_n))^{-1}\| \leq 2\|(W_t(b))^{-1}\| \leq 2C.$$

Since $\{W_t\}$ is weakly sufficient for $\mathcal{B}'$, the b_n are invertible for all $n \geq n_0$, and from Theorem 2.2.7 we infer that $\|b_n^{-1}\| \leq 2C$. Being the norm limit of a sequence of uniformly invertible elements, b is invertible. □

It turns out that the mapping $A \mapsto A_h$ provides us with a family of weakly sufficient homomorphisms. To be more precise, for $A \in \mathcal{A}_E^\$$, let $\mathcal{H}_A$ denote the set of all sequences $h \in \mathcal{H}$ for which the limit operator A_h exists. If $h \in \mathcal{H}_A$ then, by Proposition 1.2.2, the limit operators B_h exist for all operators B which belong to the smallest closed subalgebra $C^*(A)$ of $L(E)$ which contains the operators A, A^* and the identity operator I. The algebra $C^*(A)$ is a unital C^*-algebra, and the mapping $B \mapsto B_h$ is a unital *-homomorphism from $C^*(A)$ into $L(E)$. As we have seen in Proposition 1.3.2,

$$\sigma_{op}(B) = \{B_h : h \in \mathcal{H}_A\} \quad \text{for every operator } B \in C^*(A). \tag{2.28}$$

Further, the ideal $C^*(A) \cap K(E, \mathcal{P})$ lies in the kernel of the homomorphism $B \mapsto B_h$. Thus, the quotient homomorphism

$$W_h : C^*(A)/(C^*(A) \cap K(E, \mathcal{P})) \to L(E), \quad B + C^*(A) \cap K(E, \mathcal{P}) \mapsto B_h$$

is well defined for every sequence $h \in \mathcal{H}_A$, and Theorem 2.2.1 together with the inverse closedness of C^*-algebras yields that $\{W_h\}_{h \in \mathcal{H}_A}$ is a weakly sufficient family of homomorphisms for the algebra $C^*(A)/(C^*(A) \cap K(E, \mathcal{P}))$.

2.2.3 Symbol calculus for rich band-dominated operators

In terms introduced in Section 1.3.2, Theorem 2.2.1 can be restated as follows.

Theorem 2.2.9 *A rich band-dominated operator is $\mathcal{P}$-Fredholm on E if and only if its symbol is invertible in $\mathcal{F}_E$.*

Proof. If $A \in \mathcal{A}_E^\$$ is $\mathcal{P}$-Fredholm then (Propositions 1.2.6 and 1.2.10) there are an operator $B \in L^\$(E, \mathcal{P})$ and $\mathcal{P}$-compact operators $T_1, T_2 \in L^\$(E, \mathcal{P})$ such that

$$AB = I + T_1, \qquad BA = I + T_2.$$

Applying the homomorphism smb to both sides of these equalities yields invertibility of smb A because of smb T_1 = smb $T_2 = 0$. Conversely, if smb A is invertible, then all operators in $W(\text{smb}^0 A) = \sigma_{op}(A)$ are invertible, and the norms of their inverses are uniformly bounded. Thus, A is $\mathcal{P}$-Fredholm by Theorem 2.2.1. □

We complete this result by showing that the symbol mapping induces an isomorphism from $\mathcal{A}_E^\$/K(E, \mathcal{P})$ onto a closed subalgebra of $\mathcal{F}_E$.

Theorem 2.2.10 *The following estimate holds for $A \in \mathcal{A}_E^{\$}$:*

$$\|\mathrm{smb}\, A\|_{\mathcal{F}_E} \le \|\pi(A)\|_{L(E,\mathcal{P})/K(E,\mathcal{P})} \le 2^N \|\mathrm{smb}\, A\|_{\mathcal{F}_E}.$$

Proof. It is sufficient to prove the assertion for band operators. The first inequality is an immediate consequence of assertion (a) of Proposition 1.2.2 and of Proposition 1.2.6. For a proof of the second inequality, let $\varphi_{\alpha,R}$ and $\psi_{\alpha,R}$ be the functions introduced in Section 2.2.1. Since A is a band operator, we have, as in (2.19),

$$A = \sum_{\alpha\in\mathbb{Z}^n} \hat{\varphi}^2_{\alpha,R} A = \sum_{\alpha\in\mathbb{Z}^n} \hat{\varphi}^2_{\alpha,R} A \hat{\psi}_{\alpha,R} I$$

for all sufficiently large R. Given a positive real number ρ, define

$$Q_{R,\rho}(A) := \sum_{|\alpha|\ge\rho} \hat{\varphi}^2_{\alpha,R} A \hat{\psi}_{\alpha,R} I.$$

Proposition 2.2.2 implies the strong convergence of this series and the estimate

$$\begin{aligned}\|Q_{R,\rho}(A)\|_{L(E)} &\le 2^N \sup_{|\alpha|\ge\rho} \|\hat{\varphi}^2_{\alpha,R} A \hat{\psi}_{\alpha,R} I\|_{L(E)} \\ &\le 2^N \sup_{|\alpha|\ge\rho} \|\hat{\varphi}_{\alpha,R} A \hat{\psi}_{\alpha,R} I\|_{L(E)}.\end{aligned}$$

Since $\hat{\varphi}_{\alpha,R} I = V_{\alpha R} \hat{\varphi}_{0,R} V_{-\alpha R}$ we get

$$\|Q_{R,\rho}(A)\|_{L(E)} \le 2^N \sup_{|\alpha|\ge\rho} \|\hat{\varphi}_{0,R} V_{-\alpha R} A V_{\alpha R} \hat{\psi}_{0,R} I\|, \tag{2.29}$$

for all ρ and for all sufficiently large R. Now we fix a sufficiently large R and choose a strongly monotonically increasing sequence $(\rho_j)_{j\in\mathbb{N}}$. The sequence

$$\Big(\sup_{|\alpha|\ge\rho_j} \|\hat{\varphi}_{0,R} V_{-\alpha R} A V_{\alpha R} \hat{\psi}_{0,R} I\|\Big)_{j=1}^{\infty}$$

is monotonically decreasing and bounded below, hence convergent. Denote its limit by $M(A)$. From (2.29) we conclude the estimate

$$\|Q_{R,\rho}(A)\|_{L(E)} \le 2^N M(A).$$

We claim that

$$M(A) \le \|\mathrm{smb}\, A\|_{\mathcal{F}_E}.$$

Contrary to what we want, let us assume that $M(A) > \|\mathrm{smb}\, A\|_{\mathcal{F}_E}$. By the definition of $M(A)$, there is a sequence (α_k) tending to infinity such that

$$M(A) = \lim_{k\to\infty} \|\hat{\varphi}_{0,R} V_{-\alpha_k R} A V_{\alpha_k R} \hat{\psi}_{0,R} I\|.$$

Without loss of generality we can moreover suppose that the limit operator of A with respect to the sequence $h = (\alpha_k R)_{k=1}^\infty$ exists (otherwise we choose a convenient subsequence of (α_k)). Thus,

$$M(A) = \|\hat{\varphi}_{0,R} A_h \hat{\psi}_{0,R} I\|.$$

Then our assumption yields

$$\|A_h\| \geq \|\hat{\varphi}_{0,R} A_h \hat{\psi}_{0,R} I\| = M(A) > \|\mathrm{smb}\, A\|_{\mathcal{F}_E},$$

which is a contradiction since $\|\mathrm{smb}\, A\|_{\mathcal{F}_E} = \sup\{\|A_h\|,\ A_h \in \sigma(A)\}$. □

Let $\mathcal{S}_E$ denote the set of all symbols of operators in $\mathcal{A}_E^\$$.

Proposition 2.2.11 *$\mathcal{S}_E$ is a closed subalgebra of $\mathcal{F}_E$.*

Proof. The mapping smb is an algebra homomorphism. Hence, $\mathcal{S}_E$ is an algebra. To check the closedness of $\mathcal{S}_E$ in $\mathcal{F}_E$, let $(\mathrm{smb}\, A_j)_{j=1}^\infty$ be a sequence in $\mathcal{S}_E$ which converges to a certain coset $X^\sim$ in $\mathcal{F}_E$. Then $(\mathrm{smb}\, A_j)_{j=1}^\infty$ is a Cauchy sequence in $\mathcal{F}_E^\sim$, and $(\pi(A_j))_{j=1}^\infty$ is a Cauchy sequence in $L^\$(E, \mathcal{P})/K(E, \mathcal{P})$ by the preceding theorem implies. Since $\mathcal{A}_E^\$/K(E, \mathcal{P})$ is closed in $L^\$(E, \mathcal{P})/K(E, \mathcal{P})$, there is an operator $A \in \mathcal{A}_E^\$$ such that $\|\pi(A_j) - \pi(A)\| \to 0$. Again from Theorem 2.2.10 we conclude that $X^\sim = \mathrm{smb}\, A$; hence, $\mathcal{S}_E$ is closed. □

Let us further observe that the symbol $\mathrm{smb}\, A$ depends on the coset $\pi(A)$ only. Thus, the quotient mapping

$$\mathcal{A}_E^\$/K(E, \mathcal{P}) \to \mathcal{S}_E, \quad \pi(A) \mapsto \mathrm{smb}\, A \tag{2.30}$$

is well defined.

Theorem 2.2.12 *The mapping* (2.30) *is a continuous isomorphism between the algebras $\mathcal{A}_E^\$/K(E, \mathcal{P})$ and $\mathcal{S}_E$.*

Corollary 2.2.13 *Let $E = l^2(\mathbb{Z}^N, H)$ with a Hilbert space H. Then the algebras $\mathcal{A}_E^\$/K(E, \mathcal{P})$ and $\mathcal{S}_E$ are isometrically isomorphic.*

Indeed, under these assumptions, both $\mathcal{A}_E^\$/K(E, \mathcal{P})$ and $\mathcal{S}_E$ are C^*-algebras, and smb is a *-isomorphism. Thus, smb is actually an isometry. □

Observe also that, in the Hilbert space case, Theorem 2.2.12 offers another way to prove Theorem 2.2.9. Indeed, if $\mathrm{smb}\, A$ is invertible in $\mathcal{F}_E$, then it is also invertible in $\mathcal{S}_E$ due to the inverse closedness of C^*-algebras. The isomorphy of $\mathcal{S}_E$ and $\mathcal{A}_E^\$/K(E, \mathcal{P})$ implies that A is $\mathcal{P}$-Fredholm.

In case of a general space E, the Theorems 2.2.9 and 2.2.12 show that $\mathcal{S}_E$ is an inverse closed subalgebra of $\mathcal{F}_E$.

2.2.4 Appendix A: Second version of a symbol calculus

The symbol calculus introduced in the previous section associates with every band-dominated operator with rich spectrum an operator-valued function on (a subset of) the space $\mathcal{H}$ of all sequences tending to infinity. Here we will establish another symbol calculus where the symbol function is defined on the maximal ideal space of the C^*-algebra $l^\infty(\mathbb{Z}^N)$, i.e., where the limit operators are labelled by the maximal ideals of $l^\infty(\mathbb{Z}^N)$. Some basic facts concerning the Gelfand theory of commutative Banach algebras are summarized in Appendix B.

We restrict our considerations to band-dominated operators with matrix-valued coefficients. That is, we specify $X = \mathbb{C}^m$ with some positive integer m, and we agree upon writing $l^\infty_{m\times m}$ in place of $l^\infty(\mathbb{Z}^N, L(\mathbb{C}^m))$ and l^∞ in place of $l^\infty_{1\times 1}$. This particular choice implies that every band-dominated operator on E is rich.

Let $M(l^\infty)$ stand for the maximal ideal space of the Banach algebra l^∞, provided with its Gelfand topology, and let $M^\infty(l^\infty)$ denote the subset of $M(l^\infty)$ which consists of all non-trivial multiplicative functionals ξ such that $\xi(f) = 0$ for all $f \in l^\infty$ with finite support. The fiber $M^\infty(l^\infty)$ is a closed (hence, compact) subset of $M(l^\infty)$.

Our goal is to associate with each operator $A \in \mathcal{A}_E$ a function Smb A on $M^\infty(l^\infty)$ with values in $L(E)$ which is a symbol in the sense of the definition in Section 1.3.2.

Given a function $a = (a_{ij})_{i,j=1}^m \in l^\infty_{m\times m}$ and a multiplicative functional $\xi \in M^\infty(l^\infty)$, we let $\xi(a)$ denote the matrix $(\xi(a_{ij}))_{i,j=1}^m$, and we define a function $a_\xi \in l^\infty_{m\times m}$ by

$$a_\xi(x) := \xi(V_{-x} a V_x), \quad x \in \mathbb{Z}^N$$

(where we again identify a function with the operator of multiplication by this function). Further, if A be a band operator, then it can be uniquely written as

$$A = \sum_{|\alpha|\le k} a_\alpha V_\alpha \quad \text{with} \quad a_\alpha \in l^\infty_{m\times m},$$

and we define an operator A_ξ by

$$A_\xi := \sum_{|\alpha|\le k} (a_\alpha)_\xi V_\alpha.$$

Proposition 2.2.14 *Let A, $B \in L(E)$ be band operators. Then $(A+B)_\xi = A_\xi + B_\xi$, $(AB)_\xi = A_\xi B_\xi$, and $(\alpha A)_\xi = \alpha A_\xi$ for all complex α. Moreover, $\|A_\xi\| \le \|A\|$.*

Proof. Having in mind that $\xi(a+b) = \xi(a)+\xi(b)$ and $\xi(ab) = \xi(a)\xi(b)$ for all a, $b \in l^\infty_{m\times m}$, one easily verifies that the mapping $A \mapsto A_\xi$ is an algebra homomorphism. For the norm inequality observe that, given $\xi \in M^\infty(l^\infty)$, there is a directed set T and a net $h : T \to \mathbb{Z}^N$ $(= M(l^\infty) \setminus M^\infty(l^\infty))$ such that h_τ converges to ξ in the Gelfand topology (compare [58], Section 43). Thus, $a(h_\tau) \to \xi(a)$ for all $a \in l^\infty_{m\times m}$

and, consequently, $V_{-h_\tau} a V_{h_\tau} \to a_\xi$ in the $\mathcal{P}$-strong topology. For band operators $A = \sum a_\alpha V_\alpha$, this implies that

$$V_{-h_\tau} A V_{h_\tau} = \sum V_{-h_\tau} a_\alpha V_{h_\tau} V_\alpha \to A_\xi \quad \mathcal{P}\text{-strongly}.$$

By Proposition 1.1.17, $\|A_\xi\| \leq \liminf \|V_{-h_\tau} A V_{h_\tau}\| \leq \|A\|$. □

Let now $A \in \mathcal{A}_E$ be a band-dominated operator, and let $(A^{(k)})_{k=1}^\infty$ be a sequence of band operators tending to A in the norm. Then $(A^{(k)})$ is a Cauchy sequence, and the preceding proposition shows that $(A_\xi^{(k)})$ is a Cauchy sequence for every $\xi \in M^\infty(l^\infty)$. Thus, $\lim_{k\to\infty} A_\xi^{(k)}$ exists, and this limit is independent of the concrete choice of the $A^{(k)}$. We denote this limit by A_ξ.

Definition 2.2.15 *The* symbol $\operatorname{Smb} A$ *of the operator* $A \in \mathcal{A}_E$ *is the function from* $M^\infty(l^\infty)$ *into* $L(E)$ *defined by* $(\operatorname{Smb} A)(\xi) := A_\xi$.

Let $\mathcal{B}_E$ denote the set of all bounded functions on $M^\infty(l^\infty)$ with values in $L(E)$. Provided with pointwise operations and the supremum norm, $\mathcal{B}_E$ becomes a Banach algebra and, if a pointwise involution is defined in case $E = l^2(\mathbb{Z}^N, L(\mathbb{C}^m))$, a C^*-algebra.

Theorem 2.2.16 *Let* $X = \mathbb{C}^m$. *Then the mapping*

$$\operatorname{Smb} : \mathcal{A}_E \to \mathcal{B}_E, \quad A \mapsto \operatorname{Smb} A,$$

is a continuous algebra homomorphism of norm 1, *and an operator* $A \in \mathcal{A}_E$ *is* $\mathcal{P}$*-Fredholm (thus, Fredholm) if and only if its symbol* $\operatorname{Smb} A$ *is invertible in* $\mathcal{B}_E$.

Proof. The first assertion follows without effort from Proposition 2.2.14. For a proof of the second assertion we first remark that the symbol $\operatorname{Smb} K$ of a compact operator is the zero function. In order to see this, approximate K by band operators $K^{(k)} = \sum_{|\alpha| \leq k} a_\alpha V_\alpha$ where the a_α are functions with finite support. If ξ is in the fiber $M^\infty(l^\infty)$, then $\xi(a_\alpha) = 0$, whence $\operatorname{Smb} K^{(k)} = 0$. The continuity of the mapping Smb yields $\operatorname{Smb} K = 0$ for all compact operators. With this information, it is elementary to check that the $\mathcal{P}$-Fredholmness of A implies the invertibility of $\operatorname{Smb} A$ (compare the proof of Theorem 2.2.9).

For the reverse direction, it suffices to verify that

$$\{A_\xi : \xi \in M^\infty(l^\infty)\} = \sigma_{op}(A) \quad \text{for all } A \in \mathcal{A}_E. \tag{2.31}$$

Indeed, if $\operatorname{Smb} A$ is invertible, i.e., if all operators A_ξ are invertible and if the norms of their inverses are uniformly bounded, then (2.31) implies that all limit operators of A are invertible, and that their inverses are uniformly bounded, too. Thus, by Theorem 2.2.1, and since every band-dominated operator is rich under our assumptions, A is $\mathcal{P}$-Fredholm.

So we are left with verifying (2.31), for which it is clearly sufficient to consider the scalar case $m = 1$. Choose a sequence $(A^{(r)})$ of band operators

$$A^{(r)} = \sum_{k=1}^{r} a_k^{(r)} V_{h_k^{(r)}} \tag{2.32}$$

which approximates A in the norm topology, and let l_A^∞ stand for the smallest closed subalgebra of l^∞ which contains the identity function and all functions $a_k^{(r)}$ with $k = 1, \ldots, r$ and $r = 1, 2, \ldots$ together with their complex-conjugates and which, moreover, has the property that, whenever a function a is in l_A^∞, then all shifts $V_{-h}aV_h$ (with $h \in \mathbb{Z}^N$) of this function belong to l_A^∞ again. Thus, l_A^∞ is a symmetric and separable subalgebra of l^∞. Let $M(l_A^\infty)$ denote its maximal ideal space, and abbreviate $M(l_A^\infty) \setminus \mathbb{Z}^N$ to $M^\infty(l_A^\infty)$.

Repeating the above arguments, one associates an operator A_η with each $\eta \in M^\infty(l_A^\infty)$. We claim that

$$\{A_\eta : \eta \in M^\infty(l_A^\infty)\} = \sigma_{op}(A) \tag{2.33}$$

and

$$\{A_\eta : \eta \in M^\infty(l_A^\infty)\} = \{A_\xi : \xi \in M^\infty(l^\infty)\} \tag{2.34}$$

for all $A \in \mathcal{A}_E$. Obviously, (2.33) and (2.34) imply (2.31).

To verify (2.33), let $A_h \in \sigma_{op}(A)$ for some $h \in \mathcal{H}_A$. Since l_A^∞ is separable, one can make use of a Cantor diagonalization argument in order to obtain a subsequence $g = (g_m)$ of h such that the limit operator with respect to g exists for every function $f \in l_A^\infty$. Hence, the limit $\lim_{m\to\infty} f(g_m)$ exists for every $f \in l_A^\infty$, which implies the existence of a functional $\eta \in M^\infty(l_A^\infty)$ such that

$$\lim_{m\to\infty} f(g_m) = \eta(f).$$

This identity shows that $(fI)_g = (fI)_\eta$ for all $f \in l_A^\infty$, and the l_A^∞-version of Proposition 2.2.14 gives $A_g = A_\eta$, i.e.,

$$A_h = A_g \in \{A_\eta : \eta \in M^\infty(l_A^\infty)\}.$$

For the reverse inclusion, consider A_η with $\eta \in M^\infty(l_A^\infty)$. Since l_A^∞ is separable, $M(L_A^\infty)$ is metrizable, and we can approximate η by a *sequence* $h = (h_m) \subseteq \mathbb{Z}^N$ in the Gelfand topology. This sequence contains a subsequence g which also approximates η and for which the limit operator A_g exists. It is not hard to check that A_g is just the operator A_η, hence, $A_\eta \in \sigma_{op}(A)$. This proves (2.33).

For (2.34), recall that the subalgebra l_A^∞ of l^∞ induces a subdivision of $M^\infty(l^\infty)$ into fibers $M^{\infty,\eta}(l^\infty)$ over $M^\infty(l_A^\infty)$ where, by definition, ξ belongs to $M^{\infty,\eta}(l^\infty)$ if and only if the restriction of ξ onto l_A^∞ coincides with η. Since all coefficients of the band operators $A^{(r)}$ in (2.32) belong to l_A^∞, one has

$$(A^{(r)})_\xi = (A^{(r)})_\eta \quad \text{whenever } \xi \in M^{\infty,\eta}(l^\infty),$$

whence

$$A_\xi = A_\eta \quad \text{whenever } \xi \in M^{\infty,\eta}(l^\infty).$$

This verifies (2.34) and finishes the proof of Theorem 2.2.16 since $\mathcal{P}$ is a perfect approximate identity consisting of compact operators only. □

An immediate consequence of Theorem 2.2.10 and identity (2.31) is the following.

Corollary 2.2.17 *Let $X = \mathbb{C}^m$. Then, for $A \in \mathcal{A}_E$,*

$$\|\operatorname{Smb} A\|_{\mathcal{B}_E} \leq \|\pi(A)\|_{L(E)/K(E)} \leq 2^N \|\operatorname{Smb} A\|_{\mathcal{F}_E}.$$

Having this equivalence of norms at our disposal, it is not hard to derive results for the symbol mapping Smb which are analogous to Proposition 2.2.11 and Theorem 2.2.12. Indeed, the set of all functions $\operatorname{Smb} A$ with A running through $\mathcal{A}_E$ forms a closed subalgebra of $\mathcal{B}_E$, and there is a continuous isomorphism between this subalgebra and the algebra $\mathcal{A}_E/K(E, \mathcal{P})$. In the Hilbert space case, this isomorphism is an isometry.

2.2.5 Appendix B: Commutative Banach algebras

Let $\mathcal{B}$ be a commutative unital Banach algebra. An ideal of $\mathcal{B}$ is *maximal* if is not properly contained in some proper closed ideal of $\mathcal{B}$. Maximal ideals are closed. If x is a maximal ideal of $\mathcal{B}$, then the quotient $\mathcal{B}/x$ is isomorphic to the field $\mathbb{C}$ of the complex numbers. Thus, to each maximal ideal x of $\mathcal{B}$ and to each element b of $\mathcal{B}$, there is associated a complex number $\Phi_x(b)$, which is the image of the coset $b + x$ under the above-mentioned isomorphism. The mapping $\Phi_x : b \mapsto \Phi_x(b)$ is a multiplicative linear functional on $\mathcal{B}$ and, conversely, every non-trivial multiplicative linear functional on $\mathcal{B}$ is of this form. There is a one-to-one correspondence between the non-trivial multiplicative linear functionals, which are also called the *characters* of $\mathcal{B}$, and the maximal ideals of $\mathcal{B}$: the kernel of every character is a maximal ideal, and every maximal ideal is the kernel of a uniquely determined character.

Let $M(\mathcal{B})$ refer to the set of all maximal ideals of the Banach algebra $\mathcal{B}$. Given an element $b \in \mathcal{B}$, the complex-valued function

$$b^\sharp : M(\mathcal{B}) \to \mathbb{C}, \; x \mapsto \Phi_x(b)$$

is called the *Gelfand transform of* $b \in \mathcal{B}$. The set $M(\mathcal{B})$ becomes a topological space in a natural way: The *Gelfand topology* on $M(\mathcal{B})$ is the coarsest topology on $M(\mathcal{B})$ that makes all functions $b^\sharp$ with $b \in \mathcal{B}$ continuous. Equivalently, if $M(\mathcal{B})$ is thought of as a set of multiplicative functionals and, hence, as a subset of the dual Banach space $\mathcal{B}^*$ of $\mathcal{B}$, then the Gelfand topology coincides with the restriction of the *-weak topology on $\mathcal{B}^*$ onto $M(\mathcal{B})$. The set $M(\mathcal{B})$ provided with its Gelfand topology will be referred to as the *maximal ideal space* of the Banach algebra $\mathcal{B}$. Finally, the mapping

$$^\sharp : \mathcal{B} \to C(M(\mathcal{B})), \quad b \mapsto b^\sharp$$

is called the *Gelfand transformation.*

Theorem 2.2.18 *Let $\mathcal{B}$ be a commutative unital Banach algebra. Then*

(a) *the element $b \in \mathcal{B}$ is invertible if and only if $b^\sharp(x) \neq 0$ for all $x \in M(\mathcal{B})$.*

(b) *the maximal ideal space $M(\mathcal{B})$ is a Hausdorff compact, and the set $\mathcal{B}^\sharp$ of all Gelfand transforms of elements of $\mathcal{B}$ separates the points of $M(\mathcal{B})$.*

(c) *the Gelfand transformation is a continuous algebra homomorphism with norm 1.*

(d) *the kernel of the Gelfand transformation is the radical of $\mathcal{B}$.*

In our above notations, the assertions (a) and (c) are equivalent to saying that *the Gelfand transformation $^\sharp : \mathcal{B} \to C(M(\mathcal{B}))$ is a symbol mapping for $\mathcal{B}$.* Further, (c) reveals that, for $b \in \mathcal{B}$,

$$\|b^\sharp\| = \sup\{|\Phi_x(b)| : x \in M(\mathcal{B})\} \leq \|b\|, \tag{2.35}$$

but notice that neither equality holds in this estimate nor the Gelfand transformation is injective or surjective in general. However, in the C^*-case one has the following basic result.

Theorem 2.2.19 (Gelfand-Naimark) *Let $\mathcal{B}$ be a commutative unital C^*-algebra. Then the Gelfand transformation is a *-isomorphism from $\mathcal{B}$ onto $C(M(\mathcal{B}))$.*

This theorem allows us to think of the elements of a commutative unital C^*-algebra as continuous functions on a compact Hausdorff space.

A Banach algebra $\mathcal{B}$ with identity e is *singly generated* if there is an element $b \in \mathcal{B}$ such that the smallest closed subalgebra of $\mathcal{B}$ which contains e and b coincides with $\mathcal{B}$. In this case, b is called a *generator* of $\mathcal{B}$. The maximal ideal space of the Banach algebra $\mathcal{B}$ singly generated by b is homeomorphic to the spectrum $\sigma_\mathcal{B}(b)$. Under this identification, the Gelfand transform of b is just the identity mapping on $\sigma_\mathcal{B}(b)$.

Let $\mathcal{A}$ be a commutative Banach algebra with identity e, and let $\mathcal{B}$ be a closed subalgebra of $\mathcal{A}$ which contains e. Then the mapping

$$\tau : M(\mathcal{A}) \to M(\mathcal{B}), \quad x \mapsto x|_\mathcal{B},$$

which assigns to each non-trivial multiplicative functional on $\mathcal{B}$ its restriction onto $\mathcal{B}$, is continuous. For $y \in M(\mathcal{B})$, the set

$$M_y(\mathcal{A}) := \{x \in M(\mathcal{A}) : x|_\mathcal{B} = y\}$$

is referred to as the *fiber of $M(\mathcal{A})$ over $y \in M(\mathcal{B})$*. Since τ is continuous, each fiber $M_y(\mathcal{A}) = \tau^{-1}(\{y\})$ is a compact subset of $M(\mathcal{A})$. It can happen that $M_y(\mathcal{A}) = \emptyset$.

2.3 Local $\mathcal{P}$-Fredholmness: elementary theory

Up to now, we have considered limit operators with respect to sequences h which tend to infinity in the sense that, for each $R > 0$, there is an m_0 such that

$$h(m) \in \{z \in \mathbb{Z}^N : |z| > R\} \quad \text{for all } m \geq m_0.$$

It is, however, often desirable and useful to take into account the direction into which h tends. This idea is made precise in what follows by introducing local operator spectra. Moreover, local principles will enable us to weaken the uniform invertibility condition in Theorem 2.2.1.

2.3.1 Local operator spectra and local invertibility

Let S^{N-1} denote the unit sphere $\{\eta \in \mathbb{R}^N : |\eta|_2 = 1\}$ where $|\eta|_2$ stands for the Euklidean norm of η. Given a 'radius' $R > 0$, a 'direction' $\eta \in S^{N-1}$, and a neighborhood $U \subseteq S^{N-1}$ of η, define the *neighborhood at infinity of* η by

$$W_{R,U} := \{z \in \mathbb{Z}^N : |z| > R \text{ and } z/|z| \in U\}. \tag{2.36}$$

If h is a sequence which tends to infinity, then we say that h *tends into the direction of* $\eta \in S^{N-1}$ if, for every neighborhood at infinity $W_{R,U}$ of η, there is an m_0 such that

$$h(m) \in W_{R,U} \quad \text{for all } m \geq m_0.$$

Definition 2.3.1 *Let* $\eta \in S^{N-1}$ *and* $A \in L(E, \mathcal{P})$. *The* local operator spectrum $\sigma_\eta(A)$ of A at η *is the set of all limit operators* A_h *of* A *with respect to sequences* h *tending into the direction of* η.

Proposition 2.3.2 *Let* $A \in L(E, \mathcal{P})$. *Then* $\sigma_{op}(A) = \cup_{\eta \in S^{N-1}} \sigma_\eta(A)$.

Proof. It is evident that each local operator spectrum is contained in the operator spectrum. For the reverse inclusion, let $A_h \in \sigma_{op}(A)$ with respect to a certain sequence h tending to infinity. Due to the compactness of S^{N-1}, there is a subsequence g of h for which $g(m)/|g(m)|$ converges to a certain point η in S^{N-1} as $m \to \infty$. Then, obviously, the sequence g converges into the direction of η to infinity. Thus, $A_h = A_g \in \sigma_\eta(A)$. □

If A is a band-dominated operator, then the uniform invertibility of the operators in $\sigma_{op}(A)$ is equivalent to the $\mathcal{P}$-Fredholmness of A, as we know from Theorem 2.2.1. The uniform invertibility of the operators in the local operator spectrum $\sigma_\eta(A)$ is related with some kind of local invertibility at η in a similar way.

Definition 2.3.3 *Let* $\eta \in S^{N-1}$ *and* $A \in L(E)$. *The operator* A *is* locally invertible at η *if there are operators* $B, C \in L(E)$ *and a neighborhood at infinity* W *of* η *such that*

$$BA\hat{\chi}_W I = \hat{\chi}_W AC = \hat{\chi}_W I$$

where $\hat{\chi}_W$ *refers to the characteristic function of* W.

Before going into the details, we will provide an axiomatic frame for studying local invertibility problems.

2.3.2 $\mathcal{PR}$-compactness, $\mathcal{PR}$-Fredholmness

In the section, we return to the setting from the very beginning: E is a Banach space and $\mathcal{P} = (P_n)_{n=0}^\infty$ is an increasing approximate identity. Besides $\mathcal{P}$, we consider a *decreasing* approximate projection, $\mathcal{R} := (R_n)_{n=0}^\infty$, which is related with $\mathcal{P}$ as follows: For every $m \geq 0$, there is an $N(m)$ such that

$$R_n P_m = P_m R_n = 0 \quad \text{whenever } n \geq N(m). \tag{2.37}$$

Evidently, one can choose $R_n := I - P_n$, in which case we will write $\mathcal{R} := \mathcal{R}(\mathcal{P})$.

Definition 2.3.4 *Let $K(E, \mathcal{P}, \mathcal{R})$ stand for the set of all operators $K \in L(E, \mathcal{P})$ with*

$$\|KR_n\| \to 0 \quad \text{and} \quad \|R_n K\| \to 0 \quad \text{as } n \to \infty,$$

and let $L(E, \mathcal{P}, \mathcal{R})$ denote the set of all operators $A \in L(E, \mathcal{P})$ for which both AK and KA are in $K(E, \mathcal{P}, \mathcal{R})$ whenever K is in $K(E, \mathcal{P}, \mathcal{R})$.

Proposition 2.3.5

(a) *$L(E, \mathcal{P}, \mathcal{R})$ is a Banach algebra, and $K(E, \mathcal{P}, \mathcal{R})$ is a closed ideal of $L(E, \mathcal{P}, \mathcal{R})$.*

(b) *$K(E, \mathcal{P}) \subseteq K(E, \mathcal{P}, \mathcal{R})$, and $L(E, \mathcal{P}, \mathcal{R}) \subseteq L(E, \mathcal{P})$.*

(c) *$K(E, \mathcal{P}, \mathcal{R}(\mathcal{P})) = K(E, \mathcal{P})$, and $L(E, \mathcal{P}, \mathcal{R}(\mathcal{P})) = L(E, \mathcal{P})$.*

Proof. Assertion (a) can be proved as Proposition 1.1.8, and assertion (c) is obvious. For (b), let $K \in K(E, \mathcal{P})$ and $\varepsilon > 0$. Choose m such that $\|K - KP_m\| < \varepsilon$, and then choose $N(m)$ such that $P_m R_n = 0$ for all $n \geq N(m)$. Then

$$\|KR_n\| \leq \|(K - KP_m)R_n\| + \|KP_m R_n\| \leq \varepsilon \sup \|R_n\|$$

for all $n \geq N(m)$. Thus, $\|KR_n\| \to 0$ and, analogously, $\|R_n K\| \to 0$. □

Definition 2.3.6 *The operators in $K(E, \mathcal{P}, \mathcal{R})$ are called $\mathcal{PR}$-*compact*, and the operators in $L(E, \mathcal{P}, \mathcal{R})$ which are invertible modulo $K(E, \mathcal{P}, \mathcal{R})$ are called $\mathcal{PR}$-*Fredholm.

In case $\mathcal{R} = \mathcal{R}(\mathcal{P})$, the notions $\mathcal{PR}$-compact and $\mathcal{P}$-compact as well as the notions $\mathcal{PR}$-Fredholm and $\mathcal{P}$-Fredholm coincide. It is also evident that every operator P_n is $\mathcal{PR}$-compact and that every operator R_n is $\mathcal{PR}$-Fredholm.

Proposition 2.3.7 *An operator $A \in L(E, \mathcal{P}, \mathcal{R})$ is $\mathcal{PR}$-Fredholm if and only if there are operators $C, D \in L(E, \mathcal{P}, \mathcal{R})$ and $R_m \in \mathcal{R}$ such that*

$$R_m AC = DAR_m = R_m.$$

The proof is the same as that of Proposition 1.1.12. Also the following result can be proved as its predecessor Theorem 1.1.9. A decreasing approximate projection is called *uniform* if the associated increasing approximate projection is uniform.

Theorem 2.3.8 *Let $\mathcal{P}$ be a uniform approximate identity and $\mathcal{R}$ a uniform decreasing approximate projection related to each other by* (2.37). *Then the algebra $L(E, \mathcal{P}, \mathcal{R})$ is inverse closed in $L(E)$.*

Now we let again a group action $\mathcal{V} = \{V_k\}_{k\in\mathbb{Z}^N}$ enter the scene. As in Section 1.2.1, we assume that $\mathcal{V}$ is related with $\mathcal{P}$ by (1.32) and (1.33). In addition, we require that $\mathcal{V}$ is related with $\mathcal{R}$ by

$$\forall m \geq 0,\, k \in \mathbb{Z}^N\ \exists n_0 \geq 0\ \forall n \geq n_0 :\ (I - R_m)V_k Q_n = Q_n V_k (I - R_m) = 0. \quad (2.38)$$

This condition implies that $\mathcal{V} \subseteq L(E, \mathcal{P}, \mathcal{R})$. Let

$$\mathcal{H}_{\mathcal{R}} := \{h \in \mathcal{H} : \mathcal{P}\text{-}\lim_{m\to\infty} V_{-h(m)} R_n V_{h(m)} = I \text{ for all } R_n \in \mathcal{R}\}.$$

One can think of $\mathcal{H}_{\mathcal{R}}$ as the set of all sequences which tend to infinity in the direction prescribed by $\mathcal{R}$. We denote the set of all limit operators of A with respect to sequences in $\mathcal{H}_{\mathcal{R}}$ by $\sigma_{\mathcal{R}}(A)$. Clearly, $\sigma_{\mathcal{R}}(A) \subseteq \sigma_{op}(A)$.

Proposition 2.3.9 *Let $K \in K(E, \mathcal{P}, \mathcal{R})$ and $h \in \mathcal{H}_{\mathcal{R}}$. Then the limit operator K_h exists, and $K_h = 0$.*

Proof. Let $K \in K(E, \mathcal{P}, \mathcal{R})$ and $h \in \mathcal{H}_{\mathcal{R}}$. Then, for every $P_m \in \mathcal{P}$, $R_n \in \mathcal{R}$ and $k \in \mathbb{N}$,

$$\begin{aligned} \|V_{-h(k)} K V_{h(k)} P_m\| &\leq \|V_{-h(k)} K (I - R_n) V_{h(k)} P_m\| + \|V_{-h(k)} K R_n V_{h(k)} P_m\| \\ &\leq C\|V_{-h(k)} (I - R_n) V_{h(k)} P_m\| + C\|K R_n\|. \end{aligned}$$

Given $\varepsilon > 0$, choose n such that $\|KR_n\| < \varepsilon$, and fix k_0 such that $\|V_{-h(k)}(I - R_n)V_{h(k)}P_m\| \leq \varepsilon$ for all $k \geq k_0$ (which is possible because $h \in \mathcal{H}_{\mathcal{R}}$). Hence,

$$\|V_{-h(k)} K V_{h(k)} P_m\| \leq 2\varepsilon C \quad \text{for all } k \geq k_0.$$

The dual condition $\|P_m V_{-h(k)} K V_{h(k)}\| \to 0$ follows analogously. □

With this result, we get the following generalization of Proposition 1.2.9 without effort.

Proposition 2.3.10 *Let $\mathcal{P}$ be a perfect approximate identity and $\mathcal{R}$ a decreasing approximate projection. If $A \in L(E, \mathcal{P}, \mathcal{R})$ is a $\mathcal{PR}$-Fredholm operator, then all limit operators in $\sigma_{\mathcal{R}}(A)$ are invertible, and the norms of their inverses are uniformly bounded.*

There is also a generalization of Proposition 1.2.3 and its corollary, which is often helpful to determine the spectra $\sigma_{\mathcal{R}}(A)$.

Proposition 2.3.11 *The spectrum $\sigma_{\mathcal{R}}(A)$ of an operator $A \in L(E, \mathcal{P})$ is bounded and closed with respect to $\mathcal{P}$-strong convergence.*

Proof. The boundedness of $\sigma_{\mathcal{R}}(A)$ follows from Proposition 1.2.3 since $\sigma_{\mathcal{R}}(A) \subseteq \sigma_{op}(A)$. To verify the closedness of this spectrum, let $\tilde{A} \in L(E)$ be the $\mathcal{P}$-strong limit of a sequence $(A^{(k)})_{k=1}^{\infty}$ of operators in $\sigma_{\mathcal{R}}(A)$. For a fixed k, let h be a sequence in $\mathcal{H}_{\mathcal{R}}$ with $A_h = A^{(k)}$. Then, due to the definition of a limit operator and due to the definition of the class $\mathcal{H}_{\mathcal{R}}$,

$$\max_{l,m \leq k} \{\|(V_{-h(n)}AV_{h(n)} - A^{(k)})P_l\|, \|P_l(V_{-h(n)}AV_{h(n)} - A^{(k)})\|, \\ \|(V_{-h(n)}R_mV_{h(n)} - I)P_l\|, \|P_l(V_{-h(n)}R_mV_{h(n)} - I)\|\} \to 0.$$

Thus, there is an n_k such that

$$\max_{l,m \leq k} \{\|(V_{-h(n_k)}AV_{h(n_k)} - A^{(k)})P_l\|, \|P_l(V_{-h(n_k)}AV_{h(n_k)} - A^{(k)})\|, \\ \|(V_{-h(n_k)}R_mV_{h(n_k)} - I)P_l\|, \|P_l(V_{-h(n_k)}R_mV_{h(n_k)} - I)\|\} < 1/k.$$

Set $\tilde{h}(k) := h(n_k)$. As in the proof of Proposition 1.2.3, one can verify that $\tilde{A}$ is the limit operator of A with respect to the sequence $\tilde{h}$. It remains to show that $\tilde{h}$ belongs to $\mathcal{H}_{\mathcal{R}}$. This follows immediately from our construction since

$$\max\{\|(V_{-\tilde{h}(k)}R_mV_{\tilde{h}(k)} - I)P_l\|, \|P_l(V_{-\tilde{h}(k)}R_mV_{\tilde{h}(k)} - I)\|\} < 1/k$$

for all $k \geq \max\{l, m\}$. □

Let $h \in \mathcal{H}_{\mathcal{R}}$ and $s \in \mathbb{Z}^N$. Then, for every $R_m \in \mathcal{R}$,

$$V_{-(h(n)+s)}R_mV_{h(n)+s} = V_{-s}V_{-h(n)}R_mV_{h(n)}V_s \to V_{-s}V_s = I$$

$\mathcal{P}$-strongly as $n \to \infty$. Thus, the sequence $h + s$ belongs to $\mathcal{H}_{\mathcal{R}}$, too, and the arguments from Section 1.2.1 show that the spectra $\sigma_{\mathcal{R}}(A)$ are shift invariant. In combination with the preceding proposition, this observation has the following consequence.

Corollary 2.3.12 *If $B \in \sigma_{\mathcal{R}}(A)$, then $\sigma_{op}(B) \subseteq \sigma_{\mathcal{R}}(A)$.*

2.3.3 Local $\mathcal{P}$-Fredholmness of band-dominated operators

One can reify the axioms from the previous section as follows. Given $\eta \in S^{N-1}$, let $(U_n) \subseteq S^{N-1}$ be a monotonically decreasing sequence of neighborhoods of η with $\cap_n U_n = \{\eta\}$, and let $(r_n) \subseteq \mathbb{R}^+$ be a monotonically increasing sequence which tends to infinity. Then let R_n stand for the operator of multiplication by the characteristic function of the set W_{r_n,U_n}, and set $\mathcal{R} := (R_n)_{n=0}^{\infty}$. One easily checks that then all assumptions made for $\mathcal{R}$ in the previous section are satisfied. In particular, it turns out that a sequence $h \in \mathcal{H}$ tends to infinity into the direction

η if and only if it belongs to $\mathcal{H}_{\mathcal{R}}$, that the local operator spectrum $\sigma_\eta(A)$ coincides with $\sigma_{\mathcal{R}}(A)$, and that an operator is locally invertible at η if and only if it is $\mathcal{PR}$-Fredholm. Thus, we conclude from Proposition 2.3.10 that if $A \in L(E, \mathcal{P})$ is locally invertible at η, then all operators in $\sigma_\eta(A)$ are invertible, and the norms of their inverses are uniformly bounded. We will see now, that for rich band-dominated operators, the converse of this statement is true. The following theorem can be considered as a local version of Theorem 2.2.1.

Theorem 2.3.13 *The operator $A \in \mathcal{A}_E^\$$ is locally invertible at $\eta \in S^{N-1}$ if and only if all limit operators in $\sigma_\eta(A)$ are invertible and if*

$$\sup\{\|(A_h)^{-1}\| : A_h \in \sigma_\eta(A)\} < \infty.$$

The proof proceeds in the same way as that of Theorem 2.2.1, and we omit the details. Roughly speaking, one has to replace the sets $\{z \in \mathbb{Z}^N : |z| > R\}$ by the neighborhoods at infinity of η defined in (2.36). For example, the 'local' analogue of the crucial Proposition 2.2.3 reads as follows. Again, χ_W refers to the characteristic function of the set W.

Proposition 2.3.14 *Let $A \in \mathcal{A}_E$, $\eta \in S^{N-1}$, and let $\psi_{\alpha,R}$ be as in Section* 2.2.1. *Suppose there is a constant $M > 0$ such that, for all positive integers R, there is a neighborhood at infinity $W_{\rho,U}$ of η with some $\rho = \rho(R)$ such that, for all $\alpha \in W_{\rho(R),U}$, there are operators $B_{\alpha,R}$, $C_{\alpha,R} \in L(E, \mathcal{P})$ with $\|B_{\alpha,R}\|_{L(E)} \le M$, $\|C_{\alpha,R}\|_{L(E)} \le M$ and*

$$B_{\alpha,R}\, A\, \hat{\psi}_{\alpha,R} I = \hat{\psi}_{\alpha,R}\, A\, C_{\alpha,R} = \hat{\psi}_{\alpha,R} I.$$

Then the operator A is locally invertible at η, i.e., there are operators B, $C \in \mathcal{A}_E$ and a neighborhood at infinity W of η such that

$$BA\hat{\chi}_W I = \hat{\chi}_W AC = \hat{\chi}_W I. \tag{2.39}$$

2.3.4 Allan's local principle

Here we are going to explain that the theory of limit operators is compatible with another local theory, which is called *central localization* or *Allan's local principle.*

The *center* cen $\mathcal{A}$ of an algebra $\mathcal{A}$ consists of all elements $a \in \mathcal{A}$ such that $ab = ba$ for all $b \in \mathcal{A}$. Clearly, the center of a unital Banach algebra is a closed, unital, commutative, and inverse closed subalgebra of that algebra. Allan's local principle is a generalization of the Gelfand theory for commutative Banach algebras to Banach algebras which are close to the commutative ones in the sense that their centers are non-trivial.

Let $\mathcal{A}$ be a unital Banach algebra. By a *central* subalgebra $\mathcal{B}$ of $\mathcal{A}$ we mean a closed subalgebra of the center of $\mathcal{A}$ which contains the identity element. Thus, $\mathcal{B}$ is a commutative Banach algebra with compact maximal ideal space $M(\mathcal{B})$. To each maximal ideal x of $\mathcal{B}$, we associate the smallest closed two-sided ideal $\mathcal{I}_x$ of

$\mathcal{A}$ which contains x, and we let Φ_x refer to the canonical homomorphism from $\mathcal{A}$ onto the quotient algebra $\mathcal{A}/\mathcal{I}_x$. Notice that, in contrast to the commutative setting, the quotient algebras $\mathcal{A}/\mathcal{I}_x$ can differ from each other in dependence on $x \in M(\mathcal{B})$. Moreover, it may happen that $\mathcal{I}_x = \mathcal{A}$ for some points x. In this case we *define* that $\Phi_x(a)$ is invertible in $\mathcal{A}/\mathcal{I}_x$ and that $\|\Phi_x(a)\| = 0$ for each $a \in \mathcal{A}$.

The proof of Allan's local principle is based on the following observation.

Proposition 2.3.15 *Let $\mathcal{B}$ be a central subalgebra of the unital Banach algebra $\mathcal{A}$. If $\mathcal{M}$ is a maximal left, right, or two-sided ideal of $\mathcal{A}$, then $\mathcal{M} \cap \mathcal{B}$ is a (two-sided) maximal ideal of $\mathcal{B}$.*

Proof. Suppose that $\mathcal{M}$ is a maximal left ideal. Then $\mathcal{M} \cap \mathcal{B}$ is a proper (since $e \in \mathcal{B} \setminus \mathcal{M}$) closed two-sided ideal of $\mathcal{B}$, and we are left with the proof of its maximality.

Let $z \in \mathcal{B} \setminus \mathcal{M}$. Then $\mathcal{I}_z := \{l + az : l \in \mathcal{M}, a \in \mathcal{A}\}$ is a left ideal of $\mathcal{A}$ which contains $\mathcal{M}$ properly (since $z \notin \mathcal{M}$). The maximality of $\mathcal{M}$ implies $\mathcal{I}_z = \mathcal{A}$. Hence, $e \in \mathcal{I}_z$, and z has an inverse modulo $\mathcal{M}$ (note that $z \in \operatorname{cen} A$). Furthermore, $\mathcal{K}_z := \{a \in \mathcal{A} : az \in \mathcal{M}\}$ is a proper (since $e \notin \mathcal{K}_z$) left ideal of $\mathcal{A}$ containing $\mathcal{M}$. Since $\mathcal{M}$ is maximal, we have $\mathcal{K}_z = \mathcal{M}$. In particular, if y_1, y_2 are both inverses modulo $\mathcal{M}$ of z, then $y_1 - y_2 \in \mathcal{M}$. Thus, the inverses modulo $\mathcal{M}$ of z determine a unique element of the quotient space $\mathcal{A}/\mathcal{M}$.

Suppose $z - \lambda e \notin \mathcal{M}$ for all $\lambda \in \mathbb{C}$. Let $y^\pi(\lambda)$ denote the (uniquely determined) coset of $\mathcal{A}/\mathcal{M}$ containing the inverses modulo $\mathcal{M}$ of $z - \lambda e$. We claim that $y^\pi : \mathbb{C} \to \mathcal{A}/\mathcal{M}$ is an analytic function. To see this, let $\lambda_0 \in \mathbb{C}$, and let $y_0 \in y^\pi(\lambda_0)$ be an inverse modulo $\mathcal{M}$ of $z - \lambda_0 e$. Then, for $|\lambda - \lambda_0| < 1/\|y_0\|$, the element $e - (\lambda - \lambda_0)y_0$ is invertible in $\mathcal{A}$, and it is readily verified that $y_0[e - (\lambda - \lambda_0)y_0]^{-1}$ is an inverse modulo $\mathcal{M}$ of $z - \lambda e$. Thus, for $|\lambda - \lambda_0| < 1/\|y_0\|$,

$$y^\pi(\lambda) = y_0[e - (\lambda - \lambda_0)y_0]^{-1} + \mathcal{M},$$

which implies the asserted analyticity of y^π. If $|\lambda| > \|z\|$, then $z - \lambda e$ is actually invertible in $\mathcal{A}$ and

$$\|y^\pi(\lambda)\| \le \|(z - \lambda e)^{-1}\| = (1/|\lambda|)\|\sum_{n \ge 0} z^n/\lambda^n\| = o(1) \quad \text{as } |\lambda| \to \infty.$$

Therefore, by Liouville's theorem, $y^\pi(\lambda) = 0$ for all $\lambda \in \mathbb{C}$, contrary to the assumption that $\mathcal{M}$ is a proper ideal of $\mathcal{A}$ (for $y^\pi(0) = 0$ would imply that there is a $y_0 \in \mathcal{M}$ with $y_0 z - e \in \mathcal{M}$, whence $e \in \mathcal{M}$).

Hence, there is some $\lambda \in \mathbb{C}$ such that $z - \lambda e \in \mathcal{M}$ and, since $z \notin \mathcal{M}$, we have $\lambda \neq 0$. It follows that $e = \lambda^{-1} z + l$ for some $l \in \mathcal{M} \cap \mathcal{B}$.

Assume there is a two-sided ideal $\mathcal{I}$ of $\mathcal{B}$ such that $\mathcal{M} \cap \mathcal{B} \subset \mathcal{I}$ and $\mathcal{M} \cap \mathcal{B} \neq \mathcal{I}$. Then there is a $z \in \mathcal{I} \setminus (\mathcal{M} \cap \mathcal{B}) \subset \mathcal{B} \setminus \mathcal{M}$, and by what has been proved above, there exist $\lambda \in \mathbb{C} \setminus \{0\}$ and $l \in \mathcal{M} \cap \mathcal{B}$ with $e = \lambda^{-1} z + l$. But this implies that $e \in \mathcal{I}$ and, hence, $\mathcal{I} = \mathcal{B}$, which verifies the maximality of $\mathcal{M} \cap \mathcal{B}$. □

Theorem 2.3.16 (Allan) *Let $\mathcal{B}$ be a central subalgebra of the unital Banach algebra $\mathcal{A}$. Then*

(a) *an element $a \in \mathcal{A}$ is invertible if and only if the cosets $\Phi_x(a)$ are invertible in $\mathcal{A}/\mathcal{I}_x$ for every $x \in M(\mathcal{B})$.*

(b) *the mapping $M(\mathcal{B}) \to \mathbb{R}^+$, $x \mapsto \|\Phi_x(a)\|$ is upper semi-continuous for every $a \in \mathcal{A}$.*

(c) $\|a\| \geq \max_{x \in M(\mathcal{B})} \|\Phi_x(a)\|$.

(d) $\cap_{x \in M(\mathcal{B})} \mathcal{I}_x$ *lies in the radical of $\mathcal{A}$.*

Recall that a function $f : M(\mathcal{B}) \to \mathbb{R}$ is said to be *upper semi-continuous* at $x_0 \in M(\mathcal{B})$ if, for every $\varepsilon > 0$, there exists a neighborhood $U \subset M(\mathcal{B})$ of x_0 such that $f(x) < f(x_0) + \varepsilon$ for all $x \in U$. The function f is said to be upper semi-continuous on $M(\mathcal{B})$ if it is upper semi-continuous at each point $x \in M(\mathcal{B})$. Notice that the maximum in (c) is justified since every upper semi-continuous function attains its supremum.

Proof. To prove (a), we show that $a \in \mathcal{A}$ is left invertible if and only if $\Phi_x(a)$ is left invertible for all $x \in M(\mathcal{B})$. The proof for the right invertibility is analogous.

Clearly, $\Phi_x(a)$ is left invertible if a is so. To verify the reverse implication assume the contrary, i.e., suppose $\Phi_x(a)$ to be left invertible in $\mathcal{A}/\mathcal{I}_x$ for all $x \in M(\mathcal{B})$ but let a have no left inverse in $\mathcal{A}$. Denote by $\mathcal{M}$ a maximal left ideal of $\mathcal{A}$ which contains the set $\mathcal{I} := \{ba : b \in \mathcal{A}\}$ (note that $e \notin \mathcal{I}$). Put $x = \mathcal{M} \cap \mathcal{B}$. By the preceding proposition, x is a maximal ideal of $\mathcal{B}$. We claim that $\mathcal{I}_x \subseteq \mathcal{M}$. Indeed, if $l = \sum_{k=1}^n a_k x_k b_k$ where $x_k \in x$ and a_k, $b_k \in \mathcal{A}$, then $l = \sum_{k=1}^n a_k b_k x_k$ (because $\mathcal{B}$ is central), and hence $l \in \mathcal{M}$ (because $\mathcal{M}$ is a left ideal). Thus, $\mathcal{I}_x \subseteq \mathcal{M}$. By our assumption, $\Phi_x(a)$ is left invertible in $\mathcal{A}/\mathcal{I}_x$, that is, there exist a $b \in \mathcal{A}$ with $ba - e \in \mathcal{I}_x$, and since $\mathcal{I}_x \subseteq \mathcal{M}$ we have $ba - e \in \mathcal{M}$. On the other hand, $ba \in \mathcal{I} \subseteq \mathcal{M}$. This implies that $e \in \mathcal{M}$ which contradicts the maximality of $\mathcal{M}$.
For assertion (b), let $x \in M(\mathcal{B})$ and $\varepsilon > 0$. We have to show that there is a neighborhood U of x such that

$$\|\Phi_y(a)\| < \|\Phi_x(a)\| + \varepsilon \quad \text{for all } y \in U.$$

For, choose elements $a_1, \ldots, a_n \in \mathcal{A}$ and $x_1, \ldots, x_n \in x$ such that

$$\|a + \sum_{j=1}^n a_j x_j\| < \|\Phi_x(a)\| + \varepsilon/2, \tag{2.40}$$

and define an open neighborhood $U \subset M(\mathcal{B})$ of x by

$$U := \{y \in M(\mathcal{B}) : |\Phi_y(x_j)| < \varepsilon \left(2 \sum_{i=1}^n \|a_i\| + 1\right)^{-1} \quad \text{for } j = 1, \ldots, n\}.$$

Let $y \in U$ and put $y_j := x_j - \Phi_y(x_j)e$. Since the x_j's are elements of the *commutative* Banach algebra $\mathcal{B}$, the cosets $\Phi_y(x_j)$ can be identified with complex numbers, namely with the value of the Gelfand transform of x_j at the point y. In particular, $\Phi_y(y_j) = \Phi_y(x_j - \Phi_y(x_j)e) = 0$, whence $y_j \in y$ and

$$\|\Phi_y(a)\| \leq \|a + \sum a_j y_j\|. \tag{2.41}$$

The estimates (2.40) and (2.41) give

$$\begin{aligned} \|\Phi_y(a)\| - \|\Phi_x(a)\| &\leq \|a + \sum a_j y_j\| - \|a + \sum a_j x_j\| + \varepsilon/2 \\ &\leq \|\sum a_j (y_j - x_j)\| + \varepsilon/2 \\ &= \|\sum \Phi_y(x_j) a_j\| + \varepsilon/2 < \varepsilon, \end{aligned}$$

thus proving the upper semi-continuity of $y \mapsto \|\Phi_y(a)\|$ at x.
(c) By definition, $\|a\| \geq \|\Phi_x(a)\|$ for any $x \in M(\mathcal{B})$, which implies that $\|a\| \geq \max_{x \in M(\mathcal{B})} \|\Phi_x(a)\|$.
(d) Let $k \in \cap_{x \in M(\mathcal{B})} \mathcal{I}_x$. Then, for every $a \in \mathcal{A}$, $\Phi_x(e - ka) = \Phi_x(e)$, whence via (a) the invertibility of $e - ka$. Thus, k lies in the radical of $\mathcal{A}$. □

Here is a consequence of the preceding theorem which concerns local invertibility.

Proposition 2.3.17 *Let the situation be as in Theorem* 2.3.16. *If* $\Phi_x(a)$ *is invertible in* $\mathcal{A}/\mathcal{I}_x$, *then there are a neighborhood* U *of* $x \in M(\mathcal{B})$ *as well as a neighborhood* V *of* $a \in \mathcal{A}$ *such that* $\Phi_y(c)$ *is invertible in* $\mathcal{A}/I_y$ *and*

$$\|\Phi_y(c)^{-1}\| \leq 4\,\|\Phi_x(a)^{-1}\| \quad \textit{for all } y \in U \textit{ and } c \in V.$$

The number 4 can be replaced by any constant greater than 1.

Proof. If $\Phi_x(a)$ is invertible, then there is a $b \in \mathcal{A}$ such that $\Phi_x(ab - e) = \Phi_x(ba - e) = 0$. By Theorem 2.3.16 (b), the mappings

$$y \mapsto \|\Phi_y(ab - e)\| \quad \text{and} \quad y \mapsto \|\Phi_y(ba - e)\|,$$

defined on the maximal ideal space of $\mathcal{B}$, are upper semi-continuous. Hence,

$$\|\Phi_y(ab - e)\| < 1/4 \quad \text{and} \quad \|\Phi_y(ba - e)\| < 1/4$$

for all maximal ideals y in a certain neighborhood U' of x. Let further V stand for the set of all elements $c \in \mathcal{A}$ with $\|c - a\| < (4\,\|b\|)^{-1}$. Then

$$\Phi_y(c)\Phi_y(b) = \Phi_y(e) + \Phi_y(cb - e) \quad \text{and} \quad \Phi_y(b)\Phi_y(c) = \Phi_y(e) + \Phi_y(bc - e)$$

with

$$\|\Phi_y(cb - e)\| \leq \|\Phi_y(ab - e)\| + \|\Phi_y((c - a)b)\| \leq 1/4 + \|c - a\|\,\|b\| < 1/2$$

and, analogously, $\|\Phi_y(bc - e)\| \leq 1/2$. Since $\Phi(e)$ is the identity element in $\mathcal{A}/\mathcal{I}_y$, this implies via Neumann's series that $\Phi_y(c)$ is invertible in $\mathcal{A}/\mathcal{I}_y$ and that

$$\|\Phi_y(c)^{-1}\| \leq 2\,\|\Phi_y(b)\| \quad \text{for all } y \in U' \text{ and } c \in V.$$

Employing the upper semi-continuity once more, one finally gets

$$\|\Phi_y(b)\| \leq 2\,\|\Phi_x(b)\| = 2\,\|\Phi_x(a)^{-1}\|$$

for all y in a neighborhood $U \subseteq U'$ of x. □

Another supplement to Allan's local principle concerns the continuity of local spectra. Let X be a compact Hausdorff space, and let M be a mapping from X into the set of all compact subsets of the complex plane $\mathbb{C}$. Given a point $x \in X$ and a sequence $(y_n) \subseteq X$ with $y_n \to x$ as $n \to \infty$, we consider the set $L(y_n)$ of all limiting points of all sequences (λ_n) with $\lambda_n \in M(y_n)$. Then the *limes superior* $\limsup_{y\to x} M(y)$ is defined as $\cup L(y_n)$, where the union is taken over all sequences (y_n) converging to x.

Proposition 2.3.18 *Let the situation be as in Theorem* 2.3.16. *Then, for all* $a \in \mathcal{A}$ *and* $x \in M(\mathcal{B})$,

$$\limsup_{y\to x} \sigma(\Phi_y(a)) \subseteq \sigma(\Phi_x(a)).$$

Proof. Let $\lambda \in \limsup_{y\to x} \sigma(\Phi_y(a))$. By definition, then there are points $y_n \in M(\mathcal{B})$ with $y_n \to x$ and numbers $\lambda_n \in \sigma(\Phi_{y_n}(a))$ such that $\lambda_n \to \lambda$. Consider the elements $a - \lambda_n e$ which converge to $a - \lambda e$ in the norm of $\mathcal{A}$, and assume the coset $\Phi_x(a - \lambda e)$ is invertible. Then, by Proposition 2.3.17, the local cosets $\Phi_{y_n}(a - \lambda_n e)$ are invertible for all sufficiently large n, which contradicts our hypothesis. Consequently, $s \in \sigma(\Phi_x(a))$. □

2.3.5 Local $\mathcal{P}$-Fredholmness of band-dominated operators in the sense of the local principle

Consider the set of all continuous complex-valued functions f on $\mathbb{R}^N$ for which the functions

$$f_R : S^{N-1} \to \mathbb{C},\ \eta \mapsto f(\eta R),$$

which are defined for every $R > 0$, converge uniformly on S^{N-1} as $R \to \infty$. The uniform limit $f^* := \lim_{R\to\infty} f_R$ is a continuous function on S^{N-1} again. This set is a commutative unital C^*-algebra. If we let $\overline{\mathbb{R}^N}$ refer to its maximal ideal space, then the Gelfand/Naimark theorem justifies to denote this algebra by $C(\overline{\mathbb{R}^N})$.

The function $x \mapsto x/(1 + |x|)$ maps $\mathbb{R}^N$ homeomorphically onto the interior $B^N \setminus S^{N-1}$ of the unit ball $B^N := \{x \in \mathbb{R}^N : |x|_2 \leq 1\}$. This homeomorphism induces a *-isomorphism between $C(\overline{\mathbb{R}^N})$ and the algebra of all functions which are continuous on $B^N \setminus S^{N-1}$ and which admit a continuous extension onto all of B^N, i.e., between $C(\overline{\mathbb{R}^N})$ and $C(B^N)$. Thus, $\overline{\mathbb{R}^N}$ is homeomorphic to B^N, and the non-trivial multiplicative functionals of $C(\overline{\mathbb{R}^N})$ are either of the form $f \mapsto f(x)$ with some $x \in \mathbb{R}^N$ or $f \mapsto f^*(\eta)$ with some $\eta \in S^{N-1}$.

We think of $\overline{\mathbb{R}^N}$ as the compactification of $\mathbb{R}^N$ by the sphere S^{N-1} of the infinitely distant points. Evidently, a sequence $h \subset \mathbb{R}^N$, which tends to infinity, tends into the direction $\eta \in S^{N-1}$ with respect to the Euklidean topology if and only if it tends to the infinitely distant point η with respect to the Gelfand topology. This observation justifies to call the sets $W_{R,U}$ introduced in (2.36) neighborhoods at infinity of η. Notice also that, for every infinitely distant point $\eta \in S^{N-1}$, there is a sequence h in $\mathbb{Z}^N$ which tends to η in the Gelfand topology.

In what follows, we let π refer to the canonical homomorphism from $L(E, \mathcal{P})$ onto $L(E, \mathcal{P})/K(E, \mathcal{P})$.

Proposition 2.3.19

(a) *For $f \in C(\overline{\mathbb{R}^N})$, the coset $\pi(\hat{f}I)$ belongs to the center of $\mathcal{A}_E/K(E, \mathcal{P})$.*

(b) *The maximal ideal space of the algebra $\mathcal{C}_E := \{\pi(\hat{f}I) : f \in C(\overline{\mathbb{R}^N})\}$ is homeomorphic to S^{N-1}, and the homeomorphism identifies $\eta \in S^{N-1}$ with the functional $\pi(\hat{f}I) \mapsto f^*(\eta)$.*

Proof. (a) It is evident that $\hat{f}I$ commutes with all multiplication operators. Further, the commutator $\hat{f}V_\alpha - V_\alpha \hat{f}$ is $\mathcal{P}$-compact if and only if the operator $G := V_{-\alpha}\hat{f}V_\alpha - \hat{f}I$ is $\mathcal{P}$-compact. But G is just the operator of multiplication by the function $g : x \mapsto \hat{f}(x+\alpha) - \hat{f}(x)$ which belongs to $c_0(\mathbb{Z}^N, \mathbb{C})$. By Proposition 2.1.2, G is $\mathcal{P}$-compact.
(b) Let $f \in C(\overline{\mathbb{R}^N})$ and $\eta \in S^{N-1}$ be an infinitely distant point. If $h \in \mathbb{Z}^N$ is a sequence which tends to η in the Gelfand topology, then the limit operator $(\hat{f}I)_h$ exists and is equal to $f^*(\eta)I$. By Proposition 1.2.6, this limit operator depends on the coset $\pi(\hat{f}I)$ only. Thus, the mapping

$$\pi(\hat{f}I) \mapsto f^*(\eta)I \tag{2.42}$$

defines a character of $\mathcal{C}_E$ for every $\eta \in S^{N-1}$.

Let, conversely, ξ be a character of $\mathcal{C}_E$. Then $f \mapsto \xi(\pi(\hat{f}I))$ is a character of $C(\overline{\mathbb{R}^N})$ which, hence, is either of the form $f \mapsto f(x)$ with $x \in \mathbb{R}^N$ or $f \mapsto f^*(\eta)$ with $\eta \in S^{N-1}$. The first case can be ruled out since a function f which is non-zero at x and zero outside a certain bounded neighborhood of x generates a $\mathcal{P}$-compact multiplication operator. Thus, every character of $\mathcal{C}_E$ is of the form (2.42). Further, by definition, the Gelfand topology is the coarsest topology on S^{N-1} which makes all functions f^* continuous. This finally implies that the maximal ideal space of $\mathcal{C}_E$ is homeomorphic to S^{N-1} provided with its usual Euklidean topology. □

For every maximal ideal $\eta \in S^{N-1}$ of $\mathcal{C}_E$, let I_η denote the smallest closed two-sided ideal of $\mathcal{A}_E/K(E, \mathcal{P})$ which contains η, and write $\mathcal{A}_{E,\eta}$ for the quotient algebra $(\mathcal{A}_E/K(E, \mathcal{P}))/I_\eta$ and π_η for the canonical homomorphism from $\mathcal{A}_E$ onto $\mathcal{A}_{E,\eta}$. The following is the specification of Allan's local principle to the present context. Recall in this connection that $\mathcal{A}_E/K(E, \mathcal{P})$ is inverse closed in $L(E, \mathcal{P})/K(E, \mathcal{P})$ by Proposition 2.1.9 and, hence, an operator $A \in \mathcal{A}_E$ is $\mathcal{P}$-Fredholm if and only if its coset $A + K(E, \mathcal{P})$ is invertible in $\mathcal{A}_E/K(E, \mathcal{P})$.

Theorem 2.3.20 *An operator $A \in \mathcal{A}_E$ is $\mathcal{P}$-Fredholm if and only if the cosets $\pi_\eta(A)$ are invertible in $\mathcal{A}_{E,\eta}$ for all $\eta \in S^{N-1}$.*

The following theorem relates the invertibility of the coset $\pi_\eta(A)$ with the local invertibility of A at η.

Theorem 2.3.21 *The operator $A \in \mathcal{A}_E$ is locally invertible at $\eta \in S^{N-1}$ if and only if the coset $\pi_\eta(A)$ is invertible in $\mathcal{A}_{E,\eta}$.*

Proof. Let A be locally invertible at η. Then there are operators B, $C \in \mathcal{A}_E$ and a neighborhood at infinity W of η such that

$$BA\hat{\chi}_W I = \hat{\chi}_W AC = \hat{\chi}_W I \tag{2.43}$$

with χ_W referring to the characteristic function of W. Choose a function $f \in C(\overline{\mathbb{R}^N})$ with $\operatorname{supp} f \subseteq W$ and $f^*(\eta) = 1$. Multiplication of (2.43) by $\hat{f}I$ yields

$$BA\hat{f}I = \hat{f}AC = \hat{f}I,$$

and applying the local homomorphism π_η to this equality gives $\pi_\eta(B)\pi_\eta(A) = \pi_\eta(A)\pi_\eta(C) = \pi_\eta(I)$ because of $\pi_\eta(\hat{f}I) = f^*(\eta)\pi_\eta(I) = \pi_\eta(I)$.

Conversely, if $\pi_\eta(A)$ is invertible, then there are operators B, J_1, $J_2 \in \mathcal{A}_E$ with J_1, $J_2 \in I_\eta$ such that

$$BA = I + J_1, \quad AB = I + J_2.$$

Clearly, J_1 can be written as a sum $\sum_{k=1}^r A_k \hat{f}_k I + R + S$ where $A_k \in \mathcal{A}_E$, $f_k^*(\eta) = 0$, R is $\mathcal{P}$-compact, and where S is an operator in $\mathcal{A}_E$ with $\|S\| < 1/4$. Choose a neighborhood at infinity $W = W_{S,V}$ of η such that

$$\|R\chi_W I\| < 1/4 \quad \text{and} \quad \|f_k \chi_W I\| < \frac{1}{4r\,\|A_k\|} \quad \text{for all } k = 1, \ldots, r.$$

Multiplying both sides of the equality $BA = I + \sum A_k \hat{f}_k I + R + S$ by $\hat{\chi}_W I$ from the right-hand side gives

$$BA\hat{\chi}_W I = (I + \sum A_k \hat{f}_k \hat{\chi}_W I + R\hat{\chi}_W I + S)\hat{\chi}_W I. \tag{2.44}$$

Since $\|\sum A_k \hat{f}_k \hat{\chi}_W I + R\hat{\chi}_W I + S\| < 3/4$, the operator $I + \sum A_k \hat{f}_k \hat{\chi}_W I + R\hat{\chi}_W I + S$ is invertible, and from (2.44) we conclude that

$$B'A\hat{\chi}_W I = \hat{\chi}_W I \quad \text{with} \quad B' = (I + \sum A_k \hat{f}_k \hat{\chi}_W I + R\hat{\chi}_W I + S)^{-1} B.$$

Analogously, one verifies the existence of the operator C in (2.39). $\square$

Combining Theorems 2.3.20 and 2.3.21, we obtain the following.

Corollary 2.3.22 *An operator $A \in \mathcal{A}_E^\$$ is $\mathcal{P}$-Fredholm if and only if all of its limit operators are invertible and if*

$$\sup\{\|(A_h)^{-1}\| : A_h \in \sigma_\eta(A)\} < \infty \quad \textit{for each } \eta \in S^{N-1}.$$

This corollary is indeed a generalization of Theorem 2.2.1 since it does not require that the suprema are uniformly bounded with respect to $\eta \in S^{N-1}$.

Let us also mention that $\sigma_\eta(\hat{f}I) = \{0\}$ for all functions $f \in C(\overline{\mathbb{R}^N})$ with $f^*(\eta) = 0$. This implies that $\sigma_\eta(J) = \{0\}$ for all operators $J \in \mathcal{A}_E$ with $\pi_\eta(J) = 0$ and, hence, $\sigma_\eta(A)$ depends on the coset $\pi_\eta(A)$ only. Based on this observation, a symbol calculus for the invertibility in the local algebra $\mathcal{A}_{E,\eta}$ can be established in analogy to Sections 2.2.3 and 2.2.4.

2.3.6 Operators with continuous coefficients

Given a Banach space X, let $C(\overline{\mathbb{R}^N}, L(X))$ stand for the set of all continuous functions $f : \mathbb{R}^N \to L(X)$ for which the functions

$$f_R : S^{N-1} \to L(X),\ \eta \mapsto f(\eta R), \tag{2.45}$$

which are defined for $R > 0$, converge uniformly on S^{N-1} as $R \to \infty$. The uniform limit $f^* := \lim_{R\to\infty} f_R$ is a continuous $L(X)$-valued function on S^{N-1}. We write $C_{L(X)}$ for the set of all restrictions of functions in $C(\overline{\mathbb{R}^N}, L(X))$ onto $\mathbb{Z}^N$, and we let $\mathbb{C}_{L(X)}$ refer to the set of all constant functions on $\mathbb{Z}^N$ with values in $L(X)$. Both classes are closed subalgebras of $l^\infty(\mathbb{Z}^N, L(X))$. Finally, given a subset B of $l^\infty(\mathbb{Z}^N; L(X))$, we agree upon denoting the smallest closed subalgebra of $L(E)$ which contains all band operators $\sum_{|\alpha|\le k} a_\alpha V_\alpha$ with coefficients $a_\alpha \in B$ by $\mathcal{A}_E(B)$ and the intersection $\mathcal{A}_E(B) \cap L^\$(E, \mathcal{P})$ by $\mathcal{A}_E^\$(B)$.

Proposition 2.3.23

(a) *Let $f \in C_{L(X)}$ and let the sequence $h \in \mathcal{H}$ tend into the direction $\eta \in S^{N-1}$. Then the limit operator $(fI)_h$ exists, and it is equal to the constant function on $\mathbb{Z}^N$ with value $f^*(\eta) \in L(X)$.*

(b) *$\mathcal{A}_E(C_{L(X)})$ is a closed subalgebra of $\mathcal{A}_E^\$$.*

(c) *Every limit operator of an operator in $A \in \mathcal{A}_E(C_{L(X)})$ belongs to $\mathcal{A}_E(\mathbb{C}_{L(X)})$.*

(d) *For every $A \in \mathcal{A}_E(C_{L(X)})$ and $\eta \in S^{N-1}$, the local operator spectrum $\sigma_\eta(A)$ is a singleton.*

Proof. (a) Let $W_{R,U}$ be a neighborhood at infinity of $\eta \in S^{N-1}$ as defined in (2.36). Then, for every $x \in \mathbb{Z}^N$, the points $x + h(n)$ belong to $W_{R,U}$ whenever n is large enough. Thus,

$$f(x + h(n)) \to f^*(\eta) \quad \text{as } n \to \infty$$

due to the uniform convergence of the functions (2.45). This implies the $\mathcal{P}$-strong convergence of the operators $V_{-h(n)}(\hat{f}I)V_{h(n)}$ to the constant function with value $f^*(\eta)$.

For assertion (b), one has to check whether every operator of multiplication $\hat{f}I$ where f is a function in $C(\overline{\mathbb{R}^N})_{L(X)}$ is rich. Let $(\hat{f}(h(n)))_{n\in\mathbb{N}}$ be a sequence of values of f. Then there is a subsequence g of the sequence h which tends to infinity.

Moreover, g can be assumed to converge to an infinitely distant point $\eta \in S^{N-1}$ with respect to the Gelfand topology. Thus, the assertion follows from part (a). Assertions (c) and (d) are immediate consequences of (a). □

Combining assertion (d) of the preceding proposition with Corollary 2.3.22 we immediately obtain the following.

Corollary 2.3.24 *An operator in $\mathcal{A}_E(C_{L(X)})$ is $\mathcal{P}$-Fredholm if and only if all of its limit operators are invertible.*

Thus, for band-dominated operators with continuous coefficients, the *uniformness* of the invertibility of its limit operators is redundant. Another feature of these operators is the simple structure of their limit operators, which are band-dominated operators with constant coefficients (Proposition 2.3.23 (c)). For such operators, there is a simple criterion for their invertibility which we are going to derive now. This criterion will finally yield a simple and effective condition for the $\mathcal{P}$-Fredholmness of operators in $\mathcal{A}_E(C_{L(X)})$ in the Hilbert space case.

To every band operator $A = \sum_{|\alpha|\le k} a_\alpha V_\alpha$ with coefficients in $\mathbb{C}_{L(X)}$, we associate the operator-valued function $\tilde{A} : \mathbb{T}^N \to L(X)$ defined at $\xi := (\xi_1, \ldots, \xi_N)$ by

$$\tilde{A}(\xi) := \sum_{|\alpha|\le k} a_\alpha \xi^\alpha \quad \text{where} \quad \xi^\alpha := \xi_1^{\alpha_1} \cdots \xi_N^{\alpha_N}.$$

Theorem 2.3.25 *Let H be a Hilbert space and $E = l^2(\mathbb{Z}^N, H)$. Then*

(a) *the mapping $A \mapsto \tilde{A}$ extends to a contractive homomorphism from $\mathcal{A}_E(\mathbb{C}_{L(H)})$ into the algebra of the continuous functions on $\mathbb{T}^N$ with values in $L(H)$.*

(b) *an operator $A \in \mathcal{A}_E(\mathbb{C}_{L(H)})$ is invertible if and only if the function $\tilde{A}$ is invertible at each point of $\mathbb{T}^N$.*

Proof. It will be proved in Section 2.4.8 in a much more general situation that assertion (a) is true and that an operator $A \in \mathcal{A}_E(\mathbb{C}_{L(H)})$ is $\mathcal{P}$-Fredholm if and only if the function $\tilde{A}$ is invertible. It remains to show that every $\mathcal{P}$-Fredholm operator in $\mathcal{A}_E(\mathbb{C}_{L(H)})$ is already invertible.

So let $A \in \mathcal{A}_E(\mathbb{C}_{L(H)})$ be $\mathcal{P}$-Fredholm. Then there are an operator R and $\mathcal{P}$-compact operators K and L such that

$$AB = I + K \quad \text{and} \quad BA = I + L. \tag{2.46}$$

The inverse closedness of C^*-algebras implies that B belongs to the smallest closed subalgebra $\mathcal{B}$ of $L(E)$ which contains the algebra $\mathcal{A}_E(\mathbb{C}_{L(H)})$ and the ideal of the $\mathcal{P}$-compact operators. Thus, multiplying both identities in (2.46) from the left by V_{-k} and from the right by V_k, and letting $k \in \mathbb{Z}^N$ tend to infinity, we get $AC = I$ and $CA = I$ where C is the $\mathcal{P}$-strong limit of the sequence $(V_{-k}BV_k)$ as $k \to \infty$ (the existence of this strong limit is evident). Thus, A is invertible. □

2.4 Local $\mathcal{P}$-Fredholmness: advanced theory

In the previous section, we have localized the algebra $\mathcal{A}_E/K(E, \mathcal{P})$ over its central subalgebra consisting of restrictions of functions in $C(\overline{\mathbb{R}^N})$ to $\mathbb{Z}^N$. However, the center of $\mathcal{A}_E/K(E, \mathcal{P})$ proves to be much larger: actually it consists of slowly oscillating functions. This observation offers the opportunity of an essentially finer localization and, thus, of a further weakening of the uniform invertibility condition in Theorem 2.2.1.

2.4.1 Slowly oscillating functions

A function $a \in l^\infty(\mathbb{Z}^N, L(X))$ is *slowly oscillating* if

$$\lim_{x\to\infty} (a(x+k) - a(x)) = 0 \quad \text{for all } k \in \mathbb{Z}^N. \tag{2.47}$$

We denote the set of all slowly oscillating functions by $SO_{L(X)}(\mathbb{Z}^N)$ and write $SO(\mathbb{Z}^N)$ instead of $SO_{L(\mathbb{C})}(\mathbb{Z}^N)$.

Trivial examples of slowly oscillating functions are provided by the functions in $C_{L(X)}$, whereas $\mathbb{Z}^N \to \mathbb{C} : x \mapsto \sin\sqrt{|x|}$ is an example of a slowly oscillating function which is not continuous at infinity.

By Proposition 2.1.2, a function $a \in l^\infty(\mathbb{Z}^N, L(X))$ is slowly oscillating if and only if the operator $V_{-k}aV_k - aI$ is $\mathcal{P}$-compact for every $k \in \mathbb{Z}^N$ or, equivalently, if the commutator $aV_k - V_k aI = V_k(V_{-k}aV_k - aI)$ is $\mathcal{P}$-compact for every k. Since $K(E, \mathcal{P})$ is a closed ideal of $L(E, \mathcal{P})$, we conclude that $SO_{L(X)}(\mathbb{Z}^N)$ is a closed subalgebra of $l^\infty(\mathbb{Z}^N, L(X))$. If, moreover, the slowly oscillating function a is scalar-valued, then the operator of multiplication by a also commutes with every multiplication operator. Thus, the cosets $SO(\mathbb{Z}^N) + K(E, \mathcal{P})$ belong to the center of the algebra $\mathcal{A}_E^\$/K(E, \mathcal{P})$. We will see below (Theorem 2.4.2) that, conversely, every coset in the center of $\mathcal{A}_E^\$/K(E, \mathcal{P})$ is of this form.

Another special feature of slowly oscillating functions concerns the limit operators of their multiplication operators.

Proposition 2.4.1 *Let $a \in SO_{L(X)}(\mathbb{Z}^N)$. Then every limit operator of aI is a multiplication operator in $\mathbb{C}_{L(X)}$, i.e., an operator of multiplication by a constant function with values in $L(X)$.*

Proof. Let $a \in SO_{L(X)}(\mathbb{Z}^N)$. From (2.47) we conclude that

$$\lim_{k\to\infty} (a(x' + h(k)) - a(x'' + h(k))) = 0$$

for all sequences h tending to infinity and for all $x', x'' \in \mathbb{Z}^N$. Hence, if h is a sequence such that the limit operator $(aI)_h$ exists, then $\lim_{k\to\infty} a(x + h(k))$ is independent of $x \in \mathbb{Z}^N$, i.e., $(aI)_h = AI$ with an operator $A \in L(X)$. □

Again, a certain converse to this fact proves to be true.

Theorem 2.4.2 *The following assertions are equivalent for an operator $B \in \mathcal{A}_E^\$$:*

(a) $B \in SO(\mathbb{Z}^N) + K(E, \mathcal{P})$.

(b) *All limit operators of B lie in $\mathbb{C}I$.*

(c) *The coset $B + K(E, \mathcal{P})$ lies in the center of $\mathcal{A}_E^\$/K(E, \mathcal{P})$.*

For the proof we need a characterization of $\mathcal{P}$-compact operators via the asymptotic behavior of their diagonals. For, let $A \in L(E, \mathcal{P})$. Then the series

$$\sum_{i \in \mathbb{Z}^N} S_i A S_i \tag{2.48}$$

converges strongly. Indeed, let $A_n := \sum_{|i| \le n} S_i A S_i$. The sequence (A_n) is uniformly bounded since

$$\begin{aligned} \|A_n x\|^p &= \|\sum_{|i|\le n} S_i A S_i x\|^p = \sum_{|i|\le n} \|S_i A S_i x\|^p \\ &\le \|A\|^p \sum_{|i|\le n} \|S_i x\|^p = \|A\|^p \|P_n x\|^p \le \|A\|^p \|x\|^p \end{aligned} \tag{2.49}$$

for every $x \in E$, and it converges on a dense subset of $E = l^p(\mathbb{Z}^N, X)$ since $A_n S_k x = S_k A S_k x$ for all $n \ge k$. We call the operator in (2.48) the *main diagonal* of A and denote it by $D(A)$. From $A_n S_k = S_k A S_k = D(A) S_k$ for all $n \ge k$, it is also clear that the A_n converge $\mathcal{P}$-strongly to $D(A)$. Further, for $i \in \mathbb{Z}^N$, we define the jth diagonal of A by $D_j(A) := D(AV_{-j})$. We consider both $D(A)$ and the $D_j(A)$ as elements of $l^\infty(\mathbb{Z}^N, L(X))$.

Lemma 2.4.3

(a) *The mapping $D : L(E, \mathcal{P}) \to l^\infty(\mathbb{Z}^N, L(X))$ is a surjective contraction.*

(b) *If $A \in L(E, \mathcal{P})$, and if $h \in \mathcal{H}$ is a sequence such that the limit operator A_h exists, then the limit operators $D_j(A)_h$ exist for every $j \in \mathbb{Z}^N$, and*

$$D_j(A_h) = D_j(A)_h.$$

(c) *If $A \in L^\$(E, \mathcal{P})$, then $D_j(A) \in L^\$(E, \mathcal{P})$ for every $j \in \mathbb{Z}^N$.*

Proof. (a) In (2.49) we have seen that $\|A_n\| \le \|A\|$ for all $n \in \mathbb{N}$. Hence, by Banach-Steinhaus,

$$\|D(A)\| \le \liminf \|A_n\| \le \|A\|.$$

Further, D is surjective since $D(aI) = aI$ for every function $a \in l^\infty(\mathbb{Z}^N, L(X))$.
(b) First observe that if $A \in L(E, \mathcal{P})$, then both A_h and all diagonals $D_j(A)$ belong to $L(E, \mathcal{P})$. Thus, it makes sense to consider $D_j(A_h)$ and $D_j(A)_h$.

Let now $i, j \in \mathbb{Z}^N$, and let $A \in L(E, \mathcal{P})$ be an operator for which the limit operator A_h exists. Then

$$\begin{aligned}
&\|(V_{-h(n)}D_j(A)V_{h(n)} - D_j(A_h))S_i\| \\
&= \|V_{-h(n)} \sum_k S_k A V_{-j} S_k V_{h(n)} S_i - \sum_k S_k A V_{-j} S_k S_i\| \\
&= \|V_{-h(n)} S_{i+h(n)} A V_{-j} S_{i+h(n)} V_{h(n)} S_i - S_i A V_{-j} S_i\| \\
&= \|S_i V_{-h(n)} A V_{h(n)} V_{-j} S_i - S_i A V_{-j} S_i\| \\
&= \|S_i (V_{-h(n)} A V_{h(n)} - A) S_{i-j} V_{-j}\| \\
&\le \|(V_{-h(n)} A V_{h(n)} - A) S_{i-j}\| \to 0.
\end{aligned}$$

The dual assertion follows similarly.
(c) Let $A \in L^\$(E, \mathcal{P})$. Thus, every sequence $h \in \mathcal{H}$ possesses a subsequence g such that the limit operator A_g exists. Then, by part (b), the limit operator $D_j(A)_g$ exists, too. □

Proposition 2.4.4 *The following assertions are equivalent for an operator* $K \in L^\$(E, \mathcal{P})$*:*

(a) $K \in K(E, \mathcal{P})$.

(b) $D_j(K) \in c_0(\mathbb{Z}^N, L(X))$ *for every* $j \in \mathbb{Z}^N$.

Proof. Let $K \in K(E, \mathcal{P})$, $j \in \mathbb{Z}^N$, and let $h \in \mathcal{H}$ be a sequence such that the limit operator $D_j(K)_h$ exists. Then the limit operator K_h exists, too (see the proof of Proposition 1.2.6 (b)), and $D_j(K)_h = D_j(K_h)$ by the preceding lemma. Since $K_h = 0$ (Proposition 1.2.6 again), we conclude that zero is the only limit operator of $D_j(K)$. Since $D_j(K)$ is rich due to Lemma 2.4.3 (c), this implies via Theorem 2.2.12 that $D_j(K)$ is $\mathcal{P}$-compact. Thus,

$$D_j(K) \in K(E, \mathcal{P}) \cap l^\infty(\mathbb{Z}^N, L(X)) = c_0(\mathbb{Z}^N, L(X)).$$

Let now, conversely, $K \in L^\$(E, \mathcal{P})$ be an operator for which all diagonals are $\mathcal{P}$-compact, and let $h \in \mathcal{H}$ be a sequence such that the limit operator K_h exists. Then, for every $j \in \mathbb{Z}^N$, the limit operator $D_j(K)_h$ exists, too, and $D_j(K)_h = D_j(K_h)$ by the preceding lemma. Since $D_j(K)$ is $\mathcal{P}$-compact, we have $D_j(K)_h = 0$ and, thus, $D_j(K_h) = 0$ for every j. By the definition of a diagonal, this implies that $S_i K_h S_k = 0$ for every choice of $i, k \in \mathbb{Z}^N$ and, hence, $P_n K_h P_n = 0$ for every $n \in \mathbb{N}$. Letting n go to infinity we get $K_h = 0$. Thus, zero is the only limit operator of K, which implies as above that K is $\mathcal{P}$-compact. □

Proof of Theorem 2.4.2. The implication (a) $\Rightarrow$ (b) follows immediately from Propositions 1.2.6 (b) and 2.4.1. For the implication (b) $\Rightarrow$ (c) recall that the algebra $\mathcal{A}_E^\$/K(E, \mathcal{P})$ is topologically isomorphic to the symbol algebra $\mathcal{S}_E$ (Theorem 2.2.12). Clearly, (b) implies that the symbol of B belongs to the center of

$\mathcal{S}_E$. Thus, being the pre-image of smb B, the coset $B + K(E, \mathcal{P})$ lies in the center of $\mathcal{A}_E^\$/K(E, \mathcal{P})$.

It remains to show the implication (c) $\Rightarrow$ (a). Let $B + K(E, \mathcal{P})$ be in the center of $\mathcal{A}_E^\$/K(E, \mathcal{P})$. We claim that the main diagonal of B belongs to $SO(\mathbb{Z}^N)+ c_0(\mathbb{Z}^N, L(X))$, whereas all other diagonals of B are in $c_0(\mathbb{Z}^N, L(X))$.

Let us first mention that, for every $k \in \mathbb{Z}^N$, the operator $V_{-k}BV_k - B$ is $\mathcal{P}$-compact. Hence, every diagonal of $V_{-k}BV_k - B$ lies in $c_0(\mathbb{Z}^N, L(X))$ by Proposition 2.4.4, which implies via the definition of SO that every diagonal of B lies in $SO^\$(\mathbb{Z}^N, L(X))$ (with the richness being a consequence of Lemma 2.4.3 (c)).

In the next step we will prove our claim for the main diagonal of B. As we have just seen, we can consider $D := D(B)$ as a slowly oscillating function which takes at $n \in \mathbb{Z}^N$ a value $D(n)$ in $L(X)$. Let, for a moment, A be an arbitrary operator in $L(X)$. Then, for all $\lambda \in \mathbb{C}$ with $|\lambda| > 2\,\|A\|$,

$$\|A - \lambda I\| \geq |\lambda| - \|A\| > \|A\| \geq \inf_{\mu \in \mathbb{C}} \|A - \mu I\|. \tag{2.50}$$

On the other hand, the continuous function $\lambda \mapsto \|A - \lambda I\|$ attains its minimum on the compact set $\{\lambda \in \mathbb{C} : \lambda| \leq 2\,\|A\|\}$. By (2.50), this minimum is also the minimum of $\lambda \mapsto \|A - \lambda I\|$ on all of $\mathbb{C}$.

Thus, we can write every operator $D(n) \in L(X)$ as $D(n) = \alpha_n I + K_n$ where the $\alpha_n \in \mathbb{C}$ are chosen such that

$$\|D(n) - \alpha_n I\| = \min_{\lambda \in \mathbb{C}} \|D(n) - \lambda I\| \quad \text{and} \quad |\alpha_n| \leq 2\,\|D(n)\|. \tag{2.51}$$

Hence, the function $n \mapsto \alpha_n I$ is bounded by $2\|D\|_\infty$, and the diagonal operator $(\alpha_n I)$ is evidently rich. Thus, the diagonal operator $(K_n)_{n \in \mathbb{Z}^N}$ belongs to $l^\infty(\mathbb{Z}^N, L(X))^\$$. Moreover, (2.51) implies for every n that

$$\|K_n\| = \min_{\lambda \in \mathbb{C}} \|K_n + \alpha_n I - \lambda I\| = \min_{\lambda \in \mathbb{C}} \|K_n - \lambda I\|. \tag{2.52}$$

Finally, the coset $(K_n)+c_0(\mathbb{Z}^N, L(X))$ belongs to the center of the quotient algebra $l^\infty(\mathbb{Z}^N, L(X))^\$/c_0(\mathbb{Z}^N, L(X))$ which can be seen as follows. By hypothesis, we have $AB - BA \in K(E, \mathcal{P})$ for all $A \in \mathcal{A}_E^\$$. Proposition 2.4.4 implies that then $D(AB - BA) \in c_0(\mathbb{Z}^N, L(X))$ for all $A \in \mathcal{A}_E^\$$ whence, in particular,

$$D(AB - BA) = AD(B) - D(B)A \in c_0(\mathbb{Z}^N, L(X))$$

for all $A \in l^\infty(\mathbb{Z}^N, L(X))^\$$. Thus, the coset $D(B) + c_0(\mathbb{Z}^N, L(X))$ lies in the center of $l^\infty(\mathbb{Z}^N, L(X))^\$/c_0(\mathbb{Z}^N, L(X))$. Since the coset $(\alpha_n I) + c_0(\mathbb{Z}^N, L(X))$ evidently also belongs to the center of that algebra, we conclude that $(K_n) + c_0(\mathbb{Z}^N, L(X))$ is in that center, too. In particular,

$$\|K_n A - AK_n\| \to 0 \quad \text{for all} \quad A \in L(X) \tag{2.53}$$

since (K_n) commutes with all constant functions modulo $c_0(\mathbb{Z}^N, L(X))$.

We will show now that $\|K_n\| \to 0$ as $n \to \infty$. Contrary to what we want, assume there are a monotonically increasing sequence (n_k) and an $\varepsilon > 0$ such that $\|K_{n_k}\| \geq \varepsilon$ for all k. Since the diagonal operator (K_n) is rich, there are a subsequence (n_{k_l}) of (n_k) and an operator $K \in L(X)$ such that $\|K_{n_{k_l}} - K\| \to 0$ as $l \to \infty$. Then, by 2.53,

$$KA = AK \quad \text{for all} \quad A \in L(X).$$

This observation implies that K is a scalar operator: $K = \kappa I$ with $\kappa \in \mathbb{C}$ (this fact is usually referred to as Schur's lemma). On the other hand, we conclude from $\|K_{n_{k_l}} - \kappa I\| \to 0$ that

$$\|K_{n_{k_l}} - \kappa I\| < \varepsilon \leq \|K_{n_{k_l}}\|$$

for l being sufficiently large, which contradicts (2.52). This contradiction shows that $(K_n) \in c_0(\mathbb{Z}^N, L(X))$. Finally, since

$$|\alpha_{n+1} - \alpha_n| \leq \|D(n+1) - D(n)\| + \|K_{n+1}\| + \|K_n\| \to 0,$$

the scalar-valued function $n \mapsto \alpha_n$ is slowly oscillating, and the assertion for the main diagonal follows.

In the next step we show that every diagonal $D_k(B)$ with $k \neq 0$ belongs to $c_0(\mathbb{Z}^N, L(X))$. Given $k \in \mathbb{Z}^N \setminus \{0\}$, choose $r \in \mathbb{N} \setminus \{0, 1\}$ such that

$$(r\mathbb{Z}^N + k) \cap r\mathbb{Z}^N = \emptyset. \tag{2.54}$$

(Clearly, r cannot exist if $k = 0$. If $k = (k_1, \ldots, k_N) \neq 0$, every $r > \max_i |k_i|$ can be chosen.) Further, let $f \in l^\infty(\mathbb{Z}^N)$ be the function which is 1 on $r\mathbb{Z}^N$ and zero outside this set, and consider the matrix representation of fB. Clearly, if $j \in r\mathbb{Z}^N$, then the jth rows of fB and B coincide, whereas all other rows of fB are zero. The same happens with the columns of BfI. Thus, if $B = (B_{ij})_{i,\, j \in \mathbb{Z}^N}$, then all entries of B of the form

$$B_{rm+k,\, rm} \quad \text{with} \quad m \in \mathbb{Z}^N$$

lie in nonzero columns of BfI but in zero rows of fB (recall our choice in (2.54)). Moreover, these entries belong to the kth diagonal of B and, hence, of $BfI - fB$. Since this operator is $\mathcal{P}$-compact, we conclude via Proposition 2.4.4 that

$$\|B_{rm+k,\, rm}\| \to 0 \quad \text{as} \quad m \to \infty.$$

Let $b \in l^\infty(\mathbb{Z}^N, L(X))$ be defined by $b(m) := A_{m+k,\, m}$ (thus, b is the kth diagonal of B). We know already that $b \in SO^\$(\mathbb{Z}^N, L(X))$ and that $\|b(rm)\| \to 0$ as $m \to \infty$. These two facts together already imply that $b \in c_0(\mathbb{Z}^N, L(X))$. Indeed, let $s = (s_1, \ldots, s_N) \in \mathbb{Z}^N$ with

$$0 \leq s_i \leq r - 1 \quad \text{for all} \quad i. \tag{2.55}$$

Then the SO-property implies that

$$\|b(rm+s)\| \leq \|b(rm+s) - b(rm)\| + \|b(rm)\| \to 0$$

as $m \to \infty$. Since there are only finitely many N-tuples s with property (2.55), we have $\|b(m)\| \to 0$. This finishes the proof of our claim.

Now, if $B + K(E, \mathcal{P})$ is in the center of $\mathcal{A}_E^\$/K(E, \mathcal{P})$, then we can write it as the sum of a multiplication operator in $SO(\mathbb{Z}^N)$ and an operator $K \in \mathcal{A}_E^\$$ all diagonals of which belong to $c_0(\mathbb{Z}^N, L(X))$. By Proposition 2.4.4, $K \in K(E, \mathcal{P})$, and we are done. □

Thus, $SO(\mathbb{Z}^N)$ (more precisely, the image of $SO(\mathbb{Z}^N)$ in $\mathcal{A}_E^\$/K(E, \mathcal{P})$ under the canonical embedding) is *the* natural candidate for a central subalgebra with respect to which the algebra $\mathcal{A}_E^\$/K(E, \mathcal{P})$ can be localized by means of Allan's local principle. We will pursue this idea in the following.

2.4.2 The maximal ideal space of $SO(\mathbb{Z}^N)$

Let $M(SO(\mathbb{Z}^N))$ denote the maximal ideal space of the commutative C^*-algebra $SO(\mathbb{Z}^N)$, and write $M^\infty(SO(\mathbb{Z}^N))$ for the fiber of $M(SO(\mathbb{Z}^N))$ consisting of all characters $\eta \in M(SO(\mathbb{Z}^N))$ such that $\eta(a) = 0$ whenever $a \in c_0$. Every $m \in \mathbb{Z}^N$ defines a character of $SO(\mathbb{Z}^N)$ by $f \mapsto f(m)$. In this sense, $\mathbb{Z}^N$ is embedded into $M(SO(\mathbb{Z}^N))$, and $M(SO(\mathbb{Z}^N))$ is the union of its disjoint subsets $\mathbb{Z}^N$ and $M^\infty(SO(\mathbb{Z}^N))$. The following is a special case of a general result on compactifications of topological spaces, see [57], Chapter I, Theorem 8.2.

Theorem 2.4.5 *$\mathbb{Z}^N$ is densely and homeomorphically embedded into $M(SO(\mathbb{Z}^N))$ with respect to the Gelfand topology.*

Connectedness properties. We will show now that the fiber $M^\infty(SO(\mathbb{Z}^N))$ is connected if $N > 1$ and consists of two connected components if $N = 1$. In the following proposition, we let $SO(\mathbb{N})$ denote the algebra of all restrictions of functions in $SO(\mathbb{Z})$ onto $\mathbb{N}$.

Proposition 2.4.6

(a) *Let $N > 1$, and let $a \in SO(\mathbb{Z}^N)$ be a function which only takes the values $+1$ and -1. Then one of the sets $M_\pm := \{z \in \mathbb{Z}^N : a(z) = \pm 1\}$ is finite.*

(b) *The same holds for functions in $SO(\mathbb{N})$.*

Proof. Let $N > 1$. For $z \in \mathbb{Z}^N$, set $U(z) := \{y \in \mathbb{Z}^N : |z - y|_\infty \leq 1\}$, and let

$$\tilde{M}_+ := \{z \in M_+ : U(z) \cap M_- \neq \emptyset\}.$$

Suppose that $\tilde{M}_+$ is infinite. Let $v_1, \ldots, v_{3^N-1}$ stand for the unit vectors in $\mathbb{Z}^N$, i.e., for the vectors in $\mathbb{Z}^N$ with $|v_l|_\infty = 1$. Then, for each $z \in \mathbb{Z}^N$,

$$U(z) \setminus \{z\} = \{z + v_l : l = 1, \ldots, 3^N - 1\}.$$

Since $\tilde{M}_+$ is infinite by assumption, there are a sequence $x : \mathbb{N} \to \tilde{M}_+$ with $x(k) \to \infty$ as $k \to \infty$ and an index $l_0 \in \{1, \ldots, 3^N - 1\}$ such that $a(x(k)) = 1$ and $a(x(k) + v_{l_0}) = -1$ for each $k \in \mathbb{N}$. Thus, $a(x(k) + v_{l_0}) - a(x(k)) = -2$ for all $k \in \mathbb{N}$, implying that a cannot be slowly oscillating. This contradiction shows that $\tilde{M}_+$ is finite.

Since $\tilde{M}_+$ is finite, this set is contained in a ball $V := \{x \in \mathbb{Z}^N : |x|_\infty < K\}$ with radius K chosen large enough. We claim that either $B := \mathbb{Z}^N \setminus V \subseteq M_+$ or $B \subseteq M_-$. Indeed, let $z_1 \in B$ be a point which belongs to M_+. Then every point in B lies in M_+ which can be seen as follows. Given a point $z_2 \in B$, we choose points $y_k \in B$, $k = 1, \ldots, l$ such that $y_1 = z_1$, $y_l = z_2$ and $y_{k+1} \in U(y_k)$ for each $k \in \{1, \ldots, l-1\}$ (here we use that $N \geq 2$). Since $y_k \notin \tilde{M}_+$ and $y_1 \in M_+$, all points y_k belong to M_+. Hence, $z_2 \in M_+$ and $B \subseteq M_+$ in this case. If there is no point from M_+ in B, then $B \subseteq M_-$. This proves our claim, and the claim on its hand shows that either M_+ or M_- are finite. This proves assertion (a). The proof of (b) follows similarly. □

Let us mention that, in case $N = 1$, both sets $M_\pm$ can be infinite. For an example, consider the sign function which obviously belongs to $SO(\mathbb{Z})$.

Let $M^{\pm\infty}(SO(\mathbb{Z}))$ stand for the fiber of $M(SO(\mathbb{Z}))$ which contains all characters of $SO(\mathbb{Z})$ which vanish at every function $a \in l^\infty(\mathbb{Z})$ with $\lim_{x\to\pm\infty} a(x) = 0$.

Theorem 2.4.7

(a) *$M^\infty(SO(\mathbb{Z}^N))$ is connected if $N > 1$.*

(b) *$M^{+\infty}(SO(\mathbb{Z}))$ and $M^{-\infty}(SO(\mathbb{Z}))$ are connected.*

Proof. We prove assertion (a) only. Assume that $M^\infty(SO(\mathbb{Z}^N))$ is not connected. Then there exists a continuous function $\hat{a}$ on $M^\infty(SO(\mathbb{Z}^N))$ which takes exactly the values $+1$ and -1. By the Tietze extension theorem ([140], Theorem IV.11), there is a real-valued continuous function $\tilde{a}$ on $M(SO(\mathbb{Z}^N))$ which coincides with $\hat{a}$ on $M^\infty(SO(\mathbb{Z}^N))$. Hence, there is a slowly oscillating real-valued function a on $\mathbb{Z}^N$ which has $\tilde{a}$ as its Gelfand transform and for which, consequently, ± 1 are the only partial limits at infinity. Let

$$M_+ := \{z \in \mathbb{Z}^N : a(z) > 1/2\}, \quad M_- := \{z \in \mathbb{Z}^N : a(z) < -1/2\}.$$

Then $\mathbb{Z}^N \setminus (M_+ \cup M_-)$ is finite. Indeed, otherwise there would exist infinitely many points $z \in \mathbb{Z}^N$ with $-1/2 \leq a(z) \leq 1/2$. In this case, one could choose a sequence $(z_n) \subseteq \mathbb{Z}^N$ which tends to infinity, and for which $\lim a(z_n)$ exists and belongs to $[-1/2, 1/2]$. This is impossible (recall that ± 1 are the only limit points of a at infinity). Thus, $\mathbb{Z}^N \setminus (M_+ \cup M_-)$ is indeed finite, and the function

$$b(z) := \begin{cases} a(z) & z \in M_+ \cup M_- \\ -1 & z \in \mathbb{Z}^N \setminus (M_+ \cup M_-) \end{cases}$$

differs from a at a finite number of points only. Thus, the function b is slowly oscillating, it has exactly two partial limits at infinity (namely ± 1), and $|b| > 1/2$.

The latter property guarantees that $|b|$ is invertible and that $|b|^{-1} \in SO(\mathbb{Z}^N)$. Hence, the function

$$b/|b| : z \mapsto \begin{cases} 1 & z \in M_+ \\ -1 & z \in \mathbb{Z}^N \setminus M_+ \end{cases}$$

belongs to $SO(\mathbb{Z}^N)$, and both sets M_+ and $\mathbb{Z}^N \setminus M_+$ are infinite (because both $+1$ and -1 are partial limits of $b/|b|$ at infinity). By Proposition 2.4.6, this is impossible. □

A particular consequence of the preceding theorem is that the set of all partial limits at infinity of a real-valued slowly oscillating function a on $\mathbb{Z}^N$ with $N > 1$ is a closed interval. Indeed, this set is just the range of the restriction of the Gelfand transform of a onto the fiber $M^\infty(SO(\mathbb{Z}^N))$, i.e., it is the image of a compact and connected set under a continuous mapping.

Neighborhoods at infinity. Let $\eta \in M^\infty(SO)$, and let U be a neighborhood of η in $M(SO)$ with respect to the Gelfand topology. We agree upon calling the intersection $U \cap \mathbb{Z}^N$ a *neighborhood at infinity of* η. The following definition as well as Theorem 2.4.9 are taken from [161].

Definition 2.4.8

(a) *A subset $V \subseteq \mathbb{Z}^N$ is called* growing *if, for every bounded set $D \subset \mathbb{Z}^N$, there is an $x \in \mathbb{Z}^N$ such that $x + D \subseteq V$.*

(b) *An unbounded subset V_0 of a growing set V is called a* center *if, for every bounded set $D \subset \mathbb{Z}^N$, there is a bounded set M such that $(V_0 \setminus M) + D \subseteq V$.*

Theorem 2.4.9 *Let W be an unbounded subset of $\mathbb{Z}^N$ and $\eta \in \overline{W} \cap M^\infty(SO)$ (where the bar refers to the closure with respect to the Gelfand topology on $M(SO)$), and let $U \subseteq M(SO)$ be a neighborhood of η. Then $V := U \cap \mathbb{Z}^N$ is a growing set, and there is a neighborhood $U_0 \subseteq U$ of η such that $V_0 := U_0 \cap \mathbb{Z}^N$ is contained in W and a center of V.*

Proof. By Uryson's lemma, there is a continuous function $f : M(SO) \to [0, 1]$ which is 0 at η and 1 on $M(SO) \setminus U$. Since f is continuous on $M(SO)$, the restriction of f onto $\mathbb{Z}^N$ is a slowly oscillating function. Set

$$U_0' := \{x \in M(SO) : f(x) < 1/2\} \quad \text{and} \quad U_0 := U_0' \cap W,$$

and define $V := U \cap \mathbb{Z}^N$ and $V_0 := U_0 \cap \mathbb{Z}^N$. Then $V_0 \subseteq W \cap V$. Moreover, since U_0' is a neighborhood of η, the set V_0 is unbounded. We claim that, for every bounded set M, there is a bounded set D such that $(V_0 \setminus D) + M \subseteq V$. The claim implies that V is growing and that V_0 is a center of V.

Assume the claim is wrong. Then there exists a bounded set M such that $(V_0 \setminus D) + M \not\subseteq V$, hence, $V_1 := (V_0 + M) \setminus V$ is an unbounded set. So it makes sense to consider the limes superior of $|f(x)|$ when $x \in V_1$ tends to infinity. Since $V_1 \subseteq V_0 + M$ and f is slowly oscillating, we get

$$\begin{aligned} \limsup_{x \in V_1, x \to \infty} |f(x)| &\leq \limsup_{y \in V_0, y \to \infty} \max_{m \in M} |f(y+m)| \\ &\leq \limsup_{y \in V_0, y \to \infty} \max_{m \in M} |f(y+m) - f(y)| + \limsup_{y \in V_0, y \to \infty} |f(y)| \leq 0 + 1/2 = 1/2. \end{aligned}$$

This is impossible since V_1 is in the complement of U and, hence, f is 1 on V_1. □

Inadequacy of sequences. We will run into a lot of trouble when trying to realize the simple and natural idea to localize the algebra $\mathcal{A}_E^{\$}/K(E, \mathcal{P})$ over SO. The main reason for this is the following observation.

Proposition 2.4.10 *Let $\eta \in M^\infty(SO(\mathbb{Z}^N))$. Then $\eta \in \text{clos}_{M(SO(\mathbb{Z}^N))}\mathbb{Z}^N$, but there is no* sequence *in $\mathbb{Z}^N$ which tends to η with respect to the Gelfand topology of $M(SO(\mathbb{Z}^N))$.*

Proof. We know from Theorem 2.4.5 that η is in $\text{clos}_{M(SO)}\mathbb{Z}^N$ and that, hence, there is a *net* with values in $\mathbb{Z}^N$ which converges to η. Assume there is a *sequence* h with values in $\mathbb{Z}^N$ and with limit η in the Gelfand topology. Since every subsequence of h also converges to η, we can assume without loss that

$$|h(n+1)| \geq |h(n)| + 2^{n+2} \quad \text{for all } n.$$

Let $\varphi_0 : \mathbb{R}^N \to [0, 1]$ be a continuous function with support in $\{t \in \mathbb{R}^N : |t| \leq 1\}$ and with $\varphi_0(0) = 1$, and set $\varphi_n(t) := \varphi(t/2^n)$ for $n \geq 1$. Then the function

$$\varphi(t) := \sum_{n \geq 0}^{\infty} \hat{\varphi}_{2n}(t - h(2n))$$

is slowly oscillating, and $\varphi(h(2n)) = 1$ and $\varphi(h(2n+1)) = 0$ for all n. The assumed convergence of h to η implies that both sequences $(\varphi(h(2n)))$ and $(\varphi(h(2n+1)))$ converge to $\varphi(\eta)$. Contradiction. □

Thus, the topological nature of $M(SO(\mathbb{Z}^N))$ requires the use of nets rather than sequences. In the following two sections, we are going to recall and provide some facts about nets and about limit operators with respect to nets.

2.4.3 Preliminaries on nets

Nets and subnets. A set T is *directed* if there is a binary relation $\succ$ on T such that

$\forall t \in T:$	$t \succ t$	(reflexivity),
$\forall r, s, t \in T:$	$r \succ s,\ s \succ t \Rightarrow r \succ t$	(transitivity),
$\forall s, t \in T:$	$s \succ t,\ t \succ s \Rightarrow s = t$	(anti-symmetry),
$\forall r, s \in T\ \exists t \in T:$	$t \succ r$ and $t \succ s$	(inductivity).

For example, the set $\mathbb{N}$ is directed with respect to the relation $\geq$.

A mapping x from a directed set T into a topological space X is called a *net*, and this net *converges to a point* $x^* \in X$ if, for every neighborhood U of x^*, there is a $t_0 \in T$ such that $x(t) \in U$ for all $t \succ t_0$. The net $x : T \to X$ is sometimes also denoted by $(x_t)_{t\in T}$ where $x_t = x(t)$. Accordingly, if $x : T \to X$ converges to x^*, we will write

$$\lim_{t\in T} x_t = x^* \quad \text{or} \quad x_t \to x^* \text{ with respect to } T.$$

A net $(y_s)_{s\in S}$ is a *subnet* of the net $(x_t)_{t\in T}$ if there is a mapping $F : S \to T$ such that

$$\begin{aligned} \forall s \in S : &\quad y_s = x_{F(s)}, \\ \forall t \in T\, \exists\, s_0 \in S : &\quad F(s) \succ t \quad \text{for all } s \succ s_0 \end{aligned}$$

(note that we use the same symbol $\succ$ for all apparently different order relations on S and T). A subset S of a directed set T is called *cofinal* if

$$\forall t \in T\, \exists\, s \in S : \quad s \succ t.$$

Every cofinal subset S of a directed set T is again a directed set with respect to the restriction of the order relation $\succ$ onto S. If S is a cofinal subset of T, and if $(x_t)_{t\in T}$ is a net, then the restriction of $(x_t)_{t\in T}$ onto S is a subnet of $(x_t)_{t\in T}$. One is mainly interested in subnets which do *not* arise in this simple manner.

Nets tending to infinity. In what follows we will only be concerned with nets in $\mathbb{Z}^N$. A net $(x_t)_{t\in T}$ with values in $\mathbb{Z}^N$ is said to *converge to infinity* if

$$\forall k \in \mathbb{N}\, \exists\, t_0 \in T : \quad |x_t| \geq k \quad \text{for all } t \succ t_0.$$

Let $\mathcal{N}$ denote the set of all nets in $\mathbb{Z}^N$ which converge to infinity.

Lemma 2.4.11

(a) *For every net $(x_t)_{t\in T} \in \mathcal{N}$, the set $\{x_t : t \in T\}$ of its values is countably infinite.*

(b) *If $h : \mathbb{N} \to \mathbb{Z}^N$ is injective, then the sequence h belongs to $\mathcal{N}$.*

Proof. (a) Since $\mathbb{Z}^N$ is countable, $(x_t)_{t\in T} \subseteq \mathbb{Z}^N$ is an at most countable set, and since $(x_t)_{t\in T}$ tends to infinity, this set cannot be finite.
(b) Suppose the sequence h does not converge to infinity. Then

$$\exists\, k \in \mathbb{N}\, \forall n_0 \in \mathbb{N}\, \exists\, n \geq n_0 : |h(n)| \leq k.$$

Repeating this argument we get an infinite sequence $n_0 < n_1 < n_2 < \cdots$ such that $|h(n_r)| \leq k$ for all r. But h is injective. Thus, $h(n_r) \neq h(n_s)$ whenever $r \neq s$. So we have infinitely many points in $\{z \in \mathbb{Z}^N : |z| \leq k\}$, which is nonsense. □

Lemma 2.4.12 *Let $x \in \mathcal{N}$ be a net, and let h be a bijection from $\mathbb{N}$ onto the set of the values of x. Then x is a subnet of the sequence h. In particular, every net $x \in \mathcal{N}$ is a subnet of a sequence $h \in \mathcal{H}$.*

Proof. Let $x = (x_t)_{t \in T} \in \mathcal{N}$, and let $h : \mathbb{N} \to \{x_t : t \in T\}$ be a bijection. Such bijections exist by Lemma 2.4.11.

To show that x is a subnet of h, define $F : T \to \mathbb{N}$ by $F(t) := h^{-1}(x_t)$. Then, clearly, $x_t = h_{F(t)}$ for every $t \in T$, and it remains to check whether

$$\forall n \in \mathbb{N}\, \exists t_0 \in T : \quad F(t) \geq n \quad \text{for all } t \succ t_0. \tag{2.56}$$

Given $n \in \mathbb{N}$, set $k := \max\{|h_1|, \ldots, |h_n|\}$. Since $(x_t)_{t \in T}$ belongs to $\mathcal{N}$, there is a $t_0 \in T$ such that

$$|x_t| \geq k + 1 \quad \text{for all } t \geq t_0.$$

By the definition of F, this implies $F(t) \geq n$ for all $t \succ t_0$ which gives (2.56). Hence, x is a subnet of h, and this sequence belongs to $\mathcal{H}$ due to Lemma 2.4.11 (b). □

A version of Cantor's diagonalization procedure. The following result can be considered as a substitute for Cantor's diagonalization argument for sequences.

Theorem 2.4.13 *Let Z be a set, and let $(f_n)_{n \geq 1}$ be a sequence of functions $f_n : Z \to \mathbb{R}^+$ which converges uniformly on Z to a function $f : Z \to \mathbb{R}^+$. Assume further that $(x^0_{t_0})_{t_0 \in T_0}$ is a net with values in Z and with the property that, for every $n \geq 1$, there is a subnet $(x^n_{t_n})_{t_n \in T_n}$ of $(x^{n-1}_{t_{n-1}})_{t_{n-1} \in T_{n-1}}$ such that*

$$\lim_{t_n \in T_n} f_n(x^n_{t_n}) = 0. \tag{2.57}$$

Then there is a subnet $(y_w)_{w \in W}$ of $(x^0_{t_0})_{t_0 \in T_0}$ with $\lim_{w \in W} f(y_w) = 0$.

Proof. We split the proof into several steps and emphasize some partial results as lemmas. We start with a net $(x^0_{t_0})_{t_0 \in T_0}$ in Z and, for every $n \geq 1$, with a subnet $(x^n_{t_n})_{t_n \in T_n}$ of $(x^{n-1}_{t_{n-1}})_{t_{n-1} \in T_{n-1}}$ which satisfies (2.57). In particular, we have mappings $F_n : T_n \to T_{n-1}$ with $x^n_{t_n} = x^{n-1}_{F_n(t_n)}$ for all $t_n \in T_n$ and such that

$$\forall t_{n-1} \in T_{n-1}\, \exists t^0_n \in T_n : \quad F(t_n) \succ t_{n-1} \quad \text{for all } t_n \succ t^0_n. \tag{2.58}$$

Step 1. *We show that the directed sets $T_0, T_1, \ldots$ can be replaced by one and the same directed set S.*
Indeed, set $S := T_0 \times T_1 \times T_2 \times \cdots$ and provide S with the order

$$(s_0, s_1, s_2, \ldots) \succ (s'_0, s'_1, s'_2, \ldots) \quad \Longleftrightarrow \quad s_k \succ s'_k \text{ for all } k$$

which makes S to a directed set. Further, there are canonical mappings

$$G_n : S \to T_n, \quad (s_0, s_1, s_2, \ldots) \mapsto s_n.$$

For every $n \in \mathbb{N}$, define a net $(y^n_s)_{s \in S}$ by $y^n_s := x^n_{G_n(s)}$.

Lemma 2.4.14

(a) *For all $n \geq 0$, $(y^n_s)_{s \in S}$ is a subnet of $(x^n_{t_n})_{t_n \in T_n}$.*

(b) *For all $n \geq 1$, $(y^n_s)_{s \in S}$ is a subnet of $(y^{n-1}_s)_{s \in S}$.*

Proof of Lemma 2.4.14. (a) By the definition of y_s^n, what we have to check is whether

$$\forall t_n \in T_n \, \exists s^0 \in S : \quad G_n(s) \succ t_n \quad \text{for all } s \succ s^0.$$

But this is obvious: Set $s^0 := (t_0, t_1, t_2, \ldots) \in S$. Then, for $s \succ s^0$, one indeed has $G_n(s) \succ t_n$.

(b) For $n \geq 1$, define

$$H_n : S \to S, \ (s_0, s_1, s_2, \ldots) \mapsto (s_0, \ldots, s_{n-2}, F_n(s_n), s_n, s_{n+1}, \ldots)$$

with the $F_n(s_n)$ standing at the $n-1$th position. Then, for all $s = (s_0, s_1, s_2, \ldots)$ in S and all $n \geq 1$,

$$y_s^n = x^n_{G_n(s)} = x^n_{s_n} = x^{n-1}_{F_n(s_n)} = x^{n-1}_{G_{n-1}(H_n(s))} = y^{n-1}_{H_n(s)}, \tag{2.59}$$

and it remains to show that

$$\forall \hat{s} \in S \, \exists s^0 \in S : \quad H_n(s) \succ \hat{s} \quad \text{for all } s \succ s^0. \tag{2.60}$$

Let $\hat{s} = (\hat{s}_0, \hat{s}_1, \hat{s}_2, \ldots) \in S$. For $k \neq n$, set $s_k^0 := \hat{s}_k$. In case $k = n$, we first choose $s_n^{00} \in T_n$ such that

$$\forall s_n \succ s_n^{00} : \quad F_n(s_n) \succ \hat{s}_{n-1} \tag{2.61}$$

(which is possible due to (2.58)), and then we choose $s_n^0 \in T_n$ such that both $s_n^0 \succ s_n^{00}$ and $s_n^0 \succ \hat{s}_n$. Define $s^0 := (s_0^0, s_1^0, s_2^0, \ldots) \in S$. Then, for all $s = (s_0, s_1, s_2, \ldots) \succ s^0$, we have

$$\begin{array}{lllll} s_k & \succ & s_0^k & = & \hat{s}_k \quad \text{for all } 0 \leq k \leq n-2, \\ s_n & \succ & s_n^0 & \succ & s_n^{00}, \quad \text{whence } F_n(s_n) \succ \hat{s}_{n-1} \text{ due to (2.61)}, \\ s_n & \succ & s_n^0 & \succ & \hat{s}_n, \\ s_k & \succ & s_k^0 & = & \hat{s}_k \quad \text{for all } k \geq n+1. \end{array}$$

Consequently,

$$\begin{aligned} H_n(s_0, s_1, s_2, \ldots) &= (s_0, \ldots, s_{n-2}, F_n(s_n), s_n, s_{n+1}, \ldots) \\ &\succ (\hat{s}_0, \ldots, \hat{s}_{n-2}, \hat{s}_{n-1}, \hat{s}_n, \hat{s}_{n+1}, \ldots) = \hat{s}. \end{aligned}$$

This proves (2.60) and the lemma. □

Step 2. *Choice of the diagonal net.*
Let $\Omega := S \times \mathbb{N}$. This set becomes directed by the order relation

$$(s, n) \succ (s', n') \quad \Longleftrightarrow \quad s \succ s' \text{ and } n \geq n'.$$

Consider the net

$$y : \Omega \to \mathbb{Z}^N, \quad y_{(s,n)} := y_s^n. \tag{2.62}$$

Of course (and as in the standard diagonalization procedure for sequences) one cannot expect that $(y_{(s,n)})_{(s,n)\in\Omega}$ is a subnet of $(y_s^n)_{s\in S}$. But (also as for standard diagonalization) one has the following result where we write $\Omega_{n_0} := \{(s, n) \in \Omega : n > n_0\}$ for brevity. Clearly, Ω_{n_0} is a cofinal subset of Ω for every $n_0 \in \mathbb{N}$.

Lemma 2.4.15 *For all $n_0 \in \mathbb{N}$, $(y_{(s,n)})_{(s,n)\in\Omega_{n_0}}$ is a subnet of $(y_s^{n_0})_{s\in S}$.*

Proof of Lemma 2.4.15. For all $s \in S$ and all $n > n_0$, we have

$$y_{(s,n)} = y_s^n = y_{H_n(s)}^{n-1} = y_{H_{n-1}(H_n(s))}^{n-2} = \cdots = y_{(H_{n_0+1}\circ H_{n_0+2}\circ\cdots\circ H_n)(s)}^{n_0}$$

(compare (2.59)). This equality suggests to define

$$K_{n_0} : \Omega_{n_0} \to S, \quad (s,\, n) \mapsto (H_{n_0+1} \circ H_{n_0+2} \circ \cdots \circ H_n)(s).$$

Then, obviously,

$$y_{(s,n)} = y_{K_{n_0}(s,n)}^{n_0} \quad \text{for all } (s,\, n) \in \Omega_{n_0},$$

and what remains to be verified is

$$\forall \hat{s} \in S\ \exists\, (\tilde{s},\, \tilde{n}) \in \Omega_{n_0} : \quad K_{n_0}(s,n) \succ \hat{s} \quad \text{for all } (s,\, n) \succ (\tilde{s},\, \tilde{n}).$$

Set $\tilde{n} := n_0 + 1$ and construct $\tilde{s} := (\tilde{s}_0,\, \tilde{s}_1,\, \ldots)$ successively as follows. Let $\hat{s} = (\hat{s}_0,\, \hat{s}_1,\, \ldots) \in S$. We set $\tilde{s}_k := \hat{s}_k$ for $k \le n_0$. Further, by (2.58), given $\hat{s}_{n_0} \in T_{n_0}$,

$$\exists\, \overline{s}_{n_0+1} \in T_{n_0+1} : \quad F_{n_0+1}(s) \succ \hat{s}_{n_0} \quad \forall s \succ \overline{s}_{n_0+1}.$$

Then choose $\tilde{s}_{n_0+1}$ larger than both $\overline{s}_{n_0+1}$ and $\hat{s}_{n_0+1}$.

For $\tilde{s}_{n_0+1} \in T_{n_0+1}$, we choose $\overline{s}_{n_0+2} \in T_{n_0+2}$ such that

$$\forall s \succ \overline{s}_{n_0+2} : \quad F_{n_0+2}(s) \succ \tilde{s}_{n_0+1}\ (\succ \hat{s}_{n_0+1})$$

and, hence,

$$F_{n_0+1}(F_{n_0+2}(s)) \succ \hat{s}_{n_0}.$$

Then choose $\tilde{s}_{n_0+2}$ larger than both $\overline{s}_{n_0+2}$ and $\hat{s}_{n_0+2}$.

We proceed in this way, i.e., we choose $\overline{s}_{n_0+3} \in T_{n_0+3}$ such that

$$\forall s \succ \overline{s}_{n_0+3} : \quad F_{n_0+3}(s) \succ \tilde{s}_{n_0+2}\ (\succ \hat{s}_{n_0+2})$$

which implies that

$$F_{n_0+2}(F_{n_0+3}(s)) \succ \hat{s}_{n_0+1}$$

and, hence,

$$F_{n_0+1}(F_{n_0+2}(F_{n_0+3}(s))) \succ \hat{s}_{n_0}.$$

Then choose $\tilde{s}_{n_0+3}$ larger than $\overline{s}_{n_0+3}$ and $\hat{s}_{n_0+3}$.

Thus we have fixed $\tilde{s}$. Let now $s = (s_0,\, s_1,\, \ldots) \succ \tilde{s}$. Then, due to our construction,

$$\begin{aligned}
s_k &\succ \hat{s}_k && \text{for all } k \le n_0 - 1,\\
(F_{n_0+1} \circ F_{n_0+2} \circ \cdots \circ F_n)(s_n) &\succ \hat{s}_{n_0},\\
(F_{n_0+2} \circ F_{n_0+3} \circ \cdots \circ F_n)(s_n) &\succ \hat{s}_{n_0+1},\\
&\ \ \vdots\\
F_n(s_n) &\succ \hat{s}_{n-1},\\
s_k \succ \tilde{s}_k &\succ \hat{s}_k && \text{for all } k \ge n.
\end{aligned}$$

This shows that

$$K_{n_0}(s, n) = (H_{n_0+1} \circ \cdots \circ H_n)(s) \succ \hat{s}$$

since

$$H_n(s) = (s_0, \ldots, s_{n-2}, F_n(s_n), s_n, s_{n+1}, \ldots),$$

$$(H_{n-1} \circ H_n)(s) = (s_0, \ldots, s_{n-3}, F_{n-1}(F_n(s_n)), F_n(s_n), s_n, s_{n+1}, \ldots),$$

$$(H_{n-2} \circ H_{n-1} \circ H_n)(s) = \\ (s_0, \ldots, s_{n-4}, F_{n-2}(F_{n-1}(F_n(s_n))), F_{n-1}(F_n(s_n)), F_n(s_n), s_n, s_{n+1}, \ldots),$$

and so on. This finishes the proof of Lemma 2.4.15. □

Step 3. *Let $W := \Omega_0$. Then $(y_w)_{w \in W}$ is the net we are looking for.*
It is obvious from the above construction that $(y_w)_{w \in W}$ is a subnet of $(x^0_{t_0})_{t_0 \in T_0}$. So we are left with verifying that $\lim_{w \in W} f(y_w) = 0$.

Given $\varepsilon > 0$, choose and fix $n \geq 1$ such that $\|f - f_n\| < \varepsilon/2$. Then, by hypothesis,

$$\lim_{t_n \in T_n} f_n(x^n_{t_n}) = 0.$$

Since $(y_w)_{w \in \Omega_n}$ is a subnet of $(x^n_{t_n})_{t_n \in T_n}$, we also have $\lim_{w \in \Omega_n} f_n(y_w) = 0$, whence the existence of an $w_n \in \Omega_n$ with

$$|f_n(y_w)| < \varepsilon/2 \quad \text{for all } w \succ w_n. \tag{2.63}$$

Let now $w \in W$ with $w \succ w_n$. Then, evidently, $w \in \Omega_n$, and from (2.63) we conclude

$$|f(y_w)| \leq |f(y_w) - f_n(y_w)| + |f_n(y_w)| \leq \|f - f_n\|_\infty + |f_n(y_w)| < \varepsilon.$$

Hence, $\lim_{w \in W} f(y_w) = 0$ which finishes the proof of Theorem 2.4.13. □

2.4.4 Limit operators with respect to nets

We return to band-dominated operators on the sequence space E. If $y := (y_w)_{w \in W}$ is a net in $\mathcal{N}$, then we call an operator A_y *limit operator* of the operator $A \in L(E)$ if the net $(V_{-y_w} A V_{y_w})_{w \in W}$ converges $\mathcal{P}$-strongly to A_y. In general, the results derived for limit operators with respect to sequences remain valid for limit operators with respect to nets without changes. We will illustrate this point by two results where the Cantor diagonalization procedure for nets is involved.

Theorem 2.4.16 *Let $A = aI \in L(E, \mathcal{P})$ be a rich multiplication operator. Then every net $(x_t)_{t \in T} \in \mathcal{N}$ possesses a subnet $y := (y_w)_{w \in W}$ such that the limit operator A_y exists.*

Proof. Recall that A_y is the limit operator of A with respect to the net y if and only if

$$\lim_{w\in W} \|(V_{-y_w} A V_{y_w} - A_y)S_k\| = 0 \quad \text{for every } k \in \mathbb{Z}^N$$

where, as before, S_k refers to the operator of multiplication by the function which is I at $k \in \mathbb{Z}^N$ and 0 at all other points.

Set $(x^0_{t_0})_{t_0\in T_0} := (x_t)_{t\in T}$ and choose a bijection $m : \mathbb{N} \to \mathbb{Z}^N$. Since A is rich we find, for every $n \geq 1$, a subnet $(x^n_{t_n})_{t_n\in T_n}$ of $(x^{n-1}_{t_{n-1}})_{t_{n-1}} \in T_{n-1}$ as well as an operator $B_n \in L(\operatorname{Im} S_{m(n)})$ such that

$$\|(V_{-x^n_{t_n}} A V_{x^n_{t_n}} - B_n)S_{m(n)}\| \to 0. \tag{2.64}$$

Let B stand for the operator of multiplication by the function

$$\mathbb{Z}^N \to L(X), \quad k \mapsto B_{m^{-1}(k)}.$$

We claim that B is the limit operator of A with respect to the net y. For, we reify Cantor's scheme (= Theorem 2.4.13) as follows. Set $Z := \mathbb{Z}^N$. For $n \geq 1$ and $z \in \mathbb{Z}^N$, define

$$f_n(z) := \sum_{k=1}^{n} 2^{-k} \|(V_{-z} A V_z - B)S_{m(k)}\|,$$

and let

$$f(z) := \sum_{k=1}^{\infty} 2^{-k} \|(V_{-z} A V_z - B)S_{m(k)}\|.$$

Then, obviously, $\|f_n - f\|_\infty \to 0$. Further, by (2.64), we have $\lim_{t_n\in T_n} f_n(x^n_{t_n}) = 0$. Now we conclude from Theorem 2.4.13 that there is a subnet $(y_w)_{w\in W}$ of $(x_t)_{t\in T}$ such that $\lim_{w\in W} f(y_w) = 0$. This immediately implies the $\mathcal{P}$-strong convergence of the net $(V_{-y_w} A V_{y_w})_{w\in W}$ to B, whence $B = A_y$. □

Of course, a similar result holds for rich band operators. For another application of Theorem 2.4.13, consider the set of all operators $A \in L(E, \mathcal{P})$ having the following property: every net $(x_t)_{t\in T} \in \mathcal{N}$ possesses a subnet $y := (y_w)_{w\in W}$ such that the limit operator A_y exists. We denote this class by $\mathcal{L}^{nets}_E$ for a moment. As we have just remarked, every rich band operator belongs to $\mathcal{L}^{nets}_E$.

Theorem 2.4.17 *$\mathcal{L}^{nets}_E$ is norm-closed.*

Proof. Let $(A_n)_{n\geq 1} \subseteq \mathcal{L}^{nets}_E$ be a sequence with norm limit $A \in L(E)$, and let $(x^0_{t_0})_{t_0\in T_0} \in \mathcal{N}$. By hypothesis, for every $n \geq 1$, there exists a subnet $x^n := (x^n_{t_n})_{t_n\in T_n}$ of $(x^{n-1}_{t_{n-1}})_{t_{n-1}\in T_{n-1}}$ such that the limit operator A_{n,x^n} of A_n with respect to x^n exists. If $n \geq m$, then $(x^n_{t_n})_{t_n\in T_n}$ is a subnet of $(x^m_{t_m})_{t_m\in T_m}$, thus, the limit operator A_{m,x^n} also exists, and it coincides with A_{m,x^m}. Since $\|A_h\| \leq \|A\|$ for every limit operator A_h of A, we obtain

$$\|A_{n,x^n} - A_{m,x^m}\| = \|A_{n,x^n} - A_{m,x^n}\| = \|(A_n - A_m)_{x^n}\| \leq \|A_n - A_m\|$$

for all $n \geq m$. Hence, the sequence (A_{n,x^n}) converges in the norm, and we let B denote its norm limit.

Now define for all $n \geq 1$ and $z \in \mathbb{Z}^N$ (with the notations S_k and m as in the proof of Theorem 2.4.16)

$$f_n(z) := \sum_{k=1}^{\infty} 2^{-k} \, \|(V_{-z}A_nV_z - A_{n,x^n})S_{m(k)}\|$$

and

$$f(z) := \sum_{k=1}^{\infty} 2^{-k} \, \|(V_{-z}AV_z - B)S_{m(k)}\|.$$

Then again $\|f_n - f\| \to 0$ and $\lim_{t_n \in T_n} f_n(x^n_{t_n}) = 0$, whence via Theorem 2.4.13 the existence of a subnet $y = (y_w)_{w\in W}$ of $(x^0_{t_0})_{t_0 \in T_0}$ such that $\lim_{w\in W} f(y_w) = 0$. Thus, $B = A_y$. □

As a consequence we get $\mathcal{A}^\$_E \subseteq \mathcal{L}^{nets}_E$. One might ask whether one gets something new when considering limit operators with respect to nets instead of sequences. The next theorem says that the answer is *no* in some sense: every limit operator, which is defined with respect to a net, can also be reached by a sequence. (Nevertheless, it *is* useful to consider limit operators with respect to nets as it will be pointed out in the next sections.)

Theorem 2.4.18 *Let $A \in L(E, \mathcal{P})$, and let $y = (y_w)_{w\in W} \in \mathcal{N}$ be a net for which the limit operator A_y exists. Then there is a sequence $z = (z_n)_{n\in\mathbb{N}} \in \mathcal{H}$ for which the limit operator A_z of A exists, and $A_z = A_y$.*

Proof. Let $y = (y_w)_{w\in W}$ be a net for which the limit operator A_y of A exists, and define a function $f : \mathbb{Z}^N \to \mathbb{R}^+$ by

$$f(z) := \sum_{k=1}^{\infty} 2^{-k} \, \|(V_{-z}AV_z - A_y)S_{m(k)}\|,$$

with the notation as in the proof of Theorem 2.4.16. Then $\lim_{w\in W} f(y_w) = 0$. For every $n \in \mathbb{N}$, choose $w_n \in W$ such that

$$0 \leq f(y_w) < 1/n \quad \text{for all } w \succ w_n, \tag{2.65}$$

and set $z_n := y_{w_n}$. Then the sequence $z = (z_n)$ tends to infinity, and $f(z_n)$ tends to 0 as $n \to \infty$. Hence, A_y is the limit operator of A with respect to the sequence z. □

2.4.5 Local invertibility at points in $M^\infty(SO(\mathbb{Z}^N))$

After these preparations, we turn over to the analogue of Theorem 2.3.13 for localization over $M^\infty(SO(\mathbb{Z}^N))$.

Definition 2.4.19 *Let* $\eta \in M^\infty(SO(\mathbb{Z}^N))$ *and* $A \in L(E, \mathcal{P})$. *The* local operator spectrum $\sigma_\eta(A)$ of A at η *is the set of all limit operators* A_y *of* A *with respect to nets* y *which tend to* η.

By Theorem 2.4.18, $\sigma_\eta(A) \subseteq \sigma_{op}(A)$ (recall that $\sigma_{op}(A)$ is the set of all limit operators with respect to *sequences*). Let, conversely, $A_h \in \sigma_{op}(A)$ for some sequence $h \in \mathcal{H}$. Then the intersection $\operatorname{clos}_{M(SO(\mathbb{Z}^N))}\{h(m) : m \in \mathbb{Z}^N\} \cap M^\infty(SO(\mathbb{Z}^N))$ is non-empty by Theorem 2.4.5. Consequently, there is a subnet y of h which converges to a point $\eta \in M^\infty(SO(\mathbb{Z}^N))$. Clearly, the limit operator A_y exists and is equal to A_h. Hence,

$$\sigma_{op}(A) = \cup_{\eta \in M^\infty(SO(\mathbb{Z}^N))}\sigma_\eta(A) \quad \text{for every } A \in L(E).$$

Definition 2.4.20 *Let* $\eta \in M^\infty(SO(\mathbb{Z}^N))$ *and* $A \in L(E)$. *The operator* A *is* locally invertible at η *if there are operators* $B, C \in L(E)$ *and a neighborhood at infinity* W *of* η *such that*

$$BA\hat{\chi}_W I = \hat{\chi}_W AC = \hat{\chi}_W I$$

where $\hat{\chi}_W$ *refers to the characteristic function of* W.

The following result, which states the analogue of Theorem 2.3.13 with respect to the much finer localization over points in $M^\infty(SO(\mathbb{Z}^N))$ instead of points in S^{N-1}, is the main outcome of this section.

Theorem 2.4.21 *Let* $A \in \mathcal{A}_E^\$$ *and* $\eta \in M^\infty(SO(\mathbb{Z}^N))$. *Then the operator* A *is locally invertible at* η *if and only if all limit operators in* $\sigma_\eta(A)$ *are invertible and if*

$$\sup\{\|(A_h)^{-1}\| : A_h \in \sigma_\eta(A)\} < \infty.$$

The proof of this result will follow the line of the proof of Theorem 2.3.13, and we will pay our attention mainly to the differences which are implied by the non-metrizability of the topology of $M(SO(\mathbb{Z}^N))$ and, hence, by the need of using nets in place of sequences.

A basic step is the specification of Propositions 2.2.3 and 2.3.14 to the present context which reads as follows.

Proposition 2.4.22 *Let* $A \in \mathcal{A}_E$, $\eta \in M^\infty(SO(\mathbb{Z}^N))$, *and let* $\psi_{\alpha,R}$ *be as in Section* 2.2.1. *Suppose there is a constant* $M > 0$ *such that, for all positive integers* R, *there is a neighborhood at infinity* U *of* η *such that, for all* $\alpha \in U$, *there are operators* $B_{\alpha,R}, C_{\alpha,R} \in L(E, \mathcal{P})$ *with* $\|B_{\alpha,R}\|_{L(E)} \le M$, $\|C_{\alpha,R}\|_{L(E)} \le M$ *and*

$$B_{\alpha,R}\, A\, \hat{\psi}_{\alpha,R} I = \hat{\psi}_{\alpha,R}\, A\, C_{\alpha,R} = \hat{\psi}_{\alpha,R} I.$$

Then the operator A *is locally invertible at* η, *i.e., there are operators* $B, C \in \mathcal{A}_E$ *and a neighborhood at infinity* W *of* η *such that*

$$BA\hat{\chi}_W I = \hat{\chi}_W AC = \hat{\chi}_W I. \tag{2.66}$$

Proof. We follow exactly the proof of Proposition 2.2.3 where we replace the condition $|\alpha| \geq \rho(R)$ by $\alpha \in U$. What results in analogy with (2.21) is the existence of a positive integer R such that

$$(I + T_R)^{-1} B_R A = I - (I + T_R)^{-1} \sum_{\alpha \in \mathbb{Z}^N \setminus U} \hat{\varphi}_{\alpha,R} I.$$

The assertion will follow once we have shown that there is a neighborhood at infinity W of η such that $\sum_{\alpha \in \mathbb{Z}^N \setminus U} \hat{\varphi}_{\alpha,R}\, \chi_W = 0$. This will be done in the following proposition. □

Proposition 2.4.23 *Let* $0 < R \in \mathbb{Z}$, $\eta \in M^\infty(SO(\mathbb{Z}^N))$ *and* $U \subseteq M(SO(\mathbb{Z}^N))$ *a neighborhood of* η. *Then there exists a neighborhood at infinity* $\tilde{U}$ *of* η *such that* $\sum_{\alpha \in \mathbb{Z}^N \setminus U} \hat{\varphi}_{\alpha,R}\, \chi_{\tilde{U}} = 0$.

Proof. The proof requires some precise knowledge on subsets of $M(SO(\mathbb{Z}^N))$ which is provided by Theorem 2.4.9. Applying this theorem (with the set W replaced by $\mathbb{Z}^N$), we find that $V := U \cap \mathbb{Z}^N$ is a growing set and that there is a neighborhood $U_0 \subseteq U$ of η such that $V_0 := U_0 \cap \mathbb{Z}^N$ is a center of V.

The support of every function $\hat{\varphi}_{\alpha,R}$ is contained in a smallest ball with center αR and with a radius r which depends on R but not on α. From V, we remove all points z for which the ball with center z and radius r is not completely contained in V. What we get is a set $\tilde{V}$, and we set $\tilde{U} := V_0 \cap \tilde{V}$.

We claim that $\tilde{V}$ is a growing set and that $\tilde{U}$ is one of its centers. Let $D \subset \mathbb{Z}^N$ be bounded, and let B be the ball with center 0 and radius r. Then $D + B$ is a bounded set, and since V_0 is a center of V, there is a bounded set M such that

$$(V_0 \setminus M) + (D + B) \subseteq V.$$

Then, of course, $(V_0 \setminus M) + D \subseteq \tilde{V}$, whence

$$(\tilde{U} \setminus M) + D \subseteq \tilde{V}. \tag{2.67}$$

Analogously, there is a bounded set N such that $(V_0 \setminus N) + B \subseteq V$. Thus, all points in $V_0 \setminus N$ belong to $\tilde{V}$ and, consequently, also to $\tilde{U}$. This shows that $\tilde{U}$ and V_0 differ by a bounded set only:

$$V_0 \setminus N \subseteq \tilde{U} \subseteq V_0. \tag{2.68}$$

A first consequence of (2.68) is that $\tilde{U}$ is an unbounded set. Together with (2.67) this implies that $\tilde{V}$ is a growing set, and that $\tilde{U}$ is a center of $\tilde{V}$. As another consequence of (2.68) we observe that, since V_0 is a neighborhood at infinity of η, also $\tilde{U}$ is a neighborhood at infinity of η. This finishes the proof since the support of every function $\hat{\varphi}_{\alpha,R}$ with $\alpha \in \mathbb{Z}^N \setminus U$ is contained in the complement of $\tilde{V}$, hence in the complement of $\tilde{U}$. □

Proof of Theorem 2.4.21. We will only prove that the uniform invertibility of the operators in $\sigma_\eta(A)$ implies the local invertibility of A at η. Let $A \in \mathcal{A}_E^\$$ be an operator with

$$M_A := \sup\{\|A_h^{-1}\| : A_h \in \sigma_\eta(A)\} < \infty,$$

but suppose A is not locally invertible at η. Then, by Proposition 2.4.22, there is a net $(y_t)_{t\in T}$ with values in $\mathbb{Z}^N$ which converges to η in the topology of $M(SO(\mathbb{Z}^N))$ and which has the property that

$$BA\hat{\psi}_{y_t,R}I \neq \hat{\psi}_{y_t,R}I \tag{2.69}$$

for all $t \in T$ and all $B \in L(E, \mathcal{P})$ with $\|B\| \leq M_A$. Since A belongs to $\mathcal{A}_E^\$ \subseteq \mathcal{L}_E^{nets}$, the net $(y_t)_{t\in T}$ possesses a subnet $x = (x_s)_{s\in S}$ such that the limit operator A_x exists. Clearly, the net $(x_s)_{s\in S}$ still converges to η, and

$$BA\hat{\psi}_{x_s,R}I \neq \hat{\psi}_{x_s,R}I \tag{2.70}$$

for all $s \in S$ and all $B \in L(E, \mathcal{P})$ with $\|B\| \leq M_A$. From Theorem 2.4.18 we conclude that there is a sequence $z = (z_n)_{n\in\mathbb{N}}$ which tends to infinity, and for which the limit operator A_z exists and coincides with A_x. Hence, A_z belongs to $\sigma_\eta(A)$. By hypothesis, A_z is invertible, and $\|A_z^{-1}\| \leq M_A$. This yields a contradiction in the very same way as in the proof of Theorem 2.2.1 by using Proposition 2.2.4. □

Our next goal is the relationship between local invertibility at points in the fiber $M^\infty(SO(\mathbb{Z}^N))$ and localization by means of the local principle. We have seen in Theorem 2.4.2 that the center of the quotient algebra $\mathcal{A}_E^\$/K(E,\mathcal{P})$ is equal to the algebra $\mathcal{C}$ of all cosets $aI + K(E, \mathcal{P})$ with $a \in SO(\mathbb{Z}^N)$. The isomorphy

$$\begin{aligned} \mathcal{C} &\cong (SO(\mathbb{Z}^N) \cdot I + K(E, \mathcal{P}))/K(E, \mathcal{P}) \\ &\cong SO(\mathbb{Z}^N) \cdot I/(SO(\mathbb{Z}^N) \cdot I \cap K(E, \mathcal{P})) \cong SO(\mathbb{Z}^N)/c_0(\mathbb{Z}^N, L(X)) \end{aligned}$$

implies that the maximal ideal space of the algebra $\mathcal{C}$ is homeomorphic to the fiber $M^\infty(SO(\mathbb{Z}^N))$. Given $\eta \in M^\infty(SO(\mathbb{Z}^N))$, we denote the local algebra of $\mathcal{A}_E^\$/K(E,\mathcal{P})$ which is associated with η by $\mathcal{A}_{E,\eta}^\$$, and we write π_η for the canonical homomorphism from $\mathcal{A}_E^\$$ onto $\mathcal{A}_{E,\eta}^\$$. The following result can be proved as its predecessor Theorem 2.3.21.

Theorem 2.4.24 *Let $A \in \mathcal{A}_E^\$$ and $\eta \in M^\infty(SO(\mathbb{Z}^N))$. The coset $\pi_\eta(A)$ is invertible in $\mathcal{A}_{E,\eta}^\$$ if and only if A is locally invertible at η.*

Together with Allan's local principle and with Theorem 2.4.21, this result yields a further essential refinement of Theorem 2.2.1 and Corollary 2.3.22.

Corollary 2.4.25 *An operator $A \in \mathcal{A}_E^\$$ is $\mathcal{P}$-Fredholm if and only if all of its limit operators are invertible, and if*

$$\sup\{\|(A_h)^{-1}\| : A_h \in \sigma_\eta(A)\} < \infty \quad \textit{for every } \eta \in M^\infty(SO(\mathbb{Z}^N)).$$

2.4.6 Fredholmness of band-dominated operators with slowly oscillating coefficients

We will now specify Corollary 2.4.25 to band-dominated operators with slowly oscillating coefficients. Let $SO^{\$}_{L(X)}$ refer to the class of all slowly oscillating functions with values in $L(X)$ for which the associated multiplication operator is rich. Further we let $\mathcal{A}_E(SO_{L(X)})$ (resp. $\mathcal{A}_E(SO^{\$}_{L(X)})$) stand for the smallest closed subalgebra of $\mathcal{A}_E$ which contains all band operators $\sum_{|\alpha|\le k} a_\alpha V_\alpha$ with $a_\alpha \in SO_{L(X)}$ (resp. $a_\alpha \in SO^{\$}_{L(X)}$). For the limit operators of operators with slowly oscillating coefficients we have the following.

Proposition 2.4.26

(a) *If $A \in \mathcal{A}_E(SO_{L(X)})$, then every limit operator of A belongs to $\mathcal{A}_E(\mathbb{C}_{L(X)})$.*

(b) *For $A \in \mathcal{A}_E(SO^{\$}_{L(X)})$, every local operator spectrum $\sigma_\eta(A)$ with $\eta \in M^\infty(SO)$ is a singleton.*

Proof. (a) Limit operators of shift operators are shift operators and, hence, in $\mathcal{A}_E(\mathbb{C}_{L(X)})$. By Proposition 2.4.1, the same is true for operators of multiplication by slowly oscillating functions.
(b) If $a \in SO^{\$}_{L(X)}$, then $\sigma_\eta(aI)$ is not empty since $\mathcal{A}^{\$}_E \subseteq \mathcal{L}^{nets}_E$ (see Section 2.4.4), and this spectrum is clearly a singleton. With Proposition 1.2.2 we conclude first that every local spectrum of a band operator with coefficients in $SO^{\$}_{L(X)}$ is a singleton, too, and get then the assertion also in the general case. □

Since $\sigma_\eta(A)$ is a singleton, the uniform boundedness condition from Corollary 2.4.25 is redundant for operators $A \in \mathcal{A}_E(SO^{\$}_{L(X)})$.

Theorem 2.4.27 *Operators in $\mathcal{A}_E(SO^{\$}_{L(X)})$ are $\mathcal{P}$-Fredholm if and only if all of their limit operators are invertible.*

Observe that the invertibility of the limit operators can be checked be means of Theorem 2.3.25 efficiently.

Recall from Section 1.1.3 that the $\mathcal{P}$-essential spectrum of an operator $A \in L(E,\mathcal{P})$ is defined as the spectrum of the coset $A + K(E,\mathcal{P})$ in the quotient algebra $L(E,\mathcal{P})/K(E,\mathcal{P})$. We denote this spectrum by $\sigma_{\mathcal{P}\text{-}ess}(A)$. Note that in case $K(E,\mathcal{P}) = K(E)$, the $\mathcal{P}$-essential spectrum of A coincides with the usual essential spectrum of A which we will denote by $\sigma_{ess}(A)$. Finally, let $\sigma_E(A)$ refer to the spectrum of A, considered as an element of the Banach algebra $L(E)$.

Corollary 2.4.28 *Let $A \in \mathcal{A}_E(SO^{\$}_{L(X)})$. Then*

$$\sigma_{\mathcal{P}\text{-}ess}(A) = \cup\, \sigma_E(A_h)$$

where the union is taken over all limit operators A_h of A.

2.4.7 Nets vs. sequences

For $h \in \mathcal{H}$, the closure $\overline{h}$ of the set $\{h(m) : m \in \mathbb{Z}^N\}$ of the values of h in the Gelfand topology cannot consist of a single point of $M^\infty(SO(\mathbb{Z}^N))$ only (Proposition 2.4.10). Nevertheless, the sequences in $\mathcal{H}$ separate the points of $M^\infty(SO(\mathbb{Z}^N))$ in the following sense.

Proposition 2.4.29 *Given η, $\theta \in M^\infty(SO(\mathbb{Z}^N))$, there is a function $h \in \mathcal{H}$ such that $\eta \in \overline{h}$ and $\theta \notin \overline{h}$.*

Proof. Choose disjoint neighborhoods U_η and U_θ of η and θ in $M(SO(\mathbb{Z}^N))$, and let $h \in \mathcal{H}$ be a sequence such that

$$\{h(m) : m \in \mathbb{Z}^N\} = U_\eta \cap \mathbb{Z}^N.$$

(Recall that the intersection $U_\eta \cap \mathbb{Z}^N$ is not empty by Theorem 2.4.5 and, hence, countable. Thus, h can be even chosen as a bijection from $\mathbb{Z}^N$ onto $U_\eta \cap \mathbb{Z}^N$.) Since $\mathbb{Z}^N$ is dense in $M(SO(\mathbb{Z}^N))$, it is clear that $\eta \in \overline{U_\eta \cap \mathbb{Z}^N} = \overline{h}$, but θ cannot belong to $\overline{h}$ since

$$\theta \in U_\theta \subseteq M(SO(\mathbb{Z}^N)) \setminus \overline{U_\eta} = M(SO(\mathbb{Z}^N)) \setminus \overline{h},$$

i.e., θ is an interior point of the complement of $\overline{h}$. □

The Proposition 2.4.29 suggests the following definition.

Definition 2.4.30 *Let $\eta \in M^\infty(SO(\mathbb{Z}^N))$ and $A \in L(E)$. The* local operator spectrum of A at η with respect to sequences *is the set*

$$\sigma_\eta^{seq}(A) := \{A_h : h \in \mathcal{H}_A \text{ and } \eta \in \overline{h}\}.$$

If h is a sequence with $\eta \in \overline{h}$, for which the limit operator A_h exists, then there is a subnet y of h which tends to η. Further, if $A \in \mathcal{A}_E$, then there is a subnet x of y for which the limit operator A_x exists. Clearly, x also tends to η and $A_x = A_h$. Thus,

$$\sigma_\eta^{seq}(A) \subseteq \sigma_\eta(A) \quad \text{for every } A \in \mathcal{A}_E. \tag{2.71}$$

Further, for $h \in \mathcal{H}$, the intersection $\text{clos}_{M(SO(\mathbb{Z}^N))}\{h(m) : m \in \mathbb{Z}^N\} \cap M^\infty(SO)$ is non-empty by Theorem 2.4.5. So we have in analogy to Proposition 2.3.2

$$\sigma_{op}(A) = \cup_{\eta \in M^\infty(SO(\mathbb{Z}^N))} \sigma_\eta^{seq}(A) \quad \text{for every } A \in L(E).$$

We conjecture that equality holds in (2.71). Some evidence to this conjecture is given by the following result which states that sequential local spectra of operators of multiplication by slowly oscillating functions are singletons.

Proposition 2.4.31 *Let $\eta \in M^\infty(SO(\mathbb{Z}^N))$.*

(a) *If $A = aI$ with $a \in SO(\mathbb{Z}^N)$, then $\sigma_\eta^{seq}(A) = \{a(\eta)\}$ (where we use the same notation for a function in $SO(\mathbb{Z}^N)$ and its Gelfand transform).*

(b) *If $A = aI$ with $a \in SO_{L(X)}(\mathbb{Z}^N)$, then $\sigma_\eta^{seq}(A)$ contains at most one operator.*

Proof. (a) Let $h \in \mathcal{H}$ be a sequence such that $\eta \in \overline{h}$ and such that the limit operator $(aI)_h$ exists. By Proposition 2.4.1, $(aI)_h = \alpha I$ with the complex number $\alpha := \lim a(h(n))$. We claim that $\alpha = a(\eta)$.

Let $\varepsilon > 0$. Since a is continuous at η, there is an open neighborhood U of η such that

$$|a(\eta) - a(\theta)| < \varepsilon/2 \quad \text{for all } \theta \in U.$$

Further, since $\eta \in \overline{h}$, there is an infinite subsequence g of h the values of which are in U. Choose m such that $|a(g(m)) - \alpha| < \varepsilon/2$. Then

$$|a(\eta) - \alpha| \leq |a(\eta) - a(g(m))| + |a(g(m)) - \alpha| < \varepsilon.$$

This estimate holds for arbitrary $\varepsilon > 0$; hence, $a(\eta) = \alpha$.
(b) Suppose there are sequences $h_1, h_2 \in \mathcal{H}$ such that $\eta \in \overline{h_1} \cap \overline{h_2}$ and that the limit operators $(aI)_{h_1}$ and $(aI)_{h_2}$ exist, but that $(aI)_{h_1} \neq (aI)_{h_2}$. By Proposition 2.4.1, $(aI)_{h_1}$ and $(aI)_{h_2}$ are the operators of multiplication by the constant functions $x \mapsto A_1$ and $x \mapsto A_2$ with $A_1, A_2 \in L(X)$. Since $A_1 \neq A_2$, there is a functional $\varphi \in L(X)^*$ such that $\varphi(A_1) \neq \varphi(A_2)$. Consider the function $\hat{a} : \mathbb{Z}^N \to \mathbb{C} : x \mapsto \varphi(a(x))$. This function is in $SO(\mathbb{Z}^N)$:

$$|\hat{a}(x+k) - \hat{a}(x)| \leq \|\varphi\| \, \|a(x+k) - a(x)\|_{L(X)} \to 0 \quad \text{as } x \to \infty.$$

From $\|a(h_i(m)) - A_i\| \to 0$ for $i = 1, 2$ we conclude that

$$\|\hat{a}(h_i(m)) - \varphi(A_i)\| \to 0 \quad \text{for } i = 1, 2.$$

Hence, both $\varphi(A_1)I$ and $\varphi(A_2)I$ are limit operators of $\hat{a}I$ at η. This contradicts assertion (a) of this proposition, stating that $\sigma_\eta^{seq}(\hat{a}I)$ is a singleton. □

2.4.8 Appendix A: A second proof of Theorem 2.4.27

The following two sections are devoted to alternative proofs of Theorem 2.4.27 which work under more special assumptions, but which have their own merits, and which shed light upon the properties of band-dominated operators with slowly oscillating coefficients.

We let H be a Hilbert space and $E := l^2(\mathbb{Z}^N, H)$. Further, we again write $SO_{L(H)}$ and $SO^{\$}_{L(H)}$ for the algebra of all slowly oscillating functions $\mathbb{Z}^N \to L(H)$ and for the algebra of all slowly oscillating functions $\mathbb{Z}^N \to L(H)$ for which the associated multiplication operator is rich, respectively, and we let $\mathcal{A}_E(SO_{L(H)})$ and $\mathcal{A}_E(SO^{\$}_{L(H)})$ stand for the closures in $L(E)$ of the algebra of the band operators with coefficients in $SO_{L(H)}$ and in $SO^{\$}_{L(H)}$.

Generating functions. The alternative proof of Theorem 2.4.27 which will be discussed in this section is based on the notion of the generating function of a band-dominated operator. This notion is borrowed from the pseudodifferential operator calculus and adapted for our purposes. Usually, the generating function of a pseudodifferential operator is referred to as the *symbol* of that operator.

We start with defining the generating function of a band operator. For

$$A = \sum_{|\alpha| \leq M} a_\alpha V_\alpha \quad \text{with } a_\alpha \in SO_{L(H)}, \tag{2.72}$$

the *generating function* of A is

$$\text{gen}_A : \mathbb{Z}^N \times \mathbb{T}^N \to L(H), \quad (x, t) \mapsto \sum_{|\alpha| \leq M} a_\alpha(x) t^\alpha \tag{2.73}$$

where $t^\alpha := t_1^{\alpha_1} \cdots t_n^{\alpha_n}$. There is a one-to-one correspondence between band operators and their generating functions.

We denote by $C_b(\mathbb{Z}^N \times \mathbb{T}^N, L(H))$ the set of all bounded continuous functions on $\mathbb{Z}^N \times \mathbb{T}^N$ with values in $L(H)$. Provided with pointwisely defined operations and the supremum norm, this set becomes a C^*-algebra, and the set $c_0(\mathbb{Z}^N \times \mathbb{T}^N, L(H))$ of all functions $a \in C_b(\mathbb{Z}^N \times \mathbb{T}^N, L(H))$ with

$$\lim_{x \to \infty} \sup_{t \in \mathbb{T}^N} \|a(x, t)\|_{L(H)} = 0$$

is a closed ideal of $C_b(\mathbb{Z}^N \times \mathbb{T}^N, L(H))$. The quotient algebra C_b/c_0 will be abbreviated to $\widehat{C}_b$, and the coset which contains $a \in C_b(\mathbb{Z}^N \times \mathbb{T}^N, L(H))$ to $\widehat{a}$. Notice that

$$\|\widehat{a}\|_0 := \limsup_{x \to \infty} \sup_{t \in \mathbb{T}^N} \|a(x, t)\|_{L(H)}$$

is just the canonical quotient norm of the coset $\widehat{a}$ in the quotient algebra $\widehat{C}_b$.

Evidently, if A is a band operator of the form (2.72), then its generating function belongs to $C_b(\mathbb{Z}^N \times \mathbb{T}^N, L(H))$.

Proposition 2.4.32 *Let A be as in* (2.72). *Then* $\|\widehat{\text{gen}_A}\|_0 \leq \|A\|$.

Proof. Choose a sequence $(x_n) \subset \mathbb{Z}^N$ tending to infinity, a sequence $(t_n) \in \mathbb{T}^N$, and a sequence (v_n) of unit vectors in H, such that

$$\|\widehat{\text{gen}_A}\|_0 = \lim_{n \to \infty} \|\text{gen}_A(x_n, t_n) v_n\|_H.$$

Since $\mathbb{T}^N$ is compact, we can moreover assume that (t_n) is a convergent sequence with limit $t_0 \in \mathbb{T}^N$. The assertion will follow once we have shown that, given $\varepsilon > 0$, there is an n_0 such that

$$\|\text{gen}_A(x_n, t_n) v_n\|_H \leq \|A\| + \varepsilon \tag{2.74}$$

for all $n \geq n_0$.

Given vectors $v \in H$ and $u = (u_k)_{k\in\mathbb{Z}^N} \in l^2$, let $v \otimes u$ denote the sequence $(u_k v)_{k\in\mathbb{Z}^N}$ in $E = l^2(\mathbb{Z}^N, H)$. Let further $A_{n,n}$, A_n and B_n stand for the band operators with generating functions

$$(x,t) \mapsto \operatorname{gen}_A(x_n, t_n), \quad (x,t) \mapsto \operatorname{gen}_A(x + x_n, t), \quad (x,t) \mapsto \operatorname{gen}_A(x_n, t),$$

respectively. Then we have, for every unit vector $u \in l^2$,

$$\begin{aligned} \|\operatorname{gen}_A(x_n, t_n)v_n\|_H &= \|A_{n,n}(v_n \otimes u)\|_E \\ &\leq \|(A_{n,n} - B_n)(v_n \otimes u)\|_E \\ &\quad + \|(B_n - A_n)(v_n \otimes u)\|_E + \|A_n(v_n \otimes u)\|_E. \end{aligned} \tag{2.75}$$

Since $A_n = V_{-x_n} A V_{x_n}$, we get

$$\|A_n(v_n \otimes u)\|_E = \|V_{-x_n} A V_{x_n}(v_n \otimes u)\|_E \leq \|A\|$$

for the last term in (2.75). The middle term on the right-hand side of (2.75) is not greater than $\|B_n - A_n\|$, which goes to zero as $n \to \infty$ since the coefficients of A are slowly oscillating. Thus, this middle becomes less than $\varepsilon/2$ uniformly with respect to u and v_n if only n is large enough.

To estimate the first term, choose $\delta > 0$ such that

$$\sup_{x\in\mathbb{Z}^N} \|\operatorname{gen}_A(x,t) - \operatorname{gen}_A(x,t_0)\| \leq \varepsilon/4 \quad \text{for all } |t - t_0| < \delta,$$

and choose the unit vector $u = (u_k)_{k\in\mathbb{Z}^N}$ in l^2 such that the u_k are the Fourier coefficients of a continuous function $\hat{u}$ on $\mathbb{T}^N$ with support in $\{t \in \mathbb{T}^N : |t - t_0| < \delta\}$. Since $A_{n,n} - B_n$ is the operator of convolution by the function $\operatorname{gen}_{A_{n,n}} - \operatorname{gen}_{B_n}$, we get

$$\begin{aligned} \|(A_{n,n} - B_n)(v_n \otimes u)\|^2_{l^2(\mathbb{Z}^N,H)} &= \|(\operatorname{gen}_{A_{n,n}} - \operatorname{gen}_{B_n})(\hat{u}v_n)\|^2_{L^2(\mathbb{T}^N,H)} \\ &= \int_{\mathbb{T}^N} \|(\operatorname{gen}_A(x_n,t_n) - \operatorname{gen}_A(x_n,t))\hat{u}(t)v_n\|^2_H \, dt \\ &\leq \sup_{|t-t_0|<\delta} \|\operatorname{gen}_A(x_n,t_n) - \operatorname{gen}_A(x_n,t)\|^2_{L(H)} \, \|v_n \otimes u\|^2. \end{aligned}$$

Due to the choice of δ, this term becomes less than $\varepsilon/2$ if n becomes large. □

This proposition allows us to associate with every operator A in $\mathcal{A}_E(SO_{L(H)})$ a uniquely determined coset in $\widehat{C}_b$, which we denote by $\Gamma(A)$.

Proposition 2.4.33 *Γ is a *-homomorphism from $\mathcal{A}_E(SO^\$_{L(H)})$ into $\widehat{C}_b$ with kernel $K(E,\mathcal{P})$.*

Proof. It is elementary to check that Γ acts as a *-homomorphism on the algebra of all band operators with slowly oscillating coefficients. Since this algebra is dense in

$\mathcal{A}_E(SO_{L(H)})$, and since Γ is continuous on this algebra by the preceding proposition, this proves the first assertion. It is further evident that the ideal $K(E,\mathcal{P})$ lies in the kernel of Γ. Let, finally, A be an operator in $\mathcal{A}_E(SO_{L(H)})$ with $\Gamma(A) = 0$. We have to show that A lies in $K(E,\mathcal{P})$.

Let (A_n) be a sequence of band operators in $\mathcal{A}_E(SO^{\$}_{L(H)})$ which converges to A. Then, trivially, $\|\Gamma(A_n)\| \to 0$. For $\alpha = (\alpha_1, \dots, \alpha_N) \in \mathbb{Z}^N$, consider the functions $a_\alpha^{(n)} : \mathbb{Z}^N \to L(H)$ which take at $x \in \mathbb{Z}^N$ the value

$$\frac{1}{(2\pi)^N} \int_0^{2\pi} \cdots \int_0^{2\pi} \operatorname{gen}_{A_n}(x, (e^{is_1}, \dots, e^{is_N}))\, e^{-i\alpha_1 s_1} \cdots e^{-i\alpha_N s_N}\, ds_1 \cdots ds_N. \tag{2.76}$$

If the band operator A_n is of the form $\sum b_\alpha^{(n)} V_\alpha$, then its αth diagonal $b_\alpha^{(n)}$ just coincides with the function $a_\alpha^{(n)}$ given by (2.76). From (2.76) we immediately conclude that

$$\|a_\alpha^{(n)}(x)\|_{L(H)} \le \sup_{t \in \mathbb{T}^N} \|\operatorname{gen}_{A_n}(x,t)\|_\infty$$

whence, in particular,

$$\limsup_{x \to \infty} \|a_\alpha^{(n)}(x)\| \le \limsup_{x \to \infty} \sup_{t \in \mathbb{T}^N} \|\operatorname{gen}_{A_n}(x,t)\|_\infty = \|\Gamma(A_n)\| \to 0$$

as $n \to \infty$. Thus, if a_α denotes the αth diagonal of A, then

$$\begin{aligned} \limsup_{x \to \infty} \|a_\alpha(x)\| &\le \sup_{x \in \mathbb{Z}^N} \|a_\alpha(x) - a_\alpha^{(n)}(x)\| + \limsup_{x \to \infty} \|a_\alpha^{(n)}(x)\| \\ &\le \|A - A_n\| + \|\Gamma(A_n)\|. \end{aligned} \tag{2.77}$$

This shows that every diagonal of A lies in $c_0(\mathbb{Z}^N, L(H))$, which on its hand implies that all limit operators of A are 0: Indeed, let h be a sequence for which the limit operator A_h exists. Then the operators $S_i V_{-h(n)} A V_{h(n)} S_j$ converge in the norm to $S_i A_h S_j$ for every pair of indices $i, j \in \mathbb{Z}^N$. Since

$$\lim_{n \to \infty} \|S_i V_{-h(n)} A V_{h(n)} S_j\| = \lim_{n \to \infty} \|a_{i-j}(i + h(n))\| = 0$$

due to (2.77), this shows that $S_i A_h S_j = 0$ for all i and j, whence $A_h = 0$. But a rich band-dominated operator having 0 as its only limit operator lies in $K(E, \mathcal{P})$ due to Theorem 2.2.10. □

Here is the alternative proof of Theorem 2.4.27.

Theorem 2.4.34 *The following assertions are equivalent for $A \in \mathcal{A}_E(SO^{\$}_{L(H)})$:*

(a) *A is $\mathcal{P}$-Fredholm.*

(b) *$\Gamma(A)$ is invertible in $\widehat{C}_b$.*

(c) *All limit operators of A are invertible.*

Proof. The equivalence of (a) and (b) is quite obvious: If the coset $A+K(E,\mathcal{P})$ is invertible in $L(E,\mathcal{P})/K(E,\mathcal{P})$, then it is also invertible in $\mathcal{A}_E(SO^\$_{L(H)})/K(E,\mathcal{P})$ (inverse closedness of C^*-algebras). Hence, there are operators $B \in \mathcal{A}_E(SO^\$_{L(H)})$ and $K_1, K_2 \in K(E,\mathcal{P})$ such that $AB = I + K_1$ and $BA = I + K_2$. Applying the homomorphism Γ to these equalities yields invertibility of $\Gamma(A)$. If, conversely, $\Gamma(A)$ is invertible in $\widehat{C}_b$, then it is also invertible in the image of $\mathcal{A}_E(SO^\$_{L(H)})$ under the mapping Γ (again by the inverse closedness of C^*-algebras). Thus, one can find a $B \in \mathcal{A}_E(SO^\$_{L(H)})$ with $\Gamma(A)\Gamma(B) = \Gamma(B)\Gamma(A) = 1$, showing that $AB - I$ and $BA - I$ belong to $\ker\Gamma = K(E,\mathcal{P})$.

Since (a) implies (c) by Proposition 1.2.9, we are left with the implication (c) $\Rightarrow$ (b). Assume that all limit operators of $A \in \mathcal{A}_E(SO^\$_{L(H)})$ are invertible, but that $\Gamma(A)$ is not invertible in $\widehat{C}_b$. If A is not a band operator, then we let gen_A be any function in the coset $\Gamma(A)$.

Define the lower norm of an operator $C \in L(H)$ by $\nu(C) := \inf_{x\neq 0} \|Cx\|/\|x\|$. It is well known that C is invertible if both $\nu(C)$ and $\nu(C^*)$ are positive and that, conversely, invertibility of C implies $\nu(C) = \nu(C^*) = 1/\|A^{-1}\|$. Thus, if both

$$\lim_{R\to\infty} \inf_{|x|\geq R,\, t\in\mathbb{T}^N} \nu(\text{gen}_A(x,t)) > 0 \tag{2.78}$$

and

$$\lim_{R\to\infty} \inf_{|x|\geq R,\, t\in\mathbb{T}^N} \nu(\text{gen}_A(x,t)^*) > 0, \tag{2.79}$$

then the function gen_A is invertible in C_b modulo functions in c_0. Since $\Gamma(A)$ is non-invertible by assumption, one of the conditions (2.78) and (2.79) must be violated, say the first one for definiteness. Then there exist a sequence $x = (x_m)_{m\geq 1} \subset \mathbb{Z}^N$ which tends to infinity, a sequence $(t_m)_{m\geq 1} \subset \mathbb{T}^N$ which we can also suppose to be convergent to a point $t_0 \in \mathbb{T}^N$, as well as a sequence $(v_m)_{m\geq 1}$ of unit vectors in H such that

$$\|\text{gen}_A(x_m,t_m)v_m\| \to 0 \quad \text{as } m\to\infty.$$

We will further suppose without loss that the limit operator A_x of A with respect to the sequence x exists.

Let $\varepsilon < 1/(4\,\|A_x^{-1}\|)$, and let A' be a band operator with coefficients in $SO^\$_{L(H)}$ such that $\|A - A'\| < \varepsilon$. Then $\|\Gamma(A) - \Gamma(A')\|_0 < \varepsilon$, which implies that

$$\begin{aligned}
&\limsup_{m\to\infty} \|\text{gen}_{A'}(x_m,t_m)v_m\| \\
&\quad \leq \limsup_{m\to\infty} \|\text{gen}_{A'}(x_m,t_m)v_m - \text{gen}_A(x_m,t_m)v_m\| + \lim_{m\to\infty} \|\text{gen}_A(x_m,t_m)v_m\| \\
&\quad \leq \limsup_{m\to\infty} \sup_{t\in\mathbb{T}^N} \|\text{gen}_{A'}(x_m,t) - \text{gen}_A(x_m,t)\| \\
&\quad = \|\Gamma(A) - \Gamma(A')\|_0 < \varepsilon.
\end{aligned}$$

Hence, $\|\text{gen}_{A'}(x_m,t_m)v_m\| < \varepsilon$ for all sufficiently large m. We further suppose without loss that the limit operator of A' with respect to the sequence x exists (otherwise we pass to a suitable subsequence of x). As in the proof of Proposition 2.4.32, we can find a unit vector $u \in l^2$ such that

$$\|V_{-x_m} A' V_{x_m}(v_m \otimes u)\| < 2\varepsilon \quad \text{for all sufficiently large } m$$

and, according to the definition of limit operators, we further have

$$\|(V_{-x_m} A' V_{x_m} - A'_x)(v \otimes u)\| \to 0$$

uniformly with respect to the unit vectors v. Hence, $\|A'_x(v_m \otimes u)\| < 3\varepsilon$ for all sufficiently large m. Since $\|v_m \otimes u\| = 1$, we conclude that

$$\text{either } A'_x \text{ is not invertible or } \|(A'_x)^{-1}\| > 1/(3\varepsilon). \tag{2.80}$$

On the other hand,

$$\|A_x - A'_x\| \leq \|A - A'\| < \varepsilon < 1/(4\,\|A_x^{-1}\|).$$

Thus, by Neumann series, A'_x is invertible, and

$$\|(A'_x)^{-1}\| \leq \frac{\|(A_x)^{-1}\|}{1 - \|(A_x)^{-1}\|\,\|A_x - A'_x\|} \leq \frac{\|(A_x)^{-1}\|}{1 - \varepsilon\,\|(A_x)^{-1}\|}.$$

Together with (2.80), this yields

$$\frac{1}{3\varepsilon} < \frac{\|(A_x)^{-1}\|}{1 - \varepsilon\,\|(A_x)^{-1}\|}$$

or, equivalently, $\varepsilon > 1/(4\,\|(A_x)^{-1}\|$. The obtained estimate contradicts the choice of ε. □

In a similar way, the following refinement of the local Fredholm criterion (Theorems 2.3.13 and 2.3.21) can be derived.

Theorem 2.4.35 *The following assertions are equivalent for $A \in \mathcal{A}_E(SO^{\$}_{L(H)})$:*

(a) *A is locally invertible at $\eta \in S^{N-1}$.*

(b) *The local coset $\pi_\eta(A)$ is invertible.*

(c) *All operators in local operator spectrum $\sigma_\eta(A)$ of A are invertible.*

2.4.9 Appendix B: A third proof of Theorem 2.4.27

Let $E = l^2 = l^2(\mathbb{Z}^N, \mathbb{C})$, i.e., we will again work in the Hilbert space context, and we will consider scalar-valued functions only. The latter assumption is for the sake of simplicity; the proof works as well for operators with entries in $\mathbb{C}^{k\times k}$ and for operators with entries in $\mathbb{C}I + K(H)$ where $K(H)$ is the ideal of the compact operators on an infinite-dimensional separable Hilbert space H.

Let $\mathcal{A}_E(SO)$ stand for the smallest closed subalgebra of $L(E)$ which contains all shift operators and all operators of multiplication by functions in SO. We will prove only the non-trivial direction of Theorem 2.4.27: an operator $A \in \mathcal{A}_E(SO)$ is Fredholm if all limit operators $A_h \in \sigma_{op}(A)$ are invertible.

Choose and fix a sequence of band operators $A_m := \sum_{k=0}^m a_{k,m} V_{\alpha_{k,m}}$ with $a_{k,m} \in SO$ which tends to A in the norm topology. Let SO_A denote the smallest closed subalgebra of SO which contains all functions $a_{k,m}$ together with their complex-conjugates as well as all functions in c_0 and the identity function $z \mapsto 1$. The algebra SO_A is a symmetric and separable closed subalgebra of SO. Let $M(SO_A)$ denote its maximal ideal space, and let $M^\infty(SO_A)$ refer to the fiber of $M(SO_A)$ containing all functionals $\eta \in M(SO_A)$ with $\eta(f) = 0$ for $f \in c_0$.

Let $h \in \mathcal{H}$ be a sequence for which the limit operator A_h exists. A diagonalization argument verifies the existence of a subsequence g of h such that the limit operators $(fI)_g$ exist for all $f \in SO_A$. By Proposition 2.4.1, $(fI)_g \in \mathbb{C}I$. Thus, the mapping $f \mapsto (fI)_g$ is a multiplicative (and non-trivial because $I_g = I$) functional on SO_A. In particular, there is an $\eta \in M^\infty(SO_A)$ such that

$$(fI)_g = \eta(f)I = \hat{f}(\eta)I \tag{2.81}$$

with $\hat{f}$ referring to the Gelfand transform of $f \in SO_A$.

Our next goal is to prove that, conversely, given $\eta \in M^\infty(SO_A)$, there is a sequence g such that $(fI)_g$ exists for all $f \in SO_A$ and that (2.81) holds.

Suppose there is an $\eta \in M^\infty(SO_A)$ for which there exists no sequence g such that (2.81) holds. Since SO_A is a separable algebra, its maximal ideal space is metrizable. Hence, there is a function f in SO_A with Gelfand transform $\hat{f} \in C(M(SO_A))$ such that $\hat{f}(\eta) = 0$ and $\hat{f}(\mu) \neq 0$ for all $\mu \neq \eta$ (for example, take $\hat{f}(\mu) := \text{dist}\,(\eta, \mu)$ where dist is a metric on SO_A which generates the Gelfand topology). The condition $\hat{f}(\mu) \neq 0$ for all $\mu \neq \eta$ guarantees that all limit operators of fI are invertible. Since all limit operators of fI are scalar multiples of the identity operator, Proposition 2.2.5 guarantees that these limit operators are *uniformly* invertible. Then, by Theorem 2.2.1, fI is a Fredholm operator on E. Thus, the coset $fI + K(E)$ is invertible in the Calkin algebra $L(E)/K(E)$ and, due to the inverse closedness of C^*-algebras, also in $(SO_A \cdot I + K(E))/K(E)$. The isomorphy

$$(SO_A \cdot I + K(E))/K(E) \cong SO_A \cdot I/(SO_A \cdot I \cap K(E)) \cong SO_A/c_0$$

shows that the coset $fI + c_0$ is invertible in SO_A/c_0. Since the maximal ideal space of the algebra SO_A/c_0 is homeomorphic to the fiber $M^\infty(SO_A)$, this observation implies that $\hat{f}(\mu) \neq 0$ for all $\mu \in M^\infty(SO_A)$ whence, in particular, $\hat{f}(\eta) \neq 0$. This contradiction proves our claim.

Now we employ Allan's local principle to localize the algebra $\mathcal{A}_E(SO)/K(E)$ over its central subalgebra $\{\pi(fI),\, f \in SO\}$. One can check as before that the maximal ideal space of this central subalgebra is homeomorphic to the fiber $M^\infty(SO)$.

Thus, to each $\tau \in M^\infty(SO)$, we associate a local ideal I_τ, a local algebra

$$\mathcal{A}_{E,\tau}(SO) := (\mathcal{A}_E(SO)/K(E))/I_\tau,$$

and a local homomorphisms $\pi_\tau : \mathcal{A}_E(SO) \to \mathcal{A}_{E,\tau}(SO)$. Then Theorem 2.3.16 states that an operator $A \in \mathcal{A}_E(SO)$ is Fredholm if and only if all cosets $\pi_\tau(A)$ with $\tau \in M^\infty(SO)$ are invertible. What remains to prove is that if all limit operators of A are invertible, then all cosets $\pi_\tau(A)$ with $\tau \in M^\infty(SO)$ are invertible.

The maximal ideals $\eta \in M^\infty(SO_A)$ define a fibration of $M^\infty(SO)$ into fibers $M^\infty_\eta(SO)$. Let $\tau \in M^\infty_\eta(SO)$, and let g be a sequence such that $(fI)_g$ exists for all $f \in SO_A$ and which is connected with η via (2.81). Then, trivially, A_g exists, and one has

$$\pi_\tau(A) = \pi_\tau(A_g). \tag{2.82}$$

Indeed, this identity holds for all multiplication operators fI with $f \in SO_A$ in place of A, and it also holds for all shift operators V_α in place of A. Let $\mathcal{A}_E(SO_A)$ stand for the smallest closed subalgebra of $L(E)$ which is generated by all operators of multiplication by functions in SO_A and by all shift operators. Since the mappings π_τ and $A \mapsto A_g$ are continuous algebra homomorphisms on $\mathcal{A}_E(SO_A)$, (2.82) also holds for all operators which belong to this algebra and, in particular, for the operator A itself. Since A_g is invertible by assumption, the coset $\pi_\tau(A_g) = \pi_\tau(A)$ is invertible, too, and we are done. □

As a by-product, we have got a version of the symbol calculus for the Fredholmness of operators in $\mathcal{A}_E(SO)$, where the symbol of an operator $A \in \mathcal{A}_E(SO)$ is a scalar function on $M^\infty(SO) \times \mathbb{T}^N$. Given $\tau \in M^\infty(SO)$, determine $\eta \in M^\infty(SO_A)$ such that τ belongs to the fiber $M^\infty_\eta(SO)$, and choose a sequence g such that A_g and $(fI)_g$ exist for all $f \in SO_A$ and which is related to η by (2.82). Consider the function $A^\dagger : M^\infty(SO) \to \mathcal{A}_E(\mathbb{C})$, $\tau \mapsto A_g$. Since η is uniquely determined by τ, and since there is an one-to-one correspondence between the η and the g as we have checked in the preceding proof, this function is well defined. Further, for every operator $B \in \mathcal{A}_E(\mathbb{C})$, let $\hat{B} : \mathbb{T}^N \to \mathbb{C}$ denote its Gelfand transform. Then we define the *symbol* of A as the function

$$\operatorname{smb} A : M^\infty(SO) \times \mathbb{T}^N \to \mathbb{C}, \quad (\tau, \xi) \mapsto \widehat{A^\dagger(\tau)}(\xi). \tag{2.83}$$

For example, for band operators $A = \sum_{|\alpha| \le k} a_\alpha V_\alpha$, we simply have

$$(\operatorname{smb} A)(\tau, \xi) = \sum_{|\alpha| \le k} \widehat{a_\alpha}(\tau)\xi^\alpha.$$

Then, in the special setting of this section, Theorem 2.4.27 can be restated as follows:

Corollary 2.4.36 *An operator $A \in \mathcal{A}_E(SO)$ is Fredholm if and only if its symbol, defined by* (2.83)*, is invertible on $M^\infty(SO) \times \mathbb{T}^N$.*

2.5 Operators in the discrete Wiener algebra

Here, we will examine another class of band-dominated operators for which Theorem 2.2.1 is valid without requiring the uniform boundedness of the inverses of the limit operators. The entries of the operators in this class satisfy a condition of Wiener type. It is essential for our approach to take into account also operators acting on l^∞ spaces. Thus, throughout this section, we let again E^∞ stand for one of the Banach spaces l^p with $1 \le p \le \infty$ and c^0.

2.5.1 The Wiener algebra

Let $(a_\alpha)_{\alpha \in \mathbb{Z}^N}$ be a sequence of functions in $l^\infty(\mathbb{Z}^N, L(X))$ satisfying

$$\sum_{\alpha \in \mathbb{Z}^N} \|a_\alpha\|_\infty < \infty. \tag{2.84}$$

Then the series $\sum_{\alpha \in \mathbb{Z}^n} a_\alpha V_\alpha$ converges in the norm of $L(E^\infty)$, and

$$\left\| \sum_{\alpha \in \mathbb{Z}^N} a_\alpha V_\alpha \right\|_{L(E^\infty)} \le \sum_{\alpha \in \mathbb{Z}^N} \|a_\alpha\|_\infty. \tag{2.85}$$

We write $\mathcal{W}$ for the set of all operators $A = \sum_{\alpha \in \mathbb{Z}^N} a_\alpha V_\alpha$ with coefficient functions a_α satisfying (2.84). Provided with the usual operations and the norm

$$\|A\|_{\mathcal{W}} := \sum_{\alpha \in \mathbb{Z}^N} \|a_\alpha\|_\infty,$$

the set $\mathcal{W}$ becomes a Banach algebra, the so-called *Wiener algebra*. By (2.85), the Wiener algebra is continuously embedded into $L(E^\infty, \mathcal{P})$ and, hence, into $\mathcal{A}_{E^\infty}$ for all choices of E^∞. The intersection $\mathcal{W} \cap L^\$(E^\infty, \mathcal{P})$ is called the *rich Wiener algebra* and will be denoted by $\mathcal{W}^\$$. Further, given a closed subalgebra $\mathcal{B}$ of $l^\infty(\mathbb{Z}^N, L(X))$ we let $\mathcal{W}(\mathcal{B})$ refer to the set of all operators in $\mathcal{W}$ with coefficients in $\mathcal{B}$.

Later on, we will also have to deal with Wiener algebras on certain function spaces on $\mathbb{R}^N$. In this setting, we will refer to the Wiener algebra $\mathcal{W}$ on the sequence spaces as the *discrete Wiener algebra.*

Proposition 2.5.1 $\mathcal{W}^\$ = \mathcal{W}(l^\infty(\mathbb{Z}^N, L(X))^\$)$.

Proof. The inclusion $\mathcal{W}(l^\infty(\mathbb{Z}^N, L(X))^\$) \subseteq \mathcal{W}^\$$ is evident since $l^\infty(\mathbb{Z}^N, L(X))^\$ \subset L^\$(E^\infty, \mathcal{P})$ by definition and since $L^\$(E^\infty, \mathcal{P})$ is a closed algebra by Proposition 1.2.8. If, conversely, A is an operator in $\mathcal{W}^\$$, then each of its diagonals a_α must be in $l^\infty(\mathbb{Z}^N, L(X))^\$$. Because of

$$A = \lim_{k \to \infty} \sum_{|\alpha| \le k} a_\alpha V_\alpha,$$

the operator A belongs to $\mathcal{W}(l^\infty(\mathbb{Z}^N, L(X))^\$)$. □

This proposition also shows that the algebra $\mathcal{W}^\$$ is independent of the concrete choice of the underlying space E^∞ (provided that X is fixed).

Theorem 2.5.2 *The Wiener algebra $\mathcal{W}$ is inverse closed in $L(E^\infty)$.*

We prepare the proof by a multi-dimensional version of the classical Wiener theorem which holds for functions with operator-valued coefficients.

Given a Banach algebra $\mathcal{B}$ with identity e, let $C(\mathbb{T}^N, \mathcal{B})$ refer to the set of all continuous functions $f : \mathbb{T}^N \to \mathcal{B}$, and write $W(\mathbb{T}^N, \mathcal{B})$ for the set of all functions $f : \mathbb{T}^N \to \mathcal{B}$ which are given by a series

$$f(z) = \sum_{n\in\mathbb{Z}^N} z^n A_n \quad \text{with } A_n \in \mathcal{B} \text{ and } \sum_{n\in\mathbb{Z}^N} \|A_n\| < \infty.$$

Here we use the notation

$$z^n := z_1^{n_1} \cdots z_N^{n_N} \quad \text{when } z = (z_1, \dots, z_N) \text{ and } n = (n_1, \dots, n_N).$$

Clearly, $C(\mathbb{T}^N, \mathcal{B})$ is a Banach algebra with respect to pointwisely defined operations and the supremum norm, and $W(\mathbb{T}^N, \mathcal{B})$ also becomes a Banach algebra when provided with the norm $\|f\|_W := \sum_n \|A_n\|$. It is also evident that $W(\mathbb{T}^N, \mathcal{B})$ is a (non-closed) subalgebra of $C(\mathbb{T}^N, \mathcal{B})$ which contains the identity function.

Theorem 2.5.3 *The algebra $W(\mathbb{T}^N, \mathcal{B})$ is inverse closed in $C(\mathbb{T}^N, \mathcal{B})$.*

Proof. If $N > 1$, we can identify the algebras $C(\mathbb{T}^N, \mathcal{B})$ and $C(\mathbb{T}, C(\mathbb{T}^{N-1}, \mathcal{B}))$ in a canonical way: the function $f \in C(\mathbb{T}^N, \mathcal{B})$ corresponds to the function

$$z_1 \quad \mapsto \quad (z' \mapsto f(z_1, z')) \quad \text{in} \quad C(\mathbb{T}, C(\mathbb{T}^{N-1}, \mathcal{B})).$$

Here, as usual, $z = (z_1, z') \in \mathbb{T} \times \mathbb{T}^{N-1}$. Similarly, the algebras $W(\mathbb{T}^N, \mathcal{B})$ and $W(\mathbb{T}, W(\mathbb{T}^{N-1}, \mathcal{B}))$ can be identified: The function

$$z \mapsto \sum_{n\in\mathbb{Z}^N} z^n A_n \quad \text{in} \quad W(\mathbb{T}^N, \mathcal{B})$$

corresponds to the function

$$z_1 \quad \mapsto \quad \left(z' \mapsto \sum_{n'\in\mathbb{Z}^{N-1}} (z')^{n'} \sum_{z_1\in\mathbb{Z}} z_1^{n_1} A_{n_1,n'} \right) \quad \text{in} \quad W(\mathbb{T}, W(\mathbb{T}^{N-1}, \mathcal{B})).$$

Thus, we are left with proving the assertion in case $N = 1$. We will do this by having recourse to Allan's local principle (Theorem 2.3.16). The algebra $W(\mathbb{T}, \mathbb{C})$ lies in the center of $W(\mathbb{T}, \mathcal{B})$, and it is a classical result in the Gelfand theory of commutative Banach algebras that the maximal ideal space of $W(\mathbb{T}, \mathbb{C})$ is homeomorphic with $\mathbb{T}$, with $t_0 \in \mathbb{T}$ corresponding to the maximal ideal

$$\{f \in W(\mathbb{T}, \mathbb{C}) : f(t_0) = 0\} \tag{2.86}$$

of $W(\mathbb{T}, \mathbb{C})$. For $t_0 \in \mathbb{T}$, let $\mathcal{J}_{t_0}$ stand for the smallest closed ideal of $W(\mathbb{T}, \mathcal{B})$ which contains the ideal (2.86). We claim that every function $f \in W(\mathbb{T}, \mathcal{B})$ is locally equivalent at t_0 to the constant function $t \mapsto f(t_0)$, i.e., that

$$f - f(t_0) \in \mathcal{J}_{t_0}. \tag{2.87}$$

Once (2.87) is verified, the assertion follows from the local principle. Indeed, if f is invertible in $C(\mathbb{T}, \mathcal{B})$, then $f(t_0)$ is invertible in $\mathcal{B}$ for every t_0. This implies the invertibility of the constant function $t \mapsto f(t_0)$ in $W(\mathbb{T}, \mathcal{B})$ for every t_0, whence via (2.87) the local invertibility of f in $W(\mathbb{T}, \mathcal{B})$ at every point $t_0 \in \mathbb{T}$.

So we are left with verifying (2.87) for all functions $f \in W(\mathbb{T}, \mathcal{B})$. Since every function in $W(\mathbb{T}, \mathcal{B})$ can be approximated by finite sums of functions of the form

$$f(z) = z^k A \quad \text{with } k \in \mathbb{N} \text{ and } A \in \mathcal{B}, \tag{2.88}$$

and since $\mathcal{J}_{t_0}$ is closed, it suffices to verify (2.87) for functions of the form (2.88). But for these functions, the assertion (2.87) is evident: We have

$$f(z) - f(t_0) = (z^k - t_0^k)A,$$

the function $z \mapsto z^k - t_0^k$ lies in the maximal ideal (2.86), hence, the function $z \mapsto (z^k - t_0^k)A$ lies in the local ideal $\mathcal{J}_{t_0}$. □

Proof of Theorem 2.5.2. We will work with the matrix representation of an operator. In particular, E_i and R_j are as in Section 2.1.2.

Let the operator $A = \sum_{n \in \mathbb{Z}^N} a_n V_n \in \mathcal{W}$ be invertible in $L(E^\infty)$. We associate with A the operator-valued function

$$\hat{A} : \mathbb{T}^N \to L(E^\infty), \quad z \mapsto \sum_{n \in \mathbb{Z}^N} z^{-n} a_n V_n.$$

Evidently, $\hat{A} \in W(\mathbb{T}^N, L(E^\infty))$. Observe that actually $\hat{A}(z) \in \mathcal{W} \subset L(E^\infty, \mathcal{P})$, implying that every operator $\hat{A}(z)$ is uniquely determined by its matrix representation (Proposition 2.1.1). For the ijth entry of this matrix representation, we find

$$\begin{aligned} R_i S_i \hat{A}(z) S_j E_j &= R_i S_i \sum_{n \in \mathbb{Z}^N} z^{-n} a_n V_n S_j E_j \\ &= \sum_{n \in \mathbb{Z}^N} z^{-n} R_i S_i a_n V_n S_j E_j \\ &= \sum_{n \in \mathbb{Z}^N} z^{-n} R_i S_i a_n S_i S_{j+n} V_n S_j E_j \\ &= z^{j-i} R_i S_i a_{i-j} V_{i-j} S_j E_j \\ &= z^{j-i} R_i S_i A S_j E_j. \end{aligned}$$

Thus, if M_z denotes the multiplication operator

$$M_z : E^\infty \to E^\infty, \quad (u_k)_{k\in\mathbb{Z}^N} \mapsto (z^k u_k)_{k\in\mathbb{Z}^N},$$

then

$$\hat{A}(z) = M_z^{-1} A M_z. \tag{2.89}$$

For every $z \in \mathbb{T}^N$, the operator M_z is invertible on each space E^∞. Hence, the operators $\hat{A}(z)$ are invertible on $L(E^\infty)$, which yields the invertibility of the function $\hat{A}$ in $C(\mathbb{T}^N, L(E^\infty))$. Let $B(z) := \hat{A}(z)^{-1}$. By Theorem 2.5.3, the function B lies in $W(\mathbb{T}^N, L(E^\infty))$. Thus, there are operators $B_n \in L(E^\infty)$ such that

$$B(z) = \sum_{n\in\mathbb{Z}^N} z^n B_n \quad \text{and} \quad \sum_{n\in\mathbb{Z}^N} \|B_n\| < \infty. \tag{2.90}$$

We write $B_n =: C_n V_n$ and claim that all operators C_n are multiplication operators. Once this is done, the assertion will follow from

$$A^{-1} = B(\mathbf{1}) = \sum_{n\in\mathbb{Z}^N} C_n V_n \in \mathcal{W}$$

with $\mathbf{1}$ referring to the N-tuple $(1, \ldots, 1)$.

To get the claim, observe first that every operator C_n is uniquely determined by its matrix representation. To see this, we have to repeat the above arguments in more detail. The point is that all operators $\hat{A}(z)$ are band-dominated. Since $\mathcal{A}_{E^\infty}$ is an inverse-closed subalgebra of $L(E^\infty)$ by Proposition 2.1.8, these operators are invertible in $\mathcal{A}_{E^\infty}$. Hence, the function $\hat{A}$ has an inverse B in $C(\mathbb{T}^N, \mathcal{A}_{E^\infty})$, and by Theorem 2.5.3, this inverse belongs to $W(\mathbb{T}^N, \mathcal{A}_{E^\infty})$. This shows that the operators B_n and, thus, the operators C_n are band-dominated. But all band-dominated operators lie in $L(E^\infty, \mathcal{P})$, i.e., they are uniquely determined by their matrix representation due to Proposition 2.1.1. Similar arguments show that every operator $B(z)$ is uniquely determined by its matrix representation.

Now recall that the coefficients B_n of the series B in (2.90) are given by

$$B_n = \frac{1}{(2\pi i)^N} \int_{\mathbb{T}^N} z^{-\mathbf{1}-n} \hat{A}(z)^{-1}\, dz$$

([83], Chapter I, Section 2.2). Hence,

$$C_n = B_n V_{-n} = \frac{1}{(2\pi i)^N} \int_{\mathbb{T}^N} z^{-\mathbf{1}-n} M_z^{-1} B(\mathbf{1}) M_z V_{-n}\, dz.$$

If $(B_{jk})_{j,\, k\in\mathbb{Z}^N}$ is the matrix representation of $B(\mathbf{1})$, we get further

$$\begin{aligned} C_n &= \frac{1}{(2\pi i)^N} \int_{\mathbb{T}^N} z^{-\mathbf{1}-n} (z^{k-j} B_{jk})_{j,\, k\in\mathbb{Z}^N} V_{-n}\, dz \\ &= \frac{1}{(2\pi i)^N} \int_{\mathbb{T}^N} (z^{-\mathbf{1}-n+k-j} B_{jk})_{j,\, k\in\mathbb{Z}^N} V_{-n}\, dz \\ &= \frac{1}{(2\pi i)^N} \int_{\mathbb{T}^N} (z^{-\mathbf{1}+k-j} B_{j,\, k+n})_{j,\, k\in\mathbb{Z}^N}\, dz. \end{aligned}$$

Thus, all entries C_{jk} in the matrix representation of C_n with $j \neq k$ vanish. Consequently, by Proposition 2.1.3, every C_n is an operator of multiplication. $\square$

As an immediate consequence we obtain:

Corollary 2.5.4 *Let $A \in \mathcal{W}$ be invertible on one of the spaces E^∞. Then A is invertible on all of these spaces, and the norms of the corresponding inverses are uniformly bounded.*

Indeed, if A is invertible on one of the spaces E^∞, then $A^{-1} \in \mathcal{W}$ by Theorem 2.5.2, and from $\|A^{-1}\|_{L(E^\infty)} \leq \|A^{-1}\|_\mathcal{W}$ we conclude that A^{-1} is the inverse for A on each of the spaces E^∞ and that the norm of A^{-1} in $L(E^\infty)$ is bounded by $\|A^{-1}\|_\mathcal{W}$. $\square$

Corollary 2.5.5 *For each of the spaces E, the rich Wiener algebra $\mathcal{W}^\$$ is inverse closed in each of the algebras $L(E, \mathcal{P})$ and $L(E)$.*

Indeed, this follows immediately from Theorem 2.5.2 and Proposition 1.2.8.

2.5.2 Fredholmness of operators in the Wiener algebra

Here is what can be said about limit operators of rich operators in the Wiener algebra.

Proposition 2.5.6 *Let $A \in \mathcal{W}^\$$ and let $h \subseteq \mathbb{Z}^N$ be a sequence tending to infinity. Then there is a subsequence g of h such that the limit operator A_g exists with respect to all spaces E^∞. This limit operator belongs to $\mathcal{W}$, and $\|A_g\|_\mathcal{W} \leq \|A\|_\mathcal{W}$.*

Proof. Let $A = \sum_{\alpha\in\mathbb{Z}^N} a_\alpha V_\alpha$ with $\sum_{\alpha\in\mathbb{Z}^N} \|a_\alpha\| < \infty$. We know from Proposition 2.5.1 that all diagonals a_α are rich multiplication operators. Thus, a diagonal argument yields the existence of a subsequence g of h such that the limit operators $(a_\alpha I)_g$ exist with respect to E^∞ for all α. These limit operators are again operators of multiplication by certain functions $a_{\alpha,g}$ (Proposition 2.1.10) and, by Proposition 1.1.17 (a),

$$\|a_{\alpha,g}\|_\infty = \|(a_\alpha I)_g\|_{L(E^\infty)} \leq \|a_\alpha\|_\infty.$$

Thus,

$$\sum_{\alpha\in\mathbb{Z}^N} \|a_{\alpha,g}\|_\infty < \infty,$$

and the operator $A_g := \sum_{\alpha\in\mathbb{Z}^N} a_{\alpha,g} V_\alpha$ is well defined. This operator belongs to the Wiener algebra $\mathcal{W}$, and $\|A_g\|_\mathcal{W} \leq \|A\|_\mathcal{W}$. Now it is evident that A_g is indeed the limit operator of A with respect to the sequence g in each of the spaces E^∞. $\square$

The main result of this section is the following theorem which states that, for rich operators A in the Wiener algebra, the uniform boundedness condition from Theorem 2.2.1,

$$\sup\{\|(A_h)^{-1}\| : A_h \in \sigma(A)\} < \infty,$$

is automatically satisfied if all limit operators of A are invertible.

Theorem 2.5.7 *Let X be a reflexive Banach space. Then the following assertions are equivalent for every operator $A \in \mathcal{W}^\$$:*

(a) *There is a space E such that A is $\mathcal{P}$-Fredholm on E.*

(b) *There is a space E such that all limit operators of A are invertible on E.*

(c) *All limit operators of A are invertible on $l^\infty(\mathbb{Z}^N, X)$.*

(d) *All limit operators of A are invertible on $l^\infty(\mathbb{Z}^N, X)$, and the norms of their inverses are uniformly bounded.*

(e) *All limit operators of A are invertible on E^∞ for all spaces E^∞, and the norms of their inverses are uniformly bounded.*

(f) *The operator A is $\mathcal{P}$-Fredholm on all spaces E.*

Proof. (a) $\Rightarrow$ (b): This is Proposition 1.2.9.
(b) $\Rightarrow$ (c): Let A_h be a limit operator of A with respect to the Banach space E. If A_h is invertible on E, then A_h^{-1} is in the Wiener algebra $\mathcal{W}$ by Proposition 2.5.6 and Theorem 2.5.2, and $A_h^{-1} \in L(l^\infty(\mathbb{Z}^N, X))$ by Corollary 2.5.4.
(c) $\Rightarrow$ (d): Let $\chi : \mathbb{R}^N \to [0,\, 1]$ be a continuous function which is identically 1 in a certain neighborhood of 0 and which vanishes outside the cube $[-1,\, 1]^N$. Further, given a positive integer k, define the function χ_k by $\chi_k(x) := \chi(x/k)$, and let T_k refer to the operator of multiplication by the restriction of the function χ_k onto $\mathbb{Z}^N$. We claim that there are constants $C > 0$ and $k \in \mathbb{N}$ such that

$$\|u\|_\infty \leq C\,(\|Au\|_\infty + \|T_k u\|_\infty) \quad \text{for all } u \in l^\infty(\mathbb{Z}^N, X). \tag{2.91}$$

The claim is evidently equivalent to the existence of constants C, k such that

$$1/C \leq \|Au\|_\infty + \|T_k u\|_\infty \quad \text{for all unit vectors } u \in l^\infty(\mathbb{Z}^N, X).$$

Assume, such constants do not exist. Then, for all $C > 0$ and $k \in \mathbb{N}$, there exists a vector $u_{k,C} \in l^\infty(\mathbb{Z}^N, X)$ with $\|u_{k,C}\|_\infty = 1$ such that

$$1/C > \|Au_{k,C}\|_\infty + \|T_k u_{k,C}\|_\infty.$$

In particular, we can choose $C = k$, i.e., for each $k \in \mathbb{N}$, there is a $u_k \in l^\infty(\mathbb{Z}^N, X)$ with $\|u_k\|_\infty = 1$ such that

$$1/k > \|Au_k\|_\infty + \|T_k u_k\|_\infty. \tag{2.92}$$

From $\|u_k\|_\infty = 1$ and $\|T_k u_k\|_\infty < 1/k$ we conclude the existence of points $x_k \in \mathbb{Z}^N$ such that

$$\|u_k(x_k)\|_{L(X)} \geq 1/2 \quad \text{and} \quad |x_k| \to \infty.$$

Let h be the sequence $h(m) := x_m$. Since A is rich, there is a subsequence g of h for which the limit operator A_g exists. Let $v_m := V_{-g(m)} u_{g(m)}$. Then, for arbitrary

$k, m \in \mathbb{N}$,

$$\begin{aligned}
\|A_g T_k v_m\| &\leq \|(A_g - V_{-g(m)}AV_{g(m)})T_k\| \, \|v_m\| + \|V_{-g(m)}AV_{g(m)}T_k v_m\| \\
&\leq \|(A_g - V_{-g(m)}AV_{g(m)})T_k\| \\
&\quad + \|(V_{-g(m)}AV_{g(m)}T_k - T_k V_{-g(m)}AV_{g(m)})v_m\| \\
&\quad + \|T_k V_{-g(m)}AV_{g(m)}v_m\| \\
&\leq \|(A_g - V_{-g(m)}AV_{g(m)})T_k\| \\
&\quad + \|V_{-g(m)}AV_{g(m)}T_k - T_k V_{-g(m)}AV_{g(m)}\| + \|Au_{g(m)}\|.
\end{aligned} \tag{2.93}$$

Let $\varepsilon > 0$ be arbitrary. Then choose and fix k such that the second term on the right-hand side of estimate (2.93) becomes less than ε for all m, which can be done due to Theorem 2.1.6 (c). Now choose $m > 1/\varepsilon$ so large that the first term in (2.93) also becomes less than ε. Since $\|Au_m\| < 1/m$ by (2.92), then the third term in (2.93) is less than ε, too. Thus,

$$\forall \varepsilon > 0 \, \exists k, m \in \mathbb{N} : \|A_g T_k v_m\|_\infty \leq 3\varepsilon. \tag{2.94}$$

On the other hand, $\|v_m(0)\| = \|u_{g(m)}(g(m))\| \geq 1/2$, whence $\|T_k v_m\|_\infty \geq 1/2$. Thus, by (2.94), and since all limit operators of A are invertible by hypothesis,

$$1/2 \leq \|T_k v_m\|_\infty \leq \|A_g^{-1}\| \, \|A_g T_k v_m\|_\infty \leq 3\varepsilon \, \|A_g^{-1}\|$$

whence

$$\|A_g^{-1}\| \geq 1/(6\varepsilon) \quad \text{for all } \varepsilon > 0.$$

This is clearly impossible, and our claim (2.91) is proved. We will now employ (2.91) to prove the uniform boundedness of the inverses of the limit operators of A on $l^\infty(\mathbb{Z}^N, X)$. From (2.91) we conclude that, for all $u \in l^\infty(\mathbb{Z}^N, X)$, $r \in \mathbb{N}$ and $l \in \mathbb{Z}^N$,

$$\|V_l T_r u\|_\infty \leq C(\|AV_l T_r u\|_\infty + \|T_k V_l T_r u\|_\infty).$$

Let $h \in \mathcal{H}$ be a sequence for which the limit operator A_h exists. Since every V_l is an isometry, we get

$$\|T_r u\|_\infty \leq C(\|V_{-h(m)}AV_{h(m)}T_r u\|_\infty + \|V_{-h(m)}T_k V_{h(m)}T_r u\|_\infty). \tag{2.95}$$

Further, since $T_r u \in c_0(\mathbb{Z}^N, X)$ and $V_{-h(m)}T_k V_{h(m)} \to 0$ strongly on $c_0(\mathbb{Z}^N, X)$, we can pass to the limit as $m \to \infty$ in (2.95) to obtain

$$\|T_r u\|_\infty \leq C\|A_h T_r u\|_\infty \tag{2.96}$$

for all $u \in l^\infty(\mathbb{Z}^N, X)$ and $r \in \mathbb{N}$. For $r \to \infty$, the left-hand side of (2.96) goes to $\|u\|_\infty$. For the right-hand side, some more care is in order. From Theorem 2.1.6 we conclude that the right-hand side of

$$| \, \|A_h T_r u\| - \|T_r A_h u\| \, | \leq \|A_h T_r - T_r A_h\| \, \|u\|$$

tends to zero as $r \to \infty$ (recall that A_h is band dominated by Proposition 2.1.10 and use the implication (a) $\Rightarrow$ (c) of Theorem 2.1.6). Since $\|T_r A_h u\| \to \|A_h u\|$ as $r \to \infty$, this estimate implies that $\|A_h T_r u\| \to \|A_h u\|$ as $r \to \infty$. Thus, passage to the limit $r \to \infty$ in (2.96) gives

$$\|u\|_\infty \leq C\|A_h u\|_\infty \quad \text{for all } u \in l^\infty(\mathbb{Z}^N, X)$$

whence $\|A_h^{-1}\| \leq C$, i.e., the uniform boundedness of the inverses of the limit operators.

(d) $\Rightarrow$ (e): The proof of this implication is based on the possibility to associate with every operator in the Wiener algebra a naturally defined adjoint operator. To make this point clear we will indicate the dependence of the Wiener algebra from the underlying Banach space X by writing $\mathcal{W}_X$ in place of $\mathcal{W}$. For each operator $A = \sum_{\alpha \in \mathbb{Z}^N} a_\alpha V_\alpha \in \mathcal{W}_X$, we define its *Wiener adjoint* $A^\star$ as $\sum_{\alpha \in \mathbb{Z}^N} V_{-\alpha} a_\alpha^* I$, where $a_\alpha^*(x)$ is the usual Banach dual operator of $a_\alpha(x)$, acting on X^*. Clearly, $A^\star = \sum_{\alpha \in \mathbb{Z}^N} b_\alpha V_{-\alpha}$, where $b_\alpha(x) = a_\alpha^*(x + \alpha)$. Hence, $A^\star$ belongs to the Wiener algebra $\mathcal{W}_{X^*}$, and it is easy to check that the mapping $A \mapsto A^\star$ is an anti-linear isometry from $\mathcal{W}_X$ into $\mathcal{W}_{X^*}$ which satisfies $(AB)^\star = B^\star A^\star$ for all $A, B \in \mathcal{W}_X$. In particular, $I^\star = I$ and, if A is invertible in $\mathcal{W}_X$, then $A^\star$ is invertible in $\mathcal{W}_{X^*}$ and $(A^\star)^{-1} = (A^{-1})^\star$.

For the proof of the implication (d) $\Rightarrow$ (e), let now $A \in \mathcal{W}_X^\$$ be an operator with

$$C_\infty(A) := \sup\{\|A_h^{-1}\|_{L(l^\infty(\mathbb{Z}^N, X))} : A_h \in \sigma_{op}(A)\} < \infty. \tag{2.97}$$

The limit operators of $A^\star$ are just the Wiener adjoints of the limit operators of A. Thus, the invertibility of all limit operators of A implies the invertibility of all limit operators of $A^\star$. So we conclude from the already established implication (c) $\Rightarrow$ (d) that

$$C_\infty(A^\star) := \sup\{\|(A_h^\star)^{-1}\|_{L(l^\infty(\mathbb{Z}^N, X^*))} : A_h \in \sigma_{op}(A)\} < \infty.$$

Since the limit operators of $A^\star$ as well as their inverses belong to the Wiener algebra $\mathcal{W}_{X^*}$ (Proposition 2.5.6 and Theorem 2.5.2), the operators $A_h^\star$ also act as bounded and invertible operators on $c_0(\mathbb{Z}^N, X^*)$, and $\|(A_h^\star)^{-1}\|_{L(c_0(\mathbb{Z}^N, X^*))} \leq \|(A_h^\star)^{-1}\|_{L(l^\infty(\mathbb{Z}^N, X^*))}$. This shows that

$$C_0(A^\star) := \sup\{\|(A_h^\star)^{-1}\|_{L(c_0(\mathbb{Z}^N, X^*))} : A_h \in \sigma_{op}(A)\} < \infty. \tag{2.98}$$

The operator A, thought of as acting on $l^1(\mathbb{Z}^N, X)$, can be identified with the usual Banach dual operator of $A^\star \in L(c_0(\mathbb{Z}^N, X^*))$ (this is the place where we need the reflexivity of X). Hence,

$$C_1(A) := \sup\{\|A_h^{-1}\|_{L(l^1(\mathbb{Z}^N, X))} : A_h \in \sigma_{op}(A)\} = C_0(A^\star) < \infty.$$

Consequently, by the Riess-Thorin interpolation theorem (Theorem 1 and Remark 4 in Section 1.18.3 of [182]), we have for every $1 < p < \infty$ and $A_h \in \sigma(A)$,

$$\|A_h^{-1}\|^p_{L(l^p(\mathbb{Z}^N, X))} \leq \|A_h^{-1}\|^{p-1}_{L(l^\infty(\mathbb{Z}^N, X))} \|A_h^{-1}\|_{L(l^1(\mathbb{Z}^N, X))} \leq C_\infty(A)^{p-1} C_1(A),$$

which verifies the uniform boundedness of the norms of the inverses of the limit operators of A on all spaces $l^p(\mathbb{Z}^N, X)$ with $1 \le p \le \infty$. For $E^\infty = c_0(\mathbb{Z}^N, X)$, this result follows in the same way as we derived (2.98).

Finally, the implication (e) $\Rightarrow$ (f) is Theorem 2.2.1, and the implication (f) $\Rightarrow$ (a) is evident. □

Observe that the implication (c) $\Rightarrow$ (d) holds for arbitrary rich operators A and arbitrary (not necessarily reflexive) Banach spaces X.

Corollary 2.5.8 *Let X be a reflexive Banach space. Then the $\mathcal{P}$-essential spectrum of an operator $A \in \mathcal{W}^\$$, considered as an operator on E^∞, does not depend on the space E^∞, and*

$$\sigma_{\mathcal{P}-ess}(A) = \cup \sigma_{E^\infty}(A_h) = \cup \sigma_{\mathcal{W}}(A_h)$$

where the unions are taken over all limit operators A_h of A.

If the space X is finite-dimensional, then the $\mathcal{P}$-essential spectrum is the usual essential spectrum. The proof of the independence of the $\mathcal{P}$-essential spectrum of the underlying space follows from Theorem 2.5.2 and from the fact that limit operators of operators in the Wiener algebra belong to the Wiener algebra again.

2.6 Band-dominated operators with special coefficients

In this section we are going to specialize the Fredholm criteria obtained for operators in $\mathcal{A}_E^\$$ to some concrete classes of band-dominated operators.

2.6.1 Band-dominated operators with almost periodic coefficients

A function a in $l^\infty(\mathbb{Z}^N, L(X))$ is called *almost periodic* if the set of all multiplication operators $V_{-k}aV_k$ with $k \in \mathbb{Z}^N$ is relatively compact with respect to the norm topology on $L(E)$. We denote the set of all almost periodic functions by $AP_{L(X)}$. An archetypal example of an almost periodic function is $x \mapsto \exp(i\alpha x)$ where $\alpha \in \mathbb{R}$. It is easy to check that $AP_{L(X)}$ is a closed subalgebra of $l^\infty(\mathbb{Z}^N, L(X))$ which is invariant with respect to shifts, i.e., if $a \in AP_{L(X)}$, then the function $x \mapsto a(x-k)$ lies in $AP_{L(X)}$ for every $k \in \mathbb{Z}^N$.

Let $\mathcal{A}_E(AP_{L(X)})$ stand for the smallest closed subalgebra of $L(E)$ which contains all band operators with coefficients in $AP_{L(X)}$.

Proposition 2.6.1 *Let $A \in \mathcal{A}_E(AP_{L(X)})$. Then every sequence $h \in \mathcal{H}$ possesses a subsequence g for which the limit operator A_g exists and*

$$\|V_{-g(m)}AV_{g(m)} - A_g\| \to 0 \quad \text{as } m \to \infty. \tag{2.99}$$

In particular, all operators in $\mathcal{A}_E(AP_{L(X)})$ have a rich operator spectrum.

Proof. It is an immediate consequence of the definition that every operator of multiplication by an almost periodic function a has a rich operator spectrum and that every sequence $h \in \mathcal{H}$ possesses a subsequence g for which the limit operator $(aI)_g$ exists and

$$\|V_{-g(m)} a V_{g(m)} - (aI)_g\| \to 0 \quad \text{as } m \to \infty.$$

Since the set $L^\$(E, \mathcal{P})$ of all operators with rich operator spectrum is a closed subalgebra of $L(E)$ (Proposition 1.2.8), we obtain the inclusion $\mathcal{A}_E(AP_{L(X)}) \subset L^\$(E, \mathcal{P})$.

Now let $A \in \mathcal{A}_E(AP_{L(X)})$, and let $h \in \mathcal{H}$ be a sequence for which the limit operator A_h exists. Let further (A_n) be a sequence of band operators in $\mathcal{A}_E(AP_{L(X)})$ which converges in the norm of $L(E)$ to A. Then, by a standard diagonalization procedure, one finds a subsequence g of h such that the limit operators $(A_n)_g$ exist for every n and that

$$\|V_{-g(m)} A_n V_{g(m)} - (A_n)_g\| \to 0 \quad \text{as } m \to \infty.$$

The estimate

$$\begin{aligned}\|A_h - V_{-g(m)} A V_{g(m)}\| &\leq \|A_g - (A_n)_g\| + \|(A_n)_g - V_{-g(m)} A_n V_{g(m)}\| \\ &\quad + \|V_{-g(m)} A_n V_{g(m)} - V_{-g(m)} A V_{g(m)}\| \\ &\leq 2\|A - A_n\| + \|(A_n)_g - V_{-g(m)} A_n V_{g(m)}\|\end{aligned}$$

yields the assertion. □

The basic Fredholm properties of operators in $\mathcal{A}_E(AP_{L(X)})$ are consequences of the norm convergence (2.99).

Theorem 2.6.2 *For $A \in \mathcal{A}_E(AP_{L(X)})$, the following assertions are equivalent:*

(a) *A is $\mathcal{P}$-Fredholm.*

(b) *All limit operators of A are invertible.*

(c) *At least one limit operator of A is invertible.*

(d) *A is invertible.*

Theorem 2.6.3 *For $A \in \mathcal{A}_E(AP_{L(X)})$, the following assertions are equivalent:*

(a) *A is $\mathcal{P}$-compact.*

(b) *All limit operators of A are zero.*

(c) *At least one limit operator of A is zero.*

(d) *$A = 0$.*

Proof of Theorems 2.6.2 *and* 2.6.3. If A is $\mathcal{P}$-Fredholm, then all limit operators of A are invertible by Proposition 1.2.9. Let, conversely, A_h be an invertible limit operator of A. By Proposition 2.6.1, there is a subsequence g of h such that (2.99) holds. Then $A_h = A_g$ and, since the invertible operators form an open subset of $L(E)$, the operators $V_{-g(m)}AV_{g(m)}$ must be invertible for all sufficiently large m. Hence, A is invertible.

Similarly, if A is $\mathcal{P}$-compact, then all limit operators of A are zero by Proposition 1.2.6. Conversely, if 0 is a limit operator of A, then (again by Proposition 2.6.1) there is a subsequence g of h such that $\|V_{-g(m)}AV_{g(m)}\| \to 0$. Since the operators V_k are isometries, A must be the zero operator. □

2.6.2 Operators on half-spaces

For $a \in \mathbb{R}^N \setminus \{0\}$ and $\beta \in \mathbb{R}$, we set

$$\mathbb{H}_{a,\beta} := \{z \in \mathbb{Z}^N : \langle a, z\rangle \geq \beta\} \quad \text{and} \quad \mathbb{H}^0_{a,\beta} := \{z \in \mathbb{Z}^N : \langle a, z\rangle > \beta\}$$

where $\langle .,. \rangle$ refers to the standard scalar product on $\mathbb{R}^N$. We refer to these half-spaces as closed and open, respectively. Our first goal is to compute the operator spectrum of the operators $P_{a,\beta}$ and $P^0_{a,\beta}$ of multiplication by the characteristic functions of the half-spaces $\mathbb{H}_{a,\beta}$ and $\mathbb{H}^0_{a,\beta}$, respectively. Recall in this connection Proposition 1.2.3 and its Corollary 1.2.4 which are perfectly illustrated by these results. We start with the trivial case $N = 1$.

Lemma 2.6.4 *If $N = 1$, then $\sigma_{op}(P_{a,\beta}) = \sigma_{op}(P^0_{a,\beta}) = \{0, I\}$.*

In what follows, it will be convenient to consider $\mathbb{R}^N$ as a $\mathbb{Z}$-module. Thus, by a *$\mathbb{Z}$-linear combination* of the real numbers $a_1, \ldots, a_N$ we mean a number of the form $a_1z_1 + \cdots + a_Nz_N$ with $z_1, \ldots z_N \in \mathbb{Z}$, and the numbers $a_1, \ldots, a_N$ are called *$\mathbb{Z}$-linearly independent* if the equality $a_1z_1 + \cdots + a_Nz_N = 0$ with $z_1, \ldots, z_N \in \mathbb{Z}$ implies that all z_i are zero. Thus, two real numbers are $\mathbb{Z}$-linearly independent if and only if they are non-zero and if their quotient is irrational.

Theorem 2.6.5 *Let $N > 1$, $a = (a_1, \ldots, a_N) \in \mathbb{R}^N \setminus \{0\}$ and $\beta \in \mathbb{R}$. If at least two of the numbers a_i are $\mathbb{Z}$-linearly independent, then*

$$\sigma_{op}(P_{a,\beta}) = \sigma_{op}(P^0_{a,\beta}) = \{0,\, I,\, P_{a,\delta},\, P^0_{a,\delta} \quad \textit{with } \delta \in \mathbb{R}\}.$$

Theorem 2.6.6 *Let $N > 1$, $a = (a_1, \ldots, a_N) \in \mathbb{R}^N \setminus \{0\}$ and $\beta \in \mathbb{R}$. If each two of the numbers a_i are $\mathbb{Z}$-linearly dependent, then a can be written as $\alpha(b_1, \ldots, b_N)$ where $\alpha \in \mathbb{R}$ and where the b_i are integers with greatest common divisor equal to 1. In this case,*

$$\sigma_{op}(P_{a,\beta}) = \sigma_{op}(P^0_{a,\beta}) = \{0,\, I,\, P_{a,\beta+\alpha z} \quad \textit{with } z \in \mathbb{Z}^N\}.$$

Notice that the half-spaces $\mathbb{H}_{a,\beta}$ and $\mathbb{H}^0_{a,\beta}$ can coincide (which happens if β is not a $\mathbb{Z}$-linear combination of the a_i) and they can differ by exactly one point (if β is a $\mathbb{Z}$-linear combination of the a_i and if the a_i are $\mathbb{Z}$-linearly independent) or by infinitely many points (if β is a $\mathbb{Z}$-linear combination of the a_i and if the a_i are $\mathbb{Z}$-linearly dependent). We would also like to emphasize that the operator spectrum of the half-space projection is a discrete set of operators in the setting of Theorem 2.6.6 whereas we observe a continuum of limit operators in the context of Theorem 2.6.5. We prepare the proofs of Theorems 2.6.5 and 2.6.6 by a simple lemma.

Lemma 2.6.7 *Let (M_n) be a sequence of subsets of $\mathbb{Z}^N$ and let the operators $\chi_{M_n} I$ of multiplication by the characteristic functions of the sets M_n converge $\mathcal{P}$-strongly to an operator $A \in L(E)$. Then $A = \chi_M I$ with a certain subset M of $\mathbb{Z}^N$, and a point $z \in \mathbb{Z}^N$ belongs to M if and only if $z \in M_n$ for all sufficiently large n.*

Proof. It is easy to check that the $\mathcal{P}$-strong limit of a sequence of projection operators is a projection operator and that the $\mathcal{P}$-strong limit of a sequence of multiplication operators is a multiplication operator again (compare Proposition 2.1.10). Thus, A is necessarily of the form $\chi_M I$ with $M \subseteq \mathbb{Z}^N$. Further, the $\mathcal{P}$-strong convergence of $\chi_{M_n} I$ to $\chi_M I$ is equivalent to the pointwise convergence of the functions χ_{M_n} to χ_M on $\mathbb{Z}^N$ whence the second assertion. □

We verify some of the inclusions in Theorems 2.6.5 and 2.6.6 in separate lemmas.

Lemma 2.6.8 *Let $a \in \mathbb{R}^N$ and $\beta \in \mathbb{R}$. Then*

$$\sigma_{op}(P_{a,\beta}) \cup \sigma_{op}(P^0_{a,\beta}) \subseteq \{0,\, I,\, P_{a,\delta},\, P^0_{a,\delta} \quad \textit{with } \delta \in \mathbb{R}\}.$$

Proof. We prove the assertion for the case of a closed half-space $\mathbb{H}_{a,\beta}$. The proof for $\mathbb{H}^0_{a,\beta}$ runs completely parallel.

Thus, let $h \in \mathcal{H}$ be a sequence for which the limit operator $(P_{a,\beta})_h$ exists. By Lemma 2.6.7, this limit operator is of the form $\chi_M I$ with $M \subseteq \mathbb{Z}^N$. It is further evident that

$$V_{-h(n)} \chi_{\mathbb{H}_{a,\beta}} V_{h(n)} = \chi_{\mathbb{H}_{a,\beta} - h(n)} I$$

and that a point $y \in \mathbb{Z}^N$ belongs to $\mathbb{H}_{a,\beta} - h(n)$ if and only if

$$\langle a, y + h(n) \rangle \geq \beta. \tag{2.100}$$

Now we distinguish several cases: The sequence $(\langle a, h(n) \rangle)$ contains a subsequence $(\langle a, h(n_k) \rangle)_{k \in \mathbb{N}}$ which

- converges to $\gamma \in \mathbb{R}$ and for which $\langle a, h(n_k) \rangle \geq \gamma$ for all $k \in \mathbb{N}$ (case 1).
- converges to $\gamma \in \mathbb{R}$ and for which $\langle a, h(n_k) \rangle < \gamma$ for all $k \in \mathbb{N}$ (case 2).
- converges to $+\infty$ (case 3).
- converges to $-\infty$ (case 4).

In case 1, we write (2.100) as

$$\langle a, y \rangle + \langle a, h(n_k) \rangle - \gamma \geq \beta - \gamma. \tag{2.101}$$

We will show that $M = \mathbb{H}_{a,\beta-\gamma}$ in this case. If $y \in \mathbb{H}_{a,\beta-\gamma}$, i.e., if $\langle a, y \rangle \geq \beta - \gamma$, then y satisfies (2.101) for all k since $\langle a, h(n_k) \rangle - \gamma$ is non-negative. If $y \notin \mathbb{H}_{a,\beta-\gamma}$, i.e., if $\langle a, y\rangle < \beta - \gamma$, then $\langle a, y\rangle + \langle a, h(n_k)\rangle - \gamma < \beta - \gamma$ for all sufficiently large k since $\langle a, h(n_k)\rangle - \gamma \to 0$. Thus y does not satisfy (2.101) for all sufficiently large k. From Lemma 2.6.7 we conclude that the sequence $(V_{-h(n_k)}\chi_{\mathbb{H}_{a,\beta}}V_{h(n_k)}) = ((V_{-h(n_k)}P_{a,\beta}V_{h(n_k)})$ converges $\mathcal{P}$-strongly to $\chi_{\mathbb{H}_{a,\beta-\gamma}}I = P_{a,\beta-\gamma}$. Hence, in this case, the limit operator of $P_{a,\beta}$ with respect to the sequence h is of the mentioned form.

Similarly, in case 2, (2.101) cannot hold if $\langle a, y\rangle \le \beta - \gamma$. Indeed, since $\langle a, h(n_k)\rangle - \gamma < 0$, we have

$$\langle a, y\rangle + \langle a, h(n_k)\rangle - \gamma < \beta - \gamma$$

which contradicts (2.101) for every k. On the other hand, if $\langle a, y\rangle > \beta - \gamma$, then

$$\langle a, y\rangle + \langle a, h(n_k)\rangle - \gamma \ge \beta - \gamma$$

for all sufficiently large k since $\langle a, h(n_k)\rangle - \gamma \to 0$. This shows that $(P_{a,\beta})_h = P^0_{a,\beta-\gamma}$ in this case.

In case 3, y belongs to M if and only if $\langle a, y\rangle + \langle a, h(n_k)\rangle \ge \beta$ for all sufficiently large k. Since $\langle a, h(n_k)\rangle \to \infty$, this holds for all $y \in \mathbb{Z}^N$, i.e., the limit operator of $P_{a,\beta}$ is the identity operator in this case. Similarly, the limit operator of $P_{a,\beta}$ is the zero operator in case 4. □

Lemma 2.6.9 *Let $a \in \mathbb{R}^N$ and $\beta \in \mathbb{R}$. Then*

$$\{0,\, I\} \subseteq \sigma_{op}(P_{a,\beta}) \cup \sigma_{op}(P^0_{a,\beta}).$$

Proof. We show that $I \in \sigma_{op}(P_{a,\beta})$. For, it is sufficient to show that there is a sequence $h \in \mathcal{H}$ such that $\langle a, h(n)\rangle \to +\infty$. This can be done be choosing

$$h(n) := [an] = ([a_1 n], \ldots, [a_N n])$$

which implies

$$\sum_{i=1}^{N} a_i(a_i n - 1) \le \langle a, [an]\rangle < \sum_{i=1}^{N} a_i^2 n.$$

The left- and right-hand sides of this estimate tend to infinity as $n \to \infty$. The other inclusions can be checked analogously. □

Proof of Theorem 2.6.5. We still have to show that

$$\{P_{a,\beta-\gamma},\, P^0_{a,\beta-\gamma} \quad \text{with } \gamma \in \mathbb{R}\} \subseteq \sigma_{op}(P_{a,\beta}) \cap \sigma_{op}(P^0_{a,\beta}),$$

and again we will do this for the operator spectrum of the closed half-space projection only.

Let, for example, a_1 and a_2 be $\mathbb{Z}$-linearly independent. Then both a_1 and a_2 are non-zero, and the ratio $\alpha := a_1/a_2$ is irrational, implying that the set

$$\alpha\mathbb{Z} + \mathbb{Z} = \{\alpha z_1 + z_2 : z_1,\, z_2 \in \mathbb{Z}\}$$

is dense in $\mathbb{R}$ (this fact is often referred to as Kronecker's theorem). Thus, given $\gamma \in \mathbb{R}$, there are sequences $(z_1^{(n)})$ and $(z_2^{(n)})$ in $\mathbb{Z}$ such that

$$\alpha z_1^{(n)} + z_2^{(n)} \to \gamma/a_2 \quad \text{as } n \to \infty \text{ and } \quad \alpha z_1^{(n)} + z_2^{(n)} > \gamma/a_2 \quad \text{for all } n \in \mathbb{N}.$$

Moreover, $z_1^{(n)}$ and $z_2^{(n)}$ can be chosen such that

$$|z_1^{(n)}|^2 + |z_2^{(n)}|^2 \to \infty \quad \text{as } n \to \infty.$$

Then the sequence h given by $h(n) := (z_1^{(n)}, z_2^{(n)}, 0, \ldots, 0)$ tends to infinity, and

$$\langle a, h(n) \rangle = a_1 z_1^{(n)} + a_2 z_2^{(n)} = a_2(\alpha z_1^{(n)} + z_2^{(n)}) \to a_2\gamma/a_2 = \gamma.$$

The discussion of case 1 in the proof of Lemma 2.6.8 shows that the limit operator $(P_{a,\beta})_h$ exists and that it is equal to $P_{a,\beta-\gamma}$. □

Proof of Theorem 2.6.6. Let $a = (a_1, \ldots, a_N)$ and, for example, $a_1 \neq 0$. Since a_1 and a_i (with $i > 1$) are $\mathbb{Z}$-linearly dependent, there exists $(z_1, z_i) \in \mathbb{Z}^2 \setminus (0,0)$ such that $a_1 z_1 + a_i z_i = 0$. In particular, $z_i \neq 0$ for all $i > 1$ since otherwise z_1 would be zero, too. Thus,

$$a = a_1(1, a_2/a_1, \ldots, a_N/a_1) = a_1(p_1/q_1, p_2/q_2, \ldots, p_N/q_N)$$

with integers p_i and positive integers q_i. Dividing a_1 by the least common multiple m of the q_i and multiplying a_1/m by the greatest common divisor d of the $p_i m/q_i$ we get $\alpha := a_1 d/m$, whence the desired factorization of a.

Let now $h(n) := (h_1(n), \ldots, h_N(n)) \in \mathbb{Z}^N$. Then

$$\langle a, h(n) \rangle = \alpha \sum b_i h_i(n) \in \alpha\mathbb{Z}.$$

Thus, only the operators mentioned in the theorem can appear as limit operators. Conversely, each of these operators is indeed a limit operator. For, let $k \in \mathbb{Z}$. Then the equation

$$b_1 z_1 + \cdots + b_N z_N = k$$

is solvable since the greatest common divisor of the b_i is one, and there are infinitely many solutions in integers z_i. In particular, one can choose a sequence $(h(n))$ of solutions of this equation such that $|h(n)| \to \infty$. Then, for all $n \in \mathbb{N}$,

$$\langle a, h(n) \rangle = \alpha(b_1 h_1(n) + \cdots + b_N h_N(n)) = \alpha k.$$

Consequently, the limit operator $(P_{a,\beta})_h$ exists, and it is equal to $P_{a,\beta-\alpha k}$. □

Our next goal are the local operator spectra of the half-space projections.

Theorem 2.6.10 *Let $a \in \mathbb{R}^N \setminus \{0\}$ and $\beta \in \mathbb{R}$, and let $\eta \in S^{N-1}$.*

(a) *If $\langle a, \eta \rangle = 0$, then $\sigma_\eta(P_{a,\beta}) = \sigma_{op}(P_{a,\beta})$.*

(b) *If $\langle a, \eta \rangle > 0$, then $\sigma_\eta(P_{a,\beta}) = \{I\}$.*

(c) *If $\langle a, \eta \rangle < 0$, then $\sigma_\eta(P_{a,\beta}) = \{0\}$.*

These assertions remain valid if $P_{a,\beta}$ is replaced by $P^0_{a,\beta}$.

Proof. Let $h(n) := [n\eta] = ([n\eta_1], \ldots, [n\eta_N])$. Then the sequence h belongs to $\mathcal{H}$, and it tends to infinity in the direction of η. Furthermore, since

$$-\sum a_i = \sum a_i(n\eta_i - 1) \leq \langle a, h(n)\rangle \leq \sum a_i n\eta_i = 0,$$

the sequence $(\langle a, h(n)\rangle)$ is bounded. Hence, there is a subsequence g of h such that the sequence $(\langle a, g(n)\rangle)$ converges to a certain real number γ. As we have seen in the proof of Theorems 2.6.5 and 2.6.6, this implies the existence of the limit operator $(P_{a,\beta})_g$ and one of the equalities

$$(P_{a,\beta})_g = P_{a,\beta-\gamma} \quad \text{or} \quad (P_{a,\beta})_g = P^0_{a,\beta-\gamma}.$$

Now the assertion (a) follows from Corollary 2.3.12 and from Theorems 2.6.5 and 2.6.6. The remaining assertions follow analogously. □

Observe that in case $N = 1$ the sphere S^0 consists of the points 1 and -1 only. Hence, case (a) in the preceding theorem cannot occur.

In what follows, we fix $a \in \mathbb{R}^N$ and $\beta \in \mathbb{R}$, and we write $P_+ := P_{a,\beta}$ as well as $P_- := I - P_+$ for brevity. Further, for $A \in L(E)$ and $\gamma \in \mathbb{R}$, we write $A_{a,\gamma}$ and $A^0_{a,\gamma}$ for $P_{a,\gamma}AP_{a,\gamma} + (I - P_{a,\gamma})$ and $P^0_{a,\gamma}AP^0_{a,\gamma} + (I - P^0_{a,\gamma})$, respectively. In case $\gamma = \beta$, we will simply write A_+ instead of $A_{a,\beta}$.

Let $A \in L^{\$}(E, \mathcal{P})$. We consider the operator P_+AP_+ on $\operatorname{Im} P_+ = P_+E$ or, equivalently, the operator A_+ on E. Then A_+ belongs to $L^{\$}(E, \mathcal{P})$ again, and its operator spectrum $\sigma_{op}(A_+)$ is the union of the local operator spectra $\sigma_\eta(A_+)$ where $\eta \in S^{N-1}$.

Proposition 2.6.11

(a) *Let $A \in L^{\$}(E, \mathcal{P})$, and define A_+ as above. Then the operator spectrum $\sigma_{op}(A_+)$ is the union of all local spectra $\sigma_\eta(A_+)$ where $\eta \in S^{N-1}$ and $\langle a, \eta\rangle = 0$ and of all local spectra $\sigma_\eta(A)$ where $\eta \in S^{N-1}$ and $\langle a, \eta\rangle > 0$ as well as of $\{I\}$.*

(b) *Let $A \in \mathcal{A}^{\$}_E$ be $\mathcal{P}$-Fredholm. Then the operator A_+ is $\mathcal{P}$-Fredholm if and only if, for every $\eta \in S^{N-1}$ with $\langle a, \eta\rangle = 0$, all operators in $\sigma_\eta(A_+)$ are invertible and the norms of their inverses are uniformly bounded.*

For the proof of (b) recall that the $\mathcal{P}$-Fredholmness of A implies the uniform invertibility of all limit operators of A. The operator I results from the artificial term P_- in the definition of A_+. An analogous result holds for the open half-space projection. We proceed with some instances where the assertions of Proposition 2.6.11 can be essentially improved.

Corollary 2.6.12 *Let $N = 1$, $a > 0$, $\beta \in \mathbb{R}$ arbitrary and $A \in \mathcal{A}^{\$}_E$. Then $A_+ := A_{a,\beta}$ is $\mathcal{P}$-Fredholm if and only if A is locally invertible at $1 \in S^0$.*

Proof. For $A \in L^{\$}(E, \mathcal{P})$ we have $\sigma_{op}(A) = \sigma_1(A) \cup \sigma_{-1}(A)$ and $\sigma_1(A_+) = \sigma_1(A)$ as well as $\sigma_{-1}(A_+) = \{I\}$. Hence, if $A \in \mathcal{A}^{\$}_E$, then A is locally invertible at

$1 \in S^0$ if and only if every operator in $\sigma_1(A)$ is invertible and if the inverses are uniformly bounded (Theorem 2.3.13), which is equivalent to the $\mathcal{P}$-Fredholmness of A_+ (Theorem 2.2.1). □

Corollary 2.6.13 *Let $N > 1$ and let $A \in \mathcal{A}_E^\$$ be $\mathcal{P}$-Fredholm. If no two of the components a_i of the vector a are $\mathbb{Z}$-linearly independent, then the operator A_+ is $\mathcal{P}$-Fredholm if and only if, for every $\eta \in S^{N-1}$ with $\langle a, \eta \rangle = 0$, all operators of the form $P_+ A_h P_+$ with $A_h \in \sigma_\eta(A)$ are invertible and the norms of their inverses are uniformly bounded.*

Proof. Let $\eta \in S^{N-1}$ and $\langle a, \eta \rangle = 0$. From Theorem 2.6.6 it is evident that $\sigma_\eta(A_+)$ consists of limit operators of A and of operators of the form

$$P_{a,\beta+\alpha z} A_h P_{a,\beta+\alpha z} + (I - P_{a,\beta+\alpha z}) \tag{2.102}$$

where $z \in \mathbb{Z}$ and $A_h \in \sigma_\eta(A)$. The uniform invertibility of the limit operators of A is a consequence of the $\mathcal{P}$-Fredholmness of A. It is easy to see that there is a $z_0 \in \mathbb{Z}^N$ such that the operator in (2.102) is just

$$V_{-z_0} P_{a,\beta} V_{z_0} A_h V_{-z_0} P_{a,\beta} V_{z_0} + V_{-z_0}(I - P_{a,\beta}) V_{z_0},$$

and this operator is invertible if and only if the operator $P_+ V_{z_0} A_h V_{-z_0} P_+ + P_-$ is invertible. Since $V_{z_0} A_h V_{-z_0}$ belongs to $\sigma_\eta(A)$ whenever A_h belongs to this local spectrum, we get the assertion. □

Corollary 2.6.14 *Let $N > 1$ and let $A \in \mathcal{A}_E(\mathbb{C}_{L(X)})$ and $a \in \mathbb{R}^N \setminus \{0\}$. Then the following conditions are equivalent:*

(a) *the operator $A_+ = P_+ A P_+ + P_-$ is $\mathcal{P}$-Fredholm.*

(b) *one of the operators*

$$A_{a,\gamma}, \; A^0_{a,\gamma} \quad \text{with } \gamma \in \mathbb{R} \tag{2.103}$$

is invertible.

(c) *each of the operators in (2.103) is invertible.*

If one of these conditions is satisfied, then A is invertible.

Proof. The shift invariance of A implies that if the limit operator of P_+ with respect to the sequence h exists, then the limit operator of A_+ with respect to h exists, too, and $(A_+)_h = (P_+)_h A (P_+)_h + (P_-)_h$. From Theorems 2.6.5 and 2.6.6 we infer that each operator in (2.103) is a limit operator of A_+. Hence, by Proposition 1.2.2, each operator in (2.103) is invertible if only A_+ is $\mathcal{P}$-Fredholm. This settles the implications (a) ⇒ (c) ⇒ (b).

To check the implication (b) ⇒ (c), assume for instance that the operator $A_{a,\gamma}$ is invertible for some $\gamma \in \mathbb{R}$. Then, for every $\alpha \in \mathbb{Z}^N$, the operators $V_{-\alpha} A_{a,\gamma} V_\alpha$ are invertible, and the norms of their inverses are uniformly bounded. Thus, if an operator B is the $\mathcal{P}$-strong limit of a sequence of operators of the form $V_{-\alpha} A_{a,\gamma} V_\alpha$, then B is invertible.

Let, for example, $B = A_{a,\delta}$ with $\delta \in \mathbb{R}$. Since

$$V_{-\alpha}\chi_M V_\alpha = \chi_{m-\alpha} \quad \text{and} \quad \mathbb{H}_{a,\gamma} - \alpha = \mathbb{H}_{a,\gamma-\langle a,\alpha\rangle}$$

we get $V_{-\alpha}A_{a,\gamma}V_\alpha = A_{a,\gamma-\langle a,\alpha\rangle}$. Thus, if $h \in \mathcal{H}$ is a sequence with

$$\gamma - \langle a, h(n)\rangle \to \delta \quad \text{and} \quad \gamma - \langle a, h(n)\rangle \geq \delta \quad \text{for all } n,$$

then one gets as in the proof of Lemma 2.6.8 that $V_{-h(n)}A_{a,\gamma}V_{h(n)}$ converges $\mathcal{P}$-strongly to B. Analogously one verifies the invertibility of the other operators in (2.103).

Next we show that condition (c) implies the invertibility of A. For, we consider the operators $A_{a,\gamma(n)}$ where the $\gamma(n)$ are real numbers tending to $-\infty$. As $n \to \infty$, these operators converge $\mathcal{P}$-strongly to A. Since the operators $A_{a,\gamma(n)}$ are invertible with uniformly bounded inverses, this implies the invertibility of A.

The final implication (c) $\Rightarrow$ (a) follows from Proposition 2.6.11 (b). Indeed, the uniform invertibility of all operators in in (2.103) implies the invertibility of A, whence the uniform invertibility of all limit operators of A_+. □

2.6.3 Operators on polyhedral convex cones

In this section, we let $N \geq 2$. Our goal is to consider restrictions of band-dominated operators onto cones, where a cone is a subset W of $\mathbb{R}_N$ such that if $w \in W$ and $\lambda \geq 0$, then $\lambda w \in W$. To avoid technical complications, we will make some special assumptions on the location of the cone as well as on its shape. To make this precise, write the intersection of the cone W with the hyperplane $\{z = (z_1, \ldots, z_N) \in \mathbb{R}^N : z_N = 1\}$ as $\Omega \times \{1\}$. What we assume is that Ω is a compact and convex polyhedron in $\mathbb{R}^{N-1}$ with non-empty interior. If, moreover, the vertices of Ω are in $\mathbb{Z}^N$, then we call W a *rational convex cone* with *basis* Ω. Clearly, $W = \{\lambda w : \lambda \geq 0, w \in \Omega \times \{1\}\}$.

Since every cone with compact and convex polyhedral basis is the intersection of a finite number of closed half-spaces, the results of the previous section yield a complete description of the set of all limit operators of the operator of multiplication by the characteristic function of the cone. The necessity to distinguish between several locations of the half-spaces makes the description of the operator spectrum of a general cone projection quite involved. So we will restrict our attention to the examination of some typical special settings.

The case $N = 2$. Let $u_1, u_2 \in \mathbb{R}$ and let $\Omega = [u_1, u_2] \subset \mathbb{R}$. Further let $W_\Omega \subset \mathbb{R}^2$ refer to the cone determined by Ω, i.e.,

$$W_\Omega := \{(x_1, x_2) \in \mathbb{R}^2 : x_2 > 0 \text{ and } x_1/x_2 \in \Omega\} \cup \{0\},$$

and set $\hat{W}_\Omega := W_\Omega \cap \mathbb{Z}^2$. Thus,

$$\hat{W}_\Omega = \mathbb{H}_{a_1,0} \cap \mathbb{H}_{a_2,0} \quad \text{with } a_1 = (1, -u_1) \text{ and } a_2 = (-1, u_2). \tag{2.104}$$

For $A \in L(E) = L(l^p(\mathbb{Z}^2, X))$, we consider the operator

$$A_{W_\Omega} := \hat{\chi}_{W_\Omega} A \hat{\chi}_{W_\Omega} I + (1 - \hat{\chi}_{W_\Omega}) I : l^p(\mathbb{Z}^2, X) \to l^p(\mathbb{Z}^2, X). \tag{2.105}$$

The $\mathcal{P}$-Fredholmness of A_{W_Ω} is subject to Theorem 2.2.1 and Corollary 2.3.22 provided that $A \in \mathcal{A}_E^{\$}$.

Given $x \in \mathbb{R}$, we denote by η_x the point on the sphere S^1 which is determined by the ray in $\mathbb{R}^2$ starting at the origin and passing through the point $(x, 1)$. Due to (2.104), it follows from the results of the previous section that if x is in the interior of Ω and if $h \subset \mathbb{Z}^2$ is a sequence tending to infinity in the direction of η_x, then the limit operator of $\hat{\chi}_{W_\Omega} I$ with respect to h exists and is equal to I, whereas one has in case $x \in \mathbb{R} \setminus \Omega$ that $\sigma_{\eta_x}(\hat{\chi}_{W_\Omega} I) = \{0\}$. If $x = u_i$ with $i \in \{1, 2\}$ and if u_i is irrational, then

$$\sigma_{\eta_x}(\hat{\chi}_{W_\Omega} I) = \{0,\ I, P_{a_i,\beta}, P^0_{a_i,0} \text{ with } \beta \in \mathbb{R}\},$$

and if u_i is rational, then

$$\sigma_{\eta_x}(\hat{\chi}_{W_\Omega} I) = \{0,\ I, P_{a_i,\alpha_i z} \text{ with } z \in \mathbb{Z}\}$$

where $\alpha_i = 1/q_i$ if $u_i = p_i/q_i$ with p_i and q_i integers with greatest common divisor 1 and q_i positive. All limit operators of $\hat{\chi}_{W_\Omega} I$ which correspond to other points in S^1 are zero.

Thus, in case $x = u_i$ is irrational, we get a decomposition of the local spectrum $\sigma_{\eta_a}(A)$,

$$\sigma_{\eta_x}(A) = \sigma_{\eta_x,-}(A) \cup \sigma_{\eta_x,+}(A) \cup \{\sigma_{\eta_x,\beta,+}(A) : \beta \in \mathbb{R}\} \cup \{\sigma_{\eta_x,\beta,-}(A) : \beta \in \mathbb{R}\},$$

where $\sigma_{\eta_x,-}(A)$, $\sigma_{\eta_x,+}(A)$, $\sigma_{\eta_x,\beta,+}(A)$ and $\sigma_{\eta_x,\beta,-}(A)$ contain all limit operators $A_h \in \sigma_{\eta_x}(A)$ for which there is a subsequence g of h such that the limit operator of $\hat{\chi}_{W_\Omega} I$ with respect to g exists and is equal to 0, I, $P_{a_i,\beta}$ and $P^0_{a_i,\beta}$, respectively. With these notations, we have the following theorem.

Theorem 2.6.15 *Let $A \in \mathcal{A}_E^{\$}$ and let $\Omega = [u_1, u_2]$ with both u_1 and u_2 irrational. Then the operator A_{W_Ω} defined by (2.105) is $\mathcal{P}$-Fredholm if and only if*

(a) *for every x in the interior of Ω, all operators in $\sigma_{\eta_x}(A)$ are uniformly invertible,*

(b) *for $x = u_i$ and $i = 1, 2$, all operators*

$$P_{a_i,\beta} A_g P_{a_i,\beta} + (I - P_{a_i,\beta}) \quad \textit{with} \quad A_g \in \sigma_{\eta_x,\beta,+}(A),$$

all operators

$$P^0_{a_i,\beta} A_g P^0_{a_i,\beta} + (I - P^0_{a_i,\beta}) \quad \textit{with} \quad A_g \in \sigma_{\eta_x,\beta,-}(A),$$

and all operators $\sigma_{\eta_x,+}(A)$ are uniformly invertible.

The proof follows immediately from Theorem 2.2.1, Corollary 2.3.22, and from the special form of the limit operators of the operator $\hat{\chi}_{W_\Omega}I$. Theorem 2.6.15 takes a simpler form if we assume a priori that A is a $\mathcal{P}$-Fredholm operator on E.

Theorem 2.6.16 *Let $A \in \mathcal{A}_E^\$$ be $\mathcal{P}$-Fredholm, and let $\Omega = [u_1, u_2]$ with both u_1 and u_2 irrational. Then the operator A_{W_Ω} defined by (2.105) is $\mathcal{P}$-Fredholm if and only if for $x = u_i$ and $i = 1, 2$ all operators*

$$P_{a_i,\beta} A_g P_{a_i,\beta} + (I - P_{a_i,\beta}) \quad \textit{with} \quad A_g \in \sigma_{\eta_x,\beta,+}(A)$$

and all operators

$$P^0_{a_i,\beta} A_g P^0_{a_i,\beta} + (I - P^0_{a_i,\beta}) \quad \textit{with} \quad A_g \in \sigma_{\eta_x,\beta,-}(A),$$

are uniformly invertible.

Analogously, if $x = u_i$ is rational, there is a decomposition of the local spectrum $\sigma_{\eta_a}(A)$ as follows:

$$\sigma_{\eta_x}(A) = \sigma_{\eta_x,-}(A) \cup \sigma_{\eta_x,+}(A) \cup \{\sigma_{\eta_x,z}(A) : z \in \mathbb{Z}\}$$

where $\sigma_{\eta_x,-}(A)$, $\sigma_{\eta_x,+}(A)$ and $\sigma_{\eta_x,z}(A)$ contain all limit operators $A_h \in \sigma_{\eta_x}(A)$ for which there is a subsequence g of h such that the limit operator of $\hat{\chi}_{W_\Omega}I$ with respect to g exists and is equal to 0, I and $P_{a_i,\alpha_i z}$, respectively. With these notations, we have the following result.

Theorem 2.6.17 *Let $A \in \mathcal{A}_E^\$$, and let $\Omega = [u_1, u_2]$ with both u_1 and u_2 rational. Then the operator A_{W_Ω} defined by (2.105) is $\mathcal{P}$-Fredholm if and only if*

(a) *for every x in the interior of Ω, all limit operators in $\sigma_{\eta_x}(A)$ are uniformly invertible,*

(b) *for $x = u_i$ and $i = 1, 2$ all operators*

$$P_{a_i,0} A_g P_{a_i,0} + (I - P_{a_i,0}) \quad \textit{with} \quad A_g \in \sigma_{\eta_x,0}(A)$$

and all operators in $\sigma_{\eta_x,+}(A)$ are uniformly invertible.

Proof. It follows immediately from Theorem 2.2.1, Corollary 2.3.22, and from the special form of the limit operators of the operator $\hat{\chi}_{W_\Omega}I$ that the conditions of the theorem are necessary. For their sufficiency we have to show that, under the conditions of the theorem, all operators

$$P_{a_i,\alpha_i z} A_g P_{a_i,\alpha_i z} + (I - P_{a_i,\alpha_i z}) \tag{2.106}$$

with $A_g \in \sigma_{\eta_{u_i},z}(A)$, $i \in \{1, 2\}$ and $z \in \mathbb{Z}$ are uniformly invertible. Let $A_g \in \sigma_{\eta_{u_i},z}(A)$, and choose $\alpha \in \mathbb{Z}^2$ such that $V_{-\alpha} P_{a_i,\alpha_i z} V_\alpha = P_{a_i,0}$. Then the operator (2.106) is invertible if and only if the operator

$$\begin{aligned} &V_{-\alpha} P_{a_i,\alpha_i z} A_g P_{a_i,\alpha_i z} V_\alpha + (I - V_{-\alpha} P_{a_i,\alpha_i z} V_\alpha) \\ &\quad = P_{a_i,0} V_{-\alpha} A_g V_\alpha P_{a_i,0} + (I - P_{a_i,0}) \end{aligned}$$

is invertible. Evidently, $V_{-\alpha}A_gV_\alpha$ is a limit operator of A which belongs to $\sigma_{\eta_x,0}(A)$. Thus, the invertibility of every operator of the form (2.106) follows from hypothesis (b) of the theorem. □

Again, the result gets a simpler form if A is $\mathcal{P}$-Fredholm.

Theorem 2.6.18 *Let $A \in \mathcal{A}_E^\$$ be $\mathcal{P}$-Fredholm, and let $\Omega = [u_1, u_2]$ with both u_1 and u_2 rational. Then the operator A_{W_Ω} defined by (2.105) is $\mathcal{P}$-Fredholm if and only if, for $x = u_i$ and $i = 1, 2$, all operators*

$$P_{a_i,0}A_gP_{a_i,0} + (I - P_{a_i,0}) \quad \text{with} \quad A_g \in \sigma_{\eta_x,0}(A)$$

are uniformly invertible.

It is evident how these results have to be modified if the interval Ω has one rational and one irrational endpoint.

The case $N = 3$. Let now Ω be a compact and convex polygon in $\mathbb{R}^2$ with vertices $u_1, u_2, \ldots, u_k \in \mathbb{Z}^2$, and denote by $W_\Omega \subset \mathbb{R}^3$ the cone generated by Ω, that is

$$W_\Omega = \{(x_1, x_2, x_3) \in \mathbb{R}^3 : x_3 > 0 \text{ and } (x_1/x_3, x_2/x_3) \in \Omega\} \cup \{0\}.$$

Further set $u_{k+1} := u_1$ and $u_0 := u_k$ and write (u_j, u_{j+1}) for the open segment in $\mathbb{R}^2$ joining u_j to u_{j+1}. Moreover, let K_j refer to the angular domain in $\mathbb{R}^2$ with vertex at u_j which is bounded by the rays starting at u_j and passing through u_{j-1} and u_{j+1}, respectively, and which contains Ω, and let $H_j \subset \mathbb{R}^2$ refer to the half-plane, which is bounded by the straight line passing through u_j and u_{j+1} and which contains Ω. Finally, set $K_j^0 := K_j - u_j$ and $H_j^0 := H_j - u_j$ (= the algebraic differences), and introduce the wedge

$$\mathbb{K}_j := \{(x_1, x_2, x_3) \in \mathbb{R}^3 : (x_1, x_2) \in K_j^0 + x_3u_j\},$$

the half-space

$$\mathbb{H}_j := \{(x_1, x_2, x_3) \in \mathbb{R}^3 : (x_1, x_2) \in H_j^0 + x_3u_j\},$$

and the sets

$$\hat{W}_\Omega := W_\Omega \cap \mathbb{Z}^3, \quad \hat{\mathbb{K}}_j = \mathbb{K}_j \cap \mathbb{Z}^3 \quad \text{and} \quad \hat{\mathbb{H}}_j = \mathbb{H}_j \cap \mathbb{Z}^3.$$

Further, given $x \in \mathbb{R}^2$, let η_x stand for the point on the sphere S^2 which lies on the ray starting at $0 \in \mathbb{R}^3$ and passing through $(x, 1) \in \mathbb{R}^3$. A little thought shows that the limit operators of the operator of multiplication by the function $\hat{\chi}_{W_\Omega}I$ are as follows. If x is in the interior of Ω, then

$$\sigma_{\eta_x}(\hat{\chi}_{V_\Omega}I) = \{I\},$$

and for $x \in \mathbb{R}^2 \setminus \Omega$ one has

$$\sigma_{\eta_x}(\hat{\chi}_{V_\Omega}I) = \{0\}.$$

Further, if x belongs to the segment (u_j, u_{j+1}), then

$$\sigma_{\eta_x}(\hat{\chi}_{V_\Omega}I) = \{0, I, \hat{\chi}_{\mathbb{H}_j+v}I : v \in \mathbb{Z}^3\}$$

and, finally, if $x = u_j$, then

$$\sigma_{\eta_x}(\hat{\chi}_{V_\Omega} I) = \{0,\, I,\, \hat{\chi}_{\mathbb{H}_{j-1}+v} I,\, \hat{\chi}_{\mathbb{H}_j+v} I,\, \hat{\chi}_{\mathbb{K}_j+v} I : v \in \mathbb{Z}^3\}.$$

Now let $E = l^p(\mathbb{Z}^3, X)$ and $A \in \mathcal{A}_E^\$$. We are interested in the $\mathcal{P}$-Fredholmness of the restriction of A onto W_Ω or, equivalently, in the $\mathcal{P}$-Fredholmness of the operator

$$A_{W_\Omega} := \hat{\chi}_{W_\Omega} A \hat{\chi}_{W_\Omega} I + (1 - \hat{\chi}_{W_\Omega}) I \in L(E). \tag{2.107}$$

In analogy to the two-dimensional setting, every point x in the boundary of Ω induces a decomposition of the local spectrum $\sigma_{\eta_x}(A)$ as follows. If x belongs to the segment $(u_j,\, u_{j+1})$, then we have

$$\sigma_{\eta_x}(A) = \sigma_{\eta_x,-}(A) \cup \sigma_{\eta_x,+}(A) \cup \{\sigma_{\eta_x,v}(A) : v \in \mathbb{Z}^3\}$$

where $\sigma_{\eta_x,-}(A)$, $\sigma_{\eta_x,+}(A)$ and $\sigma_{\eta_x,v}(A)$ contain all limit operators $A_h \in \sigma_{\eta_x}(A)$ for which there is a subsequence g of h such that the limit operator of $\hat{\chi}_{W_\Omega} I$ with respect to g exists and is equal to 0, I and $\hat{\chi}_{\mathbb{H}_j+v} I$, respectively. Similarly, if x is the vertex u_j, then

$$\begin{aligned}\sigma_{\eta_x}(A) \quad &= \quad \sigma_{\eta_x,-}(A) \cup \sigma_{\eta_x,+}(A) \cup \{\sigma_{\eta_x,-,v}(A) : v \in \mathbb{Z}^3\} \\ &\qquad \cup \{\sigma_{\eta_x,+,v}(A) : v \in \mathbb{Z}^3\} \cup \{\sigma_{\eta_x,v}(A) : v \in \mathbb{Z}^3\}\end{aligned}$$

where $\sigma_{\eta_x,-}(A)$, $\sigma_{\eta_x,+}(A)$, $\sigma_{\eta_x,-,v}(A)$, $\sigma_{\eta_x,+,v}(A)$ and $\sigma_{\eta_x,v}(A)$ consist of all limit operators $A_h \in \sigma_{\eta_x}(A)$ for which there is a subsequence g of h such that the limit operator of $\hat{\chi}_{W_\Omega} I$ with respect to g exists and is equal to 0, I, $\hat{\chi}_{\mathbb{H}_{j-1}+v} I$, $\hat{\chi}_{\mathbb{H}_j+v} I$ and $\hat{\chi}_{\mathbb{K}_j+v} I$, respectively. With these notations we have the following result.

Theorem 2.6.19 *Let $E = l^p(\mathbb{Z}^3, X)$ and $A \in \mathcal{A}_E^\$$. Then the operator A_{W_Ω} defined by (2.107) is $\mathcal{P}$-Fredholm if and only if*

(a) *for every x in the interior of Ω, all operators in $\sigma_{\eta_x}(A)$ are uniformly invertible.*

(b) *for every j and every $x \in (u_j, u_{j+1})$, all operators*

$$\hat{\chi}_{\mathbb{H}_j} A_h \hat{\chi}_{\mathbb{H}_j} I + (1 - \hat{\chi}_{\mathbb{H}_j}) I \quad \textit{with} \quad A_h \in \sigma_{\eta_x,0}(A)$$

as well as all operators in $\sigma_{\eta_x,+}(A)$ are uniformly invertible.

(c) *for every $x = u_j$, all operators*

$$\hat{\chi}_{\mathbb{K}_j} A_h \hat{\chi}_{\mathbb{K}_j} I + (1 - \hat{\chi}_{\mathbb{K}_j}) I \quad \textit{with} \quad A_h \in \sigma_{\eta_x,0}(A),$$

all operators

$$\hat{\chi}_{\mathbb{H}_{j-1}} A_h \hat{\chi}_{\mathbb{H}_{j-1}} I + (1 - \hat{\chi}_{\mathbb{H}_{j-1}}) I \quad \textit{with} \quad A_h \in \sigma_{\eta_x,-,0}(A),$$

all operators

$$\hat{\chi}_{\mathbb{H}_j} A_h \hat{\chi}_{\mathbb{H}_j} I + (1 - \hat{\chi}_{\mathbb{H}_j}) I \quad \textit{with} \quad A_h \in \sigma_{\eta_x,+,0}(A)$$

and all operators in $\sigma_{\eta_x,+}(A)$ are uniformly invertible.

This theorem can be proved in the very same way as its predecessor Theorem 2.6.17. Observe that condition (a) as well as the last conditions in (b) and (c) are automatically satisfied if the operator A is $\mathcal{P}$-Fredholm.

2.6.4 Composed band-dominated operators on $\mathbb{Z}^2$

Consider the following quarter planes in $\mathbb{Z}^2$,

$$\mathbb{Q}_{11} := \{x \in \mathbb{Z}^2 : x_1 \geq 0,\, x_2 \geq 0\}, \qquad \mathbb{Q}_{21} := \{x \in \mathbb{Z}^2 : x_1 < 0,\, x_2 > 0\},$$
$$\mathbb{Q}_{22} := \{x \in \mathbb{Z}^2 : x_1 \leq 0,\, x_2 \leq 0\}, \qquad \mathbb{Q}_{12} := \{x \in \mathbb{Z}^2 : x_1 > 0,\, x_2 < 0\},$$

and let χ_{ij} stand for the characteristic function of $\mathbb{Q}_{ij}$. In this section, we are going to study the Fredholm properties of so-called composed band-dominated operators of the form

$$A = \sum_{1 \leq i,j \leq 2} A^{ij} \chi_{ij} I, \tag{2.108}$$

where the operators A^{ij} belong to the Wiener algebra $\mathcal{W}$ of band-dominated operators with coefficients in $L(\mathbb{C}^n)$. We consider the operator A as acting on one of the spaces $l^p(\mathbb{Z}^2, \mathbb{C}^n)$ with $1 \leq p \leq \infty$.

According to Theorem 3.5.7 and Proposition 3.3.2, we have to calculate the operator spectrum $\sigma_{op}(A) = \cup_{\eta \in S^1} \sigma_\eta(A)$. To start with, let $\eta \in S^1$ belong to one of the sets

$$\mathbb{R}_{11} := \{x \in \mathbb{R}^2 : x_1 > 0,\, x_2 > 0\}, \qquad \mathbb{R}_{21} := \{x \in \mathbb{R}^2 : x_1 < 0,\, x_2 > 0\},$$
$$\mathbb{R}_{22} := \{x \in \mathbb{R}^2 : x_1 < 0,\, x_2 < 0\}, \qquad \mathbb{R}_{12} := \{x \in \mathbb{R}^2 : x_1 > 0,\, x_2 < 0\}.$$

Then it is easy to see that $\sigma_\eta(\chi_{ij} I) = I$ if $\eta \in \mathbb{R}_{ij}$ and that $\sigma_\eta(\chi_{ij} I) = 0$ otherwise, whence

$$\sigma_\eta(A) = \sigma_\eta(A^{ij}) \quad \text{whenever} \quad \eta \in S^1 \cap \mathbb{R}_{ij}.$$

Let now $\eta = (1, 0)$. Then

$$\sigma_\eta(A) = \sigma_\eta(A^{11} \chi_{11} I + A^{12} \chi_{12} I).$$

Further, we know from Section 3.6.2 that

$$\sigma_\eta(\chi_{11} I) = \{I,\, 0,\, V_\alpha \chi_2^1 V_{-\alpha} : \alpha \in \mathbb{Z}^2\}$$

where χ_2^1 refers to the characteristic function of the half-space $\{x \in \mathbb{Z}^2 : x_2 \geq 0\}$, and that

$$\sigma_\eta(\chi_{12} I) = \{I,\, 0,\, V_\alpha \chi_2^2 V_{-\alpha} : \alpha \in \mathbb{Z}^2\}$$

where $\chi_2^2 = 1 - \chi_2^1$.

Let $j \in \{1,\, 2\}$. By $\sigma_{(1,0),+}(A^{1j})$, we denote the set of all limit operators of A^{1j} with respect to sequences h which tend to infinity into the direction of $\eta = (1,\, 0)$ and for which $(\chi_{1j} I)_h = I$. Similarly, $\sigma_{(1,0),\,\alpha}(A^{1j})$ refers to the set of all limit operators of A^{1j} which correspond to sequences h which tend to infinity into the direction of $\eta = (1, 0)$ and for which $(\chi_{1j}I)_h = V_\alpha \chi_2^j V_{-\alpha}$ with $\alpha \in \mathbb{Z}^2$. Hence,

$$\begin{aligned}\sigma_{(1,0)}(A) &= \sigma_{(1,0),+}(A^{11}) \cup \sigma_{(1,0),+}(A^{12}) \\ &\cup \left\{ A_h^{11} V_\alpha \chi_2^1 V_{-\alpha} + A_h^{12} V_\alpha \chi_2^2 V_{-\alpha} : A_h^{1j} \in \sigma_{(1,0),\alpha}(A^{1j}),\, j = 1,\, 2,\, \alpha \in \mathbb{Z}^2 \right\}.\end{aligned}$$

Since local operator spectra of band-dominated operators are invariant with respect to shifts (Section 2.3.2), we obtain that all operators in $\sigma_{(1,0)}(A)$ are invertible if and only if all operators in

$$\begin{aligned}\sigma_{(1,0)}^{red}(A) &:= \sigma_{(1,0),+}(A^{11}) \cup \sigma_{(1,0),+}(A^{12}) \\ &\cup \left\{ A_h^{11} \chi_2^1 I + A_h^{12} \chi_2^2 I : A_h^{1j} \in \sigma_{(1,0),\alpha}(A^{1j}),\, j = 1,\, 2,\, \alpha \in \mathbb{Z}^2 \right\}. \qquad (2.109)\end{aligned}$$

are invertible. The local spectra $\sigma_\eta(A)$ as well as the *reduced* local spectra $\sigma_\eta^{red}(A)$ at the points points $\eta = (0, 1), (-1, 0), (0, -1)$ can be defined and described in a completely analogous way. The following theorem is a corollary of the general Theorem 2.5.7.

Theorem 2.6.20 *Let A be an operator of the form* (2.108) *with coefficients A^{ij} in the Wiener algebra. Then the following assertions are equivalent:*

(a) *A is a Fredholm operator on one of the spaces $l^p := l^p(\mathbb{Z}^2, \mathbb{C}^n)$ with $1 \le p \le \infty$.*

(b) *A is a Fredholm operator on each of the spaces l^p.*

(c) *There is one space l^p on which all operators in*

$$\cup_{\eta \in S^1} \sigma_\eta^{red}(A), \qquad (2.110)$$

where $\sigma_\eta^{red}(A) := \sigma_\eta(A)$ for $\eta \ne (1, 0), (0, 1), (-1, 0), (0, -1)$, are invertible.

(d) *The operators in* (2.110) *are uniformly invertible on each of the spaces l^p.*

Let us now consider composed operators of the form (2.108) with coefficients A^{ij} from the class $\mathcal{W}(SO)$, i.e., with

$$A^{ij} = \sum_{\alpha \in \mathbb{Z}^2} a_\alpha^{ij} V_\alpha$$

where

$$a_\alpha^{ij} \in SO(\mathbb{Z}^2, L(\mathbb{C}^n)) \quad \text{and} \quad \sum_{\alpha \in \mathbb{Z}^2} \|a_\alpha^{ij}\|_\infty < \infty.$$

In this case, all limit operators A_h^{ij} of A^{ij} are invariant with respect to shifts, and they have the form

$$A_h^{ij} = \sum_{\alpha \in \mathbb{Z}^2} (a_\alpha^{ij})_h V_\alpha$$

where

$$(a_\alpha^{ij})_h \in L(\mathbb{C}^n) \quad \text{and} \quad \sum_{\alpha \in \mathbb{Z}^2} \|(a_\alpha^{ij})_h\|_{L(\mathbb{C}^n)} < \infty.$$

Set

$$\Sigma_{(1,0)}(A) := \left\{ A_h^{11}\chi_2^1 I + A_h^{12}\chi_2^2 I : A_h^{1j} \in \sigma_{(1,0),\alpha}(A^{1j}),\ j=1,\ 2,\ \alpha \in \mathbb{Z}^2 \right\}$$

and define $\Sigma_\eta(A)$ at $\eta = (0,\ 1),\ (-1,\ 0),\ (0,\ -1)$ analogously.

Theorem 2.6.21 *Let A be an operator of the form (2.108) with coefficients $A^{ij} \in \mathcal{W}(SO)$. Then the following assertions are equivalent:*

(a) *A is a Fredholm operator on one of the spaces $l^p := l^p(\mathbb{Z}^2, \mathbb{C}^n)$ with $1 \le p \le \infty$.*

(b) *A is a Fredholm operator on each of the spaces l^p.*

(c) *There is a space l^p on which all operators in*

$$\cup_{i,j=1,2} \cup_{\eta \in S^1 \cap \overline{\mathbb{R}_{ij}}} \sigma_\eta(A^{ij}) \tag{2.111}$$

and all operators in

$$\Sigma_{(1,0)}(A) \cup \Sigma_{(0,1)}(A) \cup \Sigma_{(-1,0)}(A) \cup \Sigma_{(0,-1)}(A) \tag{2.112}$$

are invertible.

(d) *All operators in (2.111) and (2.112) are uniformly invertible on each of the spaces l^p.*

Proof. Let A be as in (2.108) with coefficients in $\mathcal{W}(SO)$. Then all limit operators of A occur among the operators in (2.111) and (2.112). Thus, (c) implies (b) by Theorem 2.6.20.

For the reverse implication, we have to show that the Fredholmness of A implies the invertibility of all operators in (2.111) and (2.112). By Theorem 2.6.20, this is clear for the operators in (2.112) and for the operators in

$$\cup_{i,j=1,2} \cup_{\eta \in S^1 \cap \mathbb{R}_{ij}} \sigma_\eta(A^{ij}).$$

So let, for example, $B \in \sigma_{(1,0)}(A^{11})$ be a limit operator of A^{11} with respect to a sequence h which tends into the direction of $(1,0) \in S^1$ to infinity. Without loss we can assume that the limit operator of $\chi_{11}I$ exists (otherwise we pass to a suitable subsequence of h). Hence, either $(\chi_{11}I)_h$ is equal to I or to 0 or to $V_\alpha \chi_2^1 V_{-\alpha}$ with $\alpha \in \mathbb{Z}^2$. In the first two instances, B is a limit operator of A and, thus, invertible. In the third case, there is a limit operator of A of the form

$$BV_\alpha \chi_2^1 V_{-\alpha} + CV_\alpha \chi_2^2 V_{-\alpha} \tag{2.113}$$

where C is (as B) a certain shift invariant operator. Being a limit operator of a Fredholm operator, the operator (2.113) is invertible. Since its coefficients B, C are shift invariant, this easily implies the invertibility of B (and C), too. □

Finally, we will briefly discuss the invertibility of the limit operators which appear in the previous theorem. It is clearly sufficient to study their invertibility on the space $l^2(\mathbb{Z}^2, \mathbb{C}^n)$.

The operator $A_h^{ij} = \sum_{\alpha\in\mathbb{Z}^2}(a_\alpha^{ij})_h V_\alpha$ is unitarily equivalent to the operator of multiplication by the function

$$\widehat{A_h^{ij}} : \mathbb{T}^2 \to L(\mathbb{C}^n), \quad t = (t_1, t_2) \mapsto \sum_{(\alpha_1, \alpha_2)\in\mathbb{Z}^2} (a_\alpha^{ij})_h t_1^{\alpha_1} t_2^{\alpha_2}.$$

Hence, A_h^{ij} is invertible on $l^2(\mathbb{Z}^2, \mathbb{C}^n)$ if and only if

$$\inf_{t\in\mathbb{T}^2} \left| \det \widehat{A_h^{ij}}(t) \right| > 0.$$

Similarly, the discrete Fourier transform provides the equivalence of the invertibility of the operator

$$A_h^{11}\chi_2^+ I + A_h^{12}\chi_2^- I \tag{2.114}$$

at the one hand and the invertibility of the operator

$$\mathcal{A}_h^{11}\chi^+ I + \mathcal{A}_h^{12}\chi^- I, \tag{2.115}$$

on the other hand, where χ^+ is the characteristic function of $\{z \in \mathbb{Z} : z \geq 0\}$ and $\chi^- := 1 - \chi^+$, and where $\mathcal{A}_h^{1j}$ stands for the continuous operator-valued function

$$\mathbb{T} \to L(l^2(\mathbb{Z})), \ t_1 \mapsto \sum_{\alpha_1\in\mathbb{Z}} (a_{\alpha_1,\alpha_2}^{1j})_h t_1^{\alpha_1} V_{\alpha_2} \quad \text{with } V_{\alpha_2}u(x_2) := u(x_2 + \alpha_2), \ j = 1, 2.$$

It is well known (see, e.g., [59], VIII, 8, 2°) that the operator in (2.115) (hence, the operator in (2.114)) is invertible if and only if

$$\inf_{t\in\mathbb{T}^2} |\det \widehat{A_h^{1j}}(t)| > 0, \quad j = 1, 2, \tag{2.116}$$

and if the partial indices of the matrix-function

$$\mathbb{T} \to L(\mathbb{C}^n), \quad t_2 \mapsto \left(\widehat{A_h^{11}}(t_1, t_2)\right)^{-1} \widehat{A_h^{12}}(t_1, t_2)$$

are equal to 0 for each $t_1 \in \mathbb{T}$. In the scalar case $n = 1$, the condition (2.116) simply becomes

$$\inf_{t\in\mathbb{T}^2} |\widehat{A_h^{1j}}(t)| > 0, \quad j = 1, 2,$$

and the condition imposed on the partial indices can be replaced by

$$\left[\arg \left(\widehat{A_h^{11}}(t_1, t_2)\right)^{-1} \widehat{A_h^{12}}(t_1, t_2)\right]_{t_2\in\mathbb{T}} = 0 \quad \text{for every } t_1 \in \mathbb{T}.$$

In the same way, the invertibility of the other operators in (2.112) can be checked effectively.

2.6.5 Difference operators of second order

We proceed with some applications of Theorem 2.6.19 to difference operators of second order which appear, for example, as approximations of the Schrödinger operator or of the acoustic operator. Throughout this section, let $E = l^p(\mathbb{Z}^N, \mathbb{C}) =: l^p(\mathbb{Z}^N)$ with $1 < p < \infty$. Thus, an operator on E is $\mathcal{P}$-Fredholm if and only if it is Fredholm in the common sense.

For $j = 1, \ldots, N$, let $e_j = (0, \ldots, 0, 1, 0, \ldots, 0)$ (with the 1 standing at the jth position) denote the jth unit vector, and let

$$\nabla_j := V_{e_j} - I \quad \text{and} \quad \nabla_j^* := V_{-e_j} - I$$

refer to the difference analogue of the operators of partial derivative into the direction of e_j. We consider the difference operator of second order on $\mathbb{Z}^N$,

$$A = \sum_{i,j=1}^{N} \nabla_i^* a_{ij} \nabla_j + cI \tag{2.117}$$

where a_{ij} and c are bounded functions on $\mathbb{Z}^N$. Clearly, this operator acts boundedly on E for every p.

Theorem 2.6.22 *Suppose there is an $S > 0$ such that $a_{ij}(x) = \overline{a_{ji}(x)}$ for all $x \in \mathbb{Z}^N$ with $|x| > S$, and let*

$$\lim_{R\to\infty} \inf_{|x|>R} \operatorname{Im} c(x) > 0. \tag{2.118}$$

Then the operator (2.117) *is Fredholm on E.*

Proof. If $h \in \mathcal{H}$ is a sequence such that the limit operator A_h exists, then there is a subsequence g of h such that all limit operators $(a_{ij}I)_g$ and $(cI)_g$ exist, too. Hence,

$$A_h = A_g = \sum_{i,j=1}^{N} \nabla_i^* (a_{ij}I)_g \nabla_j + (cI)_g.$$

The conditions imposed on the coefficients yield

$$(a_{ij}I)_g(x) = \overline{(a_{ji}I)_g(x)} \quad \text{and} \quad \operatorname{Im}(cI)_g(x) \geq \delta > 0 \tag{2.119}$$

with a constant δ.

We first show the invertibility of A_h on $l^2(\mathbb{Z}^N)$. Let $\langle ., . \rangle$ stand for the standard inner product on $l^2(\mathbb{Z}^N)$. Then, for every $u \in l^2(\mathbb{Z}^N)$,

$$\begin{aligned} \langle A_h u, u \rangle &= \sum_{i,j=1}^{N} \langle \nabla_i^* (a_{ij}I)_g \nabla_j u, u \rangle + \langle (cI)_g u, u \rangle \\ &= \sum_{i,j=1}^{N} \langle (a_{ij}I)_g \nabla_j u, \nabla_i u \rangle + \langle (cI)_g u, u \rangle. \end{aligned}$$

The matrix $((a_{ij}I)_g)_{i,j=1}^N$ is Hermitian. Hence,

$$\operatorname{Im} \langle A_h u,\, u\rangle = \operatorname{Im} \langle (cI)_g u,\, u\rangle,$$

whence via (2.119)

$$|\langle A_h u,\, u\rangle| \geq |\operatorname{Im} \langle (cI)_g u,\, u\rangle| \geq \delta \|u\|_2^2.$$

Repeating these considerations for the adjoint operator of A, we obtain the invertibility of A_h on $l^2(\mathbb{Z}^N)$. Since A_h is a band operator, this implies the invertibility of A_h on E. By Theorem 2.5.7, A is Fredholm on E. □

We are going to apply this result to the difference analogue of the acoustic operator which describes the sound extension in perturbed stratified media (see [38] and [189]). This operator is of the form

$$A = \sum_{i=1}^N \rho \nabla_i^* \rho^{-1} \nabla_i + cI \tag{2.120}$$

where ρ and c are bounded functions.

Theorem 2.6.23 *Let $\rho \in l^\infty(\mathbb{Z}^N)$ be a real-valued and invertible (in l^∞) function, and let $c \in l^\infty(\mathbb{Z}^N)$ be a function with* $\operatorname{Im} c(x) \geq 0$ *for all $x \in \mathbb{Z}^N$ and such that* (2.118) *is satisfied. Then the operator A in* (2.120) *is invertible on E.*

Proof. Let A be the operator in (2.120). The hypotheses guarantee that the operator $\rho^{-1}A$ is subject to Theorem 2.6.22. Hence, it is Fredholm on E. This implies the Fredholmness of A on E, and it remains to verify that both the kernel and the cokernel of A consist of zero only.

For, we introduce a new inner product $\langle .,.\rangle_\rho$ on $l^2(\mathbb{Z}^N)$ by

$$\langle u,\, v\rangle_\rho := \sum_{x\in\mathbb{Z}^N} \rho^{-1}(x) u(x) \overline{v(x)} = \langle \rho^{-1} u,\, v\rangle.$$

Let u be a solution of the homogeneous equation $Au = 0$. Then

$$\begin{aligned} 0 &= \operatorname{Im} \langle Au,\, u\rangle_\rho = \operatorname{Im} \langle \rho^{-1} Au,\, u\rangle \\ &= \operatorname{Im} \sum_{i=1}^N (\langle \rho^{-1} \nabla_i u,\, \nabla_i u\rangle + \langle \rho^{-1} cu,\, u\rangle) \\ &= \operatorname{Im} \sum_{i=1}^N (\langle \nabla_i u,\, \nabla_i u\rangle_\rho + \langle cu,\, u\rangle_\rho) = \operatorname{Im} \langle cu,\, u\rangle_\rho. \end{aligned} \tag{2.121}$$

Given $R > 0$ and $x = (x_1, \ldots, x_N) \in \mathbb{Z}^N$, set

$$u_R(x) := \begin{cases} u(x) & \text{if } |x|_\infty > R, \\ 0 & \text{if } |x|_\infty \leq R. \end{cases}$$

From $\operatorname{Im} c(x) \geq 0$ we conclude that

$$\operatorname{Im} \langle cu, u\rangle_\rho \geq \operatorname{Im} \langle cu_R, u_R\rangle_\rho$$

for all R, and from (2.118) we obtain

$$\operatorname{Im}\langle cu_R, u_R\rangle_\rho \geq \delta\langle u_R, u_R\rangle_\rho$$

for some $\delta > 0$ and R large enough. Combining these estimates with (2.121) we find

$$0 = \operatorname{Im} \langle cu, u\rangle_\rho \geq \operatorname{Im} \langle cu_R, u_R\rangle_\rho \geq \delta\langle u_R, u_R\rangle_\rho.$$

Thus, if R is large enough, then $u_R = 0$ or, equivalently,

$$u(x) = 0 \quad \text{for all } x \text{ with } |x|_\infty \geq R + 1. \tag{2.122}$$

On the other hand, since u is in the kernel of A, we have

$$0 = \rho^{-1}Au = du - \sum_{j=1}^{N}(\rho^{-1}V_{e_j}u + \rho_j^{-1}V_{-e_j}u) \tag{2.123}$$

with a certain bounded function d and invertible bounded functions ρ_j which are shifts of the function ρ. We claim that this identity together with (2.122) implies $u(x) = 0$ for all $x \in \mathbb{Z}^N$. Indeed, choose a vector $x \in \mathbb{Z}^N$ with $x_{j_0} \geq R+1$ for some j_0 (the other entries of x are arbitrary). Due to (2.122) we have $u(x) = 0$ and $u(x_1, \ldots, x_{j-1}, x_j \pm 1, x_{j+1}, \ldots, x_N) = 0$ for all $j \neq j_0$ as well as $u(x_1, \ldots, x_{j_0-1}, x_{j_0} + 1, x_{j_0+1}, \ldots, x_N) = 0$. Evaluating (2.123) at this vector x yields

$$\rho^{-1}(x)\, u(x_1, \ldots, x_{j_0-1}, x_{j_0} - 1, x_{j_0+1}, \ldots, x_N) = 0$$

whence

$$u(x_1, \ldots, x_{j_0-1}, x_{j_0} - 1, x_{j_0+1}, \ldots, x_n) = 0.$$

The same argumentation with $-R-1$ in place of $R+1$ yields

$$u(x_1, \ldots, x_{j_0-1}, x_{j_0} + 1, x_{j_0+1}, \ldots, x_n) = 0$$

for all $x \in \mathbb{Z}^N$ with $x_{j_0} \leq -R-1$. Consequently, $u(x) = 0$ whenever $|x_i| \geq R$ for some i. This shows that (2.122) also holds with R in place of $R+1$. Repeating this argumentation, we finally conclude that (2.122) holds with 0 in place of $R+1$. Thus, the kernel of A is trivial. Applying similar arguments to the adjoint operator of A we get that the cokernel of A is trivial, too. Hence, A is invertible. □

2.6.6 Discrete Schrödinger operators

In this section we will determine the essential spectrum of the discrete Schrödinger operator

$$A = \sum_{j=1}^{N} \left(V_{e_j} + V_{-e_j} \right) + vI \tag{2.124}$$

where the unit vectors e_j are as in the beginning of Section 2.6.5 and where the scalar-valued function v is in $l^\infty(\mathbb{Z}^N)$. This operator is the discrete analogue of the Schrödinger operator $-\Delta + vI$, where $\Delta := \sum_{j=1}^{N} D_{x_j}^2$ is the Laplacian on $\mathbb{R}^N$ and where the potential v is supposed to be in $L^\infty(\mathbb{R}^N)$. Spectral properties of Schrödinger operators are of extreme significance in quantum mechanics.

Being band operators, discrete Schrödinger operators belong to the Wiener algebra $\mathcal{W}$. Thus, the essential spectrum of the operator (2.124) can be determined via Corollary 2.5.8. Note that the limit operator of the operator (2.124) with respect to the sequence $h \in \mathcal{H}$ exists if and only if the limit operator of vI with respect to this sequence exists and that

$$A_h = \sum_{j=1}^{N} \left(V_{e_j} + V_{-e_j} \right) + (vI)_h \tag{2.125}$$

in this case.

Theorem 2.6.24 *Let A be as in* (2.124). *Then the essential spectrum of A in the space E^∞ does not depend on the concrete choice of E^∞, and*

$$\sigma_{ess}(A) = \cup \, \sigma_{E^\infty}(A_h)$$

where A_h is as in (2.125) *and where the union is taken over all limit operators A_h of A.*

Slowly oscillating potentials. In case the potential v is slowly oscillating on $\mathbb{Z}^N$, all limit operators of vI are scalar operators. Hence, every limit operator of the discrete Schrödinger operator (2.124) is of the form

$$A_q = \sum_{j=1}^{N} \left(V_{e_j} + V_{-e_j} \right) + qI \tag{2.126}$$

where the complex number q is a partial limit of the function v as $x \to \infty$, and the operators of this form exhaust the operator spectrum of A. We denote the set of all partial limits at infinity of the function $v \in SO(\mathbb{Z}^N)$ by v_∞. This set coincides with the set of the values of the restriction of the Gelfand transform of the function v onto the fiber $M^\infty(SO)$.

By Theorem 2.3.25, the spectrum of the operator (2.126) coincides with the set of the values of the function

$$\mathbb{T}^N \to \mathbb{C}, \quad (\xi_1, \dots, \xi_N) \mapsto \sum_{j=1}^{N} (\xi_j + \overline{\xi_j}) + q = \sum_{j=1}^{N} 2\mathrm{Re}\, \xi_j + q$$

which evidently is the interval $[-2N + q,\, 2N + q]$.

Theorem 2.6.25 *Let $v \in SO(\mathbb{Z}^N)$ and let A be as in (2.124). Then the essential spectrum of A, considered as an operator on E^∞, does not depend on the space E^∞, and*

$$\sigma_{ess}(A) = \cup_{q \in v_\infty} [-2N + q,\, 2N + q] = v_\infty + [-2N,\, 2N].$$

For the following corollary, let $m_v := \liminf_{x\to\infty} v(x)$ and $M_v := \limsup_{x\to\infty} v(x)$ in case $N > 1$, and set $m_v^\pm := \liminf_{x\to\pm\infty} v(x)$ and $M_v^\pm := \limsup_{x\to\pm\infty} v(x)$ in case $N = 1$.

Corollary 2.6.26 *Let, in addition to the hypotheses of the previous theorem, the potential v be real-valued. Then*

$$\sigma_{ess}(A) = [m_v - 2N,\, M_v + 2N] \quad \textit{in case } N > 1 \tag{2.127}$$

and

$$\sigma_{ess}(A) = [m_v^+ - 2N,\, M_v^+ + 2N] \cup [m_v^- - 2N,\, M_v^- + 2N] \quad \textit{in case } N = 1. \tag{2.128}$$

Further, $\sigma(A) \subset \mathbb{R}$, and all points in $\sigma(A) \setminus \sigma_{ess}(A)$ are discrete.

Proof. Let $N > 1$. Since $M^\infty(SO(\mathbb{Z}^N))$ is compact and connected by Theorem 2.4.7, and since v is real-valued, v_∞ is the interval $[m_v,\, M_v]$, whence (2.127). Similarly, in case $N = 1$, one concludes (2.128) from the connectedness of the fibers of $M(SO(\mathbb{Z}))$ over $\pm\infty$. The discreteness of the part of the spectrum of A which lies outside the essential spectrum is a consequence of the Gohberg/Sigal theorem (see [60], Ch. XI, Corollaries 8.4 and 8.5). □

The Gohberg/Sigal theorem states moreover that all points in $\sigma(A) \setminus \sigma_{ess}(A)$ are eigenvalues of A and that these points can cluster only at the boundary points of $\sigma_{ess}(A)$.

Separately slowly oscillating potentials. Let $N = N_1 + N_2$ with positive integers N_1, N_2 and, accordingly, write $x \in \mathbb{Z}^N$ as $(x_1, x_2) \in \mathbb{Z}^{N_1} \times \mathbb{Z}^{N_2}$. We call the function $a \in l^\infty(\mathbb{Z}^N)$ *separately slowly oscillating* if

$$\lim_{x_1 \to \infty} \sup_{x_2 \in \mathbb{Z}^{N_2}} (a(x_1 + y_1,\, x_2) - a(x_1,\, x_2)) = 0 \quad \text{for each } y_1 \in \mathbb{Z}^{N_1}$$

and

$$\lim_{x_2 \to \infty} \sup_{x_1 \in \mathbb{Z}^{N_1}} (a(x_1, x_2 + y_2) - a(x_1, x_2)) = 0 \quad \text{for each } y_2 \in \mathbb{Z}^{N_2}.$$

We denote the class of all separately slowly oscillating functions by $SO(\mathbb{Z}^{N_1}, \mathbb{Z}^{N_2})$.

Our goal is to describe the structure of the essential spectra of Schrödinger operators with separately slowly oscillating and real-valued potentials. To start with, let the separately slowly oscillating potential v depend on $x_1 \in \mathbb{Z}^{N_1}$ only. Then the Schrödinger operator

$$A_v = \sum_{j=1}^{N} (V_{e_j} + V_{-e_j}) + vI \tag{2.129}$$

is invariant with respect to shifts by the vectors $h = (0, h_2) \in \mathbb{Z}^{N_1} \times \mathbb{Z}^{N_2}$, and therefore

$$\sigma_{ess}(A) = \sigma(A),$$

i.e., A has no discrete spectrum.

As before, it follows from Theorem 2.3.25, that the spectrum of the operator (2.126) coincides with the spectrum of the operator-valued function

$$\mathbb{T}^{N_2} \to L(l^2(\mathbb{Z}^{N_1})),\ (\xi_{N_1+1}, \dots, \xi_{N_1+N_2}) \mapsto \sum_{j=1}^{N_1} (V_{e_j} + V_{-e_j}) + \sum_{j=N_1+1}^{N} 2\mathrm{Re}\, \xi_j + vI.$$

Since, by Corollary 2.6.26, the spectrum of the operator

$$\sum_{j=1}^{N_1} (V_{e_j} + V_{-e_j}) + vI,$$

thought of as acting on $l^2(\mathbb{Z}^{N_1})$, is the union of the segment $[m_v - 2N_1, M_v + 2N_1]$ with a discrete set $\{\mu_1, \mu_2, \dots\} \subset \mathbb{R} \setminus [m_v - 2N_1, M_v + 2N_1]$, we get

$$\begin{aligned}\sigma_{ess}(A_v) &= \sigma(A_v) \\ &= [m_v - 2N, M_v + 2N] \cup_{j=1}^{\infty} [m_v + \mu_j - 2N_2, M_v + \mu_j + 2N_2].\end{aligned} \tag{2.130}$$

Now we turn over to the general case of a real-valued separately slowly oscillating potential v. Recall that the essential spectrum of an operator $A \in \mathcal{W}$ is given by

$$\sigma_{ess}(A) = \bigcup_{\eta \in S^{N-1}} \bigcup_{A_h \in \sigma_\eta(A)} \sigma_E(A_h). \tag{2.131}$$

Thus, we have to determine the spectra of the limit operators of A for each infinitely distant point $\eta \in S^{N-1}$. Let

$$\Sigma_1 := S^{N-1} \cap (\mathbb{R}^{N_1} \times \{0_{N_2}\}), \quad \Sigma_2 := S^{N-1} \cap (\{0_{N_1}\} \times \mathbb{R}^{N_2})$$

where 0_{N_1} and 0_{N_2} are the zero vectors in $\mathbb{R}^{N_1}$ and $\mathbb{R}^{N_2}$, respectively.

If $\eta \notin \Sigma_1 \cup \Sigma_2$, then every limit operator of A which lies in the local operator spectrum at η is of the form

$$A_h = \sum_{j=1}^{N} (V_{e_j} + V_{-e_j}) + v_h I$$

where $v_h = \lim_{m\to\infty} v(h(m)) \in \mathbb{R}$. Hence,

$$\sigma_E(A_h) = [v_h - 2N,\, v_h + 2N] \tag{2.132}$$

and

$$\bigcup_{A_h \in \sigma_\eta(A)} \sigma_E(A_h) = [m_v^\eta - 2N,\, M_v^\eta + 2N], \tag{2.133}$$

where $m_v^\eta := \liminf_{x\to\eta} v(x)$ and $M_v^\eta := \limsup_{x\to\eta} v(x)$.

Let now $\eta \in \Sigma_2$. Then the operators $A_h \in \sigma_\eta(A)$ are of the form

$$A_h = \sum_{j=1}^{N} (V_{e_j} + V_{-e_j}) + v_h^\eta I$$

where v_h^η is a slowly oscillating function depending on $x_1 \in \mathbb{Z}^{N_1}$ only. Actually, this function is given by

$$v_h^\eta(x_1) = \lim_{m\to\infty} v((x_1,\, x_2) + h(m)).$$

Due to (2.130),

$$\begin{aligned} \sigma_E(A_h) \;&=\; \left[m_{v,1}^{\eta,h} - 2N,\, M_{v,1}^{\eta,h} + 2N\right] \\ &\quad \cup_{j=1}^{\infty}[m_{v,1}^{\eta,h} + \mu_j(A_h) - 2N_2,\, M_{v,1}^{\eta,h} + \mu_j(A_h) + 2N_2], \end{aligned} \tag{2.134}$$

where the $\mu_j(A_h)$ are the eigenvalues of the operator A_h and where

$$m_{v,1}^{\eta,h} := \liminf_{x_1\to\infty} v_h^\eta(x_1), \quad M_{v,1}^{\eta,h} := \limsup_{x_1\to\infty} v_h^\eta(x_1).$$

Analogously, if $\eta \in \Sigma_1$, then the operators $A_h \in \sigma_\eta(A)$ are of the form

$$A_h = \sum_{j=1}^{N} (V_{e_j} + V_{-e_j}) + v_h^\eta I,$$

where now v_h^η is a slowly oscillating function depending only on $x_2 \in \mathbb{Z}^{N_2}$ which is given by

$$v_h^\eta(x_2) = \lim_{m\to\infty} v((x_1,\, x_2) + h(m)).$$

For the spectrum of A_h, we find due to (2.130)

$$\sigma_E(A_h) = \left[m_{v,2}^{\eta,h} - 2N,\, M_{v,2}^{\eta,h} + 2N\right] \cup_{j=1}^{\infty}[m_{v,2}^{\eta,h} + \mu_j(A_h) - 2N_1,\, M_{v,2}^{\eta,h} + \mu_j(A_h) + 2N_1], \tag{2.135}$$

where again the $\mu_j(A_h)$ are the eigenvalues of the operator A_h and where

$$m_{v,2}^{\eta,h} := \liminf_{x_2\to\infty} v_h^\eta(x_2), \quad M_{v,2}^{\eta,h} := \limsup_{x_2\to\infty} v_h^\eta(x_2).$$

It follows from (2.131)–(2.135) that

$$\begin{aligned}\sigma_{ess}(A) &= [m_v - 2N,\, M_v + 2N] \\ &\bigcup_{\eta\in\Sigma_1}\bigcup_{A_h\in\sigma_\eta(A)}\bigcup_{j=1}^{\infty}[m_{v,2}^{\eta,h} + \mu_j(A_h) - 2N_1,\, M_{v,2}^{\eta,h} + \mu_j(A_h) + 2N_1] \\ &\bigcup_{\eta\in\Sigma_2}\bigcup_{A_h\in\sigma_\eta(A)}\bigcup_{j=1}^{\infty}[m_{v,1}^{\eta,h} + \mu_j(A_h) - 2N_2,\, M_{v,1}^{\eta,h} + \mu_j(A_h) + 2N_2]\end{aligned}$$

with $m_v = \liminf_{x\to\infty} v(x)$ and $M_v = \limsup_{x\to\infty} v(x)$.

Quasi almost periodic potentials. We call a function in $l^\infty(\mathbb{Z}^N)$ *quasi almost periodic* if it can be approximated in the norm of $l^\infty(\mathbb{Z}^N)$ by finite sums of functions of the form bc where $b \in SO(\mathbb{Z}^N)$ and $c \in AP(\mathbb{Z}^N)$. The C^*-algebra of all quasi almost periodic functions will be denoted by $QAP(\mathbb{Z}^N)$.

We consider a discrete Schrödinger operator A with potential $v \in QAP(\mathbb{Z}^N)$. It follows from the definition of the class QAP that all limit operators of the operator vI are operators of multiplication by a function $v_h \in AP(\mathbb{Z}^N)$. Thus, by Theorem 2.6.24,

$$\sigma_{ess}(A) = \bigcup_{A_h\in\sigma_{op}(A)} \sigma_E(A_h),$$

where all limit operators A_h are discrete Schrödinger operators with almost periodic potentials. It follows from Theorems 2.6.2 and 2.6.23 that $\sigma_{ess}(A_h) = \sigma_E(A_h)$ and that these spectra do not depend on the underlying space E^∞.

2.7 Indices of Fredholm band-dominated operators

Recall that an operator $A \in L(X)$, acting on a Banach space X, is a Fredholm operator if its kernel $\ker A := \{x \in X : Ax = 0\}$ and its cokernel $\operatorname{coker} A := X/\operatorname{im} A$ are finite-dimensional linear spaces, and that in this case the number

$$\operatorname{ind} A := \dim\ker A - \dim\operatorname{coker} A$$

is called the Fredholm index of A. From Theorem 2.2.1 we know that a band-dominated operator on $X = l^p(\mathbb{Z}^N)$ is Fredholm if and only if its limit operators are invertible and if the norms of their inverses are uniformly bounded. It is the goal of this section to present results which allow us to determine the index of a Fredholm band-dominated operator from its operator spectrum in case when $p = 2$ and $N = 1$.

2.7.1 Main results

Let $\mathbb{Z}_+$ and $\mathbb{Z}_-$ stand for the sets of the non-negative and negative integers, and write P and Q for the orthogonal projections from $l^2(\mathbb{Z})$ onto $l^2(\mathbb{Z}_+)$ and $l^2(\mathbb{Z}_-)$, respectively. (We identify $l^2(\mathbb{Z}_+)$ and $l^2(\mathbb{Z}_-)$ with subspaces of $l^2(\mathbb{Z})$ in the obvious way.) If A is a band-dominated operator, then the operators PAQ and QAP are compact (they are of finite rank if A is a band operator). Hence, the operators $A - (PAP + Q)(P + QAQ)$ and $A - (P + QAQ)(PAP + Q)$ are compact for every band-dominated operator A, and this shows that a band-dominated operator A is Fredholm if and only if both operators $PAP + Q$ and $P + QAQ$ are Fredholm. In this case, we call $\mathrm{ind}_+ A := \mathrm{ind}\,(PAP + Q)$ and $\mathrm{ind}_- A := \mathrm{ind}\,(P + QAQ)$ the *plus-index* and the *minus-index* of A. Evidently,

$$\mathrm{ind}\, A = \mathrm{ind}_+ A + \mathrm{ind}_- A$$

for every Fredholm band-dominated operator A.

Recall further that the operator spectrum of a band-dominated operator A splits into

$$\sigma_{op}(A) = \sigma_+(A) \cup \sigma_-(A)$$

where $\sigma_+(A)$ and $\sigma_-(A)$ stand for the sets of all limit operators of A which correspond to sequences tending to $+\infty$ and to $-\infty$, respectively.

The surprisingly simple answer to the index questions posed above is as follows.

Theorem 2.7.1 *Let A be a Fredholm band-dominated operator. Then*

(a) *for all $B \in \sigma_\pm(A)$,*

$$\mathrm{ind}_\pm(B) = \mathrm{ind}_\pm(A),$$

(b) *all operators in $\sigma_+(A)$ have the same plus-index, and all operators in $\sigma_-(A)$ have the same minus-index.*

(c) *for arbitrarily chosen operators $B_+ \in \sigma_+(A)$ and $B_- \in \sigma_-(A)$,*

$$\mathrm{ind}\, A = \mathrm{ind}_+ B_+ + \mathrm{ind}_- B_-. \tag{2.136}$$

So we can think of the plus- and the minus-index of A as local indices at $+\infty$ and $-\infty$.

To mention at least one example in which the identity (2.136) implies an explicit and effective formula for the computation of the Fredholm index, we consider band-dominated operators with slowly oscillating coefficients. These are the norm limits of band operators of the form $\sum_{n=-k}^{k} a_n V_n$ where the $a_n I$ are operators of multiplication by slowly oscillating functions. By definition, a function $a \in l^\infty(\mathbb{Z})$ is *slowly oscillating* if

$$\lim_{n\to\pm\infty} |a(n+1) - a(n)| = 0.$$

From Proposition 2.4.26 we know that every limit operator A_h of a band-dominated operator A with slowly oscillating coefficients is shift invariant. Thus, there is a continuous function a_h on the unit circle $\mathbb{T}$ such that A_h is just the Laurent operator $L(a_h)$. Recall that every function $a \in C(\mathbb{T})$ induces a bounded linear Laurent operator $L(a)$ on $l^2(\mathbb{Z})$ by $(L(a)x)(n) := \sum_{k\in\mathbb{Z}} a_{n-k}x(k)$ where a_n refers to the nth Fourier coefficient of a. The Laurent operator $L(a)$ is invertible if and only if the function a is invertible in $C(\mathbb{T})$. Thus, Theorem 2.2.1 yields an effective criterion for the Fredholmness of band-dominated operators with slowly oscillating coefficients. Moreover, the compression $PL(a)P$ of the Laurent operator $L(a)$ onto $l^2(\mathbb{Z}_+)$ is the Toeplitz operator $T(a)$, which is Fredholm if and only if its generating function a is invertible in $C(\mathbb{T})$, and which has minus the winding number of a with respect to the origin as its index (see [30, 32, 59], for example). Thus, also the plus- and minus-index of Fredholm band-dominated operators with slowly oscillating coefficients can be effectively determined.

The following subsections are devoted to the proof of Theorem 2.7.1. The strategy of the proof is as follows. Let $\mathcal{A}(\mathbb{Z})$ stand for the C^*-algebra of all band-dominated operators on $l^2(\mathbb{Z})$, and let $\mathcal{J}_+$ be the closed ideal of $\mathcal{A}(\mathbb{Z})$ generated by P. If A is a Fredholm band-dominated operator, then $PAP + Q$ is a Fredholm operator in the unitization $\mathcal{A}_1$ of $\mathcal{J}_+$. We would like to show that

$$\operatorname{ind}(PAP + Q) = \operatorname{ind}(PA_hP + Q)$$

for every sequence h tending to $+\infty$ for which the limit operator A_h exists. A simple reduction shows that it is enough to prove that the right-hand side vanishes if the left-hand side does. Suppose then that $PAP + Q$ has zero index. Then it is a compact perturbation of an invertible operator in $\mathcal{A}_1$. If we knew that the group of the invertible elements in the C^*-algebra $\mathcal{A}_1$ was path connected, then we could produce a continuous path of Fredholm operators in $\mathcal{A}_1$ joining $PAP + Q$ to the identity. Taking limit operators (perhaps with respect to a suitable subsequence of h) produces a continuous path of Fredholm operators joining $PA_hP + Q$ to the identity, thus showing that the operator $PA_hP + Q$ has index zero.

In fact, we do not know whether the group of invertibles of $\mathcal{A}_1$ is connected; but we can prove that the K-theory group $K_1(\mathcal{A}_1)$ vanishes. This implies that any invertible element in $\mathcal{A}_1$ can be connected to the identity after *stabilization* (i.e., after taking the direct sum with the identity in a matrix algebra), and that is enough to carry out the argument sketched above.

This K-theory calculations use techniques which are well known in the study of index theory on open manifolds and the coarse Baum-Connes conjecture. We first show that algebra $\mathcal{A}(\mathbb{Z})$ can be identified with a crossed product of $l^\infty(\mathbb{Z})$

by the group $\mathbb{Z}$. The Pimsner-Voiculescu exact sequence allows us to compute the K_1-group of this crossed product. (This calculation is essentially due to Yu [188]; compare also Roe [152], Lecture 4.) Then we plug in this result into a Mayer-Vietoris exact sequence to obtain that the K_1-group of $\mathcal{J}_+$ is $\{0\}$.

Notice also that $\mathcal{A}(\mathbb{Z})$ is just the rough algebra of the coarse space $\mathbb{Z}$ which is discussed in [152].

2.7.2 The algebra $\mathcal{A}(\mathbb{Z})$ as a crossed product

We start with recalling some facts on crossed products and reduced crossed products where we follow [18, 43, 114]. A *C^*-dynamical system* $(\mathcal{B}, G, \alpha)$ consists of a C^*-algebra $\mathcal{A}$, a locally compact group G, and a group homomorphism $\alpha : G \to \operatorname{Aut} \mathcal{B}$, $s \mapsto \alpha_s$. A pair (π, U) constituted by a *-representation $\pi : \mathcal{B} \to L(H)$ of $\mathcal{B}$ and a unitary representation $U : G \to L(H)$, $t \mapsto U_t$ of G on the same Hilbert space H, is called a *covariant representation* of the C^*-dynamical system $(\mathcal{B}, G, \alpha)$ if the covariance condition

$$U_t \pi(B) U_t^* = \pi(\alpha_t(B)) \quad \text{for all } B \in \mathcal{B} \text{ and } t \in G$$

is satisfied. A special class of covariant representations is obtained by taking the tensor product of a *-representation of $\mathcal{B}$ by the left regular representation of G which is defined as follows. Given a *-representation $\pi : \mathcal{B} \to L(H)$ of $\mathcal{B}$, let $l^2(G, H)$ refer to the Hilbert space of all square summable functions $x : G \to H$ with norm $\|x\|^2 := \sum_{t \in G} \|x(t)\|^2$. Then one has a covariant representation $(\tilde{\pi}, U)$ of $(\mathcal{B}, G, \alpha)$ which acts at $x \in l^2(G, H)$ by

$$(\tilde{\pi}(B)x)(s) := \pi(\alpha_s^{-1}(B))(x(s)) \quad \text{and} \quad (U_t x)(s) := x(t^{-1}s)$$

for $B \in \mathcal{B}$ and $s, t \in G$. If π is a faithful representation of $\mathcal{B}$, then the smallest C^*-subalgebra of $L(l^2(G, H))$ which contains all operators $\tilde{\pi}(B)$ with $B \in \mathcal{B}$ as well as all operators U_t with $t \in G$ is independent of the concrete choice of π. This algebra is called the *reduced crossed product of $\mathcal{B}$ by G* and is denoted by $\mathcal{B} \times_{\alpha r} G$ ([114], Theorem 7.7.5). Moreover, if the group G is amenable (for example, if G is commutative), then the reduced crossed product $\mathcal{B} \times_{\alpha r} G$ coincides with the crossed product $\mathcal{B} \times_\alpha G$ ([114], Theorem 7.7.7 and [43], Corollary VII.2.2).

Now we consider the special dynamical system $(l^\infty(\mathbb{Z}), \mathbb{Z}, \alpha)$ where $\alpha_k = \alpha(k)$, $k \in \mathbb{Z}$, acts at $a \in l^\infty(\mathbb{Z})$ by

$$(\alpha_k(a))(n) = a(n-k), \quad n \in \mathbb{Z}. \tag{2.137}$$

Proposition 2.7.2 *For the dynamical system $(l^\infty(\mathbb{Z}), \mathbb{Z}, \alpha)$ with α specified by (2.137), one has*

$$l^\infty(\mathbb{Z}) \times_\alpha \mathbb{Z} = l^\infty(\mathbb{Z}) \times_{\alpha r} \mathbb{Z} \cong \mathcal{A}(\mathbb{Z}).$$

Proof. We have already mentioned that the first identity holds in general for products by amenable groups. So we are left with showing that the algebra $\mathcal{A}(\mathbb{Z})$ is *-isomorphic to the reduced crossed product $l^\infty(\mathbb{Z}) \times_{\alpha r} \mathbb{Z}$.

The mapping π which associates with every sequence $a \in l^\infty(\mathbb{Z})$ the operator $aI \in L(l^2(\mathbb{Z}))$ of multiplication by a represents the C^*-algebra $l^\infty(\mathbb{Z})$ faithfully. This representation induces a covariant representation of the dynamical system $(l^\infty(\mathbb{Z}), \mathbb{Z}, \alpha)$ on the Hilbert space $H = l^2(\mathbb{Z}, l^2(\mathbb{Z}))$ via

$$(\tilde{\pi}(a)x)(s) := \pi(\alpha_s^{-1}(a))(x(s)) \quad \text{and} \quad (U_t x)(s) := x(s-t)$$

where $a \in l^\infty(\mathbb{Z})$ and $t \in \mathbb{Z}$. We identify $l^2(\mathbb{Z}, l^2(\mathbb{Z}))$ with $l^2(\mathbb{Z} \times \mathbb{Z})$ via $x(s, n) := (x(s))(n)$. Then we can identify $\tilde{\pi}(a)$ and U_t with the operators

$$(\tilde{\pi}(a)x)(s, n) := a(n+s)x(s, n) \quad \text{and} \quad (U_t x)(s, n) := x(s-t, n). \qquad (2.138)$$

Let $\mathcal{C}$ refer to the smallest C^*-subalgebra of $L(l^2(\mathbb{Z} \times \mathbb{Z}))$ which contains all operators $\tilde{\pi}(a)$ and U_t with $a \in l^\infty(\mathbb{Z})$ and $t \in \mathbb{Z}$, given by (2.138). This algebra is *-isomorphic to the reduced crossed product $l^\infty(\mathbb{Z}) \times_{\alpha r} \mathbb{Z}$ as quoted above, and we claim that it is also *-isomorphic to the algebra $\mathcal{A}(\mathbb{Z})$ of the band-dominated operators on $l^2(\mathbb{Z})$. For $n \in \mathbb{Z}$, let

$$H_n := \{x \in l^2(\mathbb{Z} \times \mathbb{Z}) : x(s, m) = 0 \text{ whenever } m \neq n\}.$$

We identify $l^2(\mathbb{Z} \times \mathbb{Z})$ with the orthogonal sum $\oplus_{n \in \mathbb{Z}} H_n$ such that $x \in l^2(\mathbb{Z} \times \mathbb{Z})$ is identified with $\oplus h_n \in \oplus H_n$ if $x(s, n) = h_n(s)$. From (2.138) we conclude that each space H_n is invariant with respect to each operator $C \in \mathcal{C}$. Hence, each operator $C \in \mathcal{C}$ corresponds to a diagonal matrix operator $\operatorname{diag}(\ldots, C_n, C_{n+1}, \ldots)$ with respect to the decomposition of $l^2(\mathbb{Z} \times \mathbb{Z})$ into the orthogonal sum of its subspaces H_n. In particular, C_n is nothing but the restriction of C onto H_n. Let $\mathcal{C}_n$ denote the C^*-algebra of all restrictions of operators in $\mathcal{C}$ onto H_n.

It is clear that each of the spaces H_n is isometric to $l^2(\mathbb{Z})$ with the isometry given by

$$J_n : H_n \to l^2(\mathbb{Z}), \quad (J_n x)(s) := x(s, n).$$

Thus, $J_n \mathcal{C}_n J_n^{-1}$ is a C^*-subalgebra of $L(l^2(\mathbb{Z}))$ which we denote by $\mathcal{B}_n$. Clearly, for $a \in l^\infty(\mathbb{Z})$, the operator $J_n \tilde{\pi}(a) J_n^{-1}$ is just the operator $\pi(\alpha_n(a))$, whereas $J_n U_t J_n^{-1}$ is the shift operator V_t. Since $\pi(\alpha_{-n}(a)) = V_n \pi(a) V_n^*$ and $V_t = V_n V_t V_n^*$, the mapping $B \mapsto V_n B V_{-n}$ is a *-isomorphism from $\mathcal{B}_n$ onto $\mathcal{A}(\mathbb{Z})$. Consequently, the mapping

$$\mathcal{A}(\mathbb{Z}) \to \mathcal{C}, \quad A \mapsto \operatorname{diag}(\ldots, J_n^{-1} V_n^* A V_n J_n, \ldots)$$

is a *-isomorphism. □

2.7.3 The K_1-group of $\mathcal{A}(\mathbb{Z})$

To compute the K_1-group of the algebra $\mathcal{A}(\mathbb{Z})$ we will make use of the fact that $\mathcal{A}(\mathbb{Z})$ is *-isomorphic to the crossed product $l^\infty(\mathbb{Z}) \times_{\alpha r} \mathbb{Z}$ by Proposition 2.7.2. The K-theory of crossed products by $\mathbb{Z}$ is dominated by the Pimsner-Voiculescu exact sequence ([115], see also [43], Theorem VIII.5.1) which we restate below. Recall in this connection that every automorphism α of a C^*-algebra $\mathcal{B}$ induces a group homomorphism from $\mathbb{Z}$ into $\operatorname{Aut} \mathcal{B}$ by $n \mapsto \alpha^n$ which we denote by α again.

Theorem 2.7.3 (The Pimsner-Voiculescu exact sequence.) *Let α be an automorphism of the C^*-algebra $\mathcal{B}$. Then there is a cyclic six term exact sequence*

$$\begin{array}{ccccc} K_0(\mathcal{B}) & \xrightarrow{\mathrm{id}_* - \alpha_*} & K_0(\mathcal{B}) & \longrightarrow & K_0(\mathcal{B} \times_\alpha \mathbb{Z}) \\ \uparrow & & & & \downarrow \\ K_1(\mathcal{B} \times_\alpha \mathbb{Z}) & \longleftarrow & K_1(\mathcal{B}) & \xleftarrow{\mathrm{id}_* - \alpha_*} & K_1(\mathcal{B}). \end{array} \tag{2.139}$$

We wish to apply this exact sequence to the algebra $\mathcal{A}(\mathbb{Z}) = l^\infty(\mathbb{Z}) \times_\alpha \mathbb{Z}$, i. e. with $\mathcal{B} = l^\infty(\mathbb{Z})$. Since $l^\infty(\mathbb{Z})$ is a von Neumann algebra, one has $K_1(l^\infty(\mathbb{Z})) = \{0\}$ ([153], Exercise 8.14). Thus, (2.139) becomes

$$\begin{array}{ccccc} K_0(l^\infty(\mathbb{Z})) & \xrightarrow{\mathrm{id}_* - \alpha_*} & K_0(l^\infty(\mathbb{Z})) & \longrightarrow & K_0(\mathcal{A}(\mathbb{Z})) \\ \uparrow & & & & \downarrow \\ K_1(\mathcal{A}(\mathbb{Z})) & \longleftarrow & \{0\} & \longleftarrow & \{0\}. \end{array} \tag{2.140}$$

The K_0-group of $l^\infty(\mathbb{Z})$. The K_0-group of the algebra $l^\infty(\mathbb{Z})$ coincides with the group of all bounded functions from $\mathbb{Z}$ into $\mathbb{Z}$ which we denote by $\mathbb{Z}_b^{\mathbb{Z}}$. Again we start with recalling the basic steps in the definition of the K_0-groupof a C^*-algebra, where we follow [153], Chapter 3.

For n a positive integer and $\mathcal{B}$ a unital C^*-algebra, let $\mathcal{P}_n(\mathcal{B})$ stand for the set of all projections (i.e., self-adjoint idempotents) in the algebra $\mathcal{B}_{n\times n}$ of all $n \times n$ matrices with entries in $\mathcal{B}$, and set $\mathcal{P}_\infty(\mathcal{B}) := \cup_n \mathcal{P}_n(\mathcal{B})$. One defines a binary operation $\oplus$ and a relation $\sim$ on $\mathcal{P}_\infty(\mathcal{B})$ as follows. For $p \in \mathcal{P}_n(\mathcal{B})$ and $q \in \mathcal{P}_m(\mathcal{B})$, one sets

$$p \oplus q := \operatorname{diag}(p, q) \in \mathcal{P}_{n+m}(\mathcal{B}),$$

and one writes $p \sim q$ if there is an element $v \in \mathcal{B}_{m\times n}$ such that $p = v^*v$ and $q = vv^*$. Thus, if both p and q belong to $\mathcal{P}_n(\mathcal{B})$ for some n, then $p \sim q$ if and only if p and q are Murray-von Neumann equivalent. The following is Proposition 2.3.2 in [153].

Proposition 2.7.4 *Let p, q, r, p', $q' \in \mathcal{P}_\infty(\mathcal{B})$ for some unital C^*-algebra $\mathcal{B}$. Then*

(a) $p \sim p \oplus 0_{n\times n}$.

(b) *If $p \sim p'$ and $q \sim q'$, then $p \oplus q \sim p' \oplus q'$.*

(c) $p \oplus q \sim q \oplus p$.

(d) *If p, $q \in \mathcal{P}_n(\mathcal{B})$ and $pq = 0$, then $p + q \in \mathcal{P}_n(\mathcal{B})$ and $p + q \sim p \oplus q$.*

(e) $(p \oplus q) \oplus r \sim p \oplus (q \oplus r)$.

Let $D(\mathcal{B}) := \mathcal{P}_\infty(\mathcal{B})/\sim$, write $[p]_\sim$ for the coset of $p \in \mathcal{P}_\infty(\mathcal{B})$ in $D(\mathcal{B})$, and define an operation $+$ on $D(\mathcal{B})$ by $[p]_\sim + [q]_\sim := [p \oplus q]_\sim$. Then $D(\mathcal{B})$ becomes an abelian semigroup, and the Grothendieck group of $D(\mathcal{B})$ is called the K_0-group of $\mathcal{B}$.

Now we specify $\mathcal{B} = l^\infty(\mathbb{Z})$ and let $P \in \mathcal{P}_\infty(\mathcal{B})$. Then $P \in \mathcal{P}_k(l^\infty(\mathbb{Z})) = \mathcal{P}(l^\infty(\mathbb{Z})_{k\times k})$ for some k. Since $l^\infty(\mathbb{Z})_{k\times k} = l^\infty(\mathbb{Z}, \mathbb{C}_{k\times k})$, we can think of P as a sequence of projections in $\mathbb{C}_{k\times k}$. Conversely, each sequence of projections in $\mathbb{C}_{k\times k}$ determines an element of $\mathcal{P}_k(l^\infty(\mathbb{Z}))$.

For $P \in \mathcal{P}(l^\infty(\mathbb{Z}, \mathbb{C}_{k\times k})$, let $\operatorname{rank} P$ be the sequence

$$\mathbb{Z} \to \mathbb{Z}_+, \quad n \mapsto \operatorname{rank} P(n).$$

Clearly, this sequence is bounded by k and, conversely, every bounded sequence from $\mathbb{Z}$ into $\mathbb{Z}_+$ is the rank of a certain projection in $\mathcal{P}_\infty(l^\infty(\mathbb{Z}))$.

We claim that if $P, Q \in \mathcal{P}_\infty(l^\infty(\mathbb{Z}))$, then

$$P \sim Q \quad \Longleftrightarrow \quad \operatorname{rank} P = \operatorname{rank} Q. \tag{2.141}$$

Since $\sim$ is an equivalence relation and by Proposition 2.7.4 (a), we can assume without loss of generality that $P, Q \in \mathcal{P}_k(l^\infty(\mathbb{Z}))$ with some positive integer k. Then the implication $\Leftarrow$ in (2.141) can be seen as follows. If the matrices $P(n), Q(n) \in \mathcal{P}(\mathbb{C}_{k\times k})$ have rank $l \le k$, then there are unitary operators U_n and V_n such that

$$U_n^* P(n) U_n = \operatorname{diag}(\underbrace{1, \ldots, 1}_{l}, 0, \ldots, 0) = V_n^* Q(n) V_n.$$

Define $W \in l^\infty(\mathbb{Z}, \mathbb{C}_{k\times k})$ by $W(n) := V_n U_n^*$. Then W is a unitary element in $l^\infty(\mathbb{Z}, \mathbb{C}_{k\times k})$, and $P = W^*QW$. Hence, the projections P and Q are unitarily equivalent, which implies their Murray-von Neumann equivalence ([153], Proposition 2.2.2).

For the reverse implication in (2.141), let $P, Q \in \mathcal{P}(l^\infty(\mathbb{Z}, \mathbb{C}_{k\times k}))$ and $P \sim Q$. Then $P(n) \sim Q(n)$ for every $n \in \mathbb{Z}$. By elementary linear algebra, this implies that $\operatorname{rank} P(n) = \operatorname{rank} Q(n)$ and, hence, $\operatorname{rank} P = \operatorname{rank} Q$ ([153], Exercise 2.9).

This proves (2.141), and from the definition of the addition $\oplus$ in $\mathcal{P}_\infty(\mathcal{B})$ we conclude that $D(l^\infty(\mathbb{Z}))$ is isomorphic to the semigroup of all bounded sequences from $\mathbb{Z}$ into $\mathbb{Z}_+$, provided with the operation of pointwise addition. Passing to the Grothendieck group of this semigroup, we get

$$K_0(l^\infty(\mathbb{Z})) \cong \mathbb{Z}_b^{\mathbb{Z}}. \tag{2.142}$$

The mapping $\mathrm{id}_* - \alpha_*$ and its kernel. K-theory is functorial, i.e., given C^*-algebras $\mathcal{B}$ and $\mathcal{C}$ and a *-homomorphism $\varphi : \mathcal{B} \to \mathcal{C}$, there is a unique group homomorphism $\varphi_* : K_0(\mathcal{B}) \to K_0(\mathcal{C})$ such that

$$\varphi_* : [p]_\sim \mapsto [\varphi(p)]_\sim \quad \text{for } p \in \mathcal{P}_\infty.$$

Here, $\varphi(p)$ is defined as follows: the mapping φ extends to a *-homomorphism from $\mathcal{B}_{k\times k}$ into $\mathcal{C}_{k\times k}$ by

$$\varphi : (b_{ij})_{i,\,j=1}^k \mapsto (\varphi(b_{ij}))_{i,\,j=1}^k,$$

and since φ maps projections to projections, it maps $\mathcal{P}_\infty(\mathcal{B})$ into $\mathcal{P}_\infty(\mathcal{C})$. Thus, in our concrete setting, the mapping

$$\mathrm{id}_* : K_0(l^\infty(\mathbb{Z})) \to K_0(l^\infty(\mathbb{Z}))$$

which is induced by the identical mapping on $l^\infty(\mathbb{Z})$ is just the identical mapping on the associated K_0-groups. It is also clear that, still under the identification of $l^\infty(\mathbb{Z})_{k\times k}$ with $l^\infty(\mathbb{Z}, \mathbb{C}_{k\times k})$, the mapping

$$\alpha : \mathcal{P}_\infty(l^\infty(\mathbb{Z})) \to \mathcal{P}_\infty(l^\infty(\mathbb{Z})) \tag{2.143}$$

acts as the shift operator. Moreover, the equivalence (2.141) implies that, for $P, Q \in \mathcal{P}_\infty(l^\infty(\mathbb{Z}))$,

$$P \sim Q \quad \Longleftrightarrow \quad \alpha(P) \sim \alpha(Q).$$

Thus, the mapping (2.143) is compatible with the relation $\sim$, which shows that α induces the shift operator on $D(l^\infty(\mathbb{Z})) \cong \mathbb{Z}_b^{\mathbb{N}}$. This finally implies that α_* acts as the shift operator on $K_0(l^\infty(\mathbb{Z})) \cong \mathbb{Z}_b^{\mathbb{Z}}$.

Consequently, the kernel of the group homomorphism $\mathrm{id}_* - \alpha_*$ consists of all shift invariant sequences in $\mathbb{Z}_b^{\mathbb{Z}}$, i.e., of all constant sequences. The subgroup of $\mathbb{Z}_b^{\mathbb{Z}}$ of all constant sequences is isomorphic to $\mathbb{Z}$; so what we get is

$$\ker(\mathrm{id}_* - \alpha_*) \cong \mathbb{Z}. \tag{2.144}$$

Identification of $K_1(\mathcal{A}(\mathbb{Z}))$. The picture we have obtained so far is

$$\begin{array}{ccccc} \mathbb{Z}_b^{\mathbb{Z}} & \xrightarrow{\mathrm{id}_* - \alpha_*} & \mathbb{Z}_b^{\mathbb{Z}} & \longrightarrow & K_0(\mathcal{A}(\mathbb{Z})) \\ \uparrow \beta & & & & \downarrow \\ K_1(\mathcal{A}(\mathbb{Z})) & \xleftarrow{\iota} & \{0\} & \longleftarrow & \{0\}. \end{array} \tag{2.145}$$

Since group homomorphisms map the zero element to the zero element, we have $\mathrm{im}\,\iota = \{0\}$, which implies that $\ker\beta = \{0\}$ due to the exactness of (2.145) at $K_1(\mathcal{A}(\mathbb{Z}))$. Further, by (2.144) and since (2.145) is exact at its left upper corner, we have $\mathrm{im}\,\beta \cong \mathbb{Z}$. Hence, β is a injective group homomorphism on $K_1(\mathcal{A}(\mathbb{Z}))$ with range $\mathbb{Z}$. Summarizing, we have found that

$$K_1(\mathcal{A}(\mathbb{Z})) \cong \mathbb{Z}. \tag{2.146}$$

2.7.4 The K_1-group of $\mathcal{A}_\pm$

Following [75], we now split the algebra $\mathcal{A}(\mathbb{Z})$ into two subalgebras $\mathcal{A}_\pm$ which essentially contain the band-dominated operators on $\mathbb{Z}_\pm$, and we compute their respective K_1-groups. The basic device for this computation is the Mayer-Vietoris exact sequence which can be found in following form in [75] (Section 3, Lemma 1) and in [69] (Exercise 3.16), for instance.

Theorem 2.7.5 (The Mayer-Vietoris exact sequence.) *Let $\mathcal{B}$ be a C^*-algebra and let $\mathcal{I}$ and $\mathcal{J}$ be closed ideals of $\mathcal{B}$ such that $\mathcal{I} + \mathcal{J} = \mathcal{B}$. Then there is a cyclic six term exact sequence*

$$\begin{array}{ccccc} K_0(\mathcal{I} \cap \mathcal{J}) & \longrightarrow & K_0(\mathcal{I}) \oplus K_0(\mathcal{J}) & \longrightarrow & K_0(\mathcal{B}) \\ \uparrow & & & & \downarrow \\ K_1(\mathcal{B}) & \longleftarrow & K_1(\mathcal{I}) \oplus K_1(\mathcal{J}) & \longleftarrow & K_1(\mathcal{I} \cap \mathcal{J}). \end{array} \tag{2.147}$$

Let $\mathcal{J}_+$ denote the smallest closed subalgebra of the algebra $\mathcal{A}(\mathbb{Z})$ of all band-dominated operators on $\mathbb{Z}$ which contains the algebra $P\mathcal{A}(\mathbb{Z})P$ and the ideal $\mathcal{K}$ of the compact operators on $l^2(\mathbb{Z})$. Alternatively, $\mathcal{J}_+$ can be described as the closed ideal of $\mathcal{A}(\mathbb{Z})$ generated by the operator P. We further define $\mathcal{J}_-$ by replacing P by Q in that definition. Then $\mathcal{J}_+$ and $\mathcal{J}_-$ are closed ideals of $\mathcal{A}(\mathbb{Z})$ which satisfy $\mathcal{J}_+ + \mathcal{J}_- = \mathcal{A}(\mathbb{Z})$ and $\mathcal{J}_+ \cap \mathcal{J}_- = \mathcal{K}$. Let further $\mathcal{A}_+ := \mathcal{J}_- + \mathbb{C}Q$ and $\mathcal{A}_- := \mathcal{J}_- + \mathbb{C}P$. Then $\mathcal{A}_+$ and $\mathcal{A}_-$ are C^*-subalgebras of $\mathcal{A}(\mathbb{Z})$ which are *-isomorphic to the (minimal) unitizations of the ideals $\mathcal{J}_+$ and $\mathcal{J}_-$, respectively. For, one easily checks that every operator $A \in \mathcal{A}_+$ can be written as $A = PAP+K+\alpha Q$ where $PAP + K \in P\mathcal{A}(\mathbb{Z})P + \mathcal{K}$ and $\alpha \in \mathbb{C}$ are uniquely determined, and that

$$\mathcal{A}_+ \to \mathcal{J}_+ \times \mathbb{C}, \quad PAP + K + \alpha Q \mapsto (PAP + K - \alpha P,\, \alpha)$$

is a *-isomorphism from $\mathcal{A}_+$ onto the unitization $\mathcal{J}_+ \times \mathbb{C}$ of the ideal $\mathcal{J}_+$.

Thus, we can apply the Mayer-Vietoris exact sequence with $\mathcal{A}(\mathbb{Z})$, $\mathcal{J}_+$, $\mathcal{J}_-$ and $\mathcal{K}$ in place of $\mathcal{B}, \mathcal{I}$, $\mathcal{J}$ and $\mathcal{I} \cap \mathcal{J}$. The K-theory of $\mathcal{K}$ is well known,

$$K_0(\mathcal{K}) \cong \mathbb{Z} \quad \text{and} \quad K_1(\mathcal{K}) = \{0\}$$

(Corollary 6.4.2 and Example 8.2.9 in [153]). Thus, and by (2.146), the general exact sequence (2.147) specifies to

$$\begin{array}{ccccc} \mathbb{Z} & \longrightarrow & K_0(\mathcal{J}_+) \oplus K_0(\mathcal{J}_-) & \longrightarrow & K_0(\mathcal{A}(\mathbb{Z})) \\ \uparrow & & & & \downarrow \\ \mathbb{Z} & \xleftarrow{\beta} & K_1(\mathcal{J}_+) \oplus K_1(\mathcal{J}_-) & \xleftarrow{\gamma} & \{0\} \end{array}$$

with certain group homomorphisms β and γ. From $\operatorname{im} \gamma = \{0\}$ we conclude that β is injective. Hence, $K_1(\mathcal{J}_+) \oplus K_1(\mathcal{J}_-)$ is isomorphic to a subgroup of $\mathbb{Z}$. But each subgroup of $\mathbb{Z}$ is either isomorphic to $\mathbb{Z}$ or equal to $\{0\}$. Suppose for a moment that $K_1(\mathcal{J}_+) \oplus K_1(\mathcal{J}_-) \cong \mathbb{Z}$. Since the ideals $\mathcal{J}_+$ and $\mathcal{J}_-$ are *-isomorphic (a *-isomorphism is given by $K \mapsto JKJ$), their K_1-groups are isomorphic, too:

$$K_1(\mathcal{J}_+) \cong K_1(\mathcal{J}_-) =: \Gamma.$$

Thus, $\mathbb{Z}$ is isomorphic to $\Gamma \oplus \Gamma$, the direct sum of two copies of Γ. But $\mathbb{Z}$ is singly generated (by 1, for example), whereas $\Gamma \oplus \Gamma$ cannot be generated by a single element. This contradiction shows that

$$K_1(\mathcal{J}_+) \oplus K_1(\mathcal{J}_-) \cong \Gamma \oplus \Gamma = \{0\},$$

whence

$$K_1(\mathcal{J}_+) = K_1(\mathcal{J}_-) = \{0\}.$$

Finally, the K_1-groups of a C^*-algebra and of its unitization coincide (Proposition 8.1.6 and Equality (8.4) in [153]) which implies that

$$K_1(\mathcal{A}_+) \cong K_1(\mathcal{A}_-) = \{0\}. \tag{2.148}$$

2.7.5 Proof of Theorem 2.7.1

In this section, we will prove assertion (a) of Theorem 2.7.1 which has assertions (b) and (c) as its corollaries. In the course of the proof, we will make use of the following elementary properties of the plus- and minus-indices of Fredholm band-dominated operators.

Proposition 2.7.6 *Let A and B be Fredholm operators in $\mathcal{A}(\mathbb{Z})$. Then*

(a) $\text{ind}_\pm A$ *is invariant with respect to small perturbations.*

(b) $\text{ind}_\pm A$ *is invariant with respect to compact perturbations.*

(c) $\text{ind}_\pm A^* = -\text{ind}_\pm A$.

(d) $\text{ind}_\pm AB = \text{ind}_\pm A + \text{ind}_\pm A$.

The latter property follows from

$$PABP + Q = (PAP + Q)(PBP + Q) + \text{compact}.$$

For the proof of Theorem 2.7.1, we abbreviate the C^*-algebra of all $k \times k$ matrices with entries in $\mathcal{A}(\mathbb{Z})_+$ to $\mathcal{A}_k$ and write P_k and Q_k for the operators

$$\text{diag}\,(P, \ldots, P), \quad \text{diag}\,(Q, \ldots, Q) : \mathcal{A}_k \to \mathcal{A}_k.$$

It is clearly sufficient to prove the theorem for the plus-case where it reads as follows:

$$\text{ind}_+ A = \text{ind}_+ A_h \quad \text{for all } A_h \in \sigma_+(A). \tag{2.149}$$

It is further sufficient to prove (2.149) only in the case when $\text{ind}_+ A = 0$. Indeed, for the shift operator V_1 one has $\text{ind}_+ V_1 = -1$ and $\text{ind}_- V_1 = 1$. Thus, if $A \in \mathcal{A}(\mathbb{Z})$ is a Fredholm operator with plus-index r, then AV_1^r is a Fredholm operator with plus-index 0. If the identity (2.149) holds for all Fredholm operators with vanishing plus-index, then this implies that

$$\text{ind}_+(AV_1^r)_h = 0 \quad \text{for every limit operator of } AV_1^r.$$

But, evidently, every limit operator of AV_1^r is of the form $A_h V_1^r$ since V_1 is shift invariant. Thus,

$$\text{ind}_+(A_h V_1^r) = 0 \quad \text{for every } A_h \in \sigma_+(A),$$

whence, by Proposition 2.7.6 (d),

$$0 = \mathrm{ind}_+(A_h V_1^r) = \mathrm{ind}_+ A_h + \mathrm{ind}_+ V_1^r = \mathrm{ind}_+ A_h - r$$

and, finally, $\mathrm{ind}_+ A_h = r$ for every limit operator of A in $\sigma_+(A)$.

So, what we really have to check is that, for all Fredholm band-dominated operators A,

$$\mathrm{ind}_+ A = 0 \quad \Longrightarrow \quad \mathrm{ind}_+ A_h = 0 \quad \text{for all } A_h \in \sigma_+(A). \tag{2.150}$$

Let $\mathrm{ind}_+ A = 0$, i.e., $\mathrm{ind}\,(PAP + Q) = 0$. Let further K be a compact operator such that $B := PAP + Q + K \in \mathcal{A}_+$ is invertible, and let $B = UR$ be the polar decomposition of B, i.e., U is a unitary operator in $\mathcal{A}_1$, and R is a positive definite operator in $\mathcal{A}_1$. A consequence of the vanishing of the K_1-group of $\mathcal{A}_1$ (according to (2.148)) is that U is stably path connected to the identity operator (see the Definition 8.1.3 of the K_1-group in [153]). Thus, there is a positive integer k such that

$$\begin{pmatrix} U & 0 \\ 0 & I_{k-1} \end{pmatrix} \sim_h \begin{pmatrix} I & 0 \\ 0 & I_{k-1} \end{pmatrix}$$

in the set of the unitary operators of $\mathcal{A}_k$. Here, $\sim_h$ denotes homotopy equivalence, and I_{k-1} refers to the identity operator in $\mathcal{A}_{k-1}$.

Choose a continuous unitary-valued function

$$f_1 : [0,\, 1] \to \mathcal{A}_k \quad \text{with} \quad f_1(0) = \begin{pmatrix} U & 0 \\ 0 & I_{k-1} \end{pmatrix}, \quad f_1(1) = \begin{pmatrix} I & 0 \\ 0 & I_{k-1} \end{pmatrix}.$$

Further, let

$$f_2 : [0,\, 1] \to \mathcal{A}_k, \quad t \mapsto (1-t) \begin{pmatrix} R & 0 \\ 0 & I_{k-1} \end{pmatrix} + t \begin{pmatrix} I & 0 \\ 0 & I_{k-1} \end{pmatrix}$$

and

$$f_3 : [0,\, 1] \to \mathcal{A}_k, \quad t \mapsto (1-t) \begin{pmatrix} K & 0 \\ 0 & 0_{k-1} \end{pmatrix}.$$

Then f_2 is a continuous function having only positive definite operators as its values, and f_3 is a continuous function with compact values. Hence,

$$f := f_1 f_2 + f_3 : [0,\, 1] \to \mathcal{A}_k$$

is a continuous function with

$$f(0) = \begin{pmatrix} U & 0 \\ 0 & I_{k-1} \end{pmatrix} \begin{pmatrix} R & 0 \\ 0 & I_{k-1} \end{pmatrix} + \begin{pmatrix} K & 0 \\ 0 & 0_{k-1} \end{pmatrix} = \begin{pmatrix} PAP + Q & 0 \\ 0 & I_{k-1} \end{pmatrix}$$

and

$$f(1) = \begin{pmatrix} I & 0 \\ 0 & I_{k-1} \end{pmatrix},$$

and all values of that function are Fredholm operators (with index 0).

Let now $h : \mathbb{N} \to \mathbb{Z}$ be a sequence which tends to $+\infty$ and for which the limit operator A_h exists. Then, obviously, the limit operator of $f(0)$ with respect to h exists, and

$$f(0)_h = \begin{pmatrix} A_h & 0 \\ 0 & I_{k-1} \end{pmatrix}.$$

We use a Cantor diagonal argument in order to produce a subsequence g of h such that the limit operator $f(q)_g$ exists for every rational number q in $[0, 1]$. For, let $q_1, q_2, \ldots$ be an enumeration of $\mathbb{Q} \cap [0, 1]$. Then one can find a subsequence g_1 of h such that $f(q_1)_{g_1}$ exists (recall Corollary 2.1.17 (a)), further a subsequence g_2 of g_1 such that $f(q_2)_{g_2}$ exists, etc. The sequence defined by $g(n) := g_n(n)$ has the desired property.

Since $\mathbb{Q} \cap [0, 1]$ is dense in $[0, 1]$, we conclude from Proposition 1.2.2 (e) that the limit operator $f(t)_g$ exists for every $t \in [0, 1]$ and that

$$[0, 1] \to \mathcal{A}_k, \quad t \mapsto f(t)_g \tag{2.151}$$

is a continuous function with

$$f(0)_g = \begin{pmatrix} A_h & 0 \\ 0 & I_{k-1} \end{pmatrix} \quad \text{and} \quad f(1)_g = \begin{pmatrix} I & 0 \\ 0 & I_{k-1} \end{pmatrix}.$$

Moreover, all values of the function (2.151) are invertible operators (because limit operators of Fredholm operators are invertible). Thus,

$$F : [0, 1] \to \mathcal{A}_k, \quad t \mapsto P_k f(t)_g P_k + Q_k$$

is a continuous function with

$$F(0) = \begin{pmatrix} PA_hP + Q & 0 \\ 0 & I_{k-1} \end{pmatrix} \quad \text{and} \quad F(1) = \begin{pmatrix} I & 0 \\ 0 & I_{k-1} \end{pmatrix}$$

all values of which are Fredholm operators (recall that $P_k B Q_k$ and $Q_k B P_k$ are compact for all band-dominated operators B). From the continuity of the index we finally conclude that

$$\operatorname{ind} F(0) = \operatorname{ind} \begin{pmatrix} PA_hP + Q & 0 \\ 0 & I_{k-1} \end{pmatrix} = \operatorname{ind} \begin{pmatrix} I & 0 \\ 0 & I_{k-1} \end{pmatrix} = \operatorname{ind} F(1),$$

whence $\operatorname{ind}(PA_hP + Q) = \operatorname{ind}_+ A_h = 0$. □

Remark. The argument of this section can also be expressed in K-theoretic terms. Namely, consider the quotient algebra $\mathcal{A}_1/\mathcal{K}$. Since $K_1(\mathcal{A}_1) = \{0\}$, the six term exact sequence of K-theory shows that $K_1(\mathcal{A}_1/\mathcal{K}) = \mathbb{Z}$, with the isomorphism being implemented by the Fredholm index. The continuity of the limit operators expressed by Proposition 1.2.2 (e) shows that the assignment

$$U \mapsto \text{plus-index of a plus-limit operator of} \, U$$

gives a homomorphism $K_1(\mathcal{A}_1/\mathcal{K}) \to \mathbb{Z}$, and to check that it agrees with the Fredholm index it suffices to check one example, the generator of $K_1(\mathcal{A}_1/\mathcal{K})$ given by $[V_1]$.

2.7.6 Unitary band-dominated operators

Our first attempt to prove Theorem 2.7.1 was to show that the unitary group of the C^*-algebra of the band-dominated operators on $l^2(\mathbb{Z}_+)$ is path connected. (Notice that this is definitely wrong for the unitary group of the band-dominated operators on $l^2(\mathbb{Z})$. Indeed, the plus-index of the unitary operator V_1 is -1, whereas the plus-index of the identity operator is 0. Since the plus-index is a continuous function on the set of the Fredholm band-dominated operators, the operators V_1 and I cannot be connected by a continuous path in that set.) Our attempt failed (and we do not know up to now whether this group is connected), but we obtained two partial results which might be of their own interest. For, we call an operator on $l^2(\mathbb{Z}_+)$ *elementary* if its matrix representation with respect to the standard basis of $l^2(\mathbb{Z}_+)$ is of the form

$$\operatorname{diag}(A_1, A_2, A_3, \ldots)$$

with blocks A_n of $k_n \times k_n$-matrices on the main diagonal.

Theorem 2.7.7

(a) *Every unitary tridiagonal operator on $l^2(\mathbb{Z}_+)$ is elementary with blocks of size 1×1 or 2×2.*

(b) *Every unitary band operator on $l^2(\mathbb{Z}_+)$ is the product of two elementary unitary band operators.*

Observe that these results imply that every unitary *band* operator on $l^2(\mathbb{Z}_+)$ can be connected with the identity operator by a continuous path running through the set of the unitary band operators. This is a simple consequence of the path connectedness of the unitary group of the algebra of all complex $k \times k$ matrices.

Proof. For a proof of assertion (a), let A be a tridiagonal unitary operator on $l^2(\mathbb{Z}_+)$ with matrix representation

$$A = \begin{pmatrix} a_0 & b_1 & & & \\ c_1 & a_1 & b_2 & & \\ & c_2 & a_2 & b_3 & \\ & & \ddots & \ddots & \ddots \end{pmatrix}$$

with respect to the standard basis of $l^2(\mathbb{Z}_+)$.

We choose unimodular numbers u_n and v_n such that $u_1 a_0 v_1$ as well as all numbers $u_{n+1} c_n v_v$ and $u_n b_{n+1} v_{n+1}$ are non-negative, and we set

$$U := \operatorname{diag}(u_1, u_2, \ldots),$$
$$V := \operatorname{diag}(v_1, v_2, \ldots).$$

Then U and V are unitary operators, and $T := UAV$ is a unitary tridiagonal operator

$$T = \begin{pmatrix} \alpha_0 & \beta_1 & & & \\ \gamma_1 & \alpha_1 & \beta_2 & & \\ & \gamma_2 & \alpha_2 & \beta_3 & \\ & & \ddots & \ddots & \ddots \end{pmatrix}$$

with $\alpha_0, \beta_n, \gamma_n \in \mathbb{R}_+$ for all positive integers n.

Consider the entries of the main diagonals of $TT^* = I$ and $T^*T = I$. The first of these entries are equal to

$$\alpha_0^2 + \beta_1^2 = 1 = \alpha_0^2 + \gamma_1^2,$$

whence $\beta_1 = \gamma_1$ due to the non-negativity of β_1 and γ_1. The second pair of these entries is

$$\gamma_1^2 + |\alpha_1|^2 + \beta_2^2 = 1 = \beta_1^2 + |\alpha_1|^2 + \gamma_2^2,$$

whence $\beta_2 = \gamma_2$. Proceeding in this way we see that T is necessarily of the form

$$T = \begin{pmatrix} \alpha_0 & \beta_1 & & & \\ \beta_1 & \alpha_1 & \beta_2 & & \\ & \beta_2 & \alpha_2 & \beta_3 & \\ & & \ddots & \ddots & \ddots \end{pmatrix}.$$

We claim that if $\beta_1 \neq 0$, then $\beta_2 = 0$. Indeed, the 12-entry of $TT^* = I$ equals $\alpha_0\beta_1 + \beta_1\overline{\alpha_1} = 0$, whence $\overline{\alpha_1} = -\alpha_0$. Thus, the first and the second entry on the main diagonal of $TT^* = I$ are actually given by $|\alpha_0|^2 + \beta_1^2 = 1$ and $\beta_1^2 + |\alpha_1|^2 + \beta_2^2 = \beta_1^2 + |\alpha_0|^2 + \beta_2^2 = 1$, respectively. These equalities imply that $\beta_2 = 0$.

Consequently, there is either a unitary 1×1-block (if $\beta_1 = 0$) or a unitary 2×2-block (if $\beta_1 \neq 0$ and hence $\beta_2 = 0$) in the upper left corner of T. Applying the same arguments to the remaining part of T (which evidently also can be identified with a unitary tridiagonal operator on $l^2(\mathbb{Z}_+)$), we obtain assertion (a).

To prove assertion (b), let A be a unitary band operator on $l^2(\mathbb{Z}_+)$ with matrix representation

$$A = \begin{pmatrix} A_0 & B_1 & & & \\ C_1 & A_1 & B_2 & & \\ & C_2 & A_2 & B_3 & \\ & & \ddots & \ddots & \ddots \end{pmatrix}$$

with respect to the standard basis of $l^2(\mathbb{Z}_+)$ where the A_n, B_n and C_n are $k \times k$-blocks of the same block size k.

We choose unitary $k \times k$ matrices U_n and V_n such that the matrix $U_1 A_0 V_1$ and all matrices $U_{n+1} C_n V_n$ and $U_n B_n V_{n+1}$ with $n \geq 1$ become non-negative (choose $U_1 := I$ and use the polar decomposition to define successively V_1, U_2, V_2, U_3, ...) and set $U_{(1)} := \text{diag}\,(U_1, U_2, \ldots)$ and $V_{(1)} := \text{diag}\,(V_1, V_2, \ldots)$. Then $U_{(1)}$ and $V_{(1)}$

are unitary operators, and $T_1 := U_{(1)}AV_{(1)}$ is a unitary tridiagonal operator, the $k \times k$ block entries of which we denote by A_n, B_n and C_n again, i.e.,

$$T_1 = \begin{pmatrix} A_0 & B_1 & & & \\ C_1 & A_1 & B_2 & & \\ & C_2 & A_2 & B_3 & \\ & & \ddots & \ddots & \ddots \end{pmatrix},$$

where now A_0, B_n and C_n are self-adjoint and non-negative. The upper left $N \times N$ corner of $T_1 T_1^* = I$ is $A_0^2 + B_1^2 = I$. Hence, the matrix A_0 is a contraction, B_1 is equal to $S_0 := (I - A_0^2)^{1/2}$, and the operator

$$W_1 := \begin{pmatrix} A_0 & S_0 & 0 & \\ S_0 & -A_0 & 0 & \\ 0 & 0 & I & \\ & & & \ddots \end{pmatrix}$$

is unitary. Further we get as in the proof of part (a) that $C_1 = B_1$. Thus, we have

$$\begin{aligned} A^{(1)} := W_1 T_1 &= \begin{pmatrix} A_0 & S_0 & 0 & \\ S_0 & -A_0 & 0 & \\ 0 & 0 & I & \\ & & & \ddots \end{pmatrix} \begin{pmatrix} A_0 & S_0 & & & \\ S_0 & A_1 & B_2 & & \\ & C_2 & A_2 & B_3 & \\ & & \ddots & \ddots & \ddots \end{pmatrix} \\ &= \begin{pmatrix} I & A_0 S_0 + S_0 A_1 & S_0 B_2 & 0 & \\ 0 & I - A_0^2 - A_0 A_1 & -A_0 B_2 & 0 & \\ 0 & C_2 & A_2 & B_3 & \\ & & \ddots & \ddots & \ddots \end{pmatrix}. \end{aligned}$$

Being the product of unitary operators, the operator $A^{(1)}$ is unitary, too. Thus, multiplying the first row of this operator by the first column of its adjoint, we get

$$I + (A_0 S_0 + S_0 A_1)(A_0 S_0 + S_0 A_1)^* + (S_0 B_2)(S_0 B_2)^* = I$$

whence

$$A_0 S_0 + S_0 A_1 = 0 \quad \text{and} \quad S_0 B_2 = 0.$$

Thus, $A^{(1)}$ is actually a unitary operator of the form

$$\begin{pmatrix} I & 0 & 0 & 0 & \\ 0 & A_1' & B_2' & 0 & \\ 0 & C_2 & A_2 & B_3 & \\ & & \ddots & \ddots & \ddots \end{pmatrix}$$

with $A_1' := I - A_0^2 - A_0 A_1$ and $B_2' := -A_0 B_2$, and the operator A can be written as

$$A = U_{(1)}^* W_1^* A^{(1)} V_{(1)}^*.$$

Now we repeat the same arguments to the second block column of the unitary operator $A^{(1)}$. That is, we choose unitary $k \times k$ block operators $U_{(2)}$ and $V_{(2)}$ such that

$$T_2 := U_{(2)} A^{(1)} V_{(2)} = \begin{pmatrix} I & 0 & 0 & 0 & \\ 0 & A_1' & B_2' & 0 & \\ 0 & C_2 & A_2 & B_3 & \\ & & \ddots & \ddots & \ddots \end{pmatrix}$$

with non-negative matrices A_1', B_2', B_n and C_n, and we set $S_1 := (I - (A_1')^2)^{1/2}$ and

$$W_2 := \begin{pmatrix} A_1' & S_1 & 0 & \\ S_1 & -A_1' & 0 & \\ 0 & 0 & I & \\ & & & \ddots \end{pmatrix}.$$

Then W_2 is an elementary unitary operator, $A^{(2)} := T_2 W_2$ is a unitary operator of the form

$$\begin{pmatrix} I & 0 & 0 & 0 & 0 & 0 \\ 0 & I & 0 & 0 & 0 & 0 \\ 0 & 0 & A_2' & B_3 & 0 & 0 \\ 0 & 0 & C_3' & A_3 & B_4 & 0 \\ & & & \ddots & \ddots & \ddots \end{pmatrix},$$

and

$$A = U_{(1)}^* W_1^* U_{(2)}^* A^{(2)} W_2^* V_{(2)}^* V_{(1)}^*.$$

Now we deal with the third row of $A^{(2)}$ (by operating from the left-hand side again), after this with its forth column (from the right-hand side) etc. What we finally get is that $A = \tilde{U}\tilde{V}$ where $\tilde{U}$ and $\tilde{V}$ are diagonal operators

$$\tilde{U} := \operatorname{diag}(\tilde{U}_1, \tilde{U}_2, \dots) \quad \text{and} \quad \tilde{V} := \operatorname{diag}(\tilde{V}_1, \tilde{V}_2, \dots)$$

with a unitary $k \times k$ matrix $\tilde{V}_1$ and with unitary $2k \times 2k$ matrices $\tilde{U}_n$ $(n \geq 1)$ and $\tilde{V}_n$ $(n \geq 2)$. Thus, $\tilde{U}$ and $\tilde{V}$ are elementary unitary operators on $l^2(\mathbb{Z}_+)$. □

2.8 Comments and references

The investigation of several concrete classes of band and band-dominated operators has a long and interesting history the detailed explanation of which would require its own book (and it would deserve its own book). So we have to limit ourselves to mentioning a few highlights.

The simplest class of operators under consideration are the band-dominated operators with constant coefficients which are called (discrete) convolution or Laurent operators. Their restrictions onto the space $l^p(\mathbb{Z}^+)$ of one-sided sequences are called discrete Wiener-Hopf or Toeplitz operators. Discrete Wiener-Hopf operators were the topic of a fundamental paper [64] by Gohberg and Krein which initiated a series of investigations culminating in the monographs by Gohberg and Feldman [59] and Böttcher and one of the authors [30]. The next step in the general theory of band-dominated operators was done by Simonenko [166, 168], who created an abstract local principle which allowed him to reduce the examination of Fredholmness of a given operator to a family of (in general, much simpler) local Fredholm problems. Further basic papers on various aspects of one- and multi-dimensional discrete convolutions and band-dominated operators including their invertibility and their Fredholmness are [16, 48, 49, 103]. For a few applications see [10], [11], [38], [189].

The determination of the spectrum of discrete Schrödinger operators with periodic and almost periodic potentials is the subject of many papers and monographs (see, for instance, [113, 165, 13, 12] and references given there). For a particularly interesting class of Schrödinger operators with almost periodic potential, called *almost Mathieu operators*, and for its fascinating relationships with rotation C^*-algebras, see [19].

Many of the works mentioned above deal with band-dominated operators the coefficients of which stabilize and satisfy some additional conditions (they stabilize at infinity in some sense, are periodic or almost periodic, etc.). Only Ben-Artzi and Gohberg [16] and Gohberg and Kaashoek [62] allow arbitrary bounded multiplication operators in their considerations. The limit operator approach is quite different from that in [16], where Fredholmness of band matrices is characterized in terms of dichotomy.

The first application of limit operators to band-dominated operators goes back to Lange and one of the authors [93]. The results from Sections 2.1, 2.2 and 2.3 are mainly due to the authors [137, 138]. Allan's local principle is from [6]. The local theory presented in Section 2.4 has been developed in [129].

The results presented in Section 2.5.1 as well as their proofs are taken from Kurbatov [91, 92]. The Fredholm criterion for operators in the Wiener algebra has been derived in [137] (the scalar case) and in [134] (the case of operator-valued coefficients). In connection with the implication (c) $\Rightarrow$ (d) in Theorem 2.5.7 and with the remark following that theorem, one might wonder whether the Wiener algebra coincides with the algebra of the band-dominated operators on $l^\infty(\mathbb{Z}^N)$. This is not the case as the operator $A \in L(l^\infty(\mathbb{N}))$ given by its matrix representation $(a_{ij})_{i,j=0}^\infty$ with $a_{n,2n} = 1/n$ for $n \geq 1$ shows. This operator is band-dominated, but it does not belong to the Wiener algebra.

The results in Section 2.7 go back to John Roe and two of the authors, and the presentation given here follows [136]. Their generalizations to the case of $l^p(\mathbb{Z})$-spaces with $p \neq 2$ and to the multi-dimensional case of band-dominated operators on $l^2(\mathbb{Z}^N)$ with $N > 1$ are still open. Moreover, basic new aspects should occur for operators with matrix-valued coefficients, as the known results in special settings (operators with continuous and stabilizing at infinity or with slowly oscillating coefficients) indicate (see [46] and [156, 157] for indices of discrete convolution operators with slowly oscillating coefficients on general abelian groups and for indices of multi-dimensional discrete convolutions, respectively).

We conclude these comments by emphasizing the basic open problem in the (generalized) Fredholm theory of band-dominated operators via their limit operators: We do not know any example of a band-dominated operator A such that all limit operators of A are invertible, but that A fails to be Fredholm. Is there such an example? Or is the uniform boundedness condition of Theorem 2.2.1 automatically satisfied if all limit operators are invertible? Theorems 2.4.34 and 2.5.7 show that for the Fredholmness of band-dominated operators with slowly oscillating coefficients as well as of operators in the Wiener algebra the *uniform* invertibility of the limit operators is indeed redundant.

Chapter 3

Convolution Type Operators on $\mathbb{R}^N$

In this chapter, we are going to introduce a Banach algebra $\mathcal{B}_p$ of operators on $L^p(\mathbb{R}^N)$, which become band-dominated operators acting on $l^p(\mathbb{Z}^N, X)$ with $X = L^p([0, 1)^N)$ after a suitable discretization and which are, thus, subject to the general theory developed in the previous chapter. Prominent members of this class are the Fourier convolution operators, certain pseudodifferential operators as well as operators in finite differences. We employ the close relationship between operators in $\mathcal{B}_p$ and their discretizations to study the local invertibility at infinity (which is related with the Fredholmness) as well as the applicability of the finite section method for operators in $\mathcal{B}_p$.

Throughout this chapter, let $1 < p < \infty$, $q := p/(p-1)$ and N a positive integer.

3.1 Band-dominated operators on $L^p(\mathbb{R}^N)$

We start with specifying the notions of $\mathcal{P}$-strong convergence, $\mathcal{P}$-Fredholmness and $\mathcal{P}$-compactness introduced in Section 1.1 to the present context and with introducing the class of band-dominated operators on $L^p(\mathbb{R}^N)$ as well as an appropriate discretization.

3.1.1 Approximate identities and $\mathcal{P}$-Fredholmness

We start with specifying the notions of $\mathcal{P}$-strong convergence, $\mathcal{P}$-Fredholmness and $\mathcal{P}$-compactness introduced in Section 1.1 to the present context.

Let P_n stand for the operator of multiplication by the characteristic function of the cube $[-n, n]^N$ acting on $L^p(\mathbb{R}^N)$. The sequence $\mathcal{P} := (P_n)_{n=1}^{\infty}$ forms a perfect uniform approximate identity in the sense of Section 1.1.2. Let $Q_n := I - P_n$.

In accordance with Definition 1.1.7, we introduce the set $K(L^p(\mathbb{R}^N), \mathcal{P})$ of the $\mathcal{P}$-compact operators, i.e., of the operators $K \in L(L^p(\mathbb{R}^N))$ such that

$$\lim_{n\to\infty} \|KQ_n\| = \lim_{n\to\infty} \|Q_nK\| = 0,$$

and the set $L(L^p(\mathbb{R}^N), \mathcal{P})$ of all operators $A \in L(L^p(\mathbb{R}^N))$ such that AK and KA are $\mathcal{P}$-compact whenever K is $\mathcal{P}$-compact. Then $L(L^p(\mathbb{R}^N), \mathcal{P})$ is a closed unital subalgebra of $L(L^p(\mathbb{R}^N))$ which contains $K(L^p(\mathbb{R}^N), \mathcal{P})$ as its closed ideal. The algebra $L(L^p(\mathbb{R}^N), \mathcal{P})$ is inverse closed in $L(L^p(\mathbb{R}^N))$ due to Theorem 1.1.9. Further, by Theorem 1.1.3, $K(L^p(\mathbb{R}^N), \mathcal{P})$ contains the ideal of the compact operators on $L^p(\mathbb{R}^N)$ (but it is strictly larger than that ideal since the operators P_n fail to be compact).

Our earlier definitions of Fredholmness, invertibility at infinity and local invertibility at infinity specify as follows to the present context.

Definition 3.1.1 *The operator $A \in L(L^p(\mathbb{R}^N), \mathcal{P})$ is $\mathcal{P}$-*Fredholm *if the coset $A + K(L^p(\mathbb{R}^N), \mathcal{P})$ is invertible in the quotient algebra $L(L^p(\mathbb{R}^N), \mathcal{P})/K(L^p(\mathbb{R}^N), \mathcal{P})$, that is if there exist operators $B, C \in L(L^p(\mathbb{R}^N), \mathcal{P})$ such that*

$$BA - I \in K(L^p(\mathbb{R}^N), \mathcal{P}) \quad \text{and} \quad AC - I \in K(L^p(\mathbb{R}^N), \mathcal{P}). \tag{3.1}$$

Equivalently, an operator $A \in L(L^p(\mathbb{R}^N), \mathcal{P})$ is $\mathcal{P}$-Fredholm if and only if it is *invertible at infinity* in the sense that there exist an $m \in \mathbb{N}$ as well as operators $B, C \in L(L^p(\mathbb{R}^N), \mathcal{P})$ such that

$$BAQ_m = Q_m \quad \text{and} \quad Q_m AC = Q_m.$$

There is also an adequate notion of local invertibility at an infinitely distant point $\eta \in S^{N-1}$. For, we reify the general scheme presented in Section 2.3.2 as follows.

Given $\eta \in S^{N-1}$, let (U_n) be a monotonically decreasing sequence of neighborhoods of η (with respect to the Gelfand topology) such that $\cap_n U_n = \{\eta\}$, and let $(r_n) \subset \mathbb{R}^+$ be a monotonically increasing sequence which tends to infinity. Further, let R_n stand for the operator of multiplication by the characteristic function of the set $W_{r_n, U_n} := \{x \in \mathbb{R}^N : x/|x| \in U_n, \, |x| > r_n\}$, and abbreviate the sequence (R_n) by $\mathcal{R}_\eta$. Clearly, $\mathcal{R}_\eta$ is a decreasing approximate projection.

Let $K(L^p(\mathbb{R}^N), \mathcal{P}, \mathcal{R}_\eta)$ be the set of the $\mathcal{P}\mathcal{R}_\eta$-compact operators, i.e., the set of all operators $K \in L(L^p(\mathbb{R}^N))$ such that

$$\lim_{n\to\infty} \|KR_n\| = \lim_{n\to\infty} \|R_n K\| = 0.$$

Further, write $L(L^p(\mathbb{R}^N), \mathcal{P}, \mathcal{R}_\eta)$ for the set of all operators $A \in L(L^p(\mathbb{R}^N), \mathcal{P})$ such that AK and KA are $\mathcal{P}\mathcal{R}_\eta$-compact whenever K is $\mathcal{P}\mathcal{R}_\eta$-compact. Then $L(L^p(\mathbb{R}^N), \mathcal{P}, \mathcal{R}_\eta)$ is a closed unital subalgebra of $L(L^p(\mathbb{R}^N), \mathcal{P})$ which contains $K(L^p(\mathbb{R}^N), \mathcal{P}, \mathcal{R}_\eta)$ as its closed ideal, and the algebra $L(L^p(\mathbb{R}^N), \mathcal{P}, \mathcal{R}_\eta)$ proves to be inverse closed in $L(L^p(\mathbb{R}^N))$ due to Theorem 2.3.8.

Definition 3.1.2 *The operator $A \in L(L^p(\mathbb{R}^N), \mathcal{P}, \mathcal{R}_\eta)$ is locally invertible at the infinitely distant point $\eta \in S^{N-1}$ if the coset $A + K(L^p(\mathbb{R}^N), \mathcal{P}, \mathcal{R}_\eta)$ is invertible in the quotient algebra $L(L^p(\mathbb{R}^N), \mathcal{P}, \mathcal{R}_\eta)/K(L^p(\mathbb{R}^N), \mathcal{P}, \mathcal{R}_\eta)$, i.e., if there exist operators $B, C \in L(L^p(\mathbb{R}^N), \mathcal{P}, \mathcal{R}_\eta)$ such that $BA - I$ and $AC - I$ belong to $K(L^p(\mathbb{R}^N), \mathcal{P}, \mathcal{R}_\eta)$.*

Equivalently, A is locally invertible at η if and only if there exist operators $B, C \in L(L^p(\mathbb{R}^N), \mathcal{P}, \mathcal{R}_\eta)$ and $R_m \in \mathcal{R}$ such that

$$BAR_m = R_m \quad \text{and} \quad R_m AC = R_m.$$

3.1.2 Shifts and limit operators

For $\alpha \in \mathbb{Z}^N$, we consider the operator

$$U_\alpha : L^p(\mathbb{R}^N) \to L^p(\mathbb{R}^N), \quad (U_\alpha f)(t) := f(t - \alpha)$$

of shift by α. Clearly, $\mathcal{U} := \{U_\alpha\}_{\alpha \in \mathbb{Z}^N}$ is a group of isometries on $L^p(\mathbb{R}^N)$ which is compatible with the approximate identity $\mathcal{P}$ in the sense that the conditions (1.32) and (1.33) are satisfied. In accordance with the definitions from Section 1.2, we again denote by $\mathcal{H}$ the set of all sequences $h = (h(m)) \subset \mathbb{Z}^N$ which tend to infinity, and we call an operator A_h a *limit operator of* $A \in L(L^p(\mathbb{R}^N), \mathcal{P})$ with respect to $h \in \mathcal{H}$ if

$$\lim_{m\to\infty} \|(U_{-h(m)} A U_{h(m)} - A_h) P_m\| = \lim_{m\to\infty} \|P_m (U_{-h(m)} A U_{h(m)} - A_h)\| = 0$$

for every $P_m \in \mathcal{P}$. The set $\sigma_{op}(A)$ of all limit operators of A is the *operator spectrum* of A. Further we denote by $L^\$(L^p(\mathbb{R}^N), \mathcal{P})$ the subalgebra of $L(L^p(\mathbb{R}^N), \mathcal{P})$ which consists of all operators with rich operator spectrum. The latter means for an operator $A \in L(L^p(\mathbb{R}^N), \mathcal{P})$, that every sequence $h \in \mathcal{H}$ has a subsequence g such that the limit operator A_g with respect to g exists. Notice finally that, as in Section 1.3.2, there is a symbol calculus for the operators in $L^\$(L^p(\mathbb{R}^N), \mathcal{P})$.

3.1.3 Discretization

Let χ_0 denote the characteristic function of the cube $I_0 := [0, 1)^N \subseteq \mathbb{R}^N$, and set $X := L^p(I_0)$. For $1 < p < \infty$, let further $E := l^p(\mathbb{Z}^N, X)$. Then the mapping G which sends the function $f \in L^p(\mathbb{R}^N)$ to the sequence

$$Gf = ((Gf)_\alpha)_{\alpha \in \mathbb{Z}^N}, \quad (Gf)_\alpha := \chi_0 U_{-\alpha} f \tag{3.2}$$

is an isometry from $L^p(\mathbb{R}^N)$ onto $l^p(\mathbb{Z}^N, X)$, the inverse of which sends the sequence $u = (u_\alpha)_{\alpha \in \mathbb{Z}^N}$ to the function

$$G^{-1} u = \sum_{\alpha \in \mathbb{Z}^N} U_\alpha u_\alpha \chi_0 \tag{3.3}$$

where the series converges in the norm in $L^p(\mathbb{R}^N)$. Thus, the mapping

$$\Gamma : L(L^p(\mathbb{R}^N)) \to L(l^p(\mathbb{Z}^N, X)), \quad A \mapsto GAG^{-1}$$

is an isometric algebra isomorphism.

It is not hard to check how the operators in $\mathcal{P}$ and $\mathcal{U}$ translate under the discretization operator Γ. Indeed, for $m \in \mathbb{N}$, the operator $\Gamma(P_m)$ is just the operator of multiplication by the characteristic function of the discrete cube $[-n, n]^N \cap \mathbb{Z}^N$ which we denote by $\hat{P}_m$ in what follows. Observe that $\hat{\mathcal{P}} := (\hat{P}_m)_{m\in\mathbb{N}}$ is just the approximate identity we had chosen in Chapter 2. Further, for $\alpha \in \mathbb{Z}^N$, one has $\Gamma(U_\alpha) = V_\alpha$ where V_α stands, as before, for the shift by α on $l^p(\mathbb{Z}^N, X)$.

Given a family $\mathcal{M}$ of operators of multiplication by functions on $\mathbb{R}^N$, we denote by $\hat{\mathcal{M}}$ the set of all operators of multiplication by the restrictions of these functions onto $\mathbb{Z}^N$.

Proposition 3.1.3

(a) Γ *maps* $K(L^p(\mathbb{R}^N), \mathcal{P})$ *onto* $K(E, \hat{\mathcal{P}})$ *and* $L(L^p(\mathbb{R}^N), \mathcal{P})$ *onto* $L(E, \hat{\mathcal{P}})$.

(b) Γ *maps* $K(L^p(\mathbb{R}^N), \mathcal{P}, \mathcal{R}_\eta)$ *onto* $K(E, \hat{\mathcal{P}}, \hat{\mathcal{R}_\eta})$ *and* $L(L^p(\mathbb{R}^N), \mathcal{P}, \mathcal{R}_\eta)$ *onto* $L(E, \hat{\mathcal{P}}, \hat{\mathcal{R}_\eta})$.

Proof. Since

$$\|K - KP_n\| = \|\Gamma(K - KP_n)\| = \|\Gamma(K) - \Gamma(K)\hat{P}_n\|$$

and $\|K - P_nK\| = \|\Gamma(K) - \hat{P}_n\Gamma(K)\|$, we get $\Gamma(K(L^p(\mathbb{R}^N), \mathcal{P})) = K(E, \hat{\mathcal{P}})$. Similarly, the second part of assertion (a) follows if one takes into account Proposition 1.1.8 (a).

For the ideal $K(L^p(\mathbb{R}^N), \mathcal{P}, \mathcal{R}_\eta)$, the situation is a little bit more involved since $\Gamma(R_n) \neq \hat{R}_n$ in general. Every point $x \in \mathbb{R}^N$ belongs to a (uniquely determined) cube $I(x)$ of the form $z + I_0$ with $z \in \mathbb{Z}^N$. We let W'_{r_n,U_n} stand for the set of all points x with $I(x) \subset W_{r_n,U_n}$. Further, we let R'_n stand for the operator of multiplication by the characteristic function of the set W'_{r_n,U_n}, and we abbreviate the sequence (R'_n) to $\mathcal{R}'_\eta$. Clearly, $\mathcal{R}'_\eta$ is a decreasing approximate projection which is equivalent to $\mathcal{R}_\eta$. Thus,

$$K(E, \hat{\mathcal{P}}, \hat{\mathcal{R}'_\eta}) = K(E, \hat{\mathcal{P}}, \hat{\mathcal{R}_\eta}) = K(L^p(\mathbb{R}^N), \mathcal{P}, \mathcal{R}_\eta)$$

(recall Lemma 1.1.10). Since $\Gamma(R'_n) = \hat{R}_n$, the second assertion of the proposition follows as above. □

A consequence is that an operator $A \in L(L^p(\mathbb{R}^N), \mathcal{P})$ is $\mathcal{P}$-Fredholm (i.e., invertible at infinity) if and only if $\Gamma(A)$ is $\hat{\mathcal{P}}$-Fredholm. An analogous result holds for the $\mathcal{P}\mathcal{R}_\eta$-Fredholmness (i.e., for local invertibility at $\eta \in S^{N-1}$).

The next result shows that also the limit operators behave nicely under discretization.

Proposition 3.1.4 *Let* $A \in L(L^p(\mathbb{R}^N), \mathcal{P})$ *and* $h \in \mathcal{H}$. *Then the limit operator* A_h *of* A *exists (with respect to* $\mathcal{P}$*) if and only if the limit operator* $(\Gamma(A))_h$ *of* $\Gamma(A)$ *exists (with respect to* $\Gamma(\mathcal{P}) = \hat{\mathcal{P}}$*), and*

$$\Gamma(A_h) = (\Gamma(A))_h. \tag{3.4}$$

In particular, A *belongs to* $L^\$(L^p(\mathbb{R}^N), \mathcal{P})$ *if and only if* $\Gamma(A)$ *belongs to* $L^\$(E, \hat{\mathcal{P}})$.

Proof. Let the limit operator A_h of A exist, i.e., let

$$\lim_{n\to\infty} \|(U_{-h(n)}AU_{h(m)} - A_h)P_m\| = \lim_{n\to\infty} \|P_m(U_{-h(n)}AU_{h(m)} - A_h)\| = 0$$

for all m. Since $\Gamma(U_\alpha) = V_\alpha$ and $\Gamma(P_m) = \hat{P}_m$, and since Γ is an isometrical algebra isomorphism, we conclude that

$$\lim_{n\to\infty} \|\Gamma((U_{-h(n)}AU_{h(m)} - A_h)P_m)\| = \lim_{n\to\infty} \|(V_{-h(n)}\Gamma(A)V_{h(m)} - \Gamma(A_h))\hat{P}_m\| = 0$$

and, analogously, $\hat{P}_m(\|(V_{-h(n)}\Gamma(A)V_{h(m)} - \Gamma(A_h))\| \to 0$ for every m. Thus, the limit operator of $\Gamma(A)$ with respect to h exists and (3.4) holds. The reverse implication follows analogously. □

In particular, an operator B belongs to the operator spectrum of A if and only if the operator $\Gamma(B)$ belongs to the operator spectrum of $\Gamma(A)$, and an analogous relation holds between the local operator spectra at points $\eta \in S^{N-1}$.

3.1.4 Band-dominated operators on $L^p(\mathbb{R}^N)$

The following definition is the analogue of the characterization of band-dominated operators on $l^p(\mathbb{Z}^N, X)$ in Theorem 2.1.6.

Definition 3.1.5 *We say that an operator $A \in L(L^p(\mathbb{R}^N))$ is* band-dominated *if, for every function $\varphi \in BUC(\mathbb{R}^N)$,*

$$\lim_{t\to 0} \|A\varphi_{t,r}I - \varphi_{t,r}A\|_{L(L^p(\mathbb{R}^N))} = 0 \quad \textit{uniformly with respect to } r \in \mathbb{R}^N. \tag{3.5}$$

The set of all band-dominated operators in $L(L^p(\mathbb{R}^N))$ will be denoted by $\mathcal{B}_p$, and we write $\mathcal{B}_p^\$$ instead of $\mathcal{B}_p \cap L^\$(L^p(\mathbb{R}^N), \mathcal{P})$.

Here, as in Section 2.1.4, for $r, t, x \in \mathbb{R}^N$,

$$\varphi_{t,r}(x) := \varphi_t(x - r) \quad \text{and} \quad \varphi_t(x) := \varphi(tx) := \varphi(t_1x_1, \ldots, t_Nx_N).$$

Proposition 3.1.6 $\Gamma(\mathcal{B}_p)$ *coincides with the algebra $\mathcal{A}_E$ of the band-dominated operators on $E = l^p(\mathbb{Z}^N, L^p(I_0))$, and $\Gamma(\mathcal{B}_p^\$) = \mathcal{A}_E^\$$.*

Proof. If $A \in \mathcal{B}_p$ then, for every function $\varphi \in BUC(\mathbb{R}^N)$,

$$\lim_{t\to 0} \|[A, \varphi_{t,r}I]\|_{L(L^p(\mathbb{R}^N))} = 0$$

and, consequently,

$$\lim_{t\to 0} \|[\Gamma(A), \Gamma(\varphi_{t,r}I)]\|_{L(E)} = 0 \tag{3.6}$$

uniformly with respect to $r \in \mathbb{R}^N$. We claim that

$$\lim_{t\to 0} \|\hat{\varphi}_{t,r}I - \Gamma(\varphi_{t,r}I)\|_{L(E)} = 0 \tag{3.7}$$

uniformly with respect to $r \in \mathbb{R}^N$. Indeed,

$$\begin{aligned}
\sup_{r\in\mathbb{R}^N} \|(\hat{\varphi}_{t,r}I - \Gamma(\varphi_{t,r}I)\|_{L(E)} &= \sup_{r\in\mathbb{R}^N} \sup_{\alpha\in\mathbb{Z}^N} \sup_{x\in I_0} |\hat{\varphi}_{t,r}(\alpha) - (\Gamma(\varphi_{t,r}I)_\alpha)(x)| \\
&= \sup_{r\in\mathbb{R}^N} \sup_{\alpha\in\mathbb{Z}^N} \sup_{x\in I_0} |\varphi(t(\alpha - r)) - \varphi(t(x+\alpha-r))| \\
&\le \sup_{\beta\in\mathbb{R}^N} \sup_{x\in I_0} |\varphi(t\beta) - \varphi(t(x+\beta))| \to 0
\end{aligned}$$

as $t \to 0$ due to the uniform continuity of φ. By (3.7) and (3.6),

$$\lim_{t\to 0} \|[\Gamma(A), \hat{\varphi}_{t,r}I]\|_{L(E)} = 0$$

uniformly with respect to $r \in \mathbb{R}^N$. Thus, $\Gamma(\mathcal{B}_p) \subseteq \mathcal{A}_E$ by Theorem 2.1.6. The reverse inclusion follows analogously. The second assertion is a consequence of the first one, together with Proposition 3.1.4. □

As immediate consequences of Propositions 2.1.7–2.1.9 and of Theorems 2.1.6, 2.2.1 and 2.3.13 we get the following.

Proposition 3.1.7 *An operator $A \in L(L^p(\mathbb{R}^N))$ is band-dominated if and only if*

$$\lim_{t\to 0} \|A\varphi_t I - \varphi_t A\|_{L(L^p(\mathbb{R}^N))} = 0$$

for every function $\varphi \in BUC(\mathbb{R}^N)$.

Proposition 3.1.8

(a) *$\mathcal{B}_p$ is a closed unital subalgebra of $L(L^p(\mathbb{R}^N))$, and it is a symmetric subalgebra in case $p = 2$.*

(b) *$\mathcal{B}_p$ is a closed subalgebra of $L(L^p(\mathbb{R}^N), \mathcal{P})$, and $K(L^p(\mathbb{R}^N), \mathcal{P})$ is a closed two-sided ideal of $\mathcal{B}_p$.*

(c) *$\mathcal{B}_p$ is inverse closed in $L(L^p(\mathbb{R}^N))$.*

(d) *$\mathcal{B}_p/K(L^p(\mathbb{R}^N), \mathcal{P})$ is inverse closed in $L(L^p(\mathbb{R}^N), \mathcal{P})/ K(L^p(\mathbb{R}^N), \mathcal{P})$.*

Theorem 3.1.9 *Let $A \in \mathcal{B}_p^\$$. Then*

(a) *A is locally invertible at point $\eta \in S^{N-1}$ if and only if all limit operators $A_h \in \sigma_\eta(A)$ are uniformly invertible.*

(b) *A is invertible at infinity if and only if, for every $\eta \in S^{N-1}$, all limit operators $A_h \in \sigma_\eta(A)$ are uniformly invertible.*

3.2 Operators of convolution

In this section, we collect some basic facts on convolution operators on L^p-spaces. Our presentation follows [158].

Let $k \in L^1(\mathbb{R}^N)$ and $u \in L^p(\mathbb{R}^N)$. Then Young's inequality implies that the *convolution*

$$(k * u)(t) := \int_{\mathbb{R}^N} k(t-s)u(s)ds, \quad t \in \mathbb{R}^N, \tag{3.8}$$

belongs to $L^p(\mathbb{R}^N)$, and that

$$\|k * u\|_p \leq \|k\|_1 \|u\|_p \tag{3.9}$$

(see [141], IX.4). Hence, the operator $C(k)u := k * u$ of convolution by $k \in L^1(\mathbb{R}^N)$ acts boundedly on $L^p(\mathbb{R}^N)$, and

$$\|C(k)\|_{L^p(\mathbb{R}^N)} \leq \|k\|_1. \tag{3.10}$$

We let $\mathcal{C}_p$ denote the closure in $L(L^p(\mathbb{R}^N))$ of the set of all convolution operators $C(k)$ with kernels $k \in L^1(\mathbb{R}^N)$. $\mathcal{C}_p$ is a closed subalgebra of $L(L^p(\mathbb{R}^N))$.

Since the step functions lie dense in $L^1(\mathbb{R}^N)$, we conclude from (3.10) that $\mathcal{C}_p$ is the smallest closed subalgebra of $L(L^p(\mathbb{R}^N))$ which contains all operators $C(\chi_K)$ where χ_K is the characteristic function of a compact subset K of $\mathbb{R}^N$.

It follows from the commutativity of the convolution product $*$ that $\mathcal{C}_p$ is a commutative Banach algebra (without identity element). Its maximal ideal space can be identified with $\mathbb{R}^N$ (with its standard topology) in such a way that the Gelfand transform $\hat{C}$ of $C \in \mathcal{C}_p$ coincides with the Fourier transform of k if $C = C(k)$ (see, e.g., [167]). Consequently, an operator $\gamma I + C$ in the unitization $\mathbb{C}I + \mathcal{C}_p$ of $\mathcal{C}_p$ is invertible if and only if

$$\inf_{\xi \in \mathbb{R}^N} |\gamma + \hat{C}(\xi)| > 0. \tag{3.11}$$

Observe that $\gamma + \hat{C}$ is just the Gelfand transform of $\gamma I + C$ in $\mathbb{C}I + \mathcal{C}_p$.

3.2.1 Compactness of semi-commutators

By a *semi-commutator* we mean an operator of the form $aC(k)$ or $C(k)aI$ where k is in $L^1(\mathbb{R}^N)$ and $a \in L^\infty(\mathbb{R}^N)$. The main result of this subsection characterizes the functions a for which the semi-commutators $aC(k)$ and $C(k)aI$ are compact for every function $k \in L^1(\mathbb{R}^N)$.

Definition 3.2.1 *Let $Q_{SC}(\mathbb{R}^N)$ refer to the set of all functions $a \in L^\infty(\mathbb{R}^N)$ such that*

$$\limsup_{t \to \infty} \int_M |a(t+s)| \, ds = 0$$

for every compact subset M of $\mathbb{R}^N$.

Thus, $Q_{SC}(\mathbb{R}^N)$ contains the set

$$L_0^\infty(\mathbb{R}^N) := \{a \in L^\infty(\mathbb{R}^N) : \lim_{R\to\infty} \text{ess sup}_{|t|\geq R}|a(t)| = 0\} \tag{3.12}$$

which is a C^*-subalgebra of $L^\infty(\mathbb{R}^N)$. In particular, all compactly supported functions belong to $Q_{SC}(\mathbb{R}^N)$. But this class turns out to be much larger: for example, it contains the characteristic function of the set $\cup_{n\geq 2}[n - \frac{1}{n}, n + \frac{1}{n}]$ in case $N = 1$.

Theorem 3.2.2 *The following conditions are equivalent for a bounded measurable function a:*

(a) *the operators BaI and aB are compact on $L^p(\mathbb{R}^N)$ for every $B \in \mathcal{C}_p$ and every $1 < p < \infty$,*

(b) *$a \in Q_{SC}(\mathbb{R}^N)$,*

(c) *There is a bounded open set $D \subset \mathbb{R}^N$ such that $\lim_{t\to\infty} \int_D |a(t+s)|\, ds = 0$.*

The following observation, which is an immediate consequence of the preceding theorem, has also a simple direct proof.

Corollary 3.2.3 *$Q_{SC}(\mathbb{R}^N)$ is a closed ideal in $L^\infty(\mathbb{R}^N)$.*

We prepare the proof of Theorem 3.2.2 by two lemmas the first of which treats an important special case of the implication (b) $\Rightarrow$ (a).

Lemma 3.2.4 *Let K, L be compact subsets of $\mathbb{R}^N$. Then the semi-commutators $\chi_K C(\chi_L)$ and $C(\chi_L)\chi_K I$ are compact on $L^p(\mathbb{R}^N)$.*

Proof. Let M be a compact subset of $\mathbb{R}^N$ which contains K and $L - K$ in its interior. Then $\chi_K C(\chi_L) = \chi_M \chi_K C(\chi_L)\chi_M I$, such that it is sufficient to prove the compactness of $\chi_K C(\chi_L)$ on $L^p(M)$.

Let $\varepsilon > 0$, and choose a continuous function k on $M \times M$ such that

$$\int_M \int_M |\chi_K(t)\chi_L(t-s) - k(s, t)|\, ds\, dt < \varepsilon$$

which is possible since the continuous functions lie densely in $L^1(M\times M)$. Moreover we can assume that $0 \leq k \leq 1$, otherwise we replace k by the function

$$(s, t) \mapsto \max\{\min\{k(s, t), 1\}, 0\}.$$

By the Stone-Weierstrass theorem, there are continuous functions a_i and b_i such that

$$\sup_{(s,t)\in M\times M} \left|\sum_{i=1}^k a_i(s)b_i(t) - k(s, t)\right| < \min\{1, \varepsilon\,(\text{mes}\, M)^{-2}\}.$$

With this choice, one has

$$\int_M \int_M \left|\chi_K(t)\chi_L(t-s) - \sum_{i=1}^k a_i(s)b_i(t)\right|\, ds\, dt < 2\,\varepsilon$$

as well as

$$\sup_{(s,\,t)\in M\times M}\left|\chi_K(t)\chi_L(t-s)-\sum_{i=1}^{k}a_i(s)b_i(t)-k(s,\,t)\right|<3.$$

We are going to estimate the difference between the operator $\chi_K C(\chi_L)$ and the integral operator A with kernel

$$(s,\,t)\mapsto \chi_M(s)\sum_{i=1}^{k}a_i(s)b_i(t)\chi_M(t).$$

Let $q:=p/(p-1)$ and

$$l(s,\,t):=\chi_K(t)\chi_L(t-s)-\chi_M(s)\sum_{i=1}^{k}a_i(s)b_i(t)\chi_M(t).$$

The Hölder inequality yields, for $f\in L^p(\mathbb{R}^N)$,

$$\begin{aligned}
&\|(\chi_K C(\chi_L)-A)f\|^p\\
&\quad=\int_M\left|\int_M l(s,\,t)f(s)\,ds\right|^p\,dt\\
&\quad\le \|f\|^p\int_M\left(\int_M |l(s,\,t)|^q\,ds\right)^{p/q}\,dt.
\end{aligned}$$

If $p>2$, then $p/q>1$, and a further application of the Hölder inequality gives

$$\begin{aligned}
&\int_M\left(\int_M |1\cdot l(s,\,t)|^q\,ds\right)^{\frac{p}{q}}\,dt\\
&\quad\le\int_M\left(\int_M (|l(s,\,t)|^q)^{p/q}\,ds\right)^{\frac{p}{q}\frac{q}{p}}\,dt\cdot(\operatorname{mes}M)^{\frac{p-q}{p}\frac{p}{q}}\\
&\quad=\int_M\int_M |l(s,\,t)|^p\,ds\,dt\cdot(\operatorname{mes}M)^{\frac{p-q}{q}}\\
&\quad\le(\operatorname{mes}M)^{\frac{p-q}{q}}\sup_{(s,\,t)\in M\times M}|l(s,\,t)|^{p-1}\int_M\int_M |l(s,\,t)|\,ds\,dt\\
&\quad\le(\operatorname{mes}M)^{\frac{p-q}{q}}\,3^{p-1}\,2\,\varepsilon.
\end{aligned}$$

In case $p=2$, one has

$$\int_M\int_M |l(s,\,t)|^2\,ds\,dt\le\sup_{(s,\,t)\in M\times M}|l(s,\,t)|\int_M\int_M |l(s,\,t)|\,ds\,dt<6\,\varepsilon.$$

If, finally, $p < 2$, then $q/p > 1$ and

$$\begin{aligned}
&\int_M \left(\int_M |l(s,t)|^q\,ds\right)^{\frac{p}{q}} \cdot 1\,dt \\
&\quad\le \left(\int_M \left(\int_M |l(s,t)|^q\,ds\right)^{\frac{p}{q}\frac{q}{p}} dt\right)^{\frac{p}{q}} \cdot (\operatorname{mes} M)^{\frac{q-p}{q}} \\
&\quad\le (\operatorname{mes} M)^{\frac{q-p}{q}} \left(\sup_{(s,t)\in M\times M} |l(s,t)|^{q-1} \int_M\int_M |l(s,t)|\,ds\,dt\right)^{\frac{p}{q}} \\
&\quad\le \left((\operatorname{mes} M)^{\frac{q-p}{q}} 3^{q-1}\, 2\,\varepsilon\right)^{\frac{p}{q}}.
\end{aligned}$$

Since A is of finite rank and ε is arbitrary, this shows that $\chi_K C(\chi_L)$ can be approximated by finite rank operators as closely as desired. Hence, this operator is compact, and the compactness of $C(\chi_L)\chi_K I$ follows by taking adjoints. □

Lemma 3.2.5 *Let M be a compact subset of $\mathbb{R}^N$ and $k \in L^\infty(\mathbb{R}^N \times \mathbb{R}^N)$. Then the operator B, defined by*

$$(Bf)(t) := \int_{\mathbb{R}^N} \chi_M(t-s)k(t,s)f(s)\,ds, \quad t \in \mathbb{R}^N,$$

is bounded on $L^p(\mathbb{R}^N)$ and

$$\begin{aligned}
\|B\| &\le (\operatorname{mes} M)^{1/p} \operatorname{ess\,sup}_{t\in\mathbb{R}^N} \left(\int_{-M} |k(t,t+h)|^q\,dh\right)^{1/q} \\
&\le (\operatorname{mes} M) \operatorname{ess\,sup}_{(t,h)\in\mathbb{R}^N\times(-M)} |k(t,t+h)|.
\end{aligned}$$

Proof. We agree upon writing $\int$ for the integral over $\mathbb{R}^N$. With the Hölder inequality we get, for $f \in L^p(\mathbb{R}^N)$,

$$\begin{aligned}
\|Bf\|^p &\le \int |\int \chi_M(t-s)k(t,s)f(s)\,ds|^p dt \\
&\le \int (\int |\chi_M(t-s)k(t,s)| \cdot |\chi_M(t-s)f(s)|\,ds)^p dt \\
&\le \int (\int |\chi_M(t-s)k(t,s)|^q ds)^{\frac{p}{q}} (\int |\chi_M(t-s)f(s)|^p\,ds) dt \\
&\le \operatorname*{ess\,sup}_{t\in\mathbb{R}^N} (\int |\chi_M(t-s)k(t,s)|^q ds)^{\frac{p}{q}} \int\int |\chi_M(t-s)f(s)|^p\,ds\,dt \\
&\le \operatorname*{ess\,sup}_{t\in\mathbb{R}^N} (\int |\chi_{-M}(h)k(t,t+h)|^q dh)^{\frac{p}{q}} \int |f(s)|^p \int |\chi_M(t-s)| dt\,ds \\
&\le \|f\|^p (\operatorname{mes} M) \operatorname*{ess\,sup}_{t\in\mathbb{R}^N} (\int_{-M} |k(t,t+h)|^q dh)^{\frac{p}{q}} \\
&\le \|f\|^p \operatorname{mes} M\, (\operatorname{mes} M)^{\frac{p}{q}} \operatorname*{ess\,sup}_{(t,h)\in\mathbb{R}^N\times(-M)} |k(t,t+h)|^p.
\end{aligned}$$

Taking pth roots on both sides yields the assertion. □

Proof of Theorem 3.2.2. (a) $\Rightarrow$ (b): Contrary to what we want to show, assume that (b) is not satisfied. That is, there is a compact set M such that

$$\limsup_{t\to\infty} \int_M |a(t+s)|\, ds > 0.$$

Since compact sets are bounded, this property also holds with M replaced by a certain open ball D with center at 0. Thus, there is a $\delta > 0$ and a sequence (t_k) tending to infinity such that

$$\int_D |a(t_k+s)|\, ds > \delta \quad \text{for all } k.$$

The function $t \mapsto \int_M |a(t+s)|\, ds$ is uniformly continuous on $\mathbb{R}^N$ (see, e.g., [37], VIII.4., Proposition 14). Thus, there is a bounded open neighborhood U of 0 such that

$$\int_D |a(t+s)|\, ds > \delta \quad \text{for all } t \in \cup_k\,(t_k+U).$$

The Hölder inequality

$$\int_D |a(t+s)|\, ds \le \left(\int_D |a(t+s)|^2\, ds\right)^{\frac{1}{2}} (\operatorname{mes} D)^{\frac{1}{2}}$$

implies that

$$\int_D |a(t+s)|^2\, ds > \delta^2 (\operatorname{mes} D)^{-1} \quad \text{for all } t \in \cup_k\,(t_k+U).$$

Consider the functions φ_k which coincide with the complex conjugate $\overline{a}$ of a on t_k+U+D and which are zero outside this set. Clearly, the sequence (φ_k) converges weakly to zero as k tends to infinity.

The functions $C(\chi_{-M})a\varphi_k$ can be estimated as follows, where $\int$ again refers to integrals over $\mathbb{R}^N$:

$$\begin{aligned}
\|C(\chi_{-M})a\varphi_k\|^p &= \int \left| \int \chi_{-D}(t-s)a(s)\varphi_k(s)\, ds \right|^p dt \\
&= \int \left| \int_D a(r+t)\varphi_k(r+t)\, dr \right|^p dt \\
&\ge \int_{t_k+U} \left| \int_D a(r+t)\overline{a(r+t)}\, dr \right|^p dt \\
&\ge (\operatorname{mes} U) \inf_{t\in t_k+U} \left| \int_D |a(r+t)|^2\, dr \right|^p \\
&\ge (\operatorname{mes} U)\delta^{2p}/(\operatorname{mes} D)^p.
\end{aligned}$$

Thus, the operator $C(\chi_{-M})aI$ maps sequence (φ_k), which converges weakly to zero, to a sequence which does not converge to zero in the norm of $L^p(\mathbb{R}^N)$. Thus, this operator cannot be compact.

The implication (b) $\Rightarrow$ (c) is obvious. For a proof of the implication (c) $\Rightarrow$ (a), let M be a compact subset of $\mathbb{R}^N$, and let $\lim_{t\to\infty}\int_D |a(t+s)|\,ds = 0$ for a certain bounded open set $D \in \mathbb{R}^N$. Choose a finite number of points $h_1, \ldots, h_k$ such that $-M \subseteq \cup_{j=1}^k (h_j + D)$. Then

$$\begin{aligned}\lim_{t\to\infty}\int_{-M} |a(t+s)|\,ds &\le \lim_{t\to\infty}\sum_{j=1}^k \int_{h_j+D} |a(t+s)|\,ds \\ &= \sum_{j=1}^k \lim_{t-h_j\to\infty}\int_D |a(t+s)|\,ds = 0.\end{aligned}$$

Hence, given $\varepsilon > 0$, one can choose a compact set D_ε such that

$$\sup_{t\in\mathbb{R}^N\setminus D_\varepsilon}\int_{-M} |a(t+s)|\,ds < \varepsilon.$$

Applying Lemma 3.2.5, we get

$$\begin{aligned}&\|C(\chi_M)fI - \chi_{D_\varepsilon}C(\chi_M)fI\|^p = \|\chi_{\mathbb{R}^N\setminus D_\varepsilon}C(\chi_M)fI\|^p \\ &\le (\operatorname{mes} M)\operatorname{ess\,sup}_{t\in\mathbb{R}^N\setminus D_\varepsilon}\left(\int_{-M} |a(t+h)|^q\,dh\right)^{\frac{p}{q}} \\ &\le (\operatorname{mes} M)\,\|a\|_\infty \operatorname{ess\,sup}_{t\in\mathbb{R}^N\setminus D_\varepsilon}\left(\int_{-M} |a(t+h)|\,dh\right)^{\frac{p}{q}} \\ &< \varepsilon^{\frac{p}{q}}\,(\operatorname{mes} M)\,\|a\|_\infty.\end{aligned}$$

Thus, the operator $C(\chi_M)fI$ can be approximated in the operator norm as closely as desired be the compact operators $\chi_{D_\varepsilon}C(\chi_M)fI$, whence its compactness. The compactness of $fC(\chi_M)$ can be proved by passing to the adjoint. This settles the result for a family of generators of the algebra $\mathcal{C}_p$. The general assertion follows since the compact operators form a closed linear space. □

3.2.2 Compactness of commutators

Our next goal is to characterize those functions $a \in L^\infty(\mathbb{R}^N)$ for which the *commutators* $aC(k) - C(k)aI$ are compact for every function $k \in L^1(\mathbb{R}^N)$. We start with a subclass of functions which possess this property.

Definition 3.2.6 *Let $SO(\mathbb{R}^N)$ denote the set of all bounded continuous functions a on $\mathbb{R}^N$ such that, for every compact subset M of $\mathbb{R}^N$,*

$$\lim_{t\to\infty}\sup_{h\in M} |a(t) - a(t+h)| = 0. \tag{3.13}$$

It is not hard to check that a function a belongs to $SO(\mathbb{R}^N)$ if (3.13) is satisfied for $M = [0, 1]^N$ or for another compact set M with non-empty interior.

Functions in $SO(\mathbb{R}^N)$ are called *slowly oscillating* on $\mathbb{R}^N$. Examples of slowly oscillating functions are provided by the continuous functions which possess a finite limit at infinity and by the differentiable functions the derivative of which tends to zero at infinity. The latter follows from the mean value theorem.

Proposition 3.2.7 *$SO(\mathbb{R}^N)$ is a unital commutative C^*-subalgebra of $BUC(\mathbb{R}^N)$.*

Proof. Obviously, $SO(\mathbb{R}^N)$ is linear and symmetric and contains the identity function $t \mapsto 1$. If a and b are in $SO(\mathbb{R}^N)$, then the estimate

$$|(ab)(s) - (ab)(t)| \leq |a(s) - a(t)| \, \|b\|_\infty + |b(s) - b(t)| \, \|a\|_\infty$$

shows that $ab = ba \in SO(\mathbb{R}^N)$. Hence, $SO(\mathbb{R}^N)$ is a commutative algebra. Finally, let (a_n) be a sequence of functions in $SO(\mathbb{R}^N)$ which converges uniformly to a function $a \in L^\infty(\mathbb{R}^N)$. Then the estimate

$$\begin{aligned} |a(s) - a(t)| &\leq |a(s) - a_n(s)| + |a_n(s) - a_n(t)| + |a_n(t) - a(t)| \\ &\leq 2 \, \|a - a_n\|_\infty + |a_n(s) - a_n(t)| \end{aligned}$$

implies that $a \in SO(\mathbb{R}^N)$. Hence, $SO(\mathbb{R}^N)$ is closed.

It remains to show that every slowly oscillating function is uniformly continuous. Let $\varepsilon > 0$, and let B_1 be the closed unit ball in $\mathbb{R}^N$. If $a \in SO(\mathbb{R}^N)$, then

$$\lim_{t \to \infty} \sup_{h \in B_1} |a(t) - a(t+h)| = 0,$$

and we can choose T such that

$$|a(t) - a(s)| < \varepsilon \quad \text{for all } |t| \geq T \text{ and } |t - s| \leq 1.$$

Further, the continuous function a is uniformly continuous on the compact set $B_{T+1} := \{t \in \mathbb{R}^N : |t| \leq T + 1\}$. Hence, there is a $\delta > 0$ such that

$$|a(t) - a(s)| < \varepsilon \quad \text{for all } s, \, t \in B_{T+1} \text{ with } |t - s| < \delta.$$

Thus, if $s, \, t \in \mathbb{R}^N$ with $|t - s| < \min\{\delta, 1\}$, then $|a(t) - a(s)| < \varepsilon$. □

Theorem 3.2.8 *Let $a \in SO(\mathbb{R}^N)$. Then the commutator $aB - BaI$ is compact on $L^p(\mathbb{R}^N)$ for every operator $B \in \mathcal{C}_p$.*

Proof. It is sufficient to prove the result for the generators $B = C(\chi_M)$ of the algebra $\mathcal{C}_p$ where M is a compact subset of $\mathbb{R}^N$.

Let $a \in SO(\mathbb{R}^N)$. Given $\varepsilon > 0$, choose a compact subset K of $\mathbb{R}^N$ such that

$$\sup_{t \in \mathbb{R}^N \setminus K} |a(t) - a(t+h)| < \varepsilon \quad \text{for all } h \in (-M).$$

Then, by Lemma 3.2.5,

$$\begin{aligned}
&\|(aC(\chi_M) - C(\chi_M)aI) - \chi_K(aC(\chi_M) - C(\chi_M)aI)\| \\
&\quad = \|\chi_{\mathbb{R}^N\setminus K}(aC(\chi_M) - C(\chi_M)aI)\| \\
&\quad \le (\operatorname{mes} M)\, \operatorname{ess\,sup}_{(t,\,h)\in(\mathbb{R}^N\setminus K)\times(-M)} |a(t) - a(t+h)| \\
&\quad < \varepsilon\,(\operatorname{mes} M).
\end{aligned}$$

The operator

$$\chi_K(aC(\chi_M) - C(\chi_M)aI) = a\chi_K C(\chi_M) - \chi_K C(\chi_M)aI$$

is compact by Lemma 3.2.4. Hence, being a norm limit of compact operators, the commutator $aC(\chi_M) - C(\chi_M)aI$ is compact. □

Definition 3.2.9 *A function $a \in L^\infty(\mathbb{R}^N)$ belongs to the class $Q_C(\mathbb{R}^N)$ if, for every open and bounded subset M of $\mathbb{R}^N$, the function*

$$t \mapsto \int_M (a(t) - a(t+s))\, ds$$

lies in $Q_{SC}(\mathbb{R}^N)$, i.e., if there is a bounded open subset D of $\mathbb{R}^N$ such that

$$\lim_{t\to\infty} \int_D \left| \int_M (a(t+h) - a(t+s+h))\, ds \right| dh = 0.$$

The following basic result does not only solve the commutator problem; it moreover verifies the relation between the classes $Q_{SC}(\mathbb{R}^N)$, $SO(\mathbb{R}^N)$ and $Q_C(\mathbb{R}^N)$.

Theorem 3.2.10 *The following assertions are equivalent for $a \in L^\infty(\mathbb{R}^N)$:*

(a) *the operators $BaI - aB$ are compact on $L^p(\mathbb{R}^N)$ for every $B \in \mathcal{C}_p$,*

(b) *the function a belongs to $Q_C(\mathbb{R}^N)$,*

(c) *the function a belongs to $Q_{SC}(\mathbb{R}^N) + SO(\mathbb{R}^N)$.*

Proof. (a) ⇒ (b): Suppose that $a \notin Q_C(\mathbb{R}^N)$. Then there exist bounded open sets D and M, a sequence $(t_k)_{k=1}^\infty \subset \mathbb{R}^N$ which tends to infinity, and a number $\delta > 0$ such that

$$\int_D \left| \int_{-M} (a(t_k+h) - a(t_k+s+h))\, ds \right| dh > \delta.$$

Using Hölder's inequality,

$$\begin{aligned}
&\int_D \left| \int_{-M} (a(t_k+h) - a(t_k+s+h))\, ds \right| dh \\
&\quad \le \left(\int_D \left| \int_{-M} (a(t_k+h) - a(t_k+s+h))\, ds \right|^p dh \right)^{1/p} \cdot \left(\int_D dh \right)^{1/q},
\end{aligned}$$

we get the estimate

$$\int_D \left| \int_{-M} (a(t_k + h) - a(t_k + s + h))\, ds \right|^p dh > \delta^p (\text{mes}\, D)^{-p/q}.$$

Let $W := D - M$ and $\varphi_k(s) := \chi_W(s - t_k)$ for $s \in \mathbb{R}^N$ and k a positive integer. Clearly, the sequence (φ_k) tends weakly to zero. On the other hand, the sequence of the functions $(aC(\chi_M) - C(\chi_M)aI)\varphi_k$ does not converge to zero in the norm of $L^p(\mathbb{R}^N)$ as the following estimates (where again $\int$ stands for the integral over $\mathbb{R}^N$) indicate:

$$\begin{aligned}
&\|(aC(\chi_M) - C(\chi_M)aI)\varphi_k\|^p \\
&= \int \left| \int (a(t) - a(r))\chi_M(t - r)\chi_W(r - t_k)\, dr \right|^p dt \\
&= \int \left| \int (a(t) - a(t+s))\chi_{-M}(s)\chi_W(t + s - t_k)\, ds \right|^p dt \\
&= \int \left| \int (a(t_k + h) - a(t_k + h + s))\chi_{-M}(s)\chi_W(h + s)\, ds \right|^p dh \\
&= \int \left| \int_{-M} (a(t_k + h) - a(t_k + h + s))\chi_W(h + s)\, ds \right|^p dh \\
&\geq \int_D \left| \int_{-M} (a(t_k + h) - a(t_k + h + s))\chi_W(h + s)\, ds \right|^p dh \\
&= \int_D \left| \int_{-M} (a(t_k + h) - a(t_k + h + s))\, ds \right|^p dh > \delta^p (\text{mes}\, D)^{-p/q}.
\end{aligned}$$

Thus, the commutator $aC(\chi_M) - C(\chi_M)aI$ cannot be compact (otherwise it would map the sequence of the φ_k to a norm convergent sequence).
(b) $\Rightarrow$ (c): Let $a \in Q_C(\mathbb{R}^N)$ and M be an open and bounded subset of $\mathbb{R}^N$. Then, by definition, the function

$$t \mapsto \int_M (a(t) - a(t + s))\, ds$$

belongs to $Q_{SC}(\mathbb{R}^N)$. We claim that the function

$$g : t \mapsto \int_M a(t + s)\, ds$$

belongs to $SO(\mathbb{R}^N)$. Once this is shown, the assertion $a \in Q_{SC}(\mathbb{R}^N) + SO(\mathbb{R}^N)$ will follow from the identity

$$a(t) = \frac{1}{\text{mes}\, M} \left(\int_M a(t + s)\, ds + \int_M (a(t) - a(t + s))\, ds \right).$$

Assume that $g \notin SO(\mathbb{R}^N)$. Then there is a compact K, a sequence (t_k) which tends to infinity, and a $\delta > 0$ such that

$$\sup_{h \in K} |g(t_k) - g(t_k + h)| > 2\delta \quad \text{for all } k.$$

Choose $h_k \in K$ such that

$$|g(t_k) - g(t_k + h_k)| > 2\delta \quad \text{for all } k.$$

Since K is compact, we can moreover assume that the h_k converge to a point $h^* \in K$. Now we employ the uniform continuity of the function g (see [37], VIII.4., Proposition 14) in order to get

$$|g(t_k) - g(t_k + h^*)| > \delta \quad \text{for all sufficiently large } k.$$

A further use of the uniform continuity of g yields that there is a certain open neighborhood U of $0 \in \mathbb{R}^N$ such that

$$|g(t_k + u) - g(t_k + u + h^*)| > \delta$$

for all $u \in U$ and, hence,

$$\int_U |g(t_k + u) - g(t_k + u + h^*)| \, du > \delta \operatorname{mes} U$$

for all sufficiently large k. On the other hand,

$$\begin{aligned}
&\int_U |g(t_k + u) - g(t_k + u + h^*)| \, du \\
&\quad = \int_U \left| \int_M (a(t_k + u + s) - a(t_k + u + h^* + s)) \, ds \right| du \\
&\quad \le \int_U \left| \int_M (a(t_k + u + s) - a(t_k + u)) \, ds \right| du \\
&\qquad + \int_U \left| \int_M (a(t_k + u) - a(t_k + u + h^* + s)) \, ds \right| du,
\end{aligned}$$

and each term on the right-hand side of this estimate tends to zero as $k \to \infty$ since $a \in Q_C(\mathbb{R}^N)$. This contradiction proves that g is slowly oscillating.
Finally, the implication (c) $\Rightarrow$ (a) follows from Theorems 3.2.2 and 3.2.8. □

Corollary 3.2.11 *$Q_C(\mathbb{R}^N) = Q_{SC}(\mathbb{R}^N) + SO(\mathbb{R}^N)$ is a unital commutative C^*-subalgebra of $L^\infty(\mathbb{R}^N)$ and $Q_{SC}(\mathbb{R}^N)$ is a closed ideal of that algebra.*

One can moreover show that the intersection $Q_{SC}(\mathbb{R}^N) \cap SO(\mathbb{R}^N)$ consists of all continuous functions which tend to zero at infinity (see [158], Theorem 4.2).

3.3 Fredholmness of convolution type operators

As an application of the results of the preceding sections, we are now going to examine the Fredholm properties of operators on $L^p(\mathbb{R}^N)$ which belong to an algebra which is generated by compositions of convolution operators and operators of multiplication.

3.3.1 Operators of convolution type

Given a subalgebra $\mathcal{E}$ of $L^\infty(\mathbb{R}^N)$, we let $\mathcal{A}(\mathcal{E}, \mathcal{C}_p)$ denote the smallest closed subalgebra of $L(L^p(\mathbb{R}^N))$ which contains the identity operator and all operators of the form

$$aKbI \quad \text{where} \quad a,\, b \in \mathcal{E} \text{ and } K \in \mathcal{C}_p. \tag{3.14}$$

We will call the elements of $\mathcal{A}(L^\infty(\mathbb{R}^N), \mathcal{C}_p)$ *convolution type operators* in what follows. Every operator in $\mathcal{A}(L^\infty(\mathbb{R}^N), \mathcal{C}_p)$ can be approximated as closely as desired by operators of the form

$$A := \gamma I + \sum \prod a_{ij} K_{ij} b_{ij} I \tag{3.15}$$

where a_{ij}, $b_{ij} \in L^\infty(\mathbb{R}^N)$, $K_{ij} \in \mathcal{C}_p$ and $\gamma \in \mathbb{C}$, and where the sum and all products are finite. Observe that, in this definition, multiplication operators only occur in combination with convolutions. A larger algebra which also contains single multiplication operators will be considered in Chapter 7.

Let $C_0^\infty(\mathbb{R}^N)$ stand for the algebra of all infinitely differentiable functions on $\mathbb{R}^N$ with compact support.

Proposition 3.3.1 *The algebra $\mathcal{A}(C_0^\infty(\mathbb{R}^N), \mathcal{C}_p)$ contains the compact operators on $L^p(\mathbb{R}^N)$.*

Proof. It is sufficient to show that $\mathcal{A}(C_0^\infty(\mathbb{R}^N), \mathcal{C}_p)$ contains all rank one operators. Every rank one operator on $L^p(\mathbb{R}^N)$ has the form

$$(Ku)(t) = a(t) \int_{\mathbb{R}^N} b(s)u(s)ds, \quad t \in \mathbb{R}^N, \tag{3.16}$$

where $a \in L^p(\mathbb{R}^N)$ and $b \in L^q(\mathbb{R}^N)$ with $1/p + 1/q = 1$. Since $C_0^\infty(\mathbb{R}^N)$ lies densely in both $L^p(\mathbb{R}^N)$ and $L^q(\mathbb{R}^N)$ (with respect to the corresponding norms), it is further sufficient to show that every operator (3.16) with a, $b \in C_0^\infty(\mathbb{R}^N)$ belongs to $\mathcal{A}(C_0^\infty(\mathbb{R}^N), \mathcal{C}_p)$.

Let a, $b \in C_0^\infty(\mathbb{R}^N)$, and choose a function $k \in L^1(\mathbb{R}^N)$ which is 1 on the compact set $\{t - s : t \in \operatorname{supp} f,\, s \in \operatorname{supp} g\}$. Then the operator (3.16) can be written as

$$(Ku)(t) = a(t) \int_{\mathbb{R}^N} k(t-s)b(s)u(s)ds, \quad t \in \mathbb{R}^N.$$

Evidently, this operator belongs to $\mathcal{A}(C_0^\infty(\mathbb{R}^N), \mathcal{C}_p)$. □

Proposition 3.3.2 $\mathcal{A}(L^\infty(\mathbb{R}^N), \mathcal{C}_p) \subseteq \mathcal{B}_p$.

The proof is based on the following norm estimate which is known as Schur's lemma (see [178], Appendix A, Proposition 5.1).

Proposition 3.3.3 *Let l be a measurable function on $\mathbb{R}^N \times \mathbb{R}^N$ with*

$$M_1 := \sup_{x\in\mathbb{R}^N} \int_{\mathbb{R}^N} |l(x,y)|dy < \infty \quad \textit{and} \quad M_2 := \sup_{y\in\mathbb{R}^N} \int_{\mathbb{R}^N} |l(x,y)|dx < \infty.$$

Then the operator

$$(Lu)(x) := \int_{\mathbb{R}^N} l(x,y)u(y)dy, \quad x \in \mathbb{R}^N$$

acts boundedly on $L^p(\mathbb{R}^N)$, and $\|L\|_{L(L^p(\mathbb{R}^N))} \le M_1^{1/q} M_2^{1/p}$.

Proof of Proposition 3.3.2. Clearly, the algebra $\mathcal{B}_p$ contains all operators of multiplication by a bounded measurable function. Thus, and since $\mathcal{B}_p$ is a closed algebra, the result will follow once we have shown that $\mathcal{B}_p$ also contains a dense subset of $\mathcal{C}_p$. Actually, we will check that

$$\lim_{t\to 0} \sup_{h\in\mathbb{R}^N} \|[\varphi_{t,h}I, C(k)]\| = 0 \tag{3.17}$$

for every function $k \in L^1(\mathbb{R}^N)$ with compact support and every $\varphi \in BUC(\mathbb{R}^N)$. For definiteness, let the support of k be contained in a ball with center 0 and radius R. Since

$$([\varphi_{t,h}I, C(k)]u)(x) = \int_{\mathbb{R}^N} (\varphi_{t,h}(x) - \varphi_{t,h}(y))\, k(x-y)u(y)\, dy,$$

Proposition 3.3.3 implies

$$\begin{aligned} \|[\varphi_{t,h}I, C(k)]\|_{L(L^p)} &\le \|k\|_1 \sup_{x,\, y\in\mathbb{R}^N:|x-y|\le R} |\varphi_{t,h}(x) - \varphi_{t,h}(y)| \\ &= \|k\|_1 \sup_{x,\, y\in\mathbb{R}^N:|x-y|\le R} |\varphi(t(x-h)) - \varphi(t(y-h))|. \end{aligned}$$

For $|x-y| \le R$, we have

$$|t(x-h) - t(y-h)| \le |t|R \to 0 \quad \text{as } t \to 0.$$

Since $\varphi \in BUC(\mathbb{R}^N)$, we obtain (3.17). □

A striking property of operators of convolution type is that their $\mathcal{P}$-Fredholmness coincides with their common Fredholmness.

Proposition 3.3.4 *An operator in $\mathcal{A}(L^\infty(\mathbb{R}^N), \mathcal{C}_p)$ is Fredholm if and only if it is $\mathcal{P}$-Fredholm.*

Proof. Let $\mathcal{J}$ refer to the closed ideal of $\mathcal{A}(L^\infty(\mathbb{R}^N), \mathcal{C}_p)$ which contains all operators in $\mathcal{C}_p$. It is easy to check that, whenever $J \in \mathcal{J}$, the operator JP_k is compact for every k. Indeed, every operator $J \in \mathcal{J}$ can be approximated in the norm as closely as desired by a sum of products of operators of the form $aKbI$ where a and b are bounded measurable functions and $K \in \mathcal{C}_p$. The compactness of $aKbP_k = aKP_kbI$ follows from Theorem 3.2.2.

Since P_k fails to be compact, we have $I \notin \mathcal{J}$, and the algebra $\mathcal{A}(L^\infty(\mathbb{R}^N), \mathcal{C}_p)$ decomposes into the direct sum $\mathbb{C}I + \mathcal{J}$. In particular, every operator A in this algebra can be uniquely written as $\gamma_A I + K_A$ where $\gamma_A \in \mathbb{C}$ and $K_A \in \mathcal{J}$, and it turns out that the mapping $A \mapsto \gamma_A$ is a continuous algebra homomorphism.

In the next step we will show that

$$\mathcal{J} \cap K(L^p(\mathbb{R}^N), \mathcal{P}) = K(L^p(\mathbb{R}^N)).$$

The inclusion $\supseteq$ follows from Proposition 3.3.1 and from the strong convergence of the P_n to the identity operator. If, conversely, $J \in \mathcal{J} \cap K(L^p(\mathbb{R}^N), \mathcal{P})$, then JP_k is compact for every k as we have just seen. On the other hand, since $J \in K(L^p(\mathbb{R}^N), \mathcal{P})$, one has $\|J - JP_k\| \to 0$ as $k \to \infty$. Thus, being the norm limit of compact operators, the operator J is compact.

Since $K(L^p(\mathbb{R}^N)) \subseteq K(L^p(\mathbb{R}^N), \mathcal{P})$, it is clear that every Fredholm operator is also $\mathcal{P}$-Fredholm. Let, conversely, $A \in \mathcal{A}(L^\infty(\mathbb{R}^N), \mathcal{C}_p)$ be a $\mathcal{P}$-Fredholm operator. Then there are an operator $L' \in \mathcal{B}_p$ and an operator $T \in K(L^p(\mathbb{R}^N), \mathcal{P})$ such that $L'A = I + T$. We claim that then $\gamma_A \neq 0$. Contrary to what we want, assume that $\gamma_A = 0$. Then $A \in \mathcal{J}$. Choose $m > 0$ and $n \in \mathbb{Z}^N$ such that $\|P_m U_{-n} T U_n P_m\| < 1/2$ (which can be done due to Proposition 1.2.6). Then, by Neumann series, the right-hand side of

$$P_m U_{-n} L' A U_n P_m = P_m + P_m U_{-n} T U_n P_m$$

is an invertible operator acting on the range of P_m, whence

$$P_m = (P_m + P_m U_{-n} T U_n P_m)^{-1} P_m U_{-n} L' A U_n P_m. \tag{3.18}$$

Since $U_n P_m U_{-n}$ is the operator of multiplication by a compactly supported function, the operator $AU_nP_m = A(U_nP_mU_{-n})U_n$ and, hence, the operator on the right-hand side of (3.18) are compact. But P_m is not compact, and this contradiction proves the claim.

Now write A as $\gamma_A I + K_A$ and set $L := -K_A L' + I$. Then

$$LA - \gamma_A I = \gamma_A L'A - AL'A + A - \gamma_A I = (\gamma_A I - A)(L'A - I).$$

Since $L'A - I \in K(L^p(\mathbb{R}^N), \mathcal{P})$ and $\gamma_A I - A = K_A \in \mathcal{J}$, the operator $LA - \gamma_A I$ is compact. Similarly, one shows that $AR - \gamma_A I$ is compact for a certain operator $R \in \mathcal{B}_p$. Hence, and because of $\gamma_A \neq 0$, the operator A is Fredholm. □

Corollary 3.3.5 $\mathcal{A}(L^\infty(\mathbb{R}^N), \mathcal{C}_p) \cap K(L^p(\mathbb{R}^N), \mathcal{P}) = K(L^p(\mathbb{R}^N))$.

There are operators in $\mathcal{A}(L^\infty(\mathbb{R}^N),\, \mathcal{C}_p)$ which do not possess a rich operator spectrum. The next result identifies a subalgebra of $\mathcal{A}(L^\infty(\mathbb{R}^N),\, \mathcal{C}_p) \cap \mathcal{B}_p^\$$ which contains sufficiently many interesting operators.

Proposition 3.3.6 $\mathcal{A}(BUC(\mathbb{R}^N),\, \mathcal{C}_p) \subseteq \mathcal{B}_p^\$$.

Proof. Let $a \in BUC(\mathbb{R}^N)$, and let h be a sequence which tends to infinity. The family of all functions $x \mapsto a(x + h(m))$ is bounded and equicontinuous on every compact subset M of $\mathbb{R}^N$. Hence, by the Arzelà-Ascoli theorem, there are a subsequence g of h and a continuous bounded function a_h on $\mathbb{R}^N$ such that, for every compact $M \subset \mathbb{R}^N$,

$$\lim_{m\to\infty} \sup_{x\in M} |a(x + g(m)) - a_h(x)| = 0.$$

Thus, the operators $U_{-g(m)} a U_{g(m)}$ of multiplication by the function $x \mapsto a(x + g(m))$ converge *-strongly to the operator of multiplication by the function a_h.

Let A be an operator of the form (3.15), but with $a_{ij},\, b_{ij} \in BUC$. As we have just seen, given a sequence h tending to infinity, we can choose a subsequence g of h such that the operators $U_{-g(m)} a_{ij} U_{g(m)}$ and $U_{-g(m)} b_{ij} U_{g(m)}$ converge *-strongly to certain multiplication operators $(a_{ij})_h I$ and $(b_{ij})_h I$, respectively. Then

$$U_{-g(m)} A U_{g(m)} P_k = \gamma P_k + \sum \prod (U_{-g(m)} a_{ij} U_{g(m)}) K_{ij} P_k (U_{-g(m)} b_{ij} U_{g(m)})$$

converges in the norm of $L(L^p(\mathbb{R}^N))$ to

$$\gamma P_k + \sum \prod (a_{ij})_h K_{ij} P_k (b_{ij})_h I = (\gamma I + \sum \prod (a_{ij})_h K_{ij} (b_{ij})_h I) P_k$$

for every P_k (recall Theorem 1.1.3 and that the operators $K_{ij} P_k$ are compact due to Theorem 3.2.2).

Hence, all operators of the form (3.15) with $a_{ij},\, b_{ij} \in BUC$ possess a rich operator spectrum. Since these operators lie densely in $\mathcal{A}(BUC(\mathbb{R}^N),\, \mathcal{C}_p)$, and since $\mathcal{B}_p^\$$ is a closed algebra, this yields the assertion. □

3.3.2 Fredholmness

Due to Proposition 3.3.6, the operators in $\mathcal{A}(BUC(\mathbb{R}^N),\, \mathcal{C}_p)$ are subject to Theorem 3.1.9, and in combination with Proposition 3.3.4 we obtain the following result.

Theorem 3.3.7 *Let $A \in \mathcal{A}(BUC(\mathbb{R}^N),\, \mathcal{C}_p)$. Then*

(a) *A is Fredholm (i.e., locally invertible at infinity) if and only if all limit operators of A are uniformly invertible.*

(b) *A is $\mathcal{PR}_\eta$-Fredholm (i.e., locally invertible at the infinitely distant point $\eta \in S^{N-1}$) if and only if all operators in the local operator spectrum $\sigma_\eta(A)$ are uniformly invertible.*

Corollary 3.3.8 *An operator $A \in \mathcal{A}(BUC(\mathbb{R}^N), \mathcal{C}_p)$ is a Fredholm operator if and only if, for each point $\eta \in S^{N-1}$, all operators in $\sigma_\eta(A)$ are uniformly invertible.*

We are going to specialize these results to operators with coefficients in certain subalgebras of $L^\infty(\mathbb{R}^N)$.

Slowly oscillating coefficients. Slowly oscillating functions are uniformly continuous by Proposition 3.2.7. Hence, $\mathcal{A}(SO(\mathbb{R}^N), \mathcal{C}_p) \subseteq \mathcal{A}(BUC(\mathbb{R}^N), \mathcal{C}_p)$, and Theorem 3.3.7 and its corollary apply to operators in the algebra $\mathcal{A}(SO(\mathbb{R}^N), \mathcal{C}_p)$. Limit operators of operators in $\mathcal{A}(SO(\mathbb{R}^N), \mathcal{C}_p)$ are of a particularly simple form such that their invertibility can be effectively checked via (3.11).

Proposition 3.3.9 *Every limit operator of an operator in $\mathcal{A}(SO(\mathbb{R}^N), \mathcal{C}_p)$ lies in $\mathbb{C}I + \mathcal{C}_p$.*

Proof. Every operator in $\mathcal{A}(SO(\mathbb{R}^N), \mathcal{C}_p)$ can be uniformly approximated by operators of the form (3.15) where $a_{ij}, b_{ij} \in SO(\mathbb{R}^N)$. If $K \in \mathcal{C}_p$ then, clearly, the limit operator K_h exists with respect to every sequence $h \in \mathcal{H}$, and $K_h = K$. Thus, it remains to check that if $a \in SO(\mathbb{R}^N)$, and if $h \in \mathcal{H}$ is a sequence such that the operators of multiplication $U_{-h(n)} a U_{h(n)}$ converge *-strongly to $a_h I$ as $n \to \infty$, then a_h is a constant function. This can be done as in Proposition 2.4.1. □

Corollary 3.3.10 *Let A be an operator of the form* (3.15) *with $a_{ij}, b_{ij} \in SO(\mathbb{R}^N)$. Then A is Fredholm if and only if all limit operators of A are invertible.*

Thus, the uniformity of the invertibility is not required.

Proof. We conclude from the previous proposition that every limit operator of A is a linear combination of the operators $\prod_{j=1}^{n_i} K_{ij}$ with $i = 1, \ldots, n$. Thus, $\sigma_{op}(A)$ lies in a finite-dimensional subspace of $L(L^p(\mathbb{R}^N))$, and the assertion follows from Proposition 2.2.5. □

Remark A. The algebra $\mathcal{A}(Q_C(\mathbb{R}^N), \mathcal{C}_p)$ which is apparently larger than the algebra $\mathcal{A}(SO(\mathbb{R}^N), \mathcal{C}_p)$ actually coincides with the latter algebra. Indeed, by Theorem 3.2.10, every operator aK with $a \in Q_C(\mathbb{R}^N)$ and $K \in \mathcal{C}_p$ is the sum of an operator $a_1 K$ with $a_1 \in SO(\mathbb{R}^N)$ and an operator $a_2 K$ with $a_2 \in Q_{SC}(\mathbb{R}^N)$. Since slowly oscillating functions are uniformly continuous (Proposition 3.2.7) and since $a_2 K$ is compact (Theorem 3.2.2), one has $aK \in \mathcal{A}(SO(\mathbb{R}^N), \mathcal{C}_p)$ by Proposition 3.3.1.

Remark B. One can also study the $\mathcal{P}$-Fredholmness of operators in the smallest closed subalgebra of $L(L^p(\mathbb{R}^N))$ which contains the algebra $\mathcal{A}(BUC(\mathbb{R}^N), \mathcal{C}_p)$ and all shift operators U_α. We will consider such combinations of multiplication, convolution and shift operators in a more general context in Chapter 7.

Coefficients stabilizing at infinity. Theorem 3.3.7 and its corollary attain their simplest form for operators with coefficients which stabilize at infinity in the following sense.

Definition 3.3.11 *The function* $a \in L^\infty(\mathbb{R}^N)$ stabilizes at infinity *if, for every infinitely distant point* $\eta \in S^{N-1}$, *there is a constant* $y \in \mathbb{C}$ *such that, for every* $\varepsilon > 0$, *there exists a neighborhood* $U = U_{\eta,\,\varepsilon}$ *at infinity of* η *such that*

$$\text{mes}\,\{x \in U_{\eta,\,\varepsilon} : |a(x) - y| > \varepsilon\} < \varepsilon. \tag{3.19}$$

The class of all functions stabilizing at infinity will be denoted by $L^\infty_{stab}(\mathbb{R}^N)$.

If a stabilizes at infinity and η is an infinitely distant point, then the constant y which satisfies (3.19) is uniquely determined. We denote it by $\hat{a}(\eta)$.

Lemma 3.3.12 *Let* $a \in L^\infty_{stab}(\mathbb{R}^N)$ *and* $\eta \in S^{N-1}$ *be an infinitely distant point. Then* $|\hat{a}(\eta)| \leq \|a\|_\infty$.

Proof. Let $\varepsilon > 0$ and choose a neighborhood U of infinity such that

$$\text{mes}\,\{x \in U : |a(x) - \hat{a}(\eta)| > \varepsilon\} < \varepsilon.$$

Then

$$\text{mes}\,\{x \in U : |\,|a(x)| - |\hat{a}(\eta)|\,| > \varepsilon\} < \varepsilon.$$

Since the measure of U is infinite, there is a subset $M \subset U$ of measure 1 such that

$$|a(x)| - \varepsilon < |\hat{a}(\eta)| < |a(x)| + \varepsilon \quad \text{for all } x \in M.$$

This yields the assertion. □

Theorem 3.3.13 $L^\infty_{stab}(\mathbb{R}^N)$ *is a* C^**-subalgebra of* $Q_C(\mathbb{R}^N)$.

Proof. First we will show that $L^\infty_{stab}(\mathbb{R}^N)$ is closed in $L^\infty(\mathbb{R}^N)$. Let $a_n \in L^\infty_{stab}(\mathbb{R}^N)$ and $a \in L^\infty(\mathbb{R}^N)$ such that $\lim \|a_n - a\|_\infty = 0$. Fix $\varepsilon > 0$, and choose $n_0 \in \mathbb{N}$ such that

$$\|a_n - a_m\|_\infty < \varepsilon \quad \text{for all } n,\, m \geq n_0.$$

Further, let $U_{\eta,\,\varepsilon,\,n}$ be a neighborhood at infinity of η such that

$$\text{mes}\,\{x \in U_{\eta,\,\varepsilon,\,n} : |a_n(x) - \widehat{a_n}(\eta)| > \varepsilon\} < \varepsilon,$$

and set

$$U'_{\eta,\,\varepsilon,\,n} := \{x \in U_{\eta,\,\varepsilon,\,n} : |a_n(x) - \widehat{a_n}(\eta)| \leq \varepsilon\}.$$

Then, for $x \in U'_{\eta,\,\varepsilon,\,n} \cap U'_{\eta,\,\varepsilon,\,m}$ and $m,\, n > n_0$,

$$|\widehat{a_n}(\eta) - \widehat{a_m}(\eta)| \leq |\widehat{a_n}(\eta) - a_n(x)| + |a_n(x) - a_m(x)| + |a_m(x) - \widehat{a_m}(\eta)| \leq 3\,\varepsilon.$$

Thus, $(\widehat{a_n}(\eta))_{n\in\mathbb{N}}$ is a Cauchy sequence, and we let $\hat{a}(\eta)$ denote its limit.

Now we fix $n > n_0$ such that

$$\|a_n - a\|_\infty < \varepsilon/3 \quad \text{and} \quad |\widehat{a_n}(\eta) - \hat{a}(\eta)| < \varepsilon/3.$$

The estimate

$$|a_n(x) - \hat{a_n}(\eta)| \geq |a(x) - \hat{a}(\eta)| - |a(x) - a_n(x)| - |a_n(x) - \widehat{a_n}(\eta)|$$

implies that $|a_n(x) - \widehat{a_n}(\eta)| > \varepsilon/3$ whenever $|a(x) - \hat{a}(\eta)| > \varepsilon$. Since a_n stabilizes at infinity, there is a neighborhood $U_{\eta,\,\varepsilon/3,\,n}$ such that

$$\text{mes}\,\{x \in U_{\eta,\,\varepsilon/3,\,n} : |a_n(x) - \widehat{a_n}(\eta)| > \varepsilon/3\} < \varepsilon/3.$$

Thus,

$$\text{mes}\,\{x \in U_{\eta,\,\varepsilon/3,\,n} : |a(x) - \hat{a}(\eta)| > \varepsilon\} < \varepsilon/3 < \varepsilon,$$

whence $a \in L^\infty_{stab}(\mathbb{R}^N)$.

In the next step we show that $L^\infty_{stab}(\mathbb{R}^N)$ is a *-algebra. The symmetry is obvious. Let $a,\, b \in L^\infty_{stab}(\mathbb{R}^N)$, and let η be an infinitely distant point. We choose neighborhoods at infinity of η such that

$$\text{mes}\,\{x \in U_{\eta,\,\varepsilon/2,\,a} : |a(x) - \hat{a}(\eta)| > \varepsilon/2\} < \varepsilon/2 \tag{3.20}$$

and

$$\text{mes}\,\{x \in U_{\eta,\,\varepsilon/2,\,b} : |b(x) - \hat{b}(\eta)| > \varepsilon/2\} < \varepsilon/2. \tag{3.21}$$

Set $W_\eta := U_{\eta,\,\varepsilon/2,\,a} \cap U_{\eta,\,\varepsilon/2,\,b}$. Then W_η is a neighborhood at infinity of η, and it follows from

$$\begin{aligned}&\{x \in W_\eta : |a(x) + b(x) - \hat{a}(\eta) - \hat{b}(\eta)| > \varepsilon\}\\ &\qquad \subseteq \{x \in W_\eta : |a(x) - \hat{a}(\eta)| > \varepsilon/2\} \cup \{x \in W_\eta : |b(x) - \hat{b}(\eta)| > \varepsilon/2\}\end{aligned}$$

and from (3.20), (3.21) that

$$\text{mes}\,\{x \in W_\eta : |a(x) + b(x) - \hat{a}(\eta) - \hat{b}(\eta)| > \varepsilon\} < \varepsilon.$$

Thus, $a + b \in L^\infty_{stab}(\mathbb{R}^N)$ and

$$\widehat{(a+b)}(\eta) = \hat{a}(\eta) + \hat{b}(\eta) \quad \text{for all } \eta \in S^{N-1}.$$

In order to show that $ab \in L^\infty_{stab}(\mathbb{R}^N)$, too, we can assume that $a,\, b \neq 0$ (otherwise the assertion is obvious). Choose $m \in \mathbb{N}$ such that $m\|a\|_\infty > 1$ and $m\|b\|_\infty > 1$. Given an infinitely distant point η and $\varepsilon > 0$ a, choose neighborhoods at infinity of η such that

$$\text{mes}\,\{x \in U_{\eta,\,a} : |a(x) - \hat{a}(\eta)| > \varepsilon/(2m\|b\|_\infty)\} < \varepsilon/(2m\|b\|_\infty)$$

and

$$\text{mes}\,\{x \in U_{\eta,\,b} : |b(x) - \hat{b}(\eta)| > \varepsilon/(2m\|a\|_\infty)\} < \varepsilon/(2m\|a\|_\infty).$$

Set $W_\eta := U_{\eta,\,a} \cap U_{\eta,\,b}$. Then W_η is a neighborhood at infinity of η, and

$$\begin{aligned}
&\text{mes}\,\{x \in W_\eta : |(ab)(x) - \hat{a}(\eta)\hat{b}(\eta)| > \varepsilon\} \\
&\quad = \text{mes}\,\{x \in W_\eta : |(a(x) - \hat{a}(\eta))b(x) + \hat{a}(\eta))(b(x) - \hat{b}(\eta))| > \varepsilon\} \\
&\quad \le \text{mes}\,\{x \in W_\eta : |a(x) - \hat{a}(\eta)|\,\|b\|_\infty + \|a\|_\infty\,|b(x) - \hat{b}(\eta)| > \varepsilon\} \\
&\quad \le \text{mes}\,\{x \in W_\eta : |a(x) - \hat{a}(\eta)|\,\|b\|_\infty > \varepsilon/2\} \\
&\qquad\quad + \text{mes}\,\{x \in W_\eta : |b(x) - \hat{b}(\eta)|\,\|a\|_\infty > \varepsilon/2\} \\
&\quad \le \text{mes}\,\{x \in U_{\eta,\,a} : |a(x) - \hat{a}(\eta)|\,\|b\|_\infty > \varepsilon/(2m)\} \\
&\qquad\quad + \text{mes}\,\{x \in U_{\eta,\,b} : |b(x) - \hat{b}(\eta)|\,\|a\|_\infty > \varepsilon/(2m)\} \\
&\quad \le \text{mes}\,\{x \in U_{\eta,\,a} : |a(x) - \hat{a}(\eta)| > \varepsilon/(2m\|b\|_\infty)\} \\
&\qquad\quad + \text{mes}\,\{x \in U_{\eta,\,b} : |b(x) - \hat{b}(\eta)| > \varepsilon/(2m\|a\|_\infty)\} \\
&\quad < \varepsilon/(2m\|b\|_\infty) + \varepsilon/(2m\|a\|_\infty) < \varepsilon/2 + \varepsilon/2 = \varepsilon.
\end{aligned}$$

Consequently, $ab \in L^\infty_{stab}(\mathbb{R}^N)$ and

$$\widehat{(ab)}(\eta) = \hat{a}(\eta)\,\hat{b}(\eta) \quad \text{for all } \eta \in S^{N-1}.$$

It remains to show the inclusion $L^\infty_{stab}(\mathbb{R}^N) \subseteq Q_C(\mathbb{R}^N)$. Thus, if $a \in L^\infty_{stab}(\mathbb{R}^N)$, we have to show that, for every open bounded set $M \subset \mathbb{R}^N$, there is an open bounded set $D \subset \mathbb{R}^N$ such that

$$\lim_{t\to\infty} \int_D \left| \int_M (a(t+h) - a(t+h+s))\,ds \right| dh = 0 \tag{3.22}$$

(Definition 3.2.9). Let $M \subset \mathbb{R}^N$ be open and bounded, choose D as the open unit ball in $\mathbb{R}^N$, and let $d > 0$ be the radius of a ball with center 0 which contains $M + D$. Let further $\varepsilon > 0$. Then, for every infinitely distant point η, there is a neighborhood at infinity of η such that

$$\text{mes}\,\{x \in U_{\eta,\,\varepsilon} : |a(x) - \hat{a}(\eta)| > \varepsilon\} < \varepsilon.$$

Each neighborhood $U_{\eta,\,\varepsilon}$ is of the form

$$U_{\eta,\,\varepsilon} = \{y \in \mathbb{R}^N : |y| > R_{\eta,\,\varepsilon} \text{ and } y/|y| \in W_{\eta,\,\varepsilon}\}$$

where $R_{\eta,\,\varepsilon} \ge 0$ and $W_{\eta,\,\varepsilon} \subseteq S^{N-1}$ is an open neighborhood of η. In particular, $\{W_{\eta,\,\varepsilon}\}_{\eta \in S^{N-1}}$ is an open cover of the unit sphere, from which we can choose a finite subcover $\{W_{\eta_i,\,\varepsilon}\}_{i=1}^k$. Set

$$R_0 := \max\{R_{\eta_i,\,\varepsilon} : i = 1, \ldots, k\} + d.$$

Further, since the function $f : S^{N-1} \to \mathbb{R}^N$,

$$f(x) := \max\{\text{dist}\,(x,\, S^{N-1} \setminus W_{\eta_i,\,\varepsilon}) : i = 1,\, \ldots,\, k\},$$

is positive for every x (every x belongs to one of the sets $W_{\eta_i,\,\varepsilon}$) and continuous on the compact set S^{N-1}, there is a $\delta > 0$ such that $f(x) \geq \delta$ for all $x \in S^{N-1}$. Thus, for every $x \in S^{N-1}$, there is an $i \in \{1, \ldots, k\}$ such that

$$x \in W_{\eta_i,\,\varepsilon} \quad \text{and} \quad \text{dist}\,(x,\, \partial W_{\eta_i,\,\varepsilon}) \geq \delta.$$

Consequently, there is an $R_1 \geq R_0$ such that, for every $y \in \mathbb{R}^N$ with $|y| \geq R_1$, there is an $i \in \{1, \ldots, k\}$ such that

$$y \in U_{\eta_i,\,\varepsilon} \quad \text{and} \quad \text{dist}\,(y,\, \partial U_{\eta_i,\,\varepsilon}) \geq d.$$

Let now $t \in \mathbb{R}^N$ with $|t| \geq R_1$. By what we have just seen, there is an $i \in \{1, \ldots, k\}$ such that $t + D$ and $t + M + D$ are contained in $U_{\eta_i,\,\varepsilon}$. Thus,

$$\text{mes}\,\{x \in t + D : |a(x) - \hat{a}(\eta)| > \varepsilon\} < \varepsilon$$

and

$$\text{mes}\,\{x \in t + D + M : |a(x) - \hat{a}(\eta)| > \varepsilon\} < \varepsilon.$$

This implies

$$\begin{aligned}
&\int_D \left| \int_M (a(t+h) - a(t+h+s))\, ds \right| dh \\
&\quad \leq \int_D \int_M |a(t+h) - \hat{a}(\eta)|\, ds\, dh + \int_D \int_M |a(t+h+s) - \hat{a}(\eta)|\, ds\, dh \\
&\quad \leq \text{mes}\, D \int_{t+M} |a(h) - \hat{a}(\eta)|\, dh + \text{mes}\, D \int_{t+D+M} |a(h) - \hat{a}(\eta)|\, dh \\
&\quad \leq \text{mes}\, D\, (\text{mes}\, M \cdot \varepsilon + 2\varepsilon \|a\|_\infty) + \text{mes}\, D\, (\text{mes}\,(D+M) \cdot \varepsilon + 2\varepsilon \|a\|_\infty) \\
&\quad \leq \varepsilon\, \text{mes}\, D\, (\text{mes}\, M + \text{mes}\,(D+M) + 4\|a\|_\infty),
\end{aligned}$$

whence the assertion (3.22). □

Proposition 3.3.14 *Let $a \in L^\infty_{stab}(\mathbb{R}^N)$, and let h be a sequence which tends to infinity into the direction of $\eta \in S^{N-1}$. Then*

$$U_{-h(n)} a U_{h(n)} \to \hat{a}(\eta) I \quad \textit{strongly on } L^p(\mathbb{R}^N) \textit{ as } n \to \infty.$$

Proof. Given $\varepsilon > 0$, we find a neighborhood $U_{\eta,\,\varepsilon}$ at infinity of η such that

$$\text{mes}\,\{x \in U_{\eta,\,\varepsilon} : |a(x) - \hat{a}(\eta)| > \varepsilon\} < \varepsilon.$$

Let f be a continuous function with compact support. Then

$$\|(U_{-h(n)} a U_{h(n)} - \hat{a}(\eta)) f\|_p = \|(a - \hat{a}(\eta)) U_{h(n)} f\|_p.$$

Clearly, there exists an n_0 such that $\operatorname{supp}(U_{h(n)}f) \subset U_{\eta,\varepsilon}$ for all $n \geq n_0$. Thus, if $n \geq n_0$, then

$$\|(U_{-h(n)}aU_{h(n)} - \hat{a}(\eta))f\|_p \leq \varepsilon\|f\|_p + 2\|a\|_\infty\|f\|_\infty\varepsilon^{1/p}.$$

This proves the strong convergence on a dense subset of $L^p(\mathbb{R}^N)$. Since the operators $U_{-h(n)}aU_{h(n)}$ are uniformly bounded, we get the assertion. □

An obvious consequence of this proposition is that the local operator spectrum $\sigma_\eta(A)$ for operators $A \in \mathcal{A}(L^\infty_{stab}(\mathbb{R}^N), \mathcal{C}_p)$ is a singleton for every infinitely distant point $\eta \in S^{N-1}$, say $\sigma_\eta(A) = \{A_\eta\}$. Moreover, every limit operator A_η belongs to $\mathbb{C}I + \mathcal{C}_p$ since $\mathcal{A}(L^\infty_{stab}(\mathbb{R}^N), \mathcal{C}_p)$ is a subalgebra of $\mathcal{A}(SO(\mathbb{R}^N), \mathcal{C}_p)$, and by Proposition 3.3.9. Thus, the invertibility of A_η can be effectively checked via (3.11).

Corollary 3.3.15 *An operator $A \in \mathcal{A}(L^\infty_{stab}(\mathbb{R}^N), \mathcal{C}_p)$ is Fredholm if and only if every limit operator A_η (with $\eta \in S^{N-1}$) of A is invertible.*

Remark. Although we do not have a description of the algebra $L^\infty_{stab}(\mathbb{R}^N)$ in terms of other function spaces and algebras dealt with, we can at least identify the subset $\mathcal{D}$ of functions $f \in L^\infty_{stab}(\mathbb{R}^N)$ for which the operator fI of multiplication by f has a rich operator spectrum:

$$\mathcal{D} = C(\overline{\mathbb{R}^N}) + L^\infty_0(\mathbb{R}^N), \tag{3.23}$$

where $C(\overline{\mathbb{R}^N})$ and $L^\infty_0(\mathbb{R}^N)$ are introduced in Section 2.3.6 and (3.12), respectively.

To prove (3.23), it is sufficient to check the inclusion $\mathcal{D} \subseteq C(\overline{\mathbb{R}^N}) + L^\infty_0(\mathbb{R}^N)$ since the opposite inclusion is obvious. Therefore, let $f \in \mathcal{D}$, and let h be a sequence tending to infinity into the direction of $\eta \in S^{N-1}$. Then there exists a subsequence g of h such that the limit operator $(fI)_g$ exists. Proposition 3.3.14 shows that $(fI)_g = \hat{a}(\eta)I$ with some unique value $\hat{a}(\eta)$ which only depends on the direction η and not on the sequence g itself (recall that $\mathcal{P}$-convergence implies strong convergence for $1 < p < \infty$).

We show that the mapping $\hat{a} : S^{N-1} \to \mathbb{C}$, which assigns to each direction $\eta \in S^{N-1}$ the value $\hat{a}(\eta)$, is continuous. Suppose it is not continuous at some point $\eta_0 \in S^{N-1}$. Then there are a sequence $(\eta_k) \subset S^{N-1}$ and an $\varepsilon_0 > 0$ such that $\eta_k \to \eta_0$ as $k \to \infty$ (in the familiar topology of S^{N-1}) and that

$$|\hat{a}(\eta_k) - \hat{a}(\eta_0)| > \varepsilon_0.$$

By what was said above, for each $k \in \mathbb{N}$, there is a sequence $g_k : \mathbb{N} \to \mathbb{Z}^N$ which tends to infinity into the direction η_k and for which the limit operator $(fI)_{g_k} = \hat{a}(\eta_k)I$ exists. Let $U_k(\eta_k)$ denote the neighborhood of η_k (in the topology of S^{N-1}) defined by

$$U_k(\eta_k) := \{x \in S^{N-1} : |x - \eta_k| < \frac{1}{k}\}$$

and introduce

$$W_{k,\frac{1}{k}}(\eta_k) := \{x \in \mathbb{R}^N : |x| > k, \frac{x}{|x|} \in U_k(\eta_k)\},$$

which is a neighborhood at infinity of η_k. Given a unit vector $y \in L^p(\mathbb{R}^N)$ and a positive integer k, choose $l(k) := g_k(n_k)$ such that

- $\|\hat{a}(\eta_k)y - U_{-l(k)} f U_{l(k)} y\| < \frac{1}{k}$,
- $l(k) \in W_{k,\frac{1}{k}}(\eta_k)$,
- the limit operator $(fI)_l$ exists.

Because the sequence l tends to infinity into the direction η_0, we have $(fI)_l = \hat{a}(\eta_0)I$, and by Proposition 3.3.14,

$$\delta_k := \|U_{-l(k)} f U_{l(k)} y - \hat{a}(\eta_0)y\| \to 0$$

as $k \to \infty$. But now

$$\begin{aligned} \varepsilon_0 &< |\hat{a}(\eta_k) - \hat{a}(\eta_0)| = \|\hat{a}(\eta_k)y - \hat{a}(\eta_0)y\| \\ &\leq \|\hat{a}(\eta_k)y - U_{-l(k)} f U_{l(k)} y\| + \|U_{-l(k)} f U_{l(k)} y - \hat{a}(\eta_0)y\| \\ &\leq \frac{1}{k} + \delta_k \to 0 \quad \text{as } k \to \infty \end{aligned}$$

is a contradiction, whence the continuity of $\eta \mapsto \hat{a}(\eta)$.

The continuous function $\hat{a}$, thought of as defined on $\overline{\mathbb{R}^N} \setminus \mathbb{R}^N$, can be easily extended to a continuous function a defined on all of $\overline{\mathbb{R}^N}$ by Tietzes extension theorem. Then, clearly, $g := f - a$ is rich, and every limit operator of gI is the zero operator. It remains to show that these two conditions imply that $g \in L_0^\infty(\mathbb{R}^N)$.

Suppose that $g \notin L_0^\infty(\mathbb{R}^N)$. Then g has as an essential value at infinity which is not zero, say ξ. Hence, the function $\xi - g$ is not invertible in $L^\infty(\mathbb{R}^N)$, and the related multiplication operator $\xi I - gI$ is not invertible at infinity. On the other hand, all limit operators of $\xi I - gI$ are equal to $\xi I \neq 0$, whence the invertibility at infinity of $\xi I - gI$ by Theorem 3.1.9. This contradiction shows that $g \in L_0^\infty(\mathbb{R}^N)$ and, consequently, $f = a + g \in C(\overline{\mathbb{R}^N}) + L_0^\infty(\mathbb{R}^N)$. □

3.4 Compressions of convolution type operators

In this section we are going to study the Fredholm properties of compressions of operators of convolution type. If A is a bounded linear operator on $L^p(\mathbb{R}^N)$ and D is a measurable subset of $\mathbb{R}^N$, then the *compression of A onto D* is the operator

$$\chi_D A \chi_D I|_{L^p(D)} : L^p(D) \to L^p(D).$$

The archetypal example is the *Wiener-Hopf operator* $W(k)$ on $L^p(\mathbb{R}^+)$ which is the compression of the convolution operator $\gamma I + C(k)$ with $k \in L^1(\mathbb{R})$ onto $\mathbb{R}^+$. Thus,

$$W(k) = \chi_+(\gamma I + C(k))\chi_+ I|_{L^p(\mathbb{R}^+)},$$

where χ_+ refers to the characteristic function of $\mathbb{R}^+$. Clearly, this operator is Fredholm on $L^p(\mathbb{R}^+)$ if and only if the operator $\gamma I + \chi_+ C(k)\chi_+ I$ is Fredholm on $L^p(\mathbb{R})$. Let f be the function with $f(x) = 0$ if $x < 0$, $f(x) = x$ on $[0, 1]$ and $f(x) = 1$ for $x > 1$. Then the function $\chi_+ - f$ has a compact support. Thus, the operator $\chi_+ C(k)\chi_+ I - fC(k)fI$ is compact on $L^p(\mathbb{R})$, and the operator $W(k)$ is Fredholm on $L^p(\mathbb{R}^+)$ if and only if the operator $\gamma I + fC(k)fI$ is Fredholm on $L^p(\mathbb{R})$. The latter operator is subject to Corollary 3.3.10 which says that this operator (hence, the Wiener-Hopf operator $W(k)$) is Fredholm if and only if the convolution operator $\gamma I + C(k)$ is invertible.

This simple reduction is no longer possible for convolution operators on cones in $\mathbb{R}^N$ when $N > 1$ and, more general, for compressions of operators onto more involved sets.

3.4.1 Compressions of operators in $\mathcal{A}(BUC(\mathbb{R}^N), \mathcal{C}_p)$

A subset D of $\mathbb{R}^N$ with positive measure will be called *rich* if, for each sequence $h : \mathbb{N} \to \mathbb{Z}^N$, there is a subsequence g and a measurable subset D_g of $\mathbb{R}^N$ such that

$$U_{-g(n)}\chi_D U_{g(n)} \to \chi_{D_g} I \quad \text{strongly as } n \to \infty.$$

If this happens, then we call $\chi_{D_g} I$ the *strong limit operator* of $\chi_D I$ with respect to the sequence g.

We consider compressions of operators $A \in \mathcal{A}(BUC(\mathbb{R}^N), \mathcal{C}_p)$ onto rich sets D. Let $\mathcal{J}$ stand for the smallest closed ideal of $\mathcal{A}(BUC(\mathbb{R}^N), \mathcal{C}_p)$ which contains all operators in $\mathcal{C}_p$. Then A can be uniquely written as $\alpha_A I + J_A$ with a complex number α_A and with an operator $J_A \in \mathcal{J}$, and if the compression $\chi_D A \chi_D I : L^p(D) \to L^p(D)$ is Fredholm, then $\alpha_A \neq 0$. This can be checked as in the proof of Proposition 3.3.4. Consequently, the operator $\chi_D A \chi_D I : L^p(D) \to L^p(D)$ is Fredholm (invertible) if and only if the operator

$$\alpha_A(I - \chi_D I) + \chi_D A \chi_D I = \alpha_A I + \chi_D J_A \chi_D I : L^p(\mathbb{R}^N) \to L^p(\mathbb{R}^N)$$

is Fredholm (invertible).

Lemma 3.4.1 *Let $J \in \mathcal{J}$, and let D be a rich subset of $\mathbb{R}^N$. Then the operator $\chi_D J \chi_D I : L^p(\mathbb{R}^N) \to L^p(\mathbb{R}^N)$ is rich.*

Proof. We prove the assertion in case when $J = aKbI$ with BUC-functions a and b and with a convolution operator K possessing a finitely supported kernel. Sums of products of operators of this form can be treated similarly, and since the latter operators lie densely in $\mathcal{J}$, the general case follows by a simple approximation argument.

Given a sequence h tending to infinity , choose a subsequence g of h such that the limit operators $a_g I$ and $b_g I$ of aI and bI exist (which can be done by Proposition 3.3.6) and such that $U_{-g(n)}\chi_D U_{g(n)} \to \chi_{D_g} I$ strongly with a measurable set D_g (which is possible since D is rich).

For $n \in \mathbb{N}$, set $D_n := D - g(n)$, and define functions a_n and b_n by $a_n(x) := a(x + g(n))$ and $b_n(x) := b(x + g(n))$. Then $U_{-g(n)}\chi_D U_{g(n)} = \chi_{D_n} I$ as well as $U_{-g(n)} a U_{g(n)} = a_n I$ and $U_{-g(n)} b U_{g(n)} = b_n I$, and for each positive integer m one has

$$U_{-g(n)}\chi_D a K b \chi_D I U_{g(n)} P_m = \chi_{D_n} a_n K b_n \chi_{D_n} P_m = (\chi_{D_n} a_n)\,(K P_m)\,(b_n \chi_{D_n} I).$$

The sequences $(\chi_{D_n} a_n I)$ and $(\chi_{D_n} b_n I)$ converge strongly to $\chi_{D_g} a_g I$ and $\chi_{D_g} b_g I$, respectively, and the operator KP_m is compact by Theorem 3.2.2. From Theorem 1.1.3 we conclude that the operators $\chi_{D_n} a_n K P_m b_n \chi_{D_n} I$ converge in the norm of $L(L^p(\mathbb{R}^N))$ to

$$(\chi_{D_g} a_g)\,(K P_m)\,(b_g \chi_{D_g} I) = \chi_{D_g} a_g K b_g \chi_{D_g} P_m$$

as $n \to \infty$. Hence, $\chi_{D_g} a_g K b_g \chi_{D_g} I$ is the limit operator of $\chi_D a K b \chi_D I$ with respect to the sequence g. □

Together with Theorem 3.1.9 (b), the previous lemma implies the following.

Theorem 3.4.2 *Let $A \in \mathcal{A}(BUC(\mathbb{R}^N), \mathcal{C}_p)$, and let D be a rich subset of $\mathbb{R}^N$. Then the compression of A onto D is Fredholm on $L^p(D)$ if and only if, for each point $\eta \in S^{N-1}$, all limit operators in*

$$\sigma_\eta(\alpha_A(1 - \chi_D)I + \chi_D A \chi_D I) = \sigma_\eta(\alpha_A I + \chi_D J_A \chi_D I)$$

are uniformly invertible on $L^p(\mathbb{R}^N)$.

In the following subsections we will give examples of unbounded rich domains D for which the strong limit operators of the shifts of $\chi_D I$ can be explicitly calculated and, thus, for which explicit criteria for the Fredholmness of the compressions of operators from $\mathcal{A}(BUC(\mathbb{R}^N), \mathcal{C}_p)$ onto D can be derived.

3.4.2 Compressions to a half-space

Given a non-zero vector $a \in \mathbb{R}^N$, consider the half-space

$$\mathbf{H}(a) := \{x \in \mathbb{R}^N : \langle x,\, a\rangle > 0\}. \tag{3.24}$$

Let further $h \in \mathcal{H}$ be a sequence which tends to infinity into the direction of $\eta \in S^{N-1}$. We distinguish several cases.

- If $\langle \eta,\, a\rangle > 0$, then $\langle h(n),\, a\rangle \to +\infty$, and the strong limit operator of $\chi_{\mathbf{H}(a)} I$ exists and is equal to the identity operator.
- If $\langle \eta,\, a\rangle < 0$, then $\langle h(n),\, a\rangle \to -\infty$, and the strong limit operator of $\chi_{\mathbf{H}(a)} I$ exists and is equal to the zero operator.

- If $\langle \eta, a \rangle = 0$, then h has a subsequence $g \in \mathcal{H}$ such that either the numbers $\langle g(n), a \rangle$ tend to $+\infty$, or to $-\infty$, or to a finite limit $b_g \in \mathbb{R}$. In each of these cases, the strong limit operator of $\chi_{\mathbf{H}(a)} I$ with respect to g exists, and it is equal to the identity operator in the first case, to the zero operator in the second case and to the operator of multiplication by the characteristic function of the shifted half-space
$$\mathbf{H}(a, b_g) := \{x \in \mathbb{R}^N : \langle x, a \rangle > -b_g\}$$
in the third case.

Let $\mathcal{H}_\eta(A)$ stand for the set of all sequences $h \in \mathcal{H}$ which tend to infinity into the direction of $\eta \in S^{N-1}$ and for which the limit operator A_h exists. Further, we denote by $\mathcal{H}_{\eta,\,\infty}(A)$ and $\mathcal{H}_{\eta,\,b}(A)$ the set of all sequences $h \in \mathcal{H}_\eta(A)$ such that $\langle h(n), a \rangle \to \infty$ and $\langle h(n), a \rangle \to b \in \mathbb{R}^N$, respectively. Then Theorem 3.4.2 gives the following result.

Theorem 3.4.3 *Let $A \in \mathcal{A}(BUC(\mathbb{R}^N), \mathcal{C}_p)$ and $D = \mathbf{H}(a)$ with $a \in \mathbb{R}^N \setminus \{0\}$. Then the compression of A onto D is Fredholm on $L^p(D)$ if and only if the following conditions are satisfied:*

(a) *for each point $\eta \in S^{N-1}$ with $\langle \eta, a \rangle > 0$, the set $\{A_h : h \in \mathcal{H}_\eta(A)\}$ of limit operators of A is uniformly invertible.*

(b) *for each point $\eta \in S^{N-1}$ with $\langle \eta, a \rangle = 0$, the set $\{A_h : h \in \mathcal{H}_{\eta,\,\infty}(A)\}$ of limit operators of A is uniformly invertible.*

(c) *for each point $\eta \in S^{N-1}$ with $\langle \eta, a \rangle = 0$ and each $b \in \mathbb{R}^N$, the set*
$$\{(1 - \chi_{\mathbf{H}(a,\,b)})I + \chi_{\mathbf{H}(a,\,b)} A_h \chi_{\mathbf{H}(a,\,b)} I : h \in \mathcal{H}_{\eta,\,b}(A)\}$$
of extended compressions of limit operators of A is uniformly invertible.

3.4.3 Compressions to curved half-spaces

Let $N > 1$ and $f \in BUC(\mathbb{R}^{N-1})$. We consider the *curved half-space*
$$\mathbf{P}(f) := \{x = (x',\, x_N) \in \mathbb{R}^{N-1} \times \mathbb{R} : x_N > f(x')\} \subseteq \mathbb{R}^N. \tag{3.25}$$

Let further $h \in \mathcal{H}$ be a sequence which tends to infinity into the direction of $\eta = (\eta',\, \eta_N) \in S^{N-1} \subset \mathbb{R}^{N-1} \times \mathbb{R}$. Again, we distinguish several cases.

- If $\eta_N > 0$, then the strong limit operator of $\chi_{\mathbf{P}(f)} I$ exists and is equal to the identity operator.

- If $\eta_N < 0$, then the strong limit operator of $\chi_{\mathbf{P}(f)} I$ exists and is equal to the zero operator.

- Now let $\eta_N = 0$. Then h has a subsequence $g \in \mathcal{H}$ such that either the numbers $g(n)_N$ tend to $+\infty$, or to $-\infty$, or that the sequence $(g(n)_N)_{n\geq 1}$ is bounded. In the first two cases, the strong limit operator of $\chi_{\mathbf{P}(f)}I$ with respect to g exists, and it is equal to the identity operator in the first case and to the zero operator in the second case. In the third case, there exists a subsequence k of g, a real number b_k and a function $f_k : \mathbb{R}^{N-1} \to \mathbb{R}$ such that
$$\lim_{n\to\infty} k(n)_N = b_k \quad \text{and} \quad \lim_{n\to\infty} f(x' + k(n)') = f_k(x')$$
in the sense of the uniform convergence on compact subsets of $\mathbb{R}^{N-1}$. In this case, the strong limit operator of $\chi_{\mathbf{P}(f)}I$ exists, too, and it is equal to the operator of multiplication by the characteristic function of
$$\mathbf{P}(f_k - b_k) = \{x \in \mathbb{R}^N : x_N > f_k(x') - b_k\}.$$

Let $\mathcal{H}_\eta(A)$ stand for the set of all sequences $h \in \mathcal{H}$ which tend to infinity into the direction of $\eta \in S^{N-1}$ and for which the limit operator A_h exists. Further, given a real number b and a function $g : \mathbb{R}^{N-1} \to \mathbb{R}$, we denote by $\mathcal{H}_{\eta,\infty}(A)$ and $\mathcal{H}_{\eta,g,b}(A)$ the set of all sequences $h \in \mathcal{H}_\eta(A)$ such that $h(n)_N \to \infty$ and
$$h(n)_N \to b \quad \text{and} \quad f(x' + h(n)') \to g(x')$$
uniformly on compact subsets of $\mathbb{R}^{N-1}$, respectively. If the set $\mathcal{H}_{\eta,g,b}(A)$ is not empty, then we call g a *limit function* with respect to η. Then Theorem 3.4.2 implies the following result.

Theorem 3.4.4 *Let $A \in \mathcal{A}(BUC(\mathbb{R}^N), \mathcal{C}_p)$, and let $D = \mathbf{P}(f)$ with a function $f \in BUC(\mathbb{R}^{N-1})$. Then the compression of A onto D is Fredholm on $L^p(D)$ if and only if the following conditions are satisfied:*

(a) *for each point $\eta \in S^{N-1}$ with $\eta_N > 0$, the set $\{A_h : h \in \mathcal{H}_\eta(A)\}$ of limit operators of A is uniformly invertible.*

(b) *for each point $\eta \in S^{N-1}$ with $\eta_N = 0$, the set $\{A_h : h \in \mathcal{H}_{\eta,\infty}(A)\}$ of limit operators of A is uniformly invertible.*

(c) *for each point $\eta \in S^{N-1}$ with $\eta_N = 0$, each $b \in \mathbb{R}$, and each limit function $g : \mathbb{R}^{N-1} \to \mathbb{R}$, the set*
$$\{(1 - \chi_{\mathbf{P}(g-b)})I + \chi_{\mathbf{P}(g-b)}A_h\chi_{\mathbf{P}(g-b)}I : h \in \mathcal{H}_{\eta,g,b}(A)\}$$
of extended compressions of limit operators of A is uniformly invertible.

This result gets a particular simple form if $f \in SO(\mathbb{R}^{N-1})$. In the setting of assertion (c) of the theorem, this hypothesis implies that all functions g are constant (their possible values are just the partial limits of $f(x')$ as $x' \to \infty$). Thus, all possible limit domains $\mathbf{P}(g - b)$ are (uncurved) half-spaces.

3.4.4 Compressions to curved layers

Let again $N > 1$, and let $f_1, f_2 \in BUC(\mathbb{R}^{N-1})$ be such that $f_1(x') < f_2(x')$ for all $x' \in \mathbb{R}^{N-1}$. Then we call the set

$$\mathbf{L}(f_1, f_2) := \{x = (x', x_N) \in \mathbb{R}^{N-1} \times \mathbb{R} : f_1(x') < x_N < f_2(x')\} \tag{3.26}$$

a *curved layer*. Let $h \in \mathcal{H}$ be a sequence which tends to infinity into the direction of $\eta \in S^{N-1}$. If $\eta_N \neq 0$, then the strong limit operator of $\chi_{\mathbf{L}(f_1, f_2)}I$ with respect to h exists and is equal to 0. The same happens if $\eta_N = 0$ and the sequence $(h(n)_N)_{n\geq 1}$ tends to $\pm\infty$. Thus, the only non-trivial case is when $\eta_N = 0$ and the sequence $(h(n)_N)_{n\geq 1}$ is bounded. Then, as in the previous subsection, there is a subsequence k of h, a real number b_k as well as functions $f_{1k}, f_{2k} : \mathbb{R}^{N-1} \to \mathbb{R}$ such that the strong limit operator of $\chi_{\mathbf{L}(f_1, f_2)}I$ with respect to the sequence k exists and is equal to $\chi_{\mathbf{L}(f_{1k}-b_k, f_{2k}-b_k)}I$.

Let again $\mathcal{H}_\eta(A)$ stand for the set of all sequences $h \in \mathcal{H}$ which tend to infinity into the direction of $\eta \in S^{N-1}$ and for which the limit operator A_h exists, and denote by $\mathcal{H}_{\eta, g_1, g_2, b}(A)$ the set of all sequences $h \in \mathcal{H}_\eta(A)$ such that

$$h(n)_N \to b \quad \text{and} \quad f_i(x' + h(n)') \to g_i(x') \quad (i = 1, 2)$$

uniformly on compact subsets of $\mathbb{R}^{N-1}$.

Theorem 3.4.5 *Let $A \in \mathcal{A}(BUC(\mathbb{R}^N), \mathcal{C}_p)$ and $D = \mathbf{L}(f_1, f_2)$ with $f_1, f_2 \in BUC(\mathbb{R}^{N-1})$ and $f_1 < f_2$. Then the compression of A onto D is Fredholm on $L^p(D)$ if and only if, for each point $\eta \in S^{N-1}$ with $\eta_N = 0$, each $b \in \mathbb{R}$, and all limit functions $g_1, g_2 : \mathbb{R}^{N-1} \to \mathbb{R}$, the set*

$$\{(1 - \chi_{\mathbf{L}(g_1-b, g_2-b)})I + \chi_{\mathbf{L}(g_1-b, g_2-b)}A_h\chi_{\mathbf{L}(g_1-b, g_2-b)}I : h \in \mathcal{H}_{\eta, g_1, g_2, b}(A)\}$$

of extended compressions of limit operators of A is uniformly invertible.

If $f_1, f_2 \in SO(\mathbb{R}^{N-1})$, then the functions f_{1k}, f_{2k} are constant, and $\mathbf{L}(f_{1k} - b_k, f_{2k} - b_k)$ is a usual layer bounded by two parallel planes.

Corollary 3.4.6 *In addition to the hypothesis from Theorem 3.4.5, let*

$$\lim_{x'\to\infty} (f_1(x') - f_2(x')) = 0.$$

Then all strong limit operators of $\chi_{\mathbf{L}(f_1, f_2)}I$ are zero, and the compression of A onto $\mathbf{L}(f_1, f_2)$ is Fredholm on $L^p(\mathbf{L}(f_1, f_2))$.

3.4.5 Compressions to curved cylinders

Let $N > 1$, $\Omega \subset \mathbb{R}^{N-1}$ be a bounded domain and $f \in BUC(\mathbb{R})$ a positive function, and consider the *curved cylinder*

$$\mathbf{Z}_\Omega(f) := \{x = (x', x_N) \in \mathbb{R}^{N-1} \times \mathbb{R} : x' \in f(x_N)\Omega\}. \tag{3.27}$$

Let $h \in \mathcal{H}$. If $h(n)' \to \infty$, then the strong limit operator of $\chi_{\mathbf{Z}_\Omega(f)} I$ with respect to h exists and is equal to the zero operator. Thus, nontrivial strong limit operators of $\chi_{\mathbf{Z}_\Omega(f)} I$ with respect to h exist only if the sequence $(h(n)')_{n \geq 1}$ is bounded and $h(n)_N \to \pm\infty$. In this case, there is a subsequence k of h, a point $b_k \in \mathbb{Z}^{N-1}$, and a function f_k on $\mathbb{R}$ such that

$$k(n)' \to b_k \quad \text{and} \quad f(x_N + k(n)_N) \to f_k(x_N) \quad \text{as } n \to \infty$$

uniformly on compact subsets of $\mathbb{R}$. Then the strong limit operator of $\chi_{\mathbf{Z}_\Omega(f)} I$ with respect to the sequence k exists, and it is equal to the operator of multiplication by the characteristic function of the shifted curved cylinder

$$\mathbf{Z}_\Omega(f_k, b_k) := \{x = (x', x_N) \in \mathbb{R}^{N-1} \times \mathbb{R} : x' \in f_k(x_N)\Omega - b_k\}.$$

Let $\mathcal{H}_\eta(A)$ denote the set of all sequences $h \in \mathcal{H}$ which tend to infinity into the direction of $\eta \in S^{N-1}$ and for which the limit operator A_h exists, and write $\mathcal{H}_{\eta,\, g,\, b}(A)$ for the set of all sequences $h \in \mathcal{H}_\eta(A)$ such that

$$h(n)_N \to b \in \mathbb{Z}^{N-1} \quad \text{and} \quad f(x_N + h(n)_N) \to g(x_N)$$

uniformly on compact subsets of $\mathbb{R}$.

Theorem 3.4.7 *Let $A \in \mathcal{A}(BUC(\mathbb{R}^N), \mathcal{C}_p)$ and $D = \mathbf{Z}_\Omega(f)$ with $f \in BUC(\mathbb{R})$. Then the compression of A onto D is Fredholm on $L^p(D)$ if and only if, for each point $\eta \in S^{N-1}$ with $\eta' = 0$, each $b \in \mathbb{Z}^{N-1}$, and all limit functions $g : \mathbb{R} \to \mathbb{R}$, the set*

$$\{(1 - \chi_{\mathbf{Z}_\Omega(g,\, b)})I + \chi_{\mathbf{Z}_\Omega(g,\, b)} A_h \chi_{\mathbf{Z}_\Omega(g,\, b)} I : h \in \mathcal{H}_{\eta,\, g,\, b}(A)\}$$

of extended compressions of limit operators of A is uniformly invertible.

If $f \in SO(\mathbb{R})$, then every limit function g is constant and, thus, $\mathbf{Z}_\Omega(g, b)$ is a usual straight cylinder.

Corollary 3.4.8 *In addition to the hypothesis from Theorem 3.4.7, let the ends of the cylinder be cuspidal, i.e., let*

$$\lim_{x_N \to \pm\infty} f(x_N) = 0.$$

Then all strong limit operators of $\chi_{\mathbf{Z}_\Omega(f)} I$ are zero, and the compression of A onto $\mathbf{Z}_\Omega(f)$ is Fredholm on $L^p(\mathbf{Z}_\Omega(f))$.

3.4.6 Compressions to cones with smooth cross section

Let $\Omega \subseteq \mathbb{R}^N$ be an open domain with C^1-boundary $\partial\Omega$ in case $N \geq 2$ or an open interval in $\mathbb{R}^1$. By $\mathbf{C}_\Omega$, we denote the cone in $\mathbb{R}^{N+1}$ generated by Ω,

$$\mathbf{C}_\Omega := \{(y, y_{N+1}) \in \mathbb{R}^N \times [0, \infty) : y \in y_{N+1}\Omega\}. \tag{3.28}$$

Given $x \in \mathbb{R}^N$, let $\eta_x \in S^N$ be the point which lies on the ray in $\mathbb{R}^{N+1}$ starting at the origin and passing through the point $(x, 1)$, i.e.,

$$\eta_x = \frac{(x,\,1)}{\sqrt{\|x\|^2+1}}.$$

Let $h \in \mathcal{H}$ be a sequence which tends to infinity into the direction of $\eta \in S^N$. Again there are two trivial cases: If η is not of the form η_x with some $x \in \overline{\Omega}$, then the strong limit operator of $\chi_{\mathbf{C}_\Omega} I$ exists and is equal to the zero operator. If $\eta = \eta_x$ with $x \in \Omega$, then the strong limit operator of $\chi_{\mathbf{C}_\Omega} I$ exists, too, and is equal to the identity operator.

Let now $x \in \partial\Omega$ and $\eta = \eta_x$. We denote by $T_x\Omega$ the tangential space and by ν_x the interior normal unit vector to $\partial\Omega$ at x. Further, we write $\mathbb{H}_x$ for the closed half-space in $\mathbb{R}^N$ which is bounded by $T_x\Omega$ and for which ν_x is an interior normal unit vector to $\partial\mathbb{H}_x$ at x. Finally, we let $\mathbf{H}_x$ refer to the half-space in $\mathbb{R}^{N+1}$ which is generated by $\mathbb{H}_x$,

$$\mathbf{H}_x := \{(y,\, y_{N+1}) \in \mathbb{R}^N \times \mathbb{R} : y \in \mathbb{H}_x + (y_{N+1}-1)x\}.$$

Further, we write the sequence h as

$$h(n) := \alpha_n(\nu_x,\, 0) + (r_n,\, 0) + \beta_n(x,\, 1) \tag{3.29}$$

where $r_n \in T_x\Omega$ and $\alpha_n,\, \beta_n \in \mathbb{R}$. The following lemma claims the conditions under which the sequence (3.29) tends to infinity in the direction of η_x.

Lemma 3.4.9 *The sequence h defined by* (3.29) *tends to infinity into the direction of η_x if and only if $\beta_n \to +\infty$ and*

$$\alpha_n/\beta_n \to 0 \quad \textit{and} \quad r_n/\beta_n \to 0 \quad \textit{as} \quad n \to \infty. \tag{3.30}$$

Proof. The sequence h tends to infinity if and only if

$$|\alpha_n|^2 + \|r_n\|^2 + |\beta_n|^2 \to \infty, \tag{3.31}$$

and then it converges in the direction of η_x if and only if

$$\frac{(\alpha_n\nu_x + r_n + \beta_n x,\, \beta_n)}{\sqrt{\|\alpha_n\nu_x + r_n + \beta_n x\|^2 + |\beta_n|^2}} \to \frac{(x,\,1)}{\sqrt{\|x\|^2+1}}. \tag{3.32}$$

The convergence of the last component of (3.32) tells us that $\beta_n > 0$ for all sufficiently large n. Thus, (3.31) implies

$$\frac{(\frac{\alpha_n}{\beta_n}\nu_x + \frac{r_n}{\beta_n} + x,\, 1)}{\sqrt{\|\frac{\alpha_n}{\beta_n}\nu_x + \frac{r_n}{\beta_n} + x\|^2 + 1}} \to \frac{(x,\,1)}{\sqrt{\|x\|^2+1}}. \tag{3.33}$$

From the convergence of the last component of (3.33) we conclude that

$$\left\| \frac{\alpha_n}{\beta_n} \nu_x + \frac{r_n}{\beta_n} + x \right\| \to \|x\|.$$

This implies for the first component of (3.33) that

$$\frac{\alpha_n}{\beta_n} \nu_x + \frac{r_n}{\beta_n} + x \to x$$

whence (3.30) since $\nu_x \perp r_n$. Writing (3.31) as

$$\beta_n^2 \left(\left| \frac{\alpha_n}{\beta_n} \right|^2 + \left\| \frac{r_n}{\beta_n} \right\|^2 + 1 \right) \to \infty$$

and taking into account (3.30), we finally get $\beta_n \to +\infty$. The reverse implications can be checked similarly. □

In order to compute strong limit operators into the direction of η_x for $x \in \partial\Omega$, we assume for simplicity that $x = 0$ (which can be reached by shifting Ω) and that $T_x\Omega = T_0\Omega = \mathbb{R}^{N-1} \times \{0\}$ (which can be reached by rotating the shifted Ω). Then, since Ω has a C^1-boundary, there is an open neighborhood $U \subseteq \mathbb{R}^{N-1}$ of 0, an open interval $I \subseteq \mathbb{R}$ which contains 0, and a continuously differentiable function $g : U \to I$ such that

$$\partial\Omega \cap (U \times I) = \{(x,\, g(x)) \in \mathbb{R}^{N-1} \times \mathbb{R} : x \in U\}$$

and

$$\Omega \cap (U \times I) = \{(x,\, x_N) \in U \times I : x_N > g(x)\}.$$

Thus, if $\beta > 0$, then the part of the boundary of $\beta\Omega$ which lies in $\beta U \times \beta I$ is just the graph of the function

$$\beta U \to \beta I, \quad x \mapsto \beta g(x/\beta).$$

Let h be as in (3.29), and assume that the limit

$$\delta^* := \lim(\beta_n g(r_n/\beta_n) - \alpha_n) \in \mathbb{R} \cup \{\pm\infty\}$$

exists (otherwise we pass to a suitable subsequence of h). Let further $d > 0$ and $K_d^N := [-d,\, d]^N$, and set $\mathbf{C}_{n,\Omega} := U_{-h(n)}\mathbf{C}_\Omega$. We consider the intersection of the shifted cone $\mathbf{C}_{n,\Omega}$ with $\mathbb{R}^N \times \{0\}$ and identify this intersection with a subset of $\mathbb{R}^N$. Since $(y + r_n)/\beta_n \in U$ for all $y \in K_d^{N-1}$ and for all sufficiently large n, the boundary of $\mathbf{C}_{n,\Omega} \cap (\mathbb{R}^N \times \{0\})$ can be locally described as the graph of the function

$$G_n : K_d^{N-1} \to \mathbb{R}, \quad y \mapsto \beta_n g((y + r_n)/\beta_n) - \alpha_n.$$

Then, for every $y \in K_d^{N-1}$, we have

$$\lim(G_n(y) - \delta^*) = \lim(G_n(y) - G_n(0))$$

with

$$\begin{aligned} |G_n(y) - G_n(0)| &\leq \max_{\xi \in [0,\, y]} \|G'(\xi)\| \, \|y - 0\| \\ &= \max_{\xi \in [0,\, y]} \|g'((\xi + r_n)/\beta_n)\| \, \|y\|. \end{aligned}$$

Since g is continuously differentiable with $g'(0) = 0$, and since

$$\|(\xi + r_n)/\beta_n\| \leq (d + \|r_n\|)/\beta_n \to 0$$

by Lemma 3.4.9, we conclude that $G_n(y) \to \delta^*$ for every $y \in K_d^{N-1}$. Thus, if $(y,\, y_N) \in K_d^N$, then

$$\chi_{\mathbf{C}_{n,\Omega} \cap (\mathbb{R}^N \times \{0\})}(y) \to \begin{cases} 1 & \text{if} \quad y_N > \delta^* \\ 0 & \text{if} \quad y_N < \delta^* \end{cases}$$

An analogous result holds if the sequence (β_n) is replaced by $(\beta_n + \beta')$ with $\beta' \in [-d,\, d]$. This shows that

$$\chi_{\mathbf{C}_{n,\Omega}}(y) \to \begin{cases} \chi_{\mathbb{R}^{N+1}}(y) = y & \text{if} \quad \delta^* = -\infty \\ \chi_{\mathbf{H}_x + \delta^*(\nu_x,\, 0)}(y) & \text{if} \quad \delta^* \in \mathbb{R} \\ \chi_{\emptyset}(y) = 0 & \text{if} \quad \delta^* = +\infty \end{cases}$$

almost everywhere on K_d^{N+1}. By the dominated convergence theorem, this implies that

$$\chi_{\mathbf{C}_{n,\Omega}} \chi_{K_d^{N+1}} \to \begin{cases} \chi_{K_d^{N+1}} & \text{if} \quad \delta^* = -\infty \\ \chi_{\mathbf{H}_x + \delta^*(\nu_x,\, 0)} \chi_{K_d^{N+1}} & \text{if} \quad \delta^* \in \mathbb{R} \\ 0 & \text{if} \quad \delta^* = +\infty \end{cases}$$

with respect to the L^1-norm and, hence, also with respect to every L^p-norm with $1 \leq p < \infty$ (the occurring functions take values in $\{-1,\, 0,\, 1\}$ almost everywhere). Since d is arbitrary, this finally yields that

$$\chi_{\mathbf{C}_{n,\Omega}} I \to \begin{cases} I & \text{if} \quad \delta^* = -\infty \\ \chi_{\mathbf{H}_x + \delta^*(\nu_x,\, 0)} I & \text{if} \quad \delta^* \in \mathbb{R} \\ 0 & \text{if} \quad \delta^* = +\infty \end{cases}$$

strongly on $L^p(\mathbb{R}^{N+1})$.

Given $A \in \mathcal{A}(BUC(\mathbb{R}^{N+1}),\, \mathcal{C}_p(\mathbb{R}^{N+1}))$, let $\mathcal{H}_\eta(A)$ denote the set of all sequences $h \in \mathcal{H}$ which tend to infinity into the direction of $\eta \in S^N$ and for which the limit operator A_h exists. Further, write $\mathcal{H}_{\eta,-\infty}(A)$ and $\mathcal{H}_{\eta,\delta^*}(A)$ with $\delta^* \in \mathbb{R}$ for the set of all sequences $h \in \mathcal{H}_\eta(A)$ with

$$\lim(\beta_n g(r_n/\beta_n) - \alpha_n) = -\infty \quad \text{and} \quad \lim(\beta_n g(r_n/\beta_n) - \alpha_n) = \delta^*,$$

respectively. Finally, we abbreviate the shifted half-space $\mathbf{H}_x + \delta^*(\nu_x,\, 0)$ to $\mathbf{H}_{x,\delta^*}$. Then Theorem 3.4.2 has the following consequence.

Theorem 3.4.10 *Let $A \in \mathcal{A}(BUC(\mathbb{R}^{N+1}), \mathcal{C}_p(\mathbb{R}^{N+1}))$ and $D = \mathbf{C}_\Omega$ with $\Omega \in \mathbb{R}^N$ an open domain with C^1 boundary. Then the compression of A onto D is Fredholm on $L^p(D)$ if and only if the following conditions are satisfied:*

(a) *for each point $x \in \Omega$, the set $\{A_h : h \in \mathcal{H}_{\eta_x}(A)\}$ of limit operators of A is uniformly invertible.*

(b) *for each point $x \in \partial\Omega$, the set $\{A_h : h \in \mathcal{H}_{\eta_x,-\infty}(A)\}$ of limit operators of A is uniformly invertible.*

(c) *for each point $x \in \partial\Omega$, the set*

$$\{(1 - \chi_{\mathbf{H}_{x,\delta^*}})I + \chi_{\mathbf{H}_{x,\delta^*}} A_h \chi_{\mathbf{H}_{x,\delta^*}} I : h \in \mathcal{H}_{\eta,\delta^*}(A),\ \delta^* \in \mathbb{R}\}$$

of extended compressions of limit operators of A is uniformly invertible.

We still mention some special situations in which the conditions of Theorem 3.4.10 take a very simple form.

Let $A \in \mathcal{A}(SO(\mathbb{R}^{N+1}), \mathcal{C}_p(\mathbb{R}^{N+1}))$. Then all limit operators of A belong to $\mathbb{C}I + \mathcal{C}_p(\mathbb{R}^{N+1})$. In this case, the invertibility of the compressions in condition (c) can be effectively checked. Let, for example, A_h be the operator $\gamma I + C(k)$ with $\gamma \in \mathbb{C}$ and $k \in L^1(\mathbb{R}^{N+1})$. Since $C(k)$ is shift invariant, the corresponding compression (c) is invertible if and only if the operator

$$(1 - \chi_{\mathbf{H}_x})I + \chi_{\mathbf{H}_x}(\gamma I + C(k))\chi_{\mathbf{H}_x} I \tag{3.34}$$

is invertible. Further, given an orthogonal mapping S on $\mathbb{R}^{N+1}$, we write R_S for the rotation operator $(R_S f)(x) = f(Sx)$, and for $k \in L^1(\mathbb{R}^{N+1})$, we let k_S be the function $k_S(x) = k(S^T x)$ with S^T referring to the transposed of S. Then convolution operators are rotation invariant in the sense that

$$C(k)R_S = R_S C(k_S).$$

Thus, if we choose S such that it rotates $\mathbf{H}_x$ to the half-space $\mathbf{H} := \{(x_1, x) \in \mathbb{R} \times \mathbb{R}^N : x_1 \geq 0\}$, then the compression (3.34) is invertible if and only if the operator

$$(1 - \chi_{\mathbf{H}})I + \chi_{\mathbf{H}}(\gamma I + C(k_S))\chi_{\mathbf{H}} I \tag{3.35}$$

is invertible. Finally, the compression (3.35) is invertible if and only if the operator

$$(1 - \chi_{\mathbf{H}})I + (\gamma I + C(k_S))\chi_{\mathbf{H}} I$$

is invertible. This follows easily from the identities $AP+Q = (PAP+Q)(I+QAP)$ and $(I+QAP)^{-1} = I-QAP$ which hold for arbitrary operators A and idempotents P, Q with $P+Q = I$. Clearly, these results remain true if $C(k)$ is replaced by an arbitrary operator in $\mathcal{C}_p(\mathbb{R}^{N+1})$.

For the invertibility of the resulting compressions, one has the following result from [167] (Theorem 1.4).

Theorem 3.4.11 *Let $B \in \mathbb{C}I + \mathcal{C}_p(\mathbb{R}^{N+1})$. Then the operator $(1-\chi_{\mathbf{H}})I + B\chi_{\mathbf{H}}I$ is invertible on $L^p(\mathbb{R}^{N+1})$ if and only if the operator B is invertible on $L^p(\mathbb{R}^{N+1})$.*

Note once more that the invertibility of $B \in \mathbb{C}I + \mathcal{C}_p(\mathbb{R}^{N+1})$ can be efficiently checked via (3.11).

With these remarks, we get the following corollaries to Theorem 3.4.10.

Corollary 3.4.12 *Let $A \in \mathcal{A}(SO(\mathbb{R}^{N+1}), \mathcal{C}_p(\mathbb{R}^{N+1}))$ and $D = \mathbf{C}_\Omega$ with $\Omega \in \mathbb{R}^N$ an open domain with C^1 boundary. Then the compression of A onto D is Fredholm on $L^p(D)$ if and only if, for each point $x \in \overline{\Omega}$, the set $\{A_h : h \in \mathcal{H}_{\eta_x}(A)\}$ of limit operators of A (= the local operator spectrum at η_x) is uniformly invertible.*

Corollary 3.4.13 *Let $A \in \mathcal{A}(L^\infty_{stab}(\mathbb{R}^{N+1}), \mathcal{C}_p(\mathbb{R}^{N+1}))$ and $D = \mathbf{C}_\Omega$ with $\Omega \in \mathbb{R}^N$ an open domain with C^1 boundary. Then the compression of A onto D is Fredholm on $L^p(D)$ if and only if, for each point $x \in \overline{\Omega}$, the limit operator A_{η_x} of A is invertible.*

Remark. Let $f : [0, \infty) \to \mathbb{R}$ be a slowly oscillating function, and let $\Omega \in \mathbb{R}^N$ be an open domain with C^1 boundary. We consider the slowly oscillating cone,

$$\mathbf{C}_{\Omega,\, f} := \{(y,\, y_{N+1}) \in \mathbb{R}^N \times [0,\, \infty) : y \in (y_{N+1} + f(y_{N+1}))\Omega\}. \tag{3.36}$$

In a similar way as above, one can show that the strong limit operators of the multiplication operator $\chi_{\mathbf{C}_{\Omega,\, f}} I$ are the same as in case of the unperturbed cone $\mathbf{C}_\Omega$, and that the analogue of Theorem 3.4.10 holds.

3.4.7 Compressions to cones with edges

Here we are going to consider compressions of convolution operators to cones which are allowed to have a finite number of edges. For simplicity, we restrict ourselves to the case $N = 2$.

More precisely, we let Ω be an open domain in $\mathbb{R}^2$ the boundary $\partial\Omega$ of which is C^1 up to a finite set M of singular points (i.e., $\partial\Omega$ is not C^1 in any neighborhood of $x \in M$). For each point $x \in M$ we suppose that there are an open neighborhood $U_x \subseteq \mathbb{R}^2$ of x as well as two open domains $\Omega_{x,\, l}$ and $\Omega_{x,\, r}$ with C^1-boundary such that either

$$U_x \cap \Omega = U_x \cap (\Omega_{x,\, l} \cap \Omega_{x,\, r}) \tag{3.37}$$

or

$$U_x \cap \Omega = U_x \setminus (\Omega_{x,\, l} \cap \Omega_{x,\, r}). \tag{3.38}$$

If the tangent spaces $T_x\Omega_{x,\, l}$ and $T_x\Omega_{x,\, r}$ do not coincide, then we call x an outward angular point in case of (3.37) and an inward angular point in case of (3.38). If these tangent spaces coincide, then x is called an outward resp. inward cuspidal point.

As in the previous section, we consider the cone generated by Ω,

$$\mathbf{C}_\Omega := \{(y,\, y_3) \in \mathbb{R}^2 \times [0,\, \infty) : y \in y_3\Omega\} \tag{3.39}$$

and, for $x \in \mathbb{R}^2$ and $\delta \in \mathbb{R}$, the half-spaces $\mathbf{H}_x$ and $\mathbf{H}_{x,\delta}$. Further we set $\mathbf{H}_{x,-\infty} := \mathbb{R}^3$ and $\mathbf{H}_{x,+\infty} := \emptyset$ and, for $\delta, \epsilon \in \mathbb{R} \cup \{\pm\infty\}$ and $x \in M$,

$$\mathbf{K}_{x,\delta,\epsilon} := \mathbf{H}_{x,\delta,l} \cap \mathbf{H}_{x,\epsilon,r}$$

where $\mathbf{H}_{x,\delta,l}$ and $\mathbf{H}_{x,\epsilon,r}$ are half-spaces belonging to $\Omega_{x,l}$ and $\Omega_{x,r}$, respectively.

Proposition 3.4.14 *Let $x \in \mathbb{R}^2$. Then the set of the strong limit operators of $\chi_{\mathbf{C}_\Omega} I$ with respect to sequences tending to infinity into the direction of η_x is equal to*

(a) $\{0\}$ *if* $x \notin \overline{\Omega}$.

(b) $\{I\}$ *if* $x \in \Omega$.

(c) $\{\chi_{\mathbb{H}_{x,\delta}} I : \delta \in \mathbb{R} \cup \{\pm\infty\}\}$ *if* $x \in \partial\Omega \setminus M$.

(d) $\{\chi_{\mathbb{K}_{x,\delta,\epsilon}} I : \delta, \epsilon \in \mathbb{R} \cup \{\pm\infty\}\}$ *if* $x \in \partial\Omega \cap M$ *is an angular point.*

(e) $\{0\}$ *if* $x \in \partial\Omega \cap M$ *is an outward cuspidal point.*

(f) $\{I\}$ *if* $x \in \partial\Omega \cap M$ *is an inward cuspidal point.*

Proof. The proof for (a)–(c) is the same as in the previous section. The results of the previous section also show that each strong limit operator of $\chi_{\mathbf{C}_\Omega} I$ belongs to the set (d) if $x \in \partial\Omega \cap M$ is an angular point. The reverse inclusion can be seen as follows. Since x is an angular point, we can independently shift the half-space $\mathbf{H}_{x,0,l}$ by a sequence which tends into the direction of η_x and which comes closer and closer to $\partial\mathbf{H}_{x,0,r}$, and shift the half-space $\mathbf{H}_{x,0,r}$ by a sequence which also tends into the direction of η_x and which comes closer and closer to $\partial\mathbf{H}_{x,0,l}$. Since each of these sequences influences the limit operators of only one of the half-spaces we get any desired combination of shifts of the half-spaces $\mathbf{H}_{x,0,l}$ and $\mathbf{H}_{x,0,r}$ in this way. This shows (d), and (e) and (f) can be proved as in the previous section. (See the discussion before Theorem 3.4.10. The obvious point is that, in case of a cuspidal point, the half-spaces $\mathbf{H}_{x,0,l}$ and $\mathbf{H}_{x,0,r}$ cannot be shifted independently of each other.) □

Given $A \in \mathcal{A}(BUC(\mathbb{R}^3), \mathcal{C}_p(\mathbb{R}^3))$, let $\mathcal{H}_\eta(A)$ denote the set of all sequences $h \in \mathcal{H}$ which tend to infinity into the direction of $\eta \in S^2$ and for which the limit operator A_h exists. Further, if $x \in \partial\Omega \setminus M$ and $\delta \in \mathbb{R} \cup \{\pm\infty\}$, write $\mathcal{H}_{x,\delta}(A)$ for the set of all sequences $h \in \mathcal{H}_{\eta_x}(A)$ such that the strong limit operator of $\chi_{\mathbf{C}_\Omega} I$ exists and is equal to $\chi_{\mathbb{H}_{x,\delta}} I$. Finally, if $x \in \partial\Omega \cap M$ is an angular point and $\delta, \epsilon \in \mathbb{R} \cup \{\pm\infty\}$, then let $\mathcal{H}_{x,\delta,\epsilon}(A)$ stand for the set of all sequences $h \in \mathcal{H}_{\eta_x}(A)$ such that the strong limit operator of $\chi_{\mathbf{C}_\Omega} I$ exists and is equal to $\chi_{\mathbb{K}_{x,\delta,\epsilon}} I$. With these notations, we have the following consequence of Theorem 3.4.2.

Theorem 3.4.15 *Let $A \in \mathcal{A}(BUC(\mathbb{R}^3), \mathcal{C}_p(\mathbb{R}^3))$ and $D = \mathbf{C}_\Omega$ with $\Omega \in \mathbb{R}^2$ an open domain with piecewise C^1 boundary as above. Then the compression of A onto D is Fredholm on $L^p(D)$ if and only if the following conditions are satisfied:*

(a) *for each point* $x \in \Omega$, *the set* $\{A_h : h \in \mathcal{H}_{\eta_x}(A)\}$ *of limit operators of* A *is uniformly invertible.*

(b) *for each point* $x \in \partial\Omega \setminus M$, *the set*

$$\{(1 - \chi_{\mathbf{H}_{x,\delta}})I + \chi_{\mathbf{H}_{x,\,\delta}} A_h \chi_{\mathbf{H}_{x,\,\delta}} I : h \in \mathcal{H}_{x,\delta}(A),\ \delta \in \mathbb{R} \cup \{\pm\infty\}\}$$

of extended compressions of limit operators of A *is uniformly invertible.*

(c) *for each angular point* $x \in \partial\Omega \cap M$, *the set*

$$\{(1 - \chi_{\mathbf{K}_{x,\delta,\,\epsilon}})I + \chi_{\mathbf{K}_{x,\,\delta,\,\epsilon}} A_h \chi_{\mathbf{K}_{x,\,\delta,\,\epsilon}} I : h \in \mathcal{H}_{x,\,\delta,\,\epsilon}(A),\ \delta,\, \epsilon \in \mathbb{R} \cup \{\pm\infty\}\}$$

of extended compressions of limit operators of A *is uniformly invertible.*

(d) *for each inward cuspidal point* $x \in \partial\Omega \cap M$, *the set* $\{A_h : h \in \mathcal{H}_{\eta_x}(A)\}$ *of limit operators of* A *is uniformly invertible.*

Note that the conditions in (b) and (c) get a simpler form if one of the shift parameters δ and ϵ is $\pm\infty$. Let us also emphasize that if x is an outward cuspidal point, then the local invertibility at η_x of the compression of A onto D is trivially satisfied since all limit operators of $\alpha_A(1 - \chi_D)I + \chi_D A \chi_D I$ with respect to sequences which tend to infinity into the direction of η_x are equal to $\alpha_A I$.

Again we mention some special situations in which the conditions of Theorem 3.4.15 can be readily verified.

Corollary 3.4.16 *Let* $A \in \mathcal{A}(SO(\mathbb{R}^3), \mathcal{C}_p(\mathbb{R}^3))$ *and* D *be as in Theorem* 3.4.15. *Then the compression of* A *onto* D *is Fredholm on* $L^p(D)$ *if and only if the following conditions are satisfied:*

(a) *for each point* $x \in \overline{\Omega}$ *which is neither angular nor outward cuspidal, the set* $\{A_h : h \in \mathcal{H}_{\eta_x}(A)\}$ *of limit operators of* A *is uniformly invertible.*

(b) *for each angular point* $x \in M$, *the set*

$$\{(1 - \chi_{\mathbf{K}_{x,0,\,0}})I + \chi_{\mathbf{K}_{x,\,0,\,0}} A_h \chi_{\mathbf{K}_{x,\,0,\,0}} I : h \in \mathcal{H}_{x,\,\delta,\,\epsilon}(A),\ \delta,\, \epsilon \in \mathbb{R}\}$$

of extended compressions of limit operators of A *is uniformly invertible.*

(c) *for each angular point* $x \in M$, *the set*

$$\{A_h : h \in \mathcal{H}_{x,\,-\infty,\,\epsilon}(A) \cup \mathcal{H}_{x,\,\delta,\,-\infty}(A) \cup \mathcal{H}_{x,\,-\infty,\,-\infty}(A) : \delta,\, \epsilon \in \mathbb{R}\}$$

of limit operators of A *is uniformly invertible.*

Here we have used the shift invariance of the limit operators of A as well as Simonenko's Theorem 3.4.11 again.

Corollary 3.4.17 *Let* $A \in \mathcal{A}(L^\infty_{stab}(\mathbb{R}^3), \mathcal{C}_p(\mathbb{R}^3))$ *and* D *be as in Theorem* 3.4.15. *Then the compression of* A *onto* D *is Fredholm on* $L^p(D)$ *if and only if the following conditions are satisfied:*

(a) *for each point* $x \in \overline{\Omega}$ *which is neither angular nor outward cuspidal, the limit operator* A_{η_x} *of* A *is invertible.*

(b) *for each angular point* $x \in M$, *the extended compression*

$$(1 - \chi_{\mathbf{K}_{x,0,0}})I + \chi_{\mathbf{K}_{x,0,0}} A_{\eta_x} \chi_{\mathbf{K}_{x,0,0}}$$

of the limit operator A_h *of* A *is invertible.*

Indeed, this result follows from the fact that each local operator spectrum is a singleton under the hypothesis of the corollary. It can be checked as in Section 2.6.3 that the invertibility of the operator in (b) already implies the invertibility of all operators in condition (c) of Corollary 3.4.16.

3.4.8 Compressions to epigraphs of functions

We let $f : \mathbb{R} \to \mathbb{R}$ be a continuously differentiable function with

$$\lim_{t\to\pm\infty} f(t) = +\infty \quad \text{and} \quad \lim_{t\to\pm\infty} f'(t) = 0 \tag{3.40}$$

and consider its epigraph

$$\mathbf{E}_f := \{(x_1, x_2) \in \mathbb{R}^2 : x_2 > f(x_1)\}. \tag{3.41}$$

Let $h = (h_1, h_2) \in \mathcal{H}$ be a sequence which tends to infinity into the direction of $\eta = (\eta_1, \eta_2) \in S^1$. It is evident that the strong limit operator of $\chi_{\mathbf{E}_f} I$ exists and is equal to the identity operator if $\eta_2 > 0$, whereas the strong limit operator of $\chi_{\mathbf{E}_f} I$ exists and is equal to zero if $\eta_2 < 0$. Now let $\eta = (1, 0)$, and let h be a sequence for which the strong limit operator of $\chi_{\mathbf{E}_f}$ exists. We write

$$h_2(n) =: f(h_1(n)) + d_n$$

and choose a subsequence of h (which we denote by h again) such that the sequence (d_n) becomes convergent with limit $\delta \in \mathbb{R} \cup \{\pm\infty\}$. Further, for $\delta \in \mathbb{R} \cup \{\pm\infty\}$, we let

$$\mathbf{H}_\delta := \{(x_1, x_2) \in \mathbb{R}^2 : x_2 > \delta\}.$$

Then it is easy to check that the strong limit operator of $\chi_{\mathbf{E}_f} I$ coincides with $\chi_{\mathbf{H}_\delta} I$ and that, conversely, every operator of this form appears as a strong limit operator of $\chi_{\mathbf{E}_f} I$. The same holds for $\eta = (-1, 0)$.

Given $A \in \mathcal{A}(BUC(\mathbb{R}^2), \mathcal{C}_p(\mathbb{R}^2))$, let $\mathcal{H}_\eta(A)$ denote the set of all sequences $h \in \mathcal{H}$ which tend to infinity into the direction of $\eta \in S^1$ and for which the limit operator A_h exists. Further, for $\delta \in \mathbb{R} \cup \{\pm\infty\}$, write $\mathcal{H}_{\pm 1, \delta}(A)$ for the set of all sequences $h \in \mathcal{H}_{(\pm 1, 0)}(A)$ such that the strong limit operator of $\chi_{\mathbf{E}_f} I$ exists and is equal to $\chi_{\mathbb{H}_\delta} I$. Then Theorem 3.4.2 yields the following.

Theorem 3.4.18 *Let* $A \in \mathcal{A}(L^\infty_{stab}(\mathbb{R}^2), \mathcal{C}_p(\mathbb{R}^2))$, *and let* $D = \mathbf{E}_f$ *be the epigraph of the function* f *satisfying* (3.40). *Then the compression of* A *onto* D *is Fredholm on* $L^p(D)$ *if and only if, for each point* $\eta = (\eta_1, \eta_2) \in S^1$ *with* $\eta_2 \geq 0$, *the limit operator* A_η *of* A *is invertible.*

The proof is the same as for Corollaries 3.4.13 and 3.4.17. □

Finally, let $f_\pm : \mathbb{R} \to \mathbb{R}$ be continuously differentiable functions with

$$\lim_{t\to+\infty} f_\pm(t) = \lim_{t\to-\infty} f_\pm(t) = \pm\infty \quad \text{and} \quad \lim_{t\to+\infty} f'_\pm(t) = \lim_{t\to-\infty} f'_\pm(t) = 0, \tag{3.42}$$

and let

$$\mathbf{E}_{f_+,\,f_-} := \{(x_1,\, x_2) \in \mathbb{R}^2 : f_-(x_1) < x_2 < f_+(x_1)\}. \tag{3.43}$$

As before one can check that every strong limit operator of $\chi_{\mathbf{E}_{f_+,\,f_-}} I$ is of the form $\chi_{\mathbf{H}_\delta} I$ with $\delta \in \mathbb{R} \cup \{\pm\infty\}$ and that, conversely, every operator of this form is a strong limit operator of $\chi_{\mathbf{H}_\delta} I$ if h tends into the direction of $(\pm 1,\, 0) \in S^1$. If h tends to infinity into the direction of $\eta \in S^1$ with $\eta_2 \neq 0$ then, necessarily, the strong limit operator of $\chi_{\mathbf{H}_\delta} I$ with respect to h exists and is equal to the zero operator.

Theorem 3.4.19 *Let $A \in \mathcal{A}(L^\infty_{stab}(\mathbb{R}^2),\, \mathcal{C}_p(\mathbb{R}^2))$, and let $D = \mathbf{E}_{f_+,\,f_-}$ with functions $f_\pm$ satisfying* (3.42). *Then the compression of A onto D is Fredholm on $L^p(D)$ if and only if the limit operators A_η of A with $\eta = (\pm 1,\, 0) \in S^1$ are invertible.*

3.5 A Wiener algebra of convolution-type operators

Here we introduce a subalgebra of the algebra $\mathcal{A}(L^\infty(\mathbb{R}^N),\, \mathcal{C}_2)$ which consists of operators the discretization of which belongs to the Wiener algebra $\mathcal{W}$ introduced in Section 2.5. For rich operators in this algebra, the Fredholm criterion in terms of limit operators takes a simpler form since the *uniform* boundedness of the inverses of the limit operators is not required. We illustrate the efficiency of this result by describing the essential spectrum of certain Schrödinger operators with potentials possessing discontinuities along surfaces of curved half-spaces. Throughout this section, let $p = 2$.

3.5.1 Fredholmness of operators in the Wiener algebra

Let $\mathcal{W}(L^\infty(\mathbb{R}^N),\, \mathcal{C}_2)$ denote the set of all operators $A \in \mathcal{A}(L^\infty(\mathbb{R}^N),\, \mathcal{C}_2)$ such that

$$\sum_{\gamma\in\mathbb{Z}^N} \sup_{\alpha\in\mathbb{Z}^N} \|\chi_0 U_{-\alpha} A U_{\alpha-\gamma} \chi_0 I\|_{L(L^2(\mathbb{R}^N))} < \infty$$

where χ_0 again refers to the characteristic function of the cube $I_0 := [0,\, 1)^N$. It is not hard to see that $\mathcal{W}(L^\infty(\mathbb{R}^N),\, \mathcal{C}_2)$ is a (non-closed) *-subalgebra of the algebra $\mathcal{A}(L^\infty(\mathbb{R}^N),\, \mathcal{C}_2)$ (for a similar result in a more involved setting see Section 4.3.1 below). It is also evident that the discretization of an operator $A \in \mathcal{W}(L^\infty(\mathbb{R}^N),\, \mathcal{C}_2)$ leads to an operator on $l^2(\mathbb{Z}^N,\, L^2(I_0))$ which belongs to the corresponding discrete Wiener algebra $\mathcal{W}$ introduced in Section 2.5.

Combining Theorems 2.5.7, 3.1.9 and Propositions 3.3.2, 3.3.4, we immediately obtain the following.

Theorem 3.5.1 *Let $A \in \mathcal{W}(L^\infty(\mathbb{R}^N), \mathcal{C}_2)$ be a rich operator. Then A is a Fredholm operator on $L^2(\mathbb{R}^N)$ if and only if all limit operators of A are invertible.*

Corollary 3.5.2 *Let $A \in \mathcal{W}(L^\infty(\mathbb{R}^N), \mathcal{C}_2)$ be a rich operator. Then*

$$\sigma_{ess}(A) = \cup \sigma(A_h)$$

where the union is taken over all limit operators A_h of A.

In analogy with Theorem 3.4.2, and with the notations of Section 3.4.1, one gets the following result concerning the Fredholmness of compressions to rich sets of rich operators in the Wiener algebra.

Theorem 3.5.3 *Let $A \in \mathcal{W}(L^\infty(\mathbb{R}^N), \mathcal{C}_2)$ be a rich operator, and let D be a rich subset of $\mathbb{R}^N$. Then the compression of A onto D is Fredholm on $L^2(D)$ if and only if each limit operator of $\alpha_A I + \chi_D J_A \chi_D I$ is invertible on $L^2(\mathbb{R}^N)$.*

3.5.2 The essential spectrum of Schrödinger operators

Theorems 3.5.1 and 3.5.3 allow us to characterize the essential spectrum of some Schrödinger operators with potentials in $L^\infty(\mathbb{R}^N)$. We consider Schrödinger operators of the form

$$A = -\Delta + wI$$

where the potential $w \in L^\infty(\mathbb{R}^N)$ is *strongly rich* in the following sense: Each sequence $h : \mathbb{N} \to \mathbb{Z}^N$ which tends to infinity has a subsequence g such that the operator wI possesses a limit operator $w_g I$ with $w_g \in L^\infty(\mathbb{R}^N)$ with respect to g in the sense of the strong convergence on $L^2(\mathbb{R}^N)$,

$$\lim_{n\to\infty} U_{-g(n)} w U_{g(n)} u = w_g u \quad \text{for each } u \in L^2(\mathbb{R}^N).$$

Then we call w_g the limit function of w with respect to g, and we denote the set of all limit functions of w by $\operatorname{Lim}(w)$. Further, we write $A_g = -\Delta + w_g I$ for the Schrödinger operator with potential $w_g \in \operatorname{Lim}(w)$.

As usual, a point $\lambda \in \mathbb{C}$ belongs to the spectrum (the essential spectrum) of A if the operator

$$A - \lambda I : H^2(\mathbb{R}^N) \to L^2(\mathbb{R}^N)$$

is not invertible (not Fredholm). These definitions of Fredholmness and of the essential spectrum can be extended to unbounded operators on Hilbert spaces. An unbounded closed operator A acting on a Hilbert space H with domain $D(A)$ is a *Fredholm operator* if the spaces $\operatorname{Ker} A$ and $\operatorname{Coker} A := H/\operatorname{Im} A$ are finite-dimensional. For example (see [4], p. 28), if the partial differential operator

$$A := \sum_{|\alpha| \le m} a_\alpha D^\alpha, \quad a_\alpha \in L^\infty(\mathbb{R}^N)$$

with leading coefficients a_α, $|\alpha| = m$, belonging to $C_b^\infty(\mathbb{R}^N)$ is uniformly elliptic, i.e., if

$$\inf_{x\in\mathbb{R}^N} \left| \sum_{|\alpha|\le m} a_\alpha(x)\omega^\alpha \right| > 0 \quad \text{for each } \omega \in S^{N-1},$$

then A is a closed operator on $L^2(\mathbb{R}^N)$ with domain $H^m(\mathbb{R}^N)$, and the operator $A : H^m(\mathbb{R}^N) \to L^2(\mathbb{R}^N)$ is a Fredholm operator in the common sense if and only if A is a Fredholm operator when considered as an unbounded operator. The definition of the essential spectrum of A as unbounded operator is verbatim the same as in case of bounded operators.

Thus, the Schrödinger operator A with potential $w \in L^\infty(\mathbb{R}^N)$ can be considered as an unbounded closed operator on the Hilbert space $H = L^2(\mathbb{R}^N)$ with domain $D(A) = H^2(\mathbb{R}^N)$.

Theorem 3.5.4 *Let A be a Schrödinger operator with strongly rich potential $w \in L^\infty(\mathbb{R}^N)$. Then*

$$\sigma_{ess}(A) = \bigcup_{w_h \in \mathrm{Lim}\,(w)} \sigma(A_h). \tag{3.44}$$

Proof. Since $\Lambda := (I - \Delta)^{-1}$ is a unitary operator from $L^2(\mathbb{R}^N)$ onto $H^2(\mathbb{R}^N)$, the operator $A - \lambda I : H^2(\mathbb{R}^N) \to L^2(\mathbb{R}^N)$ is Fredholm if and only if the operator

$$B(\lambda) := I + (w - \lambda - 1)\Lambda : L^2(\mathbb{R}^N) \to L^2(\mathbb{R}^N)$$

is Fredholm. Further, since the operator Λ is in $\mathcal{C}_2$, the operator $B(\lambda)$ belongs to the Wiener algebra $\mathcal{W}(L^\infty, \mathcal{C}_2)$. By Theorem 3.5.3, $B(\lambda)$ is a Fredholm operator on $L^2(\mathbb{R}^N)$ if and only if all limit operators

$$B_h(\lambda) = I + (w_h - \lambda - 1)\Lambda : L^2(\mathbb{R}^N) \to L^2(\mathbb{R}^N)$$

are invertible. Hence, $A - \lambda I : H^2(\mathbb{R}^N) \to L^2(\mathbb{R}^N)$ is a Fredholm operator if and only if all limit operators $A_h - \lambda I = -\Delta + (w_h - \lambda)I : H^2(\mathbb{R}^N) \to L^2(\mathbb{R}^N)$ are invertible, which implies equality (3.44). □

Example A. Let $w' \in SO(\mathbb{R}^N)$ and $w'' \in Q_{SC}(\mathbb{R}^N)$. Then the potential $w := w' + w''$ is strongly rich. Thus, if $h : \mathbb{N} \to \mathbb{Z}^N$ is a sequence tending to infinity, we find a subsequence g of h such that

$$\text{s-lim}_{k\to\infty} U_{-g(k)} w U_{g(k)} = w_g I$$

where $w_g := \lim_{k\to\infty} w'(g(k)) \in \mathbb{C}$. Then

$$\begin{aligned} \sigma_{ess}(-\Delta + wI) &= \bigcup_{w_g \in \mathrm{Lim}\,(w)} \sigma(-\Delta + w_g I) \\ &= \bigcup_{w_g \in \mathrm{Lim}\,(w)} \{w_g + |\xi|^2 \in \mathbb{C} : \xi \in \mathbb{R}^N\} \\ &= \bigcup_{w_g \in \mathrm{Lim}\,(w)} \{w_g\} + [0, \infty). \end{aligned}$$

In particular, if $w \in SO(\mathbb{R}^N) + Q_{SC}(\mathbb{R}^N)$ is real-valued, then A is a self-adjoint operator, and since the set of partial limits of a slowly oscillating function is connected, the essential spectrum of $-\Delta + wI$ is the interval $[\liminf_{x\to\infty} w'(x), \infty)$ in that case. □

Example B. Here we will examine a more involved potential with discontinuities at infinity. Let $N > 1$ and $f \in SO(\mathbb{R}^N)$, and consider the curved half-space

$$\mathbf{P}(f) := \{x = (x', x_N) \in \mathbb{R}^{N-1} \times \mathbb{R} : x_N > f(x')\} \subset \mathbb{R}^N.$$

We are interested in discontinuous potentials of the form

$$w := a\chi_{\mathbf{P}(f)} + b(1 - \chi_{\mathbf{P}(f)})$$

where $a, b \in SO(\mathbb{R}^N)$ and where $\chi_{\mathbf{P}(f)}$ refers to the characteristic function of $\mathbf{P}(f)$. As in Section 3.4.3, we get the following characterization of the limit functions of the potential w.

- Let $h = (h', h_N) : \mathbb{N} \to \mathbb{Z}^{N-1} \times \mathbb{Z}$ be a sequence the last component h_N of which tends to $+\infty$. Then there exists a subsequence g of h such that the limit function of w defined by g exists, and this limit function is constant,

$$a_g := \lim_{n\to\infty} a(g(n)). \tag{3.45}$$

  We denote the set of all sequences g with this property by J_+.

- Similarly, if h_N of tends to $-\infty$, then there is a subsequence g of h such that the limit function of w defined by this sequence is

$$b_g := \lim_{n\to\infty} b(g(n)). \tag{3.46}$$

  We denote the set of these sequences g by J_-.

- Let, finally, h be a sequence the last component h_N of which is bounded. Then there exists a subsequence g of h such that g_N is a constant sequence, $g_N(n) =: d_h \in \mathbb{Z}$, say, and for which the limits

$$\lim_{n\to\infty} f(g'(n)) =: f_g, \quad \lim_{n\to\infty} a(g(n)) =: a_g, \quad \lim_{n\to\infty} b(g(n)) =: b_g$$

  exist. Then the limit function of w with respect to g exists, and it is equal to

$$w_g = a_g\chi_{\mathbf{P}(f_g - d_g)} + b_g(1 - \chi_{\mathbf{P}(f_g - d_g)}). \tag{3.47}$$

  We denote the set of all sequences g with this property by J_0.

Thus, the essential spectrum of A coincides with

$$\bigcup_{g\in J_+} \sigma(-\Delta + a_g I) \cup \bigcup_{g\in J_-} \sigma(-\Delta + b_g I) \cup \bigcup_{g\in J_0} \sigma(-\Delta + w_g I)$$

where a_g, b_g and w_g are defined by (3.45), (3.46) and (3.47), respectively.

Let now a and b be real-valued functions in $SO(\mathbb{R}^N)$. Then

$$\begin{aligned}\bigcup_{g\in J_+} \sigma(-\Delta + a_g I) &= [\liminf_- a,\, +\infty),\\ \bigcup_{g\in J_-} \sigma(-\Delta + b_g I) &= [\liminf_+ b,\, +\infty)\end{aligned}$$

where $\liminf_\pm a$ refers to the infimum of the set of all limits $\lim_{n\to\infty} a(x(n))$ where $x : \mathbb{N} \to \mathbb{R}^{N-1} \times \mathbb{R}$ is a sequence with $\lim_{n\to\infty} x_N(n) = \pm\infty$. In case $g \in J_0$, the operator

$$A_g := -\Delta + a_g \chi_{\mathbf{P}(f_g-d_g)} I + b_g(1 - \chi_{\mathbf{P}(f_g-d_g)})I$$

is unitarily equivalent to the operator

$$C_g := -\Delta + (a_g \chi_{\mathbb{R}^N_+} + b_g \chi_{\mathbb{R}^N_-})I$$

where $\mathbb{R}^N_\pm := \{(x',\, x_N) \in \mathbb{R}^{N-1} \times \mathbb{R} : \pm x_N > 0\}$. Applying the Fourier transform with respect to the variable $x' \in \mathbb{R}^{N-1}$ and shifting, we see that the operator C_g on its hand is unitarily equivalent to the operator

$$\hat{C}_g = mI - \frac{d^2}{dx^2} + (a_g \chi_{\mathbb{R}_+} + b_g \chi_{\mathbb{R}_-})I,$$

acting on the space $L^2(\mathbb{R}^N) = L^2(\mathbb{R}^{N-1} \times \mathbb{R})$, where m refers to the operator of multiplication $(mu)(\xi,\, x) := |\xi|^2 u(\xi,\, x)$ and where $(\chi_{\mathbb{R}_\pm} u)(\xi,\, x) := \chi_{\mathbb{R}_\pm}(x) u(\xi,\, x)$. Note that

$$L_g := -\frac{d^2}{dx^2} + (a_g \chi_{\mathbb{R}_+} + b_g \chi_{\mathbb{R}_-})I$$

is a Sturm-Liouville operator on $\mathbb{R}$. It is well known (see, for example, [187]) that this operator has a continuous spectrum only, which is

$$\sigma(L_g) = [c_g,\, +\infty) \quad \text{where } c_g := \min\{a_g,\, b_g\}.$$

Consequently, $\sigma(A_g) = [c_g,\, +\infty)$, whence

$$\bigcup_{g\in J_0} \sigma(-\Delta + w_g I) = [\inf_{g\in J_0} c_g,\, +\infty).$$

So we finally arrive at the following description of the essential spectrum of the Schrödinger operator A, which is

$$\sigma_{ess}(A) = [\gamma,\, +\infty), \tag{3.48}$$

where $\gamma := \min\{\liminf_- a,\, \liminf_+ b,\, \inf_{g\in J_0} c_g\}$. □

In the same way, invoking the calculation of limit functions for the characteristic functions of cones with smooth cross-sections performed in Section 3.4.6, one also gets a description of the essential spectrum of Schrödinger operators with potentials which have discontinuities along surfaces of smooth cones.

In this section, we restricted ourselves to consider potentials in $L^\infty(\mathbb{R}^N)$ which excludes, for instance, potentials of multi-particle systems (see [143, 41]) which are of particular practical interest. These restrictions are only for brevity, and to avoid technical complications. A slight modification of our approach also yields precise descriptions of the essential spectrum of Schrödinger operators for multi-particle systems in terms of their limit operators.

3.6 Comments and references

The Fredholmness of compressions of operators of convolution type for special classes of coefficients a, b and special domains D is a classical theme in concrete operator theory; see, e.g., [30, 49, 66, 105, 106, 107, 167]. For example, the multi-dimensional Wiener-Hopf operators

$$\chi_D(\gamma I + C(k))\chi_D I|_{L^p(D)} : L^p(D) \to L^p(D)$$

where $\gamma \in \mathbb{C}$ and $k \in L^1(\mathbb{R}^N)$ are considered in [66] for D being a half-space and in [167] in case D a cone in $\mathbb{R}^N$ with smooth cross-section, whereas the quarter plane case is the topic of [49, 105, 107]. See also Chapters 8 and 9 in [30]. Operators on 3D wedge shaped domains are studied in [106].

In Section 3.2, we have collected some basic facts on convolution operators on L^p-spaces. Theorem 3.2.2 goes back to [172], and the compactness of commutators of operators of multiplication by slowly oscillating functions with convolution operators has been verified in [40]. Our presentation follows [158], where the results mentioned in this section are proved in the more general context of operators on locally compact groups.

In Section 3.3, we follow [133]. The class of L^∞-functions which stabilize at infinity has been introduced in [81].

The results in Sections 3.4.3–3.4.8 (with the exception of the results pertaining to cones with smooth cross section) go back to the authors [133]. The material of Section 3.5 is presented here for the first time.

Many of the results in this chapter remain valid for convolution type operators acting on $L^\infty(\mathbb{R}^N)$, see [98, 99, 101].

3.6 Comments and references

Chapter 4

Pseudodifferential Operators

In this chapter, we are going to study the Fredholmness of pseudodifferential operators on $\mathbb{R}^N$. Again, our strategy is to discretize these operators in a suitable manner in order to get operators which live on a discrete l^2-space. In the situation at hand, it is convenient to perform this discretization both with respect to the variable $x \in \mathbb{R}^N$ and with respect to the co-variable $\xi \in \mathbb{R}^N$ in the Fourier image. One advantage of this discretization is that $\mathcal{P}$-Fredholm operators on the discrete l^2-space correspond to common Fredholm operators on $L^2(\mathbb{R}^N)$.

4.1 Generalities and notation

The goal of this section is to set up some notations and to summarize (mostly without proofs) some basic facts on pseudodifferential operators. Standard references are [76, 90, 164, 175, 181]. A short introduction to the material sketched in this section can be also found in [127].

4.1.1 Function spaces and Fourier transform

We denote the usual operators of first order partial differentiation on $\mathbb{R}^N$ by

$$\partial_{x_j} := \frac{\partial}{\partial x_j} \qquad \text{and} \qquad D_{x_j} := \frac{1}{i}\frac{\partial}{\partial x_j}.$$

For every multi-index $\alpha = (\alpha_1, \ldots, \alpha_N)$ with non-negative integers α_j, we write

$$\partial^\alpha := \partial_{x_1}^{\alpha_1} \cdots \partial_{x_N}^{\alpha_N} \qquad \text{and} \qquad D^\alpha := D_{x_1}^{\alpha_1} \cdots D_{x_N}^{\alpha_N}.$$

Further, we set $|\alpha| := \alpha_1 + \cdots + \alpha_N$ and, for each vector $\xi = (\xi_1, \ldots, \xi_N) \in \mathbb{R}^N$, we define $\xi^\alpha := \xi_1^{\alpha_1} \cdots \xi_N^{\alpha_N}$ and $\langle \xi \rangle := (1 + |\xi|_2^2)^{1/2}$ where $|\xi|_2$ refers to the Euklidean norm of ξ. The following classes of functions on $\mathbb{R}^N$ will be needed in what follows:

- $C^\infty(\mathbb{R}^N)$ is the linear topological space of all real-valued infinitely differentiable functions on $\mathbb{R}^N$. A sequence (φ_n) converges to zero in the topology of $C^\infty(\mathbb{R}^N)$ if, for each compact subset K of $\mathbb{R}^N$ and each multi-index α,

$$\lim_{n\to\infty} \sup_{x\in K} |\partial^\alpha \varphi_n(x)| = 0.$$

- $C_0^\infty(\mathbb{R}^N)$ is the subspace of $C^\infty(\mathbb{R}^N)$ which consists of all functions with compact support. A sequence (φ_n) converges to zero in the topology of $C_0^\infty(\mathbb{R}^N)$ if there exists a compact subset K of $\mathbb{R}^N$ such that $\operatorname{supp}\varphi_n \subseteq K$ for all $n \in \mathbb{N}$ and that, for each multi-index α,

$$\lim_{n\to\infty} \sup_{x\in K} |\partial^\alpha \varphi_n(x)| = 0.$$

- $S(\mathbb{R}^N)$ is the subspace of $C^\infty(\mathbb{R}^N)$ of all functions φ with

$$|\varphi|_{m;\,\alpha} := \sup_{x\in\mathbb{R}^N} \langle x\rangle^m |\partial^\alpha \varphi(x)| < \infty$$

  for all $m \in \mathbb{N}$ and all multi-indices α.

- $C_b^\infty(\mathbb{R}^N)$ is the subspace of $C^\infty(\mathbb{R}^N)$ of all functions φ with

$$|\varphi|_m := \sum_{|\alpha|\le m} \sup_{x\in\mathbb{R}^N} |\partial^\alpha \varphi(x)| < \infty$$

  for all $m \in \mathbb{N}$.

The convergence in $S(\mathbb{R}^N)$ and $C_b^\infty(\mathbb{R}^N)$ is defined by the families of the corresponding semi-norms $|.|_{m;\,\alpha}$ and $|.|_m$.

We will write the Fourier transform $\hat{\varphi} = F\varphi$ of the function $\varphi \in S(\mathbb{R}^N)$ in the form

$$\hat{\varphi}(\xi) = (F\varphi)(\xi) := \int_{\mathbb{R}^N} e^{-i\langle x,\,\xi\rangle} \varphi(x)\, dx, \quad \xi \in \mathbb{R}^N,$$

where $\langle x,\, \xi\rangle := x_1\xi_1 + \cdots + x_N\xi_N$. Then F acts on the Schwarz space $S(\mathbb{R}^N)$ as a continuous isomorphism the inverse of which is given by

$$(F^{-1}\varphi)(x) := (2\pi)^{-N} \int_{\mathbb{R}^N} e^{i\langle x,\,\xi\rangle} \varphi(\xi)\, d\xi, \quad x \in \mathbb{R}^N.$$

The Fourier transform can be extended continuously to a bounded and (up to a constant factor) unitary operator on $L^2(\mathbb{R}^N)$ such that Parseval's equality

$$\|\hat{\varphi}\|_{L^2(\mathbb{R}^N)} = (2\pi)^{N/2} \|\varphi\|_{L^2(\mathbb{R}^N)}$$

holds for all $\varphi \in L^2(\mathbb{R}^N)$. The Fourier transform mediates between the operators of differentiation and multiplication by polynomials as follows

$$(FD^\alpha u)(\xi) = \xi^\alpha Fu(\xi). \tag{4.1}$$

Given α, $\beta \in \mathbb{Z}^N$, let

$$(E_\alpha u)(x) := e^{i\langle \alpha,\, x\rangle} u(x) \quad \text{and} \quad (V_\beta u)(x) := u(x - \beta).$$

Both E_α and V_β are unitary operators on $L^2(\mathbb{R}^N)$, and $FV_\beta = E_{-\beta}F$ as well as $F^{-1}V_\beta = E_\beta F^{-1}$. Thus, the operators E_α act as shift operators in the Fourier image. Furthermore,

$$E_\alpha V_\beta = e^{i\langle \alpha,\, \beta\rangle} V_\beta E_\alpha. \tag{4.2}$$

4.1.2 Oscillatory integrals

For each function $p \in C^\infty(\mathbb{R}^N \times \mathbb{R}^N)$ and each multi-index α, we write ∂_x^α and ∂_ξ^α for the operator ∂^α, applied to the functions $x \mapsto p(x,\, \xi)$ and $\xi \mapsto p(x,\, \xi)$, respectively. Similarly, we let $\Delta := (\partial_{x_1})^2 + \cdots + (\partial_{x_N})^2$ be the Laplacian, and we let $\langle D_x\rangle^2$ and $\langle D_\xi\rangle^2$ refer to the operator $I - \Delta$, applied to the functions $x \mapsto p(x,\, \xi)$ and $\xi \mapsto p(x,\, \xi)$, respectively.

Definition 4.1.1 *Let $m \in \mathbb{N}$. The function $p \in C^\infty(\mathbb{R}^N \times \mathbb{R}^N)$ belongs to the* Hörmander class $S^m_{0,0}$ *if, for each r, $t \in \mathbb{N}$,*

$$|p|_{r,\,t} := \sum_{|\alpha|\le r,\, |\beta|\le t} \ \sup_{(x,\,\xi)\in\mathbb{R}^N\times\mathbb{R}^N} |\partial_\xi^\alpha \partial_x^\beta p(x, \xi)|\, \langle \xi\rangle^{-m} < \infty.$$

Thus, the Hörmander class $S^0_{0,0}$ coincides with $C_b^\infty(\mathbb{R}^N \times \mathbb{R}^N)$.

A function $\chi \in S(\mathbb{R}^N \times \mathbb{R}^N)$ is called a *cut-off function* if it is identically 1 in an open neighborhood of the origin of $\mathbb{R}^N \times \mathbb{R}^N$. For each $R > 0$, set $\chi_R(x,\, \xi) := \chi(x/R,\, \xi/R)$.

Definition 4.1.2 *Let p be a measurable function on $\mathbb{R}^N \times \mathbb{R}^N$, and let χ be a cut-off function on $\mathbb{R}^N \times \mathbb{R}^N$. If the limit*

$$\lim_{R\to\infty} (2\pi)^{-N} \int_{\mathbb{R}^N}\int_{\mathbb{R}^N} \chi_R(x,\, \xi) p(x,\, \xi) e^{-i\langle x,\, \xi\rangle}\, dx\, d\xi$$

exists, and if this limit is independent of the choice of χ, then it is called the oscillatory integral of p , *and we denote it by*

$$os \int\int_{\mathbb{R}^N} p(x,\, \xi) e^{-i\langle x,\, \xi\rangle}\, dx\, d\xi. \tag{4.3}$$

Proposition 4.1.3 *Let $p \in S^m_{0,0}$. Then the oscillatory integral* (4.3) *exists. Furthermore, for all integers $k_1 > N/2$ and $k_2 > (N+m)/2$,*

$$\begin{aligned} &os \int\int_{\mathbb{R}^N} p(x,\, \xi) e^{-i\langle x,\, \xi\rangle}\, dx\, d\xi \\ &= (2\pi)^{-N} \int_{\mathbb{R}^N}\int_{\mathbb{R}^N} e^{-i\langle x,\, \xi\rangle} \langle \xi\rangle^{-2k_2} \langle D_x\rangle^{2k_2} \langle x\rangle^{-2k_1} \langle D_\xi\rangle^{2k_1} p(x,\, \xi)\, dx\, d\xi, \end{aligned}$$

and the following estimate holds with a constant C independent of p (but depending on k_1 and k_2)

$$\left| os \int\int_{\mathbb{R}^N} p(x,\,\xi)e^{-i\langle x,\,\xi\rangle}\,dx\,d\xi \right| \le C|p|_{2k_1,\,2k_2}. \tag{4.4}$$

Proposition 4.1.4 *Let $p \in C^\infty(\mathbb{R}^N)$, and suppose there are an $m \in \mathbb{N}$ and constants C_α such that $|\partial^\alpha p(x)| \le C_\alpha \langle x\rangle^m$ for all multi-indices α. Then, for all $y \in \mathbb{R}^N$,*

$$os \int\int_{\mathbb{R}^N} p(x+y)e^{-i\langle x,\,\xi\rangle}\,dx\,d\xi = p(y).$$

4.1.3 Pseudodifferential operators

Let $p \in S^m_{0,0}$. The operator $Op(p)$ defined at $u \in S(\mathbb{R}^N)$ by

$$(Op(p)u)(x) := (2\pi)^{-N} \int_{\mathbb{R}^N} p(x,\,\xi)\hat{u}(\xi)e^{i\langle x,\,\xi\rangle}\,d\xi, \quad x \in \mathbb{R}^N \tag{4.5}$$

is called the *pseudodifferential operator with symbol* p. The class of all pseudodifferential operators with symbol in $S^m_{0,0}$ is denoted by $OPS^m_{0,0}$.

Proposition 4.1.5 *If $p \in S^m_{0,0}$, then $Op(p)$ is bounded on $S(\mathbb{R}^N)$.*

The proof follows easily from the Leibniz rule and by integration by parts in the integral (4.5). □

Writing the Fourier transform of u explicitly, one has

$$(Op(p)u)(x) = (2\pi)^{-N} \int_{\mathbb{R}^N}\int_{\mathbb{R}^N} p(x,\,\xi)u(x+y)e^{-i\langle y,\,\xi\rangle}\,dy\,d\xi.$$

This integral can be considered as the oscillatory integral

$$os \int\int_{\mathbb{R}^N} p(x,\,\xi)u(x+y)e^{-i\langle y,\,\xi\rangle}\,dy\,d\xi, \tag{4.6}$$

thus yielding an alternative definition of pseudodifferential operators via oscillatory integrals. Moreover, the oscillatory integral (4.6) makes sense also if $p \in S^m_{0,0}$, but if u is merely in $C^\infty_b(\mathbb{R}^N)$. Indeed, in this case, the function $(y,\,\xi) \mapsto p(x,\,\xi)u(x+y)$ is still in $S^m_{0,0}$ for each fixed $x \in \mathbb{R}^N$. So one can choose (4.6) as the definition of $Op(p)u$ for $u \in C^\infty_b(\mathbb{R}^N)$.

Proposition 4.1.6 *If $p \in S^m_{0,0}$, then $Op(p)$ is bounded on $C^\infty_b(\mathbb{R}^N)$.*

Examples. If $p(x,\,\xi) = a(x)$ is independent of ξ, then $Op(p)$ is just the operator of multiplication by the function a. In case $p(x,\,\xi) = b(\xi)$ is independent of x, then $Op(p)$ is the operator of convolution by the inverse Fourier transform of b. Moreover, it follows from (4.1) that also the shift operators E_α and V_β as well as the differential operators D^α can be considered as pseudodifferential operators. □

4.1.4 Formal symbols

Let $P : C_b^\infty(\mathbb{R}^N) \to C_b^\infty(\mathbb{R}^N)$ be a bounded linear operator. We define the *formal symbol* $\mathrm{sym}_P : \mathbb{R}^N \times \mathbb{R}^N \to \mathbb{R}$ of P by

$$\mathrm{sym}_P(x,\, \xi) := e^{-i\langle x,\, \xi\rangle} P(e^{i\langle \cdot,\, \xi\rangle})(x).$$

The boundedness of P ensures that the function sym_P belongs to $C_b^\infty(\mathbb{R}^N \times \mathbb{R}^N)$. Let $u \in S(\mathbb{R}^N)$. Then

$$\begin{aligned}(Pu)(x) &= (2\pi)^{-N} \int_{\mathbb{R}^N} P(e^{i\langle \cdot,\, \xi\rangle})(x)\hat{u}(\xi)\, d\xi \\ &= (2\pi)^{-N} \int_{\mathbb{R}^N} e^{i\langle x,\, \xi\rangle} \mathrm{sym}_P(x,\, \xi)\hat{u}(\xi)\, d\xi.\end{aligned}$$

Thus, if $\mathrm{sym}_P \in S^m_{0,0}$, then $P = Op(\mathrm{sym}_P)$. We claim that the correspondence $P \mapsto \mathrm{sym}_P$ is injective. In fact,

$$\int_{\mathbb{R}^N} e^{i\langle x,\, \xi\rangle} \mathrm{sym}_P(x,\, \xi)\hat{u}(\xi)\, d\xi = 0 \quad \text{for all } u \in S(\mathbb{R}^N)$$

implies that

$$\int_{\mathbb{R}^N} \mathrm{sym}_P(x,\, \xi) v(\xi)\, d\xi = 0 \quad \text{for all } v \in S(\mathbb{R}^N),$$

whence $\mathrm{sym}_P = 0$ since sym_P is a continuous function.

The following propositions describe the formal symbol of the composition and the formal adjoint of pseudodifferential operators.

Proposition 4.1.7 *Let $A = Op(a) \in OPS^{m_1}_{0,0}$ and $B = Op(b) \in OPS^{m_2}_{0,0}$. Then $AB \in OPS^{m_1+m_2}_{0,0}$, and*

$$\mathrm{sym}_{AB}(x,\, \xi) = os \int\!\!\int_{\mathbb{R}^N} a(x,\, \xi+\eta) b(x+y,\, \xi) e^{-i\langle y,\, \eta\rangle} dy\, d\eta.$$

Moreover,

$$|\mathrm{sym}_{AB}|_{m_1,\, m_2} \le C|a|_{2k_1+m_1,\, m_2} |b|_{m_1,\, 2k_2+m_2}$$

for all sufficiently large integers k_1 and k_2, where the constant C is independent of a and b (but dependent on m_1, m_2, k_1 and k_2).

For $u,\, v \in S(\mathbb{R}^N)$, let

$$(u,\, v) := \int_{\mathbb{R}^N} u(x)\overline{v(x)} dx,$$

and let $A : S(\mathbb{R}^N) \to S(\mathbb{R}^N)$ be a bounded linear operator. Then the operator $A^* : S(\mathbb{R}^N) \to S(\mathbb{R}^N)$ is called a *formal adjoint* of A if

$$(Au,\, v) = (u,\, A^* v) \quad \text{for all } u,\, v \in S(\mathbb{R}^N).$$

Clearly, every operator A has at most one formal adjoint.

Proposition 4.1.8 *Let $A = Op(a) \in OPS^m_{0,0}$. Then A possesses a formal adjoint A^* which belongs to $OPS^m_{0,0}$ again, and*

$$\text{sym}_{A^*}(x,\, \xi) = os \int\int_{\mathbb{R}^N} \bar{a}(x+y,\, \xi+\eta) e^{-i\langle y,\, \eta\rangle}\, dy\, d\eta.$$

Moreover,

$$|\text{sym}_{A^*}|_{m_1,\, m_2} \le C|a|_{2k_1+m_1,\, 2k_2+m_2}$$

for all sufficiently large integers k_1 and k_2, where the constant C is independent of a (but dependent on m_1, m_2, k_1 and k_2).

4.1.5 Pseudodifferential operators with double symbols

Occasionally, it will be convenient to work in a class of pseudodifferential operators with symbols depending on three parameters.

Definition 4.1.9 *A function $a \in C_b^\infty(\mathbb{R}^N \times \mathbb{R}^N \times \mathbb{R}^N)$ is called a* double symbol in the class $S^m_{0,0,0}$ *if*

$$|a|_{r,\, s,\, t} := \sup_{\mathbb{R}^N\times\mathbb{R}^N\times\mathbb{R}^N} \sum_{|\alpha|\le r,\, |\beta|\le s,\, |\gamma|\le t} |\partial_\xi^\alpha \partial_x^\beta \partial_y^\gamma a(x,\, y,\, \xi)| \langle \xi\rangle^{-m} < \infty$$

for each choice of r, s, $t \in \mathbb{N}$. To each $a \in S^m_{0,0,0}$, we associate the pseudodifferential operator $Op_d(a)$ with double symbol a *by*

$$(Op_d(a)u)(x) := \int_{\mathbb{R}^N}\int_{\mathbb{R}^N} a(x,\, y,\, \xi) u(y) e^{i\langle x-y,\, \xi\rangle}\, dy\, d\xi, \quad u \in S(\mathbb{R}^N).$$

The class of all operators $Op_d(a)$ with $a \in S^m_{0,0,0}$ is denoted by $OPS^m_{0,0,0}$.

Sometimes, double symbols are also referred to as *amplitudes.*

Formally, the class $OPS^m_{0,0,0}$ of the pseudodifferential operators with double symbol seems to be much larger than the class $OPS^m_{0,0}$. Surprisingly, it turns out that both classes coincide.

Proposition 4.1.10 *Let $a \in S^m_{0,0,0}$ and $A := Op_d(a)$. Then $A \in OPS^m_{0,0}$, and*

$$\text{sym}_A(x,\, \xi) = os \int\int_{\mathbb{R}^N} a(x,\, x+y,\, \xi+\eta) e^{-i\langle y,\, \eta\rangle}\, dy\, d\eta.$$

Moreover, for $k > N/2$,

$$|\text{sym}_A|_{m_1,\, m_2} \le C|a|_{2k+m_1,\, m_2,\, 2k+m_2}$$

where C is a constant independent of a (but depending on m_1, m_2 and k).

By means of pseudodifferential operators with double symbol, we get the following description of the Fourier transform of a pseudodifferential operator.

Proposition 4.1.11 *Let $A = Op(a)$ with $a \in S^0_{0,0}$, and set $\tilde{a}(x, y, \xi) := a(-\xi, y)$. Then FAF^{-1} belongs to $OPS^0_{0,0}$, and*

$$FAF^{-1} = Op_d(\tilde{a}).$$

Moreover, for $k > N/2$,

$$|\mathrm{sym}_{\hat{A}}|_{m_1, m_2} \le C|a|_{2k+m_2, 2k+m_1}$$

where C is a constant independent of a (but depending on m_1, m_2 and k).

Proof. We have

$$\begin{aligned}(FAu)(x) &= (2\pi)^{-N}\int_{\mathbb{R}^N} e^{-i\langle\xi, x\rangle}\int_{\mathbb{R}^N} a(\xi, y)e^{i\langle\xi, y\rangle}\hat{u}(y)\,dy\,d\xi \\ &= (2\pi)^{-N}\int_{\mathbb{R}^N}\int_{\mathbb{R}^N} a(-\xi, y)e^{i\langle\xi, x-y\rangle}\hat{u}(y)\,dy\,d\xi\end{aligned}$$

which obviously coincides with $(Op_d(\tilde{a})\hat{u})(x)$. The norm estimate follows from Proposition 4.1.10. □

4.1.6 Boundedness on $L^2(\mathbb{R}^N)$

The celebrated Calderon-Vaillancourt theorem states the boundedness of pseudodifferential operators on $L^2(\mathbb{R}^N)$. This result is crucial in what follows, and we provide it with a proof.

Theorem 4.1.12 (Calderon-Vaillancourt) *For $a \in S^0_{0,0}$, the operator $Op(a)$ is bounded on $L^2(\mathbb{R}^N)$, and*

$$\|Op(a)\|_{L^2} \le C|a|_{2k_1, 2k_2} \quad \text{whenever } 2k_1 > N \text{ and } 2k_2 > N,$$

where C is a constant independent of a (but depending on k_1 and k_2).

We prepare the proof by a few propositions and start with discrete versions of Proposition 3.3.3 and of estimate (3.10) on the boundedness on L^2 of convolution operators. Let H be a Hilbert space.

Proposition 4.1.13 *Let T be an operator which acts on the finitely supported sequences in $l^2(\mathbb{Z}^N, H)$ by*

$$(Tu)(\alpha) = \sum_{\beta\in\mathbb{Z}^N} t(\alpha, \beta)u(\beta), \quad \alpha \in \mathbb{Z}^N.$$

If

$$M_1 := \sup_{\alpha\in\mathbb{Z}^N}\sum_{\beta\in\mathbb{Z}^N}\|t(\alpha, \beta)\| < \infty, \quad M_2 := \sup_{\beta\in\mathbb{Z}^N}\sum_{\alpha\in\mathbb{Z}^N}\|t(\alpha, \beta)\| < \infty,$$

then T extends continuously to a bounded linear operator on $l^2(\mathbb{Z}^N, H)$, and

$$\|T\|_{L(l^2)} \le (M_1 M_2)^{1/2}.$$

Corollary 4.1.14 *Let $t \in l^1(\mathbb{Z}^N, L(H))$. Then the operator T, defined by*

$$(Tu)(\alpha) := \sum_{\beta \in \mathbb{Z}^N} t(\alpha - \beta)u(\beta), \quad \alpha \in \mathbb{Z}^N,$$

is bounded on $l^2(\mathbb{Z}^N, H)$, and $\|T\|_{L(l^2)} \le \|t\|_{l^1}$.

Proposition 4.1.15 *Let $a \in S^0_{0,0}$, and let $K \subset \mathbb{R}^N$ be a compact set such that $a(x, \xi) = 0$ if $x \notin K$. Then $Op(a)$ is a bounded operator on $L^2(\mathbb{R}^N)$, and for every $k > N/2$, one has*

$$\|Op(a)\|_{L(L^2)} \le C\,(\text{mes}\,K)\,|a|_{0,2k} \tag{4.7}$$

with a constant $C > 0$ independent of K and a (but depending on k).

Proof. Set $A := Op(a)$, and let $u \in S(\mathbb{R}^N)$. By Parseval's equality,

$$\|Au\| = (2\pi)^{N/2}\|\widehat{Au}\|,$$

where

$$(\widehat{Au})(\xi) = (2\pi)^{-N} \int_{\mathbb{R}^N} \hat{a}(\xi - \eta,\, \eta)\hat{u}(\eta)d\eta$$

and

$$\hat{a}(\eta,\, \xi) := \int_{\mathbb{R}^N} a(x,\, \xi)e^{-i\langle x,\, \eta\rangle}\, dx.$$

The function $\hat{a}$ can be estimated by

$$|\hat{a}(\eta,\, \xi)| \le \int_K |\langle D_x\rangle^{2k} a(x,\, \xi)|dx\, \langle\eta\rangle^{-2k}$$

whence $|\hat{a}(\eta,\, \xi)| \le (\text{mes}\,K)|a|_{0,2k}\langle\eta\rangle^{-2k}$ for every $k \in \mathbb{N}$. If $2k > N$, then (3.10) applies, and it yields the estimate (4.7). □

Let $f \in C_0^\infty(\mathbb{R}^N)$ be a non-negative function with $f(x) = f(-x)$ for all x, $f(x) = 1$ if $|x_i| \le 2/3$ for all $i = 1, \ldots, N$ and with $f(x) = 0$ if $|x_i| \ge 3/4$ for at least one i. Define a non-negative function φ by

$$\varphi^2(x) := \frac{f(x)}{\sum_{\beta \in \mathbb{Z}^N} f(x - \beta)}, \quad x \in \mathbb{R}^N,$$

and set $\varphi_\alpha(x) := \varphi(x - \alpha)$ for $\alpha \in \mathbb{Z}^N$. The family (φ_α) forms a partition of unit on $\mathbb{R}^N$ in the sense that

$$\sum_{\alpha \in \mathbb{Z}^N} \varphi_\alpha^2(x) = 1 \quad \text{for all } x \in \mathbb{R}^N. \tag{4.8}$$

Proposition 4.1.16 *Let $A = Op(a) \in OPS^0_{0,0}$, and let (φ_α) be a partition of unit satisfying* (4.8)*. Then, for all $\alpha, \beta \in \mathbb{Z}^N$ and $k_1, k_2 > N/2$,*

$$\|\varphi_\alpha A \varphi_\beta I\|_{L(L^2(\mathbb{R}^N))} \leq C\langle \beta - \alpha\rangle^{-2k_1} |a|_{2k_1,\, 2k_2} \tag{4.9}$$

with a constant $C > 0$ independent of α, β and a (but depending on k_1 and k_2).

Proof. Let $\alpha, \beta \in \mathbb{Z}^N$. Since shift operators act as isometries, we have

$$\|\varphi_\alpha A \varphi_\beta I\| = \|V_{-\alpha}\varphi_\alpha A \varphi_\beta V_\alpha\| = \|\varphi_0 V_{-\alpha} A V_\alpha \varphi_{\beta-\alpha} I\|.$$

It is easy to see that $A_\alpha := V_{-\alpha} A V_\alpha = Op(a_\alpha)$ with $a_\alpha(x, \xi) = a(x+\alpha, \xi)$. The formal symbol of the operator $\varphi_0 A_\alpha \varphi_{\beta-\alpha} I$ has its support with respect to x in the cube $[-1, 1]^N$. In order to apply Proposition 4.1.15, it remains to estimate $|\mathrm{sym}_{\varphi_0 A_\alpha \varphi_{\beta-\alpha} I}|_{0,2k}$. From

$$\begin{aligned}
&\mathrm{sym}_{\varphi_0 A_\alpha \varphi_{\beta-\alpha} I}(x, \xi)\\
&= \varphi_0(x)\, os \int\int_{\mathbb{R}^N} a(x+\alpha, \xi+\eta)\varphi_{\beta-\alpha}(x+y) e^{-i\langle y, \eta\rangle}\, dy\, d\eta\\
&= \varphi_0(x)(2\pi)^{-N} \int_{\mathbb{R}^N}\int_{\mathbb{R}^N} e^{-i\langle y, \eta\rangle} \langle y\rangle^{-2k_1} \langle D_\xi\rangle^{2k_1} \langle \xi\rangle^{-2k_2} \langle D_y\rangle^{2k_2}\\
&\qquad\qquad a(x+\alpha, \xi+\eta)\varphi_{\beta-\alpha}(x+y)\, dy\, d\eta
\end{aligned}$$

we get

$$\begin{aligned}
&|\mathrm{sym}_{\varphi_0 A_\alpha \varphi_{\beta-\alpha} I}(x, \xi)|\\
&\leq C|a|_{2k_1,0}\varphi_0(x) \int_{\mathbb{R}^N} \langle y\rangle^{-2k_1} |\langle D_y\rangle^{2k_2} \varphi_0(x+y+\alpha-\beta)|\, dy\\
&= C|a|_{2k_1,0}\varphi_0(x) \int_{\mathbb{R}^N} \langle z+\beta-\alpha\rangle^{-2k_1} |\langle D_z\rangle^{2k_2} \varphi_0(x+z)|\, dz\\
&\leq C|a|_{2k_1,0}\langle \beta-\alpha\rangle^{-2k_1}\varphi_0(x) \int_{\mathbb{R}^N} \langle z\rangle^{-2k_1} |\langle D_z\rangle^{2k_2} \varphi_0(x+z)|\, dz,
\end{aligned}$$

where we applied Peetre's inequality

$$\langle x-y\rangle^r \leq 2^{|r|/2}\langle x\rangle^r \langle y\rangle^{|r|}$$

(holding for arbitrary $r \in \mathbb{R}$ and $x, y \in \mathbb{R}^N$) in the last estimate. Because of $\operatorname{supp}\varphi_0 \subset [-1, 1]^N$, the function

$$x \mapsto \varphi_0(x) \int_{\mathbb{R}^N} \langle z\rangle^{-2k_1} |\langle D_z\rangle^{2k_2} \varphi_0(x+z)|\, dz$$

is bounded. Thus, we arrive at the estimate

$$|\sigma_{\varphi_0 A_\alpha \varphi_{\beta-\alpha}}(x, \xi)| \leq C|a|_{2k_1,0}\langle \beta-\alpha\rangle^{-2k_1}$$

for all $2k_1 > N$. In the same way, we obtain

$$|\sigma_{\varphi_0 A_\alpha \varphi_{\beta-\alpha}}|_{0,\,2k_2} \le C|a|_{2k_1,\,2k_2}\langle\beta-\alpha\rangle^{-2k_1}$$

for all $2k_1 > N$ and $2k_2 > N$ and with a constant C depending on k_1 and k_2 only. The latter estimate implies (1.7) via Proposition 4.1.15. □

Proof of the Calderon-Vaillancourt Theorem. Let $A = Op(a) \in OPS^0_{0,0}$ and $u \in S(\mathbb{R}^N)$, and let (φ_α) be a partition of unity satisfying (4.8). Then, by (4.9),

$$\begin{aligned}\|\varphi_\alpha Au\| &= \|\sum_{\beta\in\mathbb{Z}^N}\varphi_\alpha A\varphi_\beta\varphi_\beta u\| \le \sum_{\beta\in\mathbb{Z}^N}\|\varphi_\alpha A\varphi_\beta I\|\,\|\varphi_\beta u\| \\ &\le C|a|_{2k_1,\,2k_2}\sum_{\beta\in\mathbb{Z}^N}\langle\beta-\alpha\rangle^{-2k_1}\|\varphi_\beta u\|.\end{aligned}$$

Thus, Corollary 4.1.14 implies

$$\begin{aligned}\|Au\|^2 &= \sum_{\alpha\in\mathbb{Z}^N}\|\varphi_\alpha Au\|^2 \\ &\le C|a|^2_{2k_1,\,2k_2}\sum_{\alpha\in\mathbb{Z}^N}\left(\sum_{\beta\in\mathbb{Z}^N}\langle\beta-\alpha\rangle^{-2k_1}\|\varphi_\beta u\|\right)^2 \\ &\le C|a|^2_{2k_1,\,2k_2}\sum_{\beta\in\mathbb{Z}^N}\|\varphi_\beta u\|^2 = C|a|^2_{2k_1,\,2k_2}\|u\|^2\end{aligned}$$

which finishes the proof. □

4.1.7 Consequences of the Calderon-Vaillancourt theorem

Adjoints. Let $a \in S^0_{0,0}$. Since $Op(a)$ is bounded on $L^2(\mathbb{R}^N)$ by Theorem 4.1.12, and since $S(\mathbb{R}^N)$ is dense in $L^2(\mathbb{R}^N)$, the formal adjoint of $Op(a)$ can be extended boundedly to an operator on $L^2(\mathbb{R}^N)$, and this extension coincides with the usual Hilbert space adjoint of $Op(a) \in L(L^2(\mathbb{R}^N))$.

Compactness. For the compactness of pseudodifferential operators on L^2 one has the following result.

Proposition 4.1.17 *Let $a \in S^0_{0,0}$. Then $Op(a)$ is compact on $L^2(\mathbb{R}^N)$ if and only if*

$$\lim_{(x,\,\xi)\to\infty} a(x,\,\xi) = 0.$$

Proof. Let $\varphi \in C_0^\infty(\mathbb{R}^N)$ be equal to 1 in a neighborhood of the origin. For $R > 0$, set $\varphi_R(x) := \varphi(x/R)$ and $\psi_R(x,\,\xi) := \varphi_R(x)\varphi_R(\xi)$. The Calderon-Vaillancourt theorem implies

$$\lim_{R\to\infty}\|Op(a) - Op(a)Op(\psi_R)\| = 0.$$

The operator $Op(\psi_R)$ is the product of the multiplication operator $\varphi_R I$ with the convolution operator $F^{-1}\varphi_R F$. By Theorem 3.2.2, this operator is compact. Being a norm limit of compact operators, the operator $Op(a)$ is compact, too. This settles the proof of the *if* direction. The proof of the reverse direction will be postponed to the end of Section 4.3.3 where it will be a simple corollary of Theorem 4.3.16. □

Boundedness on Sobolev spaces. For $s \in \mathbb{R}$, we let $H^s(\mathbb{R}^N)$ stand for the space of all distributions $u \in S'(\mathbb{R}^N)$ which have a measurable Fourier transform $\hat{u}$ that satisfies

$$\|u\|^2_{H^s(\mathbb{R}^N)} := \int_{\mathbb{R}^N} |\hat{u}(\xi)|^2 \langle\xi\rangle^{2s} d\xi < \infty.$$

Evidently, this norm is equivalent to

$$\|u\|^2_{H^s(\mathbb{R}^N)} = \int_{\mathbb{R}^N} |((\langle D\rangle^s u)(x)|^2 dx,$$

where $\langle D\rangle^s$ is the pseudodifferential operator $Op(a) \in S^s_{0,0}$ with symbol $a(x, \xi) := \langle\xi\rangle^s$. Thus, the operator $\langle D\rangle^s$ acts as an isometry from $H^s(\mathbb{R}^N)$ onto $L^2(\mathbb{R}^N)$.

Theorem 4.1.18 *Let $a \in S^m_{0,0}$. Then $Op(a)$ is a bounded operator from $H^s(\mathbb{R}^N)$ into $H^{s-m}(\mathbb{R}^N)$ for every $s \in \mathbb{R}$. Further, if the integers k_1 and k_2 are chosen sufficiently large, then*

$$\|Op(a)\|_{L(H^s(\mathbb{R}^N),\, H^{s-m}(\mathbb{R}^N))} \le C|a|_{k_1,\, k_2}$$

with a constant C independent of a (but depending on k_1 and k_2).

The proof follows immediately from the Calderon-Vaillancourt Theorem and from Proposition 4.1.7. We conclude with a result on inverses of pseudodifferential operators the proof of which can be found in [14].

Theorem 4.1.19 *Let $A \in OPS^m_{0,0}$ act as an invertible operator from $H^s(\mathbb{R}^N)$ onto $H^{s-m}(\mathbb{R}^N)$ for some s. Then A^{-1} is in $OPS^{-m}_{0,0}$.*

4.2 Bi-discretization of operators on $L^2(\mathbb{R}^N)$

The discretization of operators on $L^2(\mathbb{R}^N)$ which we are going to introduce here is finer than the discretization used in the preceding chapter. Here we will not only discretize with respect to the variable $x \in \mathbb{R}^N$, but simultaneously also with respect to the co-variable $\xi \in \mathbb{R}^N$ in the Fourier image.

4.2.1 Bi-discretization

Our point of departure is the partition of unit $(\varphi_\alpha)_{\alpha\in\mathbb{Z}^N}$ introduced in Section 4.1.6. That is, we have $\varphi_\alpha(x) = \varphi(x - \alpha)$ with a suitably chosen C_0^∞-function φ, and (4.8) holds. For $\gamma := (\alpha, \beta) \in \mathbb{Z}^N \times \mathbb{Z}^N$, we set $\phi_\gamma(x, \xi) := \varphi_\alpha(x)\varphi_\beta(\xi)$

and $\Phi_\gamma := Op(\phi_\gamma)$. These operators are compact by Theorem 3.2.2. Further, due to (4.8),

$$\begin{aligned}\sum_{\gamma\in\mathbb{Z}^{2N}} \Phi_\gamma^*\Phi_\gamma u &= \sum_{\beta\in\mathbb{Z}^N} Op(\varphi_\beta) \sum_{\alpha\in\mathbb{Z}^N} \varphi_\alpha^2 Op(\varphi_\beta) u \\ &= \sum_{\beta\in\mathbb{Z}^N} Op(\varphi_\beta)^2 u = F^{-1} \sum_{\beta\in\mathbb{Z}^N} \varphi_\beta^2 F u = u\end{aligned}$$

for all $u \in L^2(\mathbb{R}^N)$. Thus, the family $(\Phi_\gamma)_{\gamma\in\mathbb{Z}^{2N}}$ forms a partition of unit in the sense that

$$\sum_{\gamma\in\mathbb{Z}^{2N}} \Phi_\gamma^*\Phi_\gamma = I \tag{4.10}$$

where the series converges strongly on $L^2(\mathbb{R}^N)$. Analogously, one checks that $\sum_\gamma \Phi_\gamma\Phi_\gamma^* = I$. Moreover,

$$\|u\|_{L^2}^2 = \sum_{\gamma\in\mathbb{Z}^{2N}} \|\Phi_\gamma u\|_{L^2}^2 = \sum_{\gamma\in\mathbb{Z}^{2N}} \|\Phi_\gamma^* u\|_{L^2}^2 \tag{4.11}$$

for every $u \in L^2(\mathbb{R}^N)$ which follows easily from (4.10):

$$\|u\|_{L^2}^2 = \sum_{\gamma\in\mathbb{Z}^{2N}} \langle \Phi_\gamma^*\Phi_\gamma u,\, u\rangle = \sum_{\gamma\in\mathbb{Z}^{2N}} \langle \Phi_\gamma u,\, \Phi_\gamma u\rangle = \sum_{\gamma\in\mathbb{Z}^{2N}} \|\Phi_\gamma u\|_{L^2}^2.$$

We write each vector $\gamma \in \mathbb{Z}^{2N}$ as $(\gamma_1, \gamma_2) \in \mathbb{Z}^N \times \mathbb{Z}^N$ and set $U_\gamma := V_{\gamma_1} E_{\gamma_2} \in L^2(\mathbb{R}^N)$ where V_α and E_β are defined as in Section 4.1.1. The operators U_γ are unitary, and one easily checks that $\Phi_\gamma = U_\gamma \Phi_0 U_\gamma^*$. Note that the operators U_γ, together with the scalar unitary operators $e^{ir} I$ with $r \in \mathbb{Z}$, form a non-commutative group, the so-called *discrete Heisenberg group*. In particular, one has the following identities, which follow easily from (4.2),

$$U_\alpha^* = e^{i\langle\alpha_2,\,\alpha_1\rangle} U_{-\alpha}, \quad U_\alpha U_\beta = e^{i\langle\alpha_2,\,\beta_1\rangle} U_{\alpha+\beta}, \tag{4.12}$$

$$U_\alpha^* U_\beta = e^{i\langle\alpha_2,\,\alpha_1-\beta_1\rangle} U_{\beta-\alpha} = e^{i\langle\beta_2,\,\alpha_1-\beta_1\rangle} U_{\alpha-\beta}^* \tag{4.13}$$

where $\alpha := (\alpha_1, \alpha_2)$, $\beta := (\beta_1, \beta_2) \in \mathbb{Z}^N \times \mathbb{Z}^N$.

We define the *bi-discretization* Gu of a function $u \in L^2(\mathbb{R}^N)$ by

$$(Gu)_\gamma := \Phi_0 U_\gamma^* u, \quad \gamma \in \mathbb{Z}^{2N},$$

i.e., we consider Gu as a vector-valued function on $\mathbb{Z}^{2N}$ with values in $L^2(\mathbb{R}^N)$.

Proposition 4.2.1 *The operator $G : L^2(\mathbb{R}^N) \to l^2(\mathbb{Z}^{2N}, L^2(\mathbb{R}^N))$ is an isometry. Its adjoint is given by*

$$G^* f = \sum_{\gamma\in\mathbb{Z}^{2N}} U_\gamma \Phi_0^* f_\gamma$$

where the series converges in $L^2(\mathbb{R}^N)$.

Proof. The isometry of G follows from (4.11) since

$$\|Gu\|_{l^2}^2 = \sum_{\gamma\in\mathbb{Z}^{2N}} \|\Phi_0 U_\gamma^* u\|_{L^2}^2 = \sum_{\gamma\in\mathbb{Z}^{2N}} \|U_\gamma \Phi_0 U_\gamma^* u\|_{L^2}^2 = \sum_{\gamma\in\mathbb{Z}^{2N}} \|\Phi_\gamma u\|_{L^2}^2 = \|u\|_{L^2}^2.$$

Further one has

$$\begin{aligned} \langle Gu,\, f\rangle_{l^2} &= \sum_{\gamma\in\mathbb{Z}^{2N}} \langle (Gu)_\gamma,\, f_\gamma\rangle_{L^2} = \sum_{\gamma\in\mathbb{Z}^{2N}} \langle \Phi_0 U_\gamma^* u,\, f_\gamma\rangle_{L^2} \\ &= \sum_{\gamma\in\mathbb{Z}^{2N}} \langle u,\, U_\gamma \Phi_0^* f\rangle_{L^2} = \langle u,\, G^* f\rangle_{L^2} \end{aligned}$$

for every $u \in L^2(\mathbb{R}^N)$ and $f \in l^2(\mathbb{Z}^{2N}, L^2(\mathbb{R}^N))$. □

Thus, $G^*G = I$, and the operator $Q := GG^*$ is an orthogonal projection on $l^2(\mathbb{Z}^{2N}, L^2(\mathbb{R}^N))$. We let $\operatorname{Im} Q$ denote its range. Then

$$G : L^2(\mathbb{R}^N) \to \operatorname{Im} Q$$

is a unitary operator, and every operator $A \in L(L^2(\mathbb{R}^N))$ is unitarily equivalent to the operator

$$A_G := GAG^*|_{\operatorname{Im} Q}.$$

We extend A_G to an operator $\Gamma(A)$ acting on all of $l^2(\mathbb{Z}^{2N}, L^2(\mathbb{R}^N))$ by setting

$$\Gamma(A) := A_G Q + I - Q = GAG^* + I - Q$$

and consider $\Gamma(A)$ as the bi-discretization of A. Clearly,

$$G^*\Gamma(A)G = G^*(GAG^* + I - GG^*)G = A.$$

4.2.2 Bi-discretization and Fredholmness

For the following result, recall the definition of the approximate identity on the Hilbert space $l^2(\mathbb{Z}^{2N}, L^2(\mathbb{R}^N))$ given in Section 2.1.1. Here we denote this identity by $\hat{\mathcal{P}} = (\hat{P}_n)_{n\in\mathbb{N}}$ in order to distinguish it from an approximate identity on $L^2(\mathbb{R}^N)$ which will be introduced later. The approximate identity $\hat{\mathcal{P}}$ is perfect, hence, every compact operator on $l^2(\mathbb{Z}^{2N}, L^2(\mathbb{R}^N))$ is $\hat{\mathcal{P}}$-compact.

Proposition 4.2.2

(a) *The operators $\hat{P}_k Q$ and $Q\hat{P}_k$ are compact for every $k \in \mathbb{N}$.*

(b) *The projection Q belongs to $L(l^2(\mathbb{Z}^{2N}, L^2(\mathbb{R}^N)), \hat{\mathcal{P}})$.*

(c) *For $A \in L(L^2(\mathbb{R}^N))$, the operator $\Gamma(A)$ is in $L(l^2(\mathbb{Z}^{2N}, L^2(\mathbb{R}^N)), \hat{\mathcal{P}})$.*

(d) *Let $K \in L(l^2(\mathbb{Z}^{2N}, L^2(\mathbb{R}^N)))$ be a $\hat{\mathcal{P}}$-compact operator of the form $K = QKQ$. Then G^*KG is compact.*

(e) *The operator $A \in L(L^2(\mathbb{R}^N))$ is invertible (Fredholm) if and only if the operator $\Gamma(A) \in L(l^2(\mathbb{Z}^{2N}, L^2(\mathbb{R}^N)))$ is invertible ($\hat{\mathcal{P}}$-Fredholm).*

Proof. (a) For a moment, let the operators S_γ, E_γ, $R_\gamma \in L(l^2(\mathbb{Z}^{2N}, L^2(\mathbb{R}^N)))$ with $\gamma \in \mathbb{Z}^{2N}$ be defined as in Section 2.1.2. It is clearly sufficient to verify the compactness of all operators $S_\gamma Q$ and QS_γ. A straightforward calculation yields that

$$S_\gamma Q = \sum_{\beta \in \mathbb{Z}^{2N}} E_\gamma \Phi_0 U_\gamma^* U_\beta \Phi_0^* R_\beta. \tag{4.14}$$

Since, with certain constants $c_{\gamma\beta}$,

$$\Phi_0 U_\gamma^* U_\beta \Phi_0^* = c_{\gamma\beta} \Phi_0 U_{\gamma-\beta}^* \Phi_0^* = c_{\gamma\beta} U_{\gamma-\beta}^* \Phi_{\gamma-\beta} \Phi_0^* = 0$$

if β is sufficiently large, the sum (4.14) has only a finite number of non-vanishing items. Each of these items is compact because Φ_0 is compact by Theorem 3.2.2. Thus, $S_\gamma Q$ and $QS_\gamma = (S_\gamma Q)^*$ are compact.

(b) Recall from Proposition 1.1.8 that Q belongs to $L(l^2(\mathbb{Z}^{2N}, L^2(\mathbb{R}^N)), \hat{\mathcal{P}})$ if and only if, for every $k \in \mathbb{N}$,

$$\|\hat{P}_k Q(I - \hat{P}_n)\| \to 0 \quad \text{and} \quad \|(I - \hat{P}_n)Q\hat{P}_k\| \to 0$$

as $n \to \infty$. These conditions follow immediately from the compactness of $\hat{P}_k Q$ and $Q\hat{P}_k$ and from the *-strong convergence of the $\hat{P}_n$ to the identity.

(c) As in the previous step, we have to show that, for every $k \in \mathbb{N}$,

$$\|\hat{P}_k \Gamma(A)(I - \hat{P}_n)\| \to 0 \quad \text{and} \quad \|(I - \hat{P}_n)\Gamma(A)\hat{P}_k\| \to 0$$

as $n \to \infty$. Let us check the first condition. We have

$$\hat{P}_k \Gamma(A)(I - \hat{P}_n) = \hat{P}_k QGAG^*(I - \hat{P}_n) + \hat{P}_k(I - \hat{P}_n) - \hat{P}_k Q(I - \hat{P}_n).$$

The first and the third term in this sum tend to zero in the norm since $\hat{P}_k Q$ is compact and since the $I - \hat{P}_n$ converge strongly to 0. The second term is zero whenever $n > k$.

(d) If K is $\hat{\mathcal{P}}$-compact, then $\|K(I - \hat{P}_n)\| \to 0$. Consequently,

$$\|G^* K(I - \hat{P}_n)G\| = \|G^* KGG^*(I - \hat{P}_n)G\| = \|G^* KG(I - G^* \hat{P}_n G)\| \to 0.$$

Since

$$G^* \hat{P}_n G = \sum_{\alpha \in [-n, n]^{2N} \cap \mathbb{Z}^{2N}} \Phi_\alpha^* \Phi_\alpha$$

and $\Phi_\alpha^* \Phi_\alpha$ is compact, the operator $G^* KG$ is the norm limit of compact operators and, hence, compact.

(e) Since A and A_G are unitarily equivalent, the operator A is invertible (Fredholm) if and only if A_G is invertible (Fredholm). We claim that the latter happens if and only if the operator $\Gamma(A)$ is invertible ($\hat{\mathcal{P}}$-Fredholm).

Let A_G be invertible on $\operatorname{Im} Q$, and let B be its inverse. Then, clearly, $QBQ + I - Q$ is the inverse of $\Gamma(A)$. Conversely, if C is the inverse of $\Gamma(A)$, then QCQ is the inverse of A_G, since $\Gamma(A)Q = Q\Gamma(A)Q = Q\Gamma(A)$.

Let now A_G be Fredholm, and let B be a regularizer of A_G, i.e., the operators $A_G B - I$ and $BA_G - I$ are compact. Then the operators

$$\begin{aligned}&\Gamma(A)(QBQ + I - Q) - I\\ &\quad = (QA_GQ + I - Q)(QBQ + I - Q) - I = QA_GBQ - Q = Q(A_GB - I)Q\end{aligned}$$

and $(QBQ+I-Q)\Gamma(A) - I$ are compact and, hence, also $\hat{\mathcal{P}}$-compact, whence the $\hat{\mathcal{P}}$-Fredholmness of $\Gamma(A)$. Let, conversely, $\Gamma(A)$ be a $\hat{\mathcal{P}}$-Fredholm operator. Thus, there are an operator $B \in L(l^2(\mathbb{Z}^{2N}, L^2(\mathbb{R}^N)), \hat{\mathcal{P}})$ and $\hat{\mathcal{P}}$-compact operators K, L such that

$$\Gamma(A)B = I + K \quad \text{and} \quad B\Gamma(A) = I + L.$$

We multiply both equalities from both sides by Q. Since $\Gamma(A)$ commutes with Q, we get

$$Q\Gamma(A)QBQ = Q + K' \quad \text{and} \quad QBQ\Gamma(A)Q = Q + L' \tag{4.15}$$

with $\hat{\mathcal{P}}$-compact operators K' and L' satisfying

$$K' = QK'Q \quad \text{and} \quad L' = QL'Q.$$

Multiplying (4.15) by G^* from the left-hand side and by G from the right-hand side we find

$$AG^*BG = I + G^*K'G \quad \text{and} \quad G^*BGA = I + G^*L'G.$$

The operators $G^*K'G$ and $G^*L'G$ are compact by assertion (d). □

4.2.3 Bi-discretization and limit operators

For $n \in \mathbb{N}$, set

$$P_n := \sum_{\alpha \in [-n,\, n]^{2N} \cap \mathbb{Z}^{2N}} \Phi_\alpha^* \Phi_\alpha.$$

As we have seen in Section 4.2.1, the sequence $\mathcal{P} := (P_n)_{n\in\mathbb{N}}$ forms a perfect approximate identity on $L^2(\mathbb{R}^N)$ which consists of compact operators only.

Let the unitary operators U_α be as in Section 4.2.1, and let $\mathcal{H}$ denote the set of all sequences in $\mathbb{Z}^{2N}$ which tend to infinity. In accordance with the notations from Section 1.2.1, we call the operator $A_h \in L(L^2(\mathbb{R}^N))$ the limit operator of $A \in L(L^2(\mathbb{R}^N))$ with respect to the sequence $h \in \mathcal{H}$ if $A_h = \mathcal{P}\text{-}\lim_{m\to\infty} U_{h(m)}^* A U_{h(m)}$. Equivalently, A_h is the limit operator of A with respect to h iff

$$\lim_{m\to\infty} \|(U_{h(m)}^* A U_{h(m)} - A_h)P_n\| = \lim_{m\to\infty} \|P_n(U_{h(m)}^* A U_{h(m)} - A_h)\| = 0$$

for every $n \in \mathbb{N}$. Since the operators P_n are compact and tend *-strongly to the identity operator, these conditions are equivalent to the *-strong convergence of $U_{h(m)}^* A U_{h(m)}$ to A_h as m tending to infinity.

Our goal is to relate the limit operators of operators A on $L^2(\mathbb{R}^N)$, which are defined by means of the unitary operators U_γ, with the limit operators of the discretization $\Gamma(A)$ on $L(l^2(\mathbb{Z}^{2N}, L^2(\mathbb{R}^N)))$, which are defined with respect to the unitary shift operators

$$(\hat{V}_\gamma u)_\alpha := u_{\alpha-\gamma}, \quad \alpha \in \mathbb{Z}^{2N}$$

where $\gamma \in \mathbb{Z}^{2N}$. For, we need a few prerequisites. Given $\gamma = (\gamma_1, \gamma_2) \in \mathbb{Z}^{2N} = \mathbb{Z}^N \times \mathbb{Z}^N$, we define a unitary operator $\hat{T}_\gamma$ on $l^2(\mathbb{Z}^{2N}, L^2(\mathbb{R}^N))$ by $(\hat{T}_\gamma u)_\alpha := e^{i\langle\gamma_2, \alpha_1\rangle} u_\alpha$.

Lemma 4.2.3 *Let $\gamma \in \mathbb{Z}^{2N}$. Then*

$$\hat{V}_{-\gamma} G = \hat{T}_\gamma G U_\gamma^* \quad \textit{and} \quad G^* \hat{V}_\gamma = U_\gamma G^* \hat{T}_\gamma^*$$

on $L^2(\mathbb{R}^N)$ and on $l^2(\mathbb{Z}^{2N}, L^2(\mathbb{R}^N))$, respectively.

Proof. Let $f \in L^2(\mathbb{R}^N)$ and $\alpha \in \mathbb{Z}^N \times \mathbb{Z}^N$. Then

$$\begin{aligned}(\hat{V}_{-\gamma} G U_\gamma f)_\alpha &= (G U_\gamma f)_{\alpha+\gamma} = \Phi_0 U_{\alpha+\gamma}^* U_\gamma f \\ &= e^{i\langle\gamma_2, \alpha_1\rangle} \Phi_0 U_\gamma f = e^{i\langle\gamma_2, \alpha_1\rangle} (Gf)_\alpha \\ &= (\hat{T}_\gamma G f)_\alpha\end{aligned}$$

where we used (4.13). Hence, $\hat{V}_{-\gamma} G U_\gamma = \hat{T}_\gamma G$ on $L^2(\mathbb{R}^N)$, which implies the assertions. □

Lemma 4.2.4 *Every sequence $h \in \mathcal{H}$ possesses a subsequence g such that the functions*

$$f_m : \mathbb{Z}^N \to \mathbb{T}, \quad \alpha \mapsto e^{i\langle g(m), \alpha\rangle} \tag{4.16}$$

converge uniformly on $\mathbb{Z}^N$ as $m \to \infty$.

Proof. Set $r_{-1} := h$, and let $\gamma : \mathbb{N} \to \mathbb{Z}^N$ be an enumeration of $\mathbb{Z}^N$. By the compactness of the unit circle $\mathbb{T}$, there is a subsequence r_0 of r_{-1} such that

$$e^{i\langle r_0(m), \gamma_0\rangle} \to f(\gamma_0) \in \mathbb{T} \quad \text{as } m \to \infty$$

and

$$|e^{i\langle r_0(m), \gamma_0\rangle} - f(\gamma_0)| < 2 \quad \text{for all } m \in \mathbb{Z}^N.$$

We proceed in this way and get, for every positive integer n, a subsequence r_n of r_{n-1} such that

$$e^{i\langle r_n(m), \gamma_n\rangle} \to f(\gamma_n) \in \mathbb{T} \quad \text{as } m \to \infty$$

and

$$|e^{i\langle r_n(m), \gamma_n\rangle} - f(\gamma_n)| < 2^{-n} \quad \text{for all } m \in \mathbb{Z}^N.$$

Set $g(n) := r_n(n)$. Since g is (with exception of a finite number of entries) a subsequence of each sequence r_n, we have $g \in \mathcal{H}$,

$$e^{i\langle g(m), \gamma_n\rangle} \to f(\gamma_n) \quad \text{as } m \to \infty$$

and

$$|e^{i\langle g(m),\,\gamma_n\rangle} - f(\gamma_n)| < 2^{-n} \quad \text{for all } m \in \mathbb{Z}^N \text{ and } n \in \mathbb{N}.$$

We claim that the functions f_m converge uniformly to the function $f : \mathbb{Z}^N \to \mathbb{T}$ defined in this way. Given $\varepsilon > 0$, choose $K \in \mathbb{N}$ such that $2^{-K} < \varepsilon$, and then choose $M \in \mathbb{N}$ such that

$$|e^{i\langle g(m),\,\gamma_n\rangle} - f(\gamma_n)| < \varepsilon \quad \text{for all } m \geq M \text{ and } n \leq K.$$

Then $|e^{i\langle g(m),\,\alpha\rangle} - f(\alpha)| < \varepsilon$ for all $m \geq M$ and $\alpha \in \mathbb{Z}^N$. □

Proposition 4.2.5 *Let $A \in L(L^2(\mathbb{R}^N))$ be such that the limit operator A_h with respect to the sequence $h \in \mathcal{H}$ exists. Then there is a subsequence g of h such that the limit operator $\Gamma(A)_g$ of $\Gamma(A)$ exists and that the operators $\Gamma(A)_g$ and $\Gamma(A_h)$ are unitarily equivalent.*

Proof. Let $h \in \mathcal{H}$ be a sequence such that the limit operator A_h exists. By the preceding lemma, there is a subsequence g of h such that the functions (4.16) converge uniformly on $\mathbb{Z}^{2N}$ to a certain function $f_g : \mathbb{Z}^{2N} \to \mathbb{T}$. Let the operator $T_g : l^2(\mathbb{Z}^{2N}, L^2(\mathbb{R}^N)) \to l^2(\mathbb{Z}^{2N}, L^2(\mathbb{R}^N))$ be defined by $(T_g u)_\alpha := f_g(\alpha_1)u_\alpha$. Since all values of f_g are unimodular, the operator T_g is unitary. Moreover, from the uniform convergence of the functions (4.16) to f_g we conclude that

$$\|\hat{T}_{g(m)} - T_g\| = \sup_{\alpha \in \mathbb{Z}^{2N}} |e^{i\langle g(m),\,\alpha\rangle} - f_g(\alpha)| \to 0 \quad \text{as } m \to \infty.$$

Now, by Lemma 4.2.3,

$$\hat{V}_{-g(m)} GAG^* \hat{V}_{g(m)} = \hat{T}_{g(m)} G U^*_{g(m)} A U_{g(m)} G^* \hat{T}_{g(m)},$$

and the right-hand side of this equality converges *-strongly to $T_g G A_h G^* T_g^*$. Hence, the limit operator $(GAG^*)_g$ exists, and

$$(GAG^*)_g = T_g G A_h G^* T_g^*. \tag{4.17}$$

Choosing $A = I$, we see in particular that every sequence h which tends to infinity possesses a subsequence g such that the limit operator Q_g of $Q = GG^*$ exists and that this limit operator is equal to $T_g Q T_g^*$. Of course, one can choose the same subsequence g as in (4.17). Consequently, the limit operator of $\Gamma(A) = GAG^* + I - Q$ with respect to g also exists, and

$$\begin{aligned} \Gamma(A)_g &= (GAG^*)_g + (I-Q)_g \\ &= T_g G A_h G^* T_g^* + T_g(I-Q)T_g^* = T_g \Gamma(A_h) T_g^*. \end{aligned} \tag{4.18}$$

This proves the assertion. □

4.3 Fredholmness of pseudodifferential operators

Here we are going to single out a class of operators which become band-dominated operators in the rich Wiener algebra after bi-discretization. This will enable us to derive Fredholm criteria for these operators. Particular examples of operators which belong to this class are provided by the pseudodifferential operators with symbol in $S^0_{0,0}$.

4.3.1 A Wiener algebra on $L^2(\mathbb{R}^N)$

We define a Wiener algebra of operators on $L^2(\mathbb{R}^N)$ by imposing conditions on the decay of the norms $\|\Phi_\alpha A \Phi^*_{\alpha-\gamma}\|$.

Definition 4.3.1 *Let A be a linear (not necessarily bounded) operator on $L^2(\mathbb{R}^N)$. We say that A belongs to $\mathcal{W}(L^2(\mathbb{R}^N))$ if*

$$\|A\|_{\mathcal{W}(L^2(\mathbb{R}^N))} := \sum_{\gamma\in\mathbb{Z}^{2N}} \sup_{\alpha\in\mathbb{Z}^{2N}} \|\Phi_\alpha A \Phi^*_{\alpha-\gamma}\|_{L(L^2(\mathbb{R}^N))} < \infty.$$

The class $\mathcal{W}(L^2(\mathbb{R}^N))$ contains sufficiently many interesting operators. Actually we will show that all pseudodifferential operators with symbol in $S^0_{0,0}$ belong to $\mathcal{W}(L^2(\mathbb{R}^N))$. To check this, we need the following analogue of Proposition 4.1.16.

Proposition 4.3.2 *Let $A = Op(a) \in OPS^0_{0,0}$ and $\alpha = (\alpha_1, \alpha_2)$, $\beta = (\beta_1, \beta_2) \in \mathbb{Z}^N \times \mathbb{Z}^N$. Then*

$$\|\Phi_\alpha A \Phi^*_\beta\|_{L(L^2(\mathbb{R}^N))} \le C|a|_{2k+2m,\,2k+2m}\langle\alpha_1-\beta_1\rangle^{-k}\langle\alpha_2-\beta_2\rangle^{-k}$$

whenever $2k > N$, $m \in \mathbb{N}$ is large enough, and with a constant $C > 0$ independent of a and of α and β (but depending on k and m).

Proof. Applying Proposition 4.1.16 to the operator $B := Op(\varphi_{\alpha_2})AOp(\varphi_{\beta_2})$, we get

$$\begin{aligned}\|\Phi_\alpha A \Phi^*_\beta\|_{L(L^2(\mathbb{R}^N))} &= \|\varphi_{\alpha_1} Op(\varphi_{\alpha_2})AOp(\varphi_{\beta_2})\varphi_{\beta_1} I\|_{L(L^2(\mathbb{R}^N))}\\ &\le C\langle\alpha_1-\beta_1\rangle^{-2k}|\mathrm{sym}_B|_{2k,\,2k}\end{aligned}$$

for all $2k > N$. By Proposition 4.1.7, $|\mathrm{sym}_B|_{2k,\,2k} \le C|a|_{2k+2m,\,2k+2m}$ whenever $2m > N$. Thus,

$$\|\Phi_\alpha A \Phi^*_\beta\|_{L(L^2(\mathbb{R}^N))} \le C\langle\alpha_1-\beta_1\rangle^{-2k}|a|_{2k+2m,\,2k+2m}. \tag{4.19}$$

Similarly, applying Proposition 4.1.11 to the right-hand side of the estimate

$$\begin{aligned}\|\Phi_\alpha A \Phi^*_\beta\|_{L(L^2(\mathbb{R}^N))} &= \|F\Phi_\alpha A \Phi^*_\beta F^{-1}\|_{L(L^2(\mathbb{R}^N))}\\ &= \|Op(\varphi_{\alpha_1})\varphi_{\alpha_2} FAF^{-1}\varphi_{\beta_2} Op(\varphi_{\alpha_2})\|_{L(L^2(\mathbb{R}^N))}\\ &\le \|\varphi_{\alpha_2} FAF^{-1}\varphi_{\beta_2} I\|_{L(L^2(\mathbb{R}^N))}\end{aligned}$$

we obtain

$$\|\Phi_\alpha A \Phi_\beta^*\|_{L(L^2(\mathbb{R}^N))} \le C \langle \alpha_2 - \beta_2 \rangle^{-2k} |a|_{2k+2m,\, 2k+2m} \tag{4.20}$$

for every $2k > N$ and for every m which is sufficiently large (recall that φ is an even function by hypothesis). Multiplying (4.19) by (4.20) and taking square roots, we get the assertion. □

Corollary 4.3.3 $OPS^0_{0,0} \subseteq \mathcal{W}(L^2(\mathbb{R}^N))$.

Indeed, for $A \in OPS^0_{0,0}$, and with $\gamma := (\gamma_1, \gamma_2)$ and $\alpha := (\alpha_1, \alpha_2)$, the preceding proposition implies

$$\sum_{\gamma \in \mathbb{Z}^{2N}} \sup_{\alpha \in \mathbb{Z}^{2N}} \|\Phi_\alpha A \Phi^*_{\alpha-\gamma}\|_{L(L^2(\mathbb{R}^N))} \le C |a|_{2k+2m,\, 2k+2m} \sum_{\gamma \in \mathbb{Z}^{2N}} \langle \gamma_1 \rangle^{-k} \langle \gamma_2 \rangle^{-k},$$

which is finite if k is chosen large enough. □

The following proposition summarizes some basic properties of $\mathcal{W}(L^2(\mathbb{R}^N))$.

Proposition 4.3.4

(a) $\mathcal{W}(L^2(\mathbb{R}^N)) \subset L(L^2(\mathbb{R}^N))$, *and*

$$\|A\|_{L(L^2(\mathbb{R}^N))} \le \|A\|_{\mathcal{W}(L^2(\mathbb{R}^N))} \quad \text{for all } A \in \mathcal{W}(L^2(\mathbb{R}^N)).$$

(b) *Provided with the norm* $A \mapsto \|A\|_{\mathcal{W}(L^2(\mathbb{R}^N))}$ *and with the involution* $A \mapsto A^*$ (*= the Hilbert space adjoint of* A), *the set* $\mathcal{W}(L^2(\mathbb{R}^N))$ *becomes a unital involutive Banach algebra.*

Proof. (a) The boundedness of $A \in \mathcal{W}(L^2(\mathbb{R}^N))$ as well as the norm estimate can be obtained as follows, where we employ (4.10) and (4.11) several times:

$$\begin{aligned}
\|Au\|^2 &= \sum_{\gamma \in \mathbb{Z}^{2N}} \|\Phi_\gamma A u\|^2 \\
&= \sum_{\gamma \in \mathbb{Z}^{2N}} \left\| \Phi_\gamma A \sum_{\delta \in \mathbb{Z}^{2N}} \Phi_\delta^* \Phi_\delta u \right\|^2 \\
&\le \sum_{\gamma \in \mathbb{Z}^{2N}} \left(\sum_{\alpha \in \mathbb{Z}^{2N}} \|\Phi_\gamma A \Phi^*_{\gamma-\alpha}\| \, \|\Phi_{\gamma-\alpha} u\| \right)^2 \\
&\le \sum_{\gamma \in \mathbb{Z}^{2N}} \left(\sum_{\alpha \in \mathbb{Z}^{2N}} k_A(\alpha) \|\Phi_{\gamma-\alpha} u\| \right)^2 \\
&\le \sum_{\gamma \in \mathbb{Z}^{2N}} \left(\sum_{\alpha \in \mathbb{Z}^{2N}} k_A(\gamma-\alpha) \|\Phi_\alpha u\| \right)^2
\end{aligned}$$

with $k_A(\alpha) := \sup_{\gamma\in\mathbb{Z}^N} \|\Phi_\gamma A \Phi^*_{\gamma-\alpha}\|$. Since k_A is in $l^1(\mathbb{Z}^N)$, Corollary 4.1.14 yields

$$\|Au\|^2 \leq \left(\sum_{\gamma\in\mathbb{Z}^{2N}} k_A(\gamma) \right)^2 \sum_{\alpha\in\mathbb{Z}^{2N}} \|\Phi_\alpha u\|^2 = \|A\|^2_{\mathcal{W}(L^2(\mathbb{R}^N))} \|u\|^2,$$

whence assertion (a).
(b) Let A, $B \in \mathcal{W}(L^2(\mathbb{R}^N))$. Then, clearly,

$$\|\alpha A\|_{\mathcal{W}(L^2(\mathbb{R}^N))} = |\alpha| \, \|A\|_{\mathcal{W}(L^2(\mathbb{R}^N))}$$

and

$$\|A + B\|_{\mathcal{W}(L^2(\mathbb{R}^N))} \leq \|A\|_{\mathcal{W}(L^2(\mathbb{R}^N))} + \|B\|_{\mathcal{W}(L^2(\mathbb{R}^N))}.$$

For the product AB, one finds

$$\begin{aligned}
\|AB\|_{\mathcal{W}(L^2(\mathbb{R}^N))} &= \sum_{\gamma\in\mathbb{Z}^{2N}} \sup_{\alpha\in\mathbb{Z}^{2N}} \|\Phi_\alpha A B \Phi^*_{\alpha-\gamma}\| \\
&= \sum_{\gamma\in\mathbb{Z}^{2N}} \sup_{\alpha\in\mathbb{Z}^{2N}} \left\| \sum_{\theta\in\mathbb{Z}^{2N}} \Phi_\alpha A \Phi^*_{\alpha-\theta} \Phi_{\alpha-\theta} B \Phi^*_{\alpha-\gamma} \right\| \\
&\leq \sum_{\gamma\in\mathbb{Z}^{2N}} \sum_{\theta\in\mathbb{Z}^{2N}} k_A(\theta) k_B(\gamma-\theta) \\
&\leq \|A\|_{\mathcal{W}(L^2(\mathbb{R}^N))} \|B\|_{\mathcal{W}(L^2(\mathbb{R}^N))}.
\end{aligned}$$

Further, since $\|\Phi_\gamma A \Phi^*_\delta\| = \|\Phi_\delta A^* \Phi^*_\gamma\|$, the operators A and A^* belong to the Wiener algebra $\mathcal{W}(L^2(\mathbb{R}^N))$ only simultaneously, and

$$\|A\|_{\mathcal{W}(L^2(\mathbb{R}^N))} = \|A^*\|_{\mathcal{W}(L^2(\mathbb{R}^N))}$$

in this case. That the identity operator belongs to $\mathcal{W}(L^2(\mathbb{R}^N))$ follows from Corollary 4.3.3.

Finally, if (A_n) is a Cauchy sequence in $\mathcal{W}(L^2(\mathbb{R}^N))$ then, by part (a), it is also a Cauchy sequence in $L(L^2(\mathbb{R}^N))$, hence convergent. Let $A \in L(L^2(\mathbb{R}^N))$ denote the limit of this sequence. Given $\varepsilon > 0$, choose M such that $\|A_n - A_m\|_{\mathcal{W}(L^2(\mathbb{R}^N))} < \varepsilon$ for all m, $n \geq M$. Letting m go to infinity in this inequality, we get the convergence of the A_m to A with respect to the norm in the Wiener algebra. □

Next we consider bi-discretizations of operators in the Wiener algebra $\mathcal{W}(L^2(\mathbb{R}^N))$. To distinguish this algebra from the Wiener algebra of operators on the discrete Hilbert space $l^2(\mathbb{Z}^{2N}, L^2(\mathbb{R}^N))$ introduced in Section 2.5, we denote the latter by $\mathcal{W}(l^2(\mathbb{Z}^{2N}))$ in what follows.

Proposition 4.3.5

(a) *Let $A \in \mathcal{W}(L^2(\mathbb{R}^N))$. Then the operators GAG^* and $\Gamma(A)$ belong to the Wiener algebra $\mathcal{W}(l^2(\mathbb{Z}^{2N}))$.*

(b) *Let $B \in \mathcal{W}(l^2(\mathbb{Z}^{2N}))$. Then the operator G^*BG belongs to the Wiener algebra $\mathcal{W}(L^2(\mathbb{R}^N))$.*

Proof. (a) Let $u \in l^2(\mathbb{Z}^{2N}, L^2(\mathbb{R}^N))$ and $\alpha \in \mathbb{Z}^{2N}$. Then

$$\begin{aligned}(GAG^*u)_\alpha &= (GA\sum_{\gamma\in\mathbb{Z}^{2N}} U_\gamma\Phi_0^*u_\gamma)_\alpha = \Phi_0U_\alpha^*A\sum_{\gamma\in\mathbb{Z}^{2N}} U_\gamma\Phi_0^*u_\gamma\\ &= \sum_{\gamma\in\mathbb{Z}^{2N}} \Phi_0U_\alpha^*AU_{\alpha-\gamma}\Phi_0^*u_{\alpha-\gamma} = \sum_{\gamma\in\mathbb{Z}^{2N}} \Phi_0U_\alpha^*AU_{\alpha-\gamma}\Phi_0^*(\hat{V}_\gamma u)_\alpha,\end{aligned}$$

which shows that $GAG^* \in \mathcal{W}(l^2(\mathbb{Z}^{2N}))$. When applied to the operator $A = I$ (which is in $\mathcal{W}(L^2(\mathbb{R}^N))$ by Proposition 4.3.4), this inclusion implies in particular that $Q = GG^* \in \mathcal{W}(l^2(\mathbb{Z}^{2N}))$. Clearly, the discrete Wiener algebra also contains the identity operator, whence the first assertion.
(b) Let $B \in \mathcal{W}(l^2(\mathbb{Z}^{2N}))$ be given by

$$B = \sum_{\beta\in\mathbb{Z}^{2N}} b_\beta\hat{V}_\beta \quad\text{with}\quad \|B\|_{\mathcal{W}(l^2(\mathbb{Z}^{2N}))} = \sum_{\beta\in\mathbb{Z}^{2N}} \|b_\beta\| < \infty$$

with multiplication operators b_β. Further, let $\alpha, \gamma \in \mathbb{Z}^{2N}$ and $u \in L^2(\mathbb{R}^N)$. Then

$$\begin{aligned}\Phi_\alpha G^*BG\Phi_{\alpha-\gamma}^*u &= \sum_{\delta\in\mathbb{Z}^{2N}} \Phi_\alpha U_\delta\Phi_0^*(BG\Phi_{\alpha-\gamma}^*u)_\delta\\ &= \sum_{\delta\in\mathbb{Z}^{2N}} \Phi_\alpha U_\delta\Phi_0^* \sum_{\beta\in\mathbb{Z}^{2N}} b_\beta(\delta)(G\Phi_{\alpha-\gamma}^*u)_{\delta-\beta}\\ &= \sum_{\delta\in\mathbb{Z}^{2N}} \Phi_\alpha U_\delta\Phi_0^* \sum_{\beta\in\mathbb{Z}^{2N}} b_\beta(\delta)\Phi_0U_{\delta-\beta}^*\Phi_{alpha-\gamma}^*u\\ &= \sum_{\delta\in\mathbb{Z}^{2N}} \Phi_\alpha\Phi_\delta^* \sum_{\beta\in\mathbb{Z}^{2N}} U_\delta b_\beta(\delta)U_{\delta-\beta}^*\Phi_{\delta-\beta}\Phi_{\alpha-\gamma}^*u\end{aligned}$$

whence

$$\begin{aligned}\|\Phi_\alpha G^*BG\Phi_{\alpha-\gamma}^*\| &\le \sum_{\delta\in\mathbb{Z}^{2N}} \|\Phi_\alpha\Phi_\delta^*\| \sum_{\beta\in\mathbb{Z}^{2N}} \|b_\beta\|\,\|\Phi_{\delta-\beta}\Phi_{\alpha-\gamma}^*\|\\ &= \sum_{\beta\in\mathbb{Z}^{2N}} \|b_\beta\| \sum_{\delta\in\mathbb{Z}^{2N}} \|\Phi_\alpha\Phi_\delta^*\|\,\|\Phi_{\delta-\beta}\Phi_{\alpha-\gamma}^*\|.\end{aligned}$$

We write all indices as $\alpha = (\alpha_1, \alpha_2) \in \mathbb{Z}^N \times \mathbb{Z}^N$ and use Proposition 4.3.2 to get

$$\begin{aligned}&\sum_{\delta\in\mathbb{Z}^{2N}} \|\Phi_\alpha\Phi_\delta^*\|\,\|\Phi_{\delta-\beta}\Phi_{\alpha-\gamma}^*\|\\ &\quad\le C\sum_{\delta\in\mathbb{Z}^{2N}} \langle\alpha_1-\delta_1\rangle^{-k}\langle\alpha_2-\delta_2\rangle^{-k}\langle\gamma_1+\delta_1-\alpha_1-\beta_1\rangle^{-k}\times\\ &\qquad\qquad\times\langle\gamma_2+\delta_2-\alpha_2-\beta_2\rangle^{-k}\\ &\quad= C\sum_{\delta_1\in\mathbb{Z}^N} \langle\alpha_1-\delta_1\rangle^{-k}\langle\gamma_1+\delta_1-\alpha_1-\beta_1\rangle^{-k}\times\\ &\qquad\qquad\times\sum_{\delta_2\in\mathbb{Z}^N} \langle\alpha_2-\delta_2\rangle^{-k}\langle\gamma_2+\delta_2-\alpha_2-\beta_2\rangle^{-k}.\end{aligned}$$

If k is large enough, then the sequence $x \mapsto \langle x \rangle^{-k}$ belongs to $l^1(\mathbb{Z}^N)$. Since $l^1(\mathbb{Z}^N)$ is closed under convolution, there is a sequence $f \in l^1(\mathbb{Z}^N)$ such that

$$\|\Phi_\alpha G^* BG\Phi^*_{\alpha-\gamma}\| \leq C \sum_{\beta \in \mathbb{Z}^{2N}} \|b_\beta\| f(\gamma_1 - \beta_1)\, f(\gamma_2 - \beta_2).$$

The sequence $g : (x_1, x_2) \mapsto f(x_1)f(x_2)$ belongs to $l^1(\mathbb{Z}^{2N})$. Hence, by the convolution theorem again,

$$\|\Phi_\alpha G^* BG\Phi^*_{\alpha-\gamma}\| \leq C \sum_{\beta \in \mathbb{Z}^{2N}} \|b_\beta\|\, g(\gamma - \beta) = h(\gamma)$$

with a certain function $h \in l^1(\mathbb{Z}^{2N})$ independent of α and γ. This estimate implies the assertion (b). □

Proposition 4.3.6 *The algebra $\mathcal{W}(L^2(\mathbb{R}^N))$ is inverse closed in $L(L^2(\mathbb{R}^N))$.*

Proof. Let $A \in \mathcal{W}(L^2(\mathbb{R}^N))$ be invertible on $L^2(\mathbb{R}^N)$. Then $\Gamma(A)$ belongs to $\mathcal{W}(l^2(\mathbb{Z}^{2N}))$ by Proposition 4.3.5 (a), and it is invertible in $L(l^2(\mathbb{Z}^{2N}, L^2(\mathbb{R}^N)))$ by Proposition 4.2.2. The inverse closedness of the discrete Wiener algebra (Theorem 2.5.2) implies $\Gamma(A)^{-1} \in \mathcal{W}(l^2(\mathbb{Z}^{2N}))$. Since

$$G^*\Gamma(A)^{-1}GA = G^*\Gamma(A)^{-1}GAG^*G = G^*\Gamma(A)^{-1}\Gamma(A)QG = I,$$

one has $G^*\Gamma(A)^{-1}G = A^{-1} \in \mathcal{W}(L^2(\mathbb{R}^N))$ by Proposition 4.3.5 (b). □

4.3.2 Fredholmness of operators in $\mathcal{W}^\$(L^2(\mathbb{R}^N))$

Operators on $L^2(\mathbb{R}^N)$ which possess a rich operator spectrum are defined in complete analogy to the discrete setting. Recall in this connection the definition of the rich discrete Wiener algebra $\mathcal{W}^\$ =: \mathcal{W}^\$(l^2(\mathbb{Z}^{2N}))$ given in Section 2.5.1.

Definition 4.3.7 *Let $\mathcal{W}^\$(L^2(\mathbb{R}^N))$ stand for the set of all operators A in the Wiener algebra $\mathcal{W}(L^2(\mathbb{R}^N))$ with the following property: every sequence $h \in \mathcal{H}$ possesses a subsequence g such that the limit operator A_g with respect to this sequence exists.*

It can be easily checked that $\mathcal{W}^\$(L^2(\mathbb{R}^N))$ is a closed and unital subalgebra of $\mathcal{W}(L^2(\mathbb{R}^N))$.

Proposition 4.3.8 *Let $A \in \mathcal{W}^\$(L^2(\mathbb{R}^N))$. Then GAG^* and $\Gamma(A)$ belong to the algebra $\mathcal{W}^\$(l^2(\mathbb{Z}^{2N}))$, and*

$$\begin{aligned} \sigma_{op}(GAG^*) &= \{T_g GA_h G^* T_g^* : A_h \in \sigma_{op}(A)\}, \\ \sigma_{op}(\Gamma(A)) &= \{T_g \Gamma(A_h) T_g^* : A_h \in \sigma_{op}(A)\}. \end{aligned}$$

Proof. Let $k \in \mathcal{H}$. Since A has a rich operator spectrum, there is a subsequence h of k such that A_h exists. By the Proposition 4.2.5, there is a subsequence g of h such that the limit operators $(GAG^*)_g$ and $\Gamma(A)_g$ exist. Hence, GAG^* and $\Gamma(A)$ are rich, too. The description of the corresponding operator spectra follows immediately from (4.17) and (4.18). □

The following is the main result of this section.

Theorem 4.3.9 *Let $A \in \mathcal{W}^\$(L^2(\mathbb{R}^N))$. Then A is a Fredholm operator if and only if all limit operators of A are invertible.*

Proof. If A is a Fredholm operator, then all limit operators of A are invertible by (the analogue of) Proposition 1.2.9. Let, conversely, all limit operators of A be invertible. Then, by Propositions 4.3.8 and 4.2.2 (e), all limit operators of $\Gamma(A)$ are invertible. Consequently, $\Gamma(A)$ is a $\mathcal{P}$-Fredholm operator by Theorem 2.5.7. By Proposition 4.2.2 (e) again, A is a Fredholm operator. □

Definition 4.3.10 *Let $\mathcal{A}^\$(L^2(\mathbb{R}^N))$ denote the closure in $L(L^2(\mathbb{R}^N))$ of the rich Wiener algebra $\mathcal{W}^\$(L^2(\mathbb{R}^N))$.*

Theorem 4.3.11 *An operator $A \in \mathcal{A}^\$(L^2(\mathbb{R}^N))$ is Fredholm on $L^2(\mathbb{R}^N)$ if and only if all limit operators of A are uniformly invertible on $L^2(\mathbb{R}^N)$.*

Proof. Let (A_n) be a sequence of operators in $\mathcal{W}^\$(L^2(\mathbb{R}^N))$ which converges to A in the norm. By $\mathcal{B}$ we denote the smallest C^*-subalgebra of $L(L^2(\mathbb{R}^N))$ which contains all operators A_n and the ideal $K(L^2(\mathbb{R}^N))$ of the compact operators. Then the mappings

$$W_h : \mathcal{A}/K(L^2(\mathbb{R}^N)) \to L(L^2(\mathbb{R}^N)), \quad A + K(L^2(\mathbb{R}^N)) \mapsto A_h$$

are well defined for $h \in \mathcal{H}_\mathcal{B}$ and, by Theorem 4.3.9, the collection $\{W_h\}_{h\in\mathcal{H}_\mathcal{B}}$ forms a weakly sufficient family for the dense subalgebra of $\mathcal{B}/K(L^2(\mathbb{R}^N))$ which is generated by the cosets of the operators A_n. Thus, $\{W_h\}_{h\in\mathcal{H}_\mathcal{B}}$ is a weakly sufficient family for the algebra $\mathcal{B}/K(L^2(\mathbb{R}^N))$ by Proposition 2.2.8. □

Corollary 4.3.12 *Let $A \in \mathcal{A}^\$(L^2(\mathbb{R}^N))$. Then*

$$\|A\|_{ess} := \|A + K(L^2(\mathbb{R}^N))\| = \sup\{\|A_h\| : A_h \in \sigma_{op}(A)\}.$$

Proof. The family $\{W_h\}_{h\in\mathcal{H}_\mathcal{B}}$ considered in the previous proof is weakly sufficient for the algebra $\mathcal{B}/K(L^2(\mathbb{R}^N))$. Thus, the result follows from Theorem 2.2.7. □

There are also local versions of Theorems 4.3.9 and 4.3.11. For, we call an operator $A \in L(L^2(\mathbb{R}^N))$ *locally invertible at the infinitely distant point* $\eta \in S^{N-1}$ if there exist a neighborhood at infinity W of η (defined analogously to (2.36)) as well as operators R, $L \in L(L^2(\mathbb{R}^N))$ such that

$$LA\chi_W I = \chi_W AR = \chi_W I.$$

We denote by $\sigma_{op,\eta}(A)$ the set of all limit operators of $A \in B(L^2(\mathbb{R}^N))$ which are defined by sequences $h = (h_1, h_2) : \mathbb{N} \to \mathbb{Z}^N \times \mathbb{Z}^N$ for which h_1 tends to infinity into the direction of η.

Theorem 4.3.13 *Let $A \in \mathcal{W}^\$(L^2(\mathbb{R}^N))$. Then A is locally invertible at the infinitely distant point $\eta \in S^{N-1}$ if and only if all limit operators $A_h \in \sigma_{op,\eta}(A)$ are invertible.*

The proof is similar to the proof of Theorem 4.3.9. An analogous result (with the invertibility of all limit operators in the local operator spectrum replaced by their uniform invertibility) holds for operators in $\mathcal{A}^\$(L^2(\mathbb{R}^N))$.

Finally, we say that $\lambda \in \mathbb{C}$ belongs to the local spectrum $\sigma_\eta(A)$ of the operator A at η if $A - \lambda I$ is not locally invertible at the infinitely distant point $\eta \in S^{N-1}$. The following is a corollary of Theorem 4.3.13.

Theorem 4.3.14 *Let $A \in \mathcal{W}^\$(L^2(\mathbb{R}^N))$. Then*

$$\sigma_\eta(A) = \bigcup_{A_h \in \sigma_{op,\eta}(A)} \sigma(A_h).$$

4.3.3 Fredholm properties of pseudodifferential operators in $OPS^0_{0,0}$

We have seen in Corollary 4.3.3, that every pseudodifferential operator with symbol in $S^0_{0,0}$ belongs to the Wiener algebra. Now we will show that, moreover, these pseudodifferential operators possess a rich operator spectrum. Thus, they become subject to Theorem 4.3.9.

Theorem 4.3.15 $OPS^0_{0,0} \subseteq \mathcal{W}^\$(L^2(\mathbb{R}^N))$.

Proof. Let $a \in S^0_{0,0}$ and $A := Op(a)$, and let $h \in \mathcal{H}$. For $k = (k_1, k_2) \in \mathbb{Z}^N \times \mathbb{Z}^N$, we consider the functions

$$a^{(k)} : \mathbb{R}^N \times \mathbb{R}^N \to \mathbb{R}, \quad (x_1, x_2) \mapsto a(x_1 + k_1, x_2 + k_2).$$

Clearly, $U^*_{h(m)} A U_{h(m)} = Op(a^{(h(m))})$. The sequence $(a^{(h(m))})_{m \in \mathbb{N}} \subseteq C^\infty(\mathbb{R}^N \times \mathbb{R}^N)$ is bounded with respect to the supremum norm. Hence, by the Arzelà-Ascoli theorem, there exists a subsequence g of h such that the functions $a^{(g(m))}$ converge in the topology of $C^\infty(\mathbb{R}^N \times \mathbb{R}^N)$ to a function a_g. It is easy to see that the limit function a_g belongs to $S^0_{0,0}$ and that

$$|a_g|_{k,l} \le |a|_{k,l} \qquad \text{for all } k,\, l \in \mathbb{N}.$$

We set $A_g := Op(a_g)$ and claim that A_g is the limit operator of A with respect to the sequence g, i.e., we claim that

$$\text{s-lim}_{m\to\infty} U^*_{g(m)} A U_{g(m)} = A_g \quad \text{and} \quad \text{s-lim}_{m\to\infty} U^*_{g(m)} A^* U_{g(m)} = A^*_g. \tag{4.21}$$

For the first assertion of (4.21), choose a function $\varphi \in C_0^\infty(\mathbb{R}^N)$ which is equal to 1 in a neighborhood of the origin. Further, for $R > 0$, set $\varphi_R(x) := \varphi(x/R)$, and consider the cut-off functions $\psi_R(x, \xi) := \varphi_R(x)\varphi_R(\xi)$ on $\mathbb{R}^N \times \mathbb{R}^N$. Evidently,

$$\text{s-lim}_{R\to\infty} Op(\psi_R) = I. \tag{4.22}$$

By Proposition 4.1.7, the operator $Op(a)Op(\psi_R)$ is a pseudodifferential operator with formal symbol $c_R \in S^0_{0,0}$, given by the oscillatory integral

$$c_R(x, \xi) = os \int\int_{\mathbb{R}^N} a(x, \xi + \eta)\psi_R(x + y, \xi)e^{-i\langle y, \eta\rangle} dy\, d\eta. \tag{4.23}$$

By means of the Lagrange formula, we write

$$\psi_R(x + y, \xi) = \psi_R(x, \xi) + q_R(x, y, \xi)$$

where $q_R(x, y, \xi) := \sum_{j=1}^N l_{j,R}(x, y, \xi)y_j$ and

$$l_{j,R}(x, y, \xi) := \int_0^1 (\partial_{x_j}\psi_R)(x + \theta y, \xi)\, d\theta.$$

Then we obtain via Proposition 4.1.4

$$os \int\int_{\mathbb{R}^N} a(x, \xi + \eta)e^{-i\langle y, \eta\rangle} dy\, d\eta = p(x, \xi),$$

such that (4.23) can be written as

$$c_R(x, \xi) = a(x, \xi)\psi_R(x, \xi) + t_R(x, \xi)$$

where

$$t_R(x, \xi) = (2\pi)^{-N} \sum_{j=1}^N \int_{\mathbb{R}^N}\int_{\mathbb{R}^N} l_{j,R}(x, y, \xi)(i\partial_{y_j})a(x, \xi + \eta)e^{-i\langle y, \eta\rangle} dy\, d\eta.$$

Simple manipulations yield the estimates

$$\left|\partial_x^\alpha \partial_\xi^\beta t_R(x + g_1(m), \xi + g_2(m))\right| \le C_{\alpha,\beta}\, |a|_{2k_1+|\alpha|, 2k_2+|\beta|}\, (1 + R)^{-1}$$

for all $2k_1 > N$ and $2k_2 > N$, and with a constant $C_{\alpha,\beta}$ independent of a. By the Calderon-Vaillancourt Theorem,

$$Op(t_R^{(g(m))}) \le C|a|_{N_1,N_2}(1 + R)^{-1} \tag{4.24}$$

whenever N_1 and N_2 are sufficiently large. Here we used again the convention

$$t_R^{(g(m))}(x, \xi) := t_R(x + g_1(m), \xi + g_2(m)).$$

Let now $u \in L^2(\mathbb{R}^N)$ and $\varepsilon > 0$. Due to (4.22) and (4.24), we can choose $R_0 > 0$ such that, for all $R > R_0$,

$$\|u - Op(\psi_R)u\| \leq \frac{\varepsilon}{6\|u\|} \quad \text{and} \quad \sup_{m\in\mathbb{N}} \|Op(t_R^{(g(m))})\| \leq \frac{\varepsilon}{3\|u\|}.$$

Thus, for all $m \in \mathbb{N}$,

$$\begin{aligned} \|(U^*_{g(m)} A U_{g(m)} - A_g)u\| &\leq \|(U^*_{g(m)} A U_{g(m)} - A_g) Op(\psi_R)u\| + \varepsilon/3 \\ &\leq \|Op((a^{(g(m))} - a_g)\psi_R)u\| + 2\varepsilon/3. \end{aligned} \tag{4.25}$$

Since the functions $a^{(g(m))} - a_g$ tend to zero in the topology of $C^\infty(\mathbb{R}^N \times \mathbb{R}^N)$, the sequence of the functions $(a^{(g(m))} - a_g)\psi_R$ tends uniformly to zero together with their derivatives. Hence, by the Calderon-Vaillancourt Theorem, there exist an m_0 such that, for all $m > m_0$

$$\|Op((a^{(g(m))} - a_g)\psi_R)\| \leq \frac{\varepsilon}{3\|u\|}. \tag{4.26}$$

Estimates (4.25) and (4.26) imply that, for arbitrary $u \in L^2(\mathbb{R}^N)$ and $\varepsilon > 0$, there exists an m_0 such that

$$\|(U^*_{g(m)} A U_{g(m)} - A_g)u\| < \varepsilon \quad \text{for all } m > m_0.$$

This settles the first assertion of (4.21). For the second one, recall that, by Proposition 4.1.8, the symbol of the adjoint operator is given by the oscillatory integral

$$\mathrm{sym}_{A^*}(x, \xi) = os \int\int_{\mathbb{R}^N} \bar{a}(x + y, \xi + \eta) e^{-i\langle y, \eta\rangle} dy\, d\eta.$$

This implies that, since $a^{(g(m))} \to a_g$ in the topology of $C^\infty(\mathbb{R}^N \times \mathbb{R}^N)$,

$$\begin{aligned} &\mathrm{sym}_{A^*}(x + g_1(m), \xi + g_2(m)) \\ &\quad = os \int\int_{\mathbb{R}^N} \bar{a}(x + g_1(m) + y, \xi + g_2(m) + \eta) e^{-i\langle y, \eta\rangle} dy\, d\eta \\ &\quad \to os \int\int_{\mathbb{R}^N} \bar{a}_g(x + y, \xi + \eta) e^{-i\langle y, \eta\rangle} dy\, d\eta. \end{aligned}$$

Hence, the formal symbols $\mathrm{sym}_A^{(g(m))}$ converge to the formal symbol of the adjoint limit operator A_g^* in the topology of $C^\infty(\mathbb{R}^N \times \mathbb{R}^N)$ as $m \to \infty$. Repeating the above arguments, we obtain also the second assertion of (4.21). □

Remark. Let $a \in S^0_{0,0,0}$. As we know from Proposition 4.1.10, the pseudodifferential operator $A := Op_d(a)$ with double symbol a can be considered as an operator in $OPS^0_{0,0}$. Thus, the results of the previous theorem apply to the operator A, and they yield the following. For $k = (k_1, k_2) \in \mathbb{Z}^N \times \mathbb{Z}^N$, we set

$$a^{(k)}(x, y, \xi) := a(x + k_1, y + k_1, \xi + k_2).$$

Then $U^*_{h(m)} A U_{h(m)} = Op(a^{(h(m))})$, and the sequence h has a subsequence g such that the functions $a^{(g(m))}$ converge to a function a_g in the topology of $C^\infty(\mathbb{R}^N \times \mathbb{R}^N \times \mathbb{R}^N)$ as $m \to \infty$. The limit function a_g belongs to $S^0_{0,0,0}$, and the limit operator of A with respect to the sequence g exists and is equal to $Op(a_g)$. □

Due to Theorem 4.3.15 and the preceding remark, the following results are straightforward consequences of Theorems 4.3.11 and 4.3.14 and of Corollary 4.3.12.

Theorem 4.3.16 *A pseudodifferential operator $A \in OPS^0_{0,0}$ (resp. a pseudodifferential operator $A \in OPS^0_{0,0,0}$ with double symbol) is Fredholm on $L^2(\mathbb{R}^N)$ if and only if all limit operators of A are invertible on $L^2(\mathbb{R}^N)$. Thus,*

$$\sigma_{ess}(A) := \sigma(A + K(L^2(\mathbb{R}^N))) = \cup_{A_h \in \sigma_{op}(A)} \sigma(A_h)$$

and, moreover,

$$\|A\|_{ess} := \|A + K(L^2(\mathbb{R}^N))\| = \inf_{K \in K(L^2(\mathbb{R}^N))} \|A - K\| = \sup_{A_h \in \sigma_{op}(A)} \|A_h\|. \quad (4.27)$$

Theorem 4.3.17 *An operator $A \in OPS^0_{0,0}$ (resp. in $OPS^0_{0,0,0}$) is locally invertible at the infinitely distant point $\eta \in S^{N-1}$ if and only if all limit operators of A in $\sigma_{op,\eta}(A)$ are invertible. In particular,*

$$\sigma_\eta(A) = \cup_{A_h \in \sigma_{op,\eta}(A)} \sigma(A_h).$$

Now we are also in a position to finish the proof of Proposition 4.1.17. Suppose that, contrary to what we want, the operator $Op(a) \in OPS^0_{0,0}$ is compact, but that $\lim_{(x,\,\xi) \to \infty} a(x, \xi) \neq 0$. Then there are sequences (x_m) and (ξ_m) in $\mathbb{R}^N$ which tend to infinity, and for which

$$\sup_{m \in \mathbb{N}} |a(x_m, \xi_m)| \geq \alpha > 0.$$

Let $[y]$ and $\{y\}$ stand for the integer and for the fractional part of $y \in \mathbb{R}^N$, respectively. Without loss we can assume that there are points x_0 and ξ_0 in $[0, 1]^N$ such that

$$(\{x_m\}, \{\xi_m\}) \to (x_0, \xi_0) \quad \text{as } m \to \infty$$

(otherwise we pass to suitable subsequences). Setting $h(m) = (h_1(m), h_2(m)) := ([x_m], [\xi_m])$, we thus get a sequence $h : \mathbb{N} \to \mathbb{Z}^N \times \mathbb{Z}^N$ which tends to infinity and for which

$$\sup_{m \in \mathbb{N}} |a(x_0 + h_1(m), \xi_0 + h_2(m))| \geq \alpha/2 > 0.$$

Let g be a subsequence of h such that the functions $a^{(g(m))}$ converge to a function a_g in the topology of $C^\infty(\mathbb{R}^N \times \mathbb{R}^N)$. Then the limit operator of A with respect to the sequence g exists, and $A_g = Op(a_g)$. But A_g cannot be the zero operator since

$$|a_g(x_0, \xi_0)| = \lim_{m \to \infty} |a(x_0 + g_1(m), \xi_0 + g_2(m) a_0)| \geq \alpha/2 > 0.$$

Thus, by (4.27), the essential norm of $Op(a)$ is positive, i.e., this operator cannot be compact. Contradiction. □

4.4 Applications

Here we specify the Fredholm results obtained in the previous section to pseudodifferential operators with symbols in particular function spaces.

4.4.1 Operators with slowly oscillating symbols

A symbol $a \in S^0_{0,0}$ is called *slowly oscillating with respect to* x if

$$\lim_{x\to\infty} \sup_{\xi\in\mathbb{R}^N} |\partial_{x_j} a(x,\, \xi)| = 0 \quad \text{for all } j = 1,\, \ldots,\, N,$$

and a is *slowly oscillating with respect to* ξ if

$$\lim_{\xi\to\infty} \sup_{x\in\mathbb{R}^N} |\partial_{\xi_j} a(x,\, \xi)| = 0 \quad \text{for all } j = 1,\, \ldots,\, N.$$

Proposition 4.4.1 *Let the function $a \in S^0_{0,0}$ be slowly oscillating with respect to x. Then every limit operator of $A := Op(a)$, which is defined with respect to a sequence $h = (h_1,\, h_2) : \mathbb{N} \to \mathbb{Z}^N \times \mathbb{Z}^N$ with $h_1(m) \to \infty$ as $m \to \infty$, is a pseudodifferential operator $Op(a_h)$ with a symbol independent of x. In particular, $Op(a_h)$ is shift invariant and, thus, a convolution operator. Similarly, if a is slowly oscillating with respect to ξ, and if $h_2(m) \to \infty$ as $m \to \infty$, then the limit operator $Op(a_h)$ has a symbol independent of ξ and is, thus, a multiplication operator.*

Proof. We will prove the first assertion only. Let a be slowly oscillating with respect to x. As we have seen in the proof of Theorem 4.3.15, the symbol a_h of the limit operator is the $C^\infty(\mathbb{R}^N \times \mathbb{R}^N)$-limit of the functions

$$a^{(h(m))} : \mathbb{R}^N \times \mathbb{R}^N \to \mathbb{R}, \quad (x,\, \xi) \mapsto a(x + h_1(m),\, \xi + h_2(m)).$$

Since, for fixed $x',\, x'' \in \mathbb{R}^N$,

$$\begin{aligned} &|a^{(h(m))}(x',\, \xi) - a^{(h(m))}(x'',\, \xi)| \\ &\quad \le \sum_{j=1}^N |x'_j - x''_j| \int_0^1 |\partial_{x_j}(a^{(h(m))}((1-t)x' + tx'',\, \xi)|\, dt \to 0 \end{aligned}$$

as $m \to \infty$, the function a_h does not depend on x. □

The most simple (and, perhaps, most important) pseudodifferential operators with slowly oscillating symbols are those whose symbols are slowly oscillating with respect to x and ξ simultaneously. We denote this class of symbols by $SO^0_{0,0}$ and the corresponding of pseudodifferential operators by $OPSO^0_{0,0}$. For operators in this class, all limit operators are operators of convolution or operators of multiplication (indeed, if the sequence $h = (h_1,\, h_2)$ tends to infinity, then at least one of the sequences h_1 and h_2 goes to infinity, too). For both kinds of limit operators, their invertibility can be easily checked.

Theorem 4.4.2 *Let $a \in SO^0_{0,0}$. Then all limit operators of $Op(a)$ are invertible if and only if*

$$\lim_{R\to\infty} \inf_{|x|+|\xi|\geq R} |a(x,\,\xi)| > 0. \tag{4.28}$$

Proof. Let condition (4.28) be satisfied, and let $h = (h_1,\, h_2)$ be a sequence which defines a limit operator of $Op(a)$. Further assume for definiteness that the sequence h_1 tends to infinity (the case when $h_2 \to \infty$ can be treated similarly). Then, as we have seen in Proposition 4.4.1, the limit operator $Op(a)_h$ is shift invariant, i.e., there is a function a_h in $S^0_{0,0}$ which is independent of x such that $Op(a)_h = Op(a_h)$. Moreover, the functions

$$a^{(h(m))} : (x,\,\xi) \mapsto a(x + h_1(m),\, \xi + h_2(m))$$

converge to the function $(x,\,\xi) \mapsto a_h(\xi)$ in the topology of $C^\infty(\mathbb{R}^N \times \mathbb{R}^N)$ as $m \to \infty$. Thus, for each $L > 0$,

$$\lim_{m\to\infty} \sup_{|x|+|\xi|\leq L} |a(x + h_1(m),\, \xi + h_2(m)) - a_h(\xi)| = 0. \tag{4.29}$$

From (4.28) and (4.29) we conclude that $\inf_\xi |a_h(\xi)| > 0$, i.e., the limit operator $Op(a)_h$ is invertible.

To prove the reverse statement, suppose that all limit operators of $Op(a)$ are invertible, but that condition (4.28) is not fulfilled. Then there exists a sequence $h = (h_1,\, h_2) : \mathbb{N} \to \mathbb{Z}^N \times \mathbb{Z}^N$ which tends to infinity and for which $a(h_1(m),\, h_2(m)) \to 0$. Without loss we can assume that the limit operator of $Op(a)$ with respect to h exists (otherwise we choose a suitable subsequence of h). We further assume for definiteness that $h_1 \to \infty$ (the case when $h_2 \to \infty$ follows similarly). Then, as before, $Op(a)_h = Op(a_h)$ with a function a_h independent of x and such that the functions $a^{(h(m))}$ converge to a_h in $C^\infty(\mathbb{R}^N \times \mathbb{R}^N)$. It follows from $a(h_1(m),\, h_2(m)) \to 0$ that $a_h(0) = 0$ which contradicts the invertibility of $Op(a_h)$. □

Corollary 4.4.3 *An operator $Op(a) \in OPSO^0_{0,0}$ is Fredholm if and only if condition (4.28) holds. Moreover,*

$$\|Op(a)\|_{ess} = \lim_{R\to\infty} \sup_{|x|+|\xi|\geq R} |a(x,\,\xi)|.$$

The proof of the first assertion follows from the previous result and from Theorem 4.3.16. For the second assertion, recall the proof of Corollary 4.3.12. □

These results admit generalizations to pseudodifferential operators with double symbols. For, we call the double symbol $a \in S^0_{0,0,0}$ *slowly oscillating* and write $a \in SO^0_{0,0,0}$ if, for arbitrary compact sets $K \subset \mathbb{R}^N$ and for all $j = 1, \ldots, N$,

$$\lim_{x\to\infty} \sup_{(y,\,\xi)\in K\times\mathbb{R}^N} |\partial_{x_j} a(x,\, x+y,\, \xi)| = 0$$

and

$$\lim_{\xi\to\infty} \sup_{(x,\,y)\in\mathbb{R}^N\times\mathbb{R}^N} |\partial_{\xi_j} a(x,\, y,\, \xi)| = 0.$$

Proposition 4.4.4

(a) *Let $a \in SO^0_{0,0,0}$, and let $h = (h_1, h_2)$ be a sequence with $h_1 \to \infty$ for which the limit operator $Op_d(a)_h$ exists. Then this limit operator is of the form $Op(a_h)$ where a_h is the limit in the topology of $C^\infty(\mathbb{R}^N \times \mathbb{R}^N)$ of the functions*

$$(x, \xi) \mapsto a(x + h_1(m), x + h_1(m), \xi + h_2(m))$$

as $m \to \infty$. The function a_h is independent of x in this case.

(b) *Let $a \in SO^0_{0,0,0}$, and let $h = (h_1, h_2)$ be a sequence with $h_2 \to \infty$ for which the limit operator $Op_d(a)_h$ exists. Then this limit operator is of the form $Op(a_h)$ where a_h is the limit in the topology of $C^\infty(\mathbb{R}^N \times \mathbb{R}^N)$ of the functions*

$$(x, \xi) \mapsto a(x + h_1(m), x + h_1(m), \xi + h_2(m))$$

as $m \to \infty$. The function a_h is independent of ξ in this case.

Proof. Let us check assertion (b) for example. The symbol a_h of the limit operator of $Op_d(a)$ with respect to h is defined as the limit as $m \to \infty$ of the oscillatory integrals

$$os \int\int_{\mathbb{R}^N} a(x + h_1(m), x + h_1(m) + y, \xi + h_2(m) + \eta) e^{-i\langle y, \eta\rangle} \, dy \, d\eta.$$

Thus,

$$a_h(x, x) = os \int\int_{\mathbb{R}^N} a_h(x, x + y) e^{-i\langle y, \eta\rangle} dy \, d\eta$$

by Proposition 4.1.4. □

One can also prove as in Theorem 4.4.2 and its Corollary 4.4.3 that if $a \in SO^0_{0,0,0}$, then all limit operators of $Op_d(a)$ are invertible if and only if

$$\lim_{R \to \infty} \sup_{|x|+|\xi| \geq R} |a(x, x, \xi)| > 0. \tag{4.30}$$

Hence, condition (4.30) is necessary and sufficient for Fredholmness of $Op_d(a)$, and

$$\|Op_d(a)\|_{ess} = \lim_{R \to \infty} \inf_{|x|+|\xi| \geq R} |a(x, x, \xi)|.$$

4.4.2 Operators with almost periodic symbols

A function a in $C_b(\mathbb{R}^N)$ is called *almost periodic* if the set $\{V_r a : r \in \mathbb{R}^N\}$ of all shifts of a is relatively compact in $C_b(\mathbb{R}^N)$, i.e., if every sequence in this set has a norm convergent subsequence. Here, $V_r a$ stands for the function $x \mapsto a(x - r)$. The class of all almost periodic functions will be denoted by $AP(\mathbb{R}^N)$. Note that $AP(\mathbb{R}^N)$ is a C^*-algebra with respect to the supremum norm. Nice references to this class are [97, 102].

We further set $AP^\infty(\mathbb{R}^N) := AP(\mathbb{R}^N) \cap C_b^\infty(\mathbb{R}^N)$, and we denote by $\mathfrak{A}^0_{0,0}$ the closure in $S^0_{0,0}$ of the class of all functions of the form

$$a(x, \xi) = \sum_{j=1}^{J} c_j(x) b_j(\xi) \tag{4.31}$$

where $J \in \mathbb{N}$, $c_j \in AP^\infty(\mathbb{R}^N)$ and $b_j \in SO^0_{0,0}$. Pseudodifferential operators with symbols in this class possess limit operators with respect to the shifts V_k where the convergence is in the operator norm.

Proposition 4.4.5 *Let $A \in OP\mathfrak{A}^0_{0,0}$. Then each sequence $h : \mathbb{N} \to \mathbb{Z}^N$ which tends to infinity has a subsequence g such that there exists an operator $A_g \in OP\mathfrak{A}^0_{0,0}$ with*

$$\lim_{m\to\infty} \|V_{-g(m)} A V_{g(m)} - A_g\| = 0.$$

Proof. To start with, let $A = Op(a)$ where $a \in \mathfrak{A}^0_{0,0}$ is a symbol of the form (4.31), and let $h \in \mathcal{H}$. Since the functions c_k are almost periodic (and by a simple diagonal argument), there are a subsequence g of h as well as functions $c_{jg} \in AP(\mathbb{R}^N)$ such that

$$\lim_{m\to\infty} \sup_{x\in\mathbb{R}^N} |c_j(x + g(m)) - c_{jg}(x)| = 0 \tag{4.32}$$

for $1 \le j \le J$. Applying the inequality

$$\sup_{\mathbb{R}^N} \sum_{|\alpha|=1} |\partial^\alpha a(x)| \le C \left(\sup_{x\in\mathbb{R}^N} |a(x)| \left(\sup_{x\in\mathbb{R}^N} |a(x)| + \sup_{x\in\mathbb{R}^N} \sum_{|\alpha|=2} |\partial^\alpha a(x)| \right) \right)^{1/2}$$

(see, for instance, [164], p. 22), one obtains that the sequence of the shifted functions $V_{g(m)} c_j$ converges to c_{jg} in the topology of $C_b^\infty(\mathbb{R}^N)$, which implies that $c_{jg} \in AP^\infty(\mathbb{R}^N)$. Now set

$$A_g := Op(a_g) \quad \text{with} \quad a_g(x, \xi) := \sum_{j=1}^{J} c_{jg}(x) b_j(\xi).$$

Then it follows from (4.32) that indeed

$$\lim_{m\to\infty} \|V_{-g(m)} A V_{g(m)} - A_g\| = 0.$$

This settles the assertion for operators $A = Op(a)$ where a is of the form (4.31). The general case follows straightforwardly by a Cantor diagonalization procedure and standard continuity arguments. □

One can also easily check that $A_g \in OP\mathfrak{A}^0_{0,0}$ again and that A_g is a limit operator of A defined by the sequence $h : m \mapsto (g(m), 0) \in \mathbb{Z}^N \times \mathbb{Z}^N$ and with respect to the shift operators $U_{h(m)}$ (cf. Section 4.2.3).

The following results can be proved as their discrete counterparts Theorem 2.6.2 and 2.6.3.

Theorem 4.4.6 *Let $A \in OP\mathfrak{A}^0_{0,0}$. Then the following assertions are equivalent:*

(a) *A is a Fredholm operator.*

(b) *All limit operators of A are invertible.*

(c) *At least one limit operator of A is invertible.*

(d) *A is an invertible operator.*

Theorem 4.4.7 *The smallest closed subalgebra of $L(L^2(\mathbb{R}^N))$ which contains all operators in $OP\mathfrak{A}^0_{0,0}$ does not contain nonzero compact operators.*

We are now going to sketch briefly how these results specialize to symbols in a subclass of $\mathfrak{A}^0_{0,0}$, in which case the Fredholmness of the operator together with its uniform ellipticity and a certain index condition yields the invertibility of the operator.

We say that the function $a \in S^0_{0,0}$ belongs to $S^0_{1,0}$ if

$$|a|_l := \sum_{|\alpha|+|\beta|\le l} \sup_{(x,\,\xi)\in\mathbb{R}^N\times\mathbb{R}^N} |\partial_x^\beta \partial_\xi^\alpha a(x,\,\xi)|\langle\xi\rangle^{|\alpha|} < \infty$$

for all non-negative integers l. The semi-norms $|.|_l$ define the topology of $S^0_{1,0}$. Further, we consider the class $\mathfrak{A}^0_{1,0}$ which is the closure in $S^0_{1,0}$ of the set of all symbols of form (4.31) where the c_j satisfy the estimates

$$|\partial^\alpha c_j(\xi)| \le C_{\alpha,\,k}\langle\xi\rangle^{-|\alpha|}$$

for all multi-indices α. Finally, an operator $Op(a) \in OP\mathfrak{A}^0_{1,0}$ is called *uniformly elliptic* if

$$\lim_{R\to\infty} \inf_{x,\,\xi\in\mathbb{R}^N,\,|\xi|>R} |a(x,\,\xi)| > 0. \tag{4.33}$$

It is easy to see that an operator $Op(a) \in OP\mathfrak{A}^0_{1,0}$ is uniformly elliptic if and only if all limit operators of A defined by sequences $(g_1,\, g_2) : \mathbb{N} \to \mathbb{Z}^N \times \mathbb{Z}^N$ with $g_2 \to \infty$ are invertible. Thus, the uniform ellipticity is a *necessary* condition for the invertibility of $Op(a)$. An analogous result holds for almost periodic operators with matrix-valued symbols, where one has to replace the value $a(x,\,\xi)$ in (4.33) by $\det a(x,\,\xi)$.

Let now $A \in OP\mathfrak{A}^0_{1,0}$ be a uniformly elliptic operator with $M \times M$-matrix-valued coefficients. Then the difference between its Fredholmness and its invertibility is measured by its *almost periodic index* $\kappa(A)$. This index has been introduced in [39] (see also [54]) by means of Breuer's Fredholm theory for II_∞ factors. In distinction to the usual (Fredholm) index, $\kappa(A)$ can be an arbitrary real number. We will not go into the details and restrict ourselves to rephrasing a few basic properties:

- If $A,\, B \in OP\mathfrak{A}^0_{1,0}$ are uniformly elliptic operators, then $\kappa(AB) = \kappa(A)\kappa(B)$.
- The almost periodic index is stable in the following sense. Given a uniformly elliptic operator $Op(a) \in OP\mathfrak{A}^0_{1,0}$, there exists an $\varepsilon > 0$ such that $\kappa(Op(b)) =$

$\kappa(Op(a))$ for all operators $Op(b) \in OP\mathfrak{A}^0_{1,0}$ with

$$\lim_{R\to\infty} \sup_{x,\,\xi\in\mathbb{R}^N,\,|\xi|>R} \|a(x,\,\xi) - b(x,\,\xi)\|_{L(\mathbb{C}^M)} < \varepsilon.$$

- If $A \in OP\mathfrak{A}^0_{1,0}$ is invertible, then $\kappa(A) = 0$.
- Let $A \in OP\mathfrak{A}^0_{1,0}$ be uniformly elliptic and $\kappa(A) = 0$. Then A is invertible if and only if
$$\nu(A) := \inf_{\|\varphi\|\le 1} \|A\varphi\| > 0.$$
- Let $A \in OP\mathfrak{A}^0_{1,0}$ be a scalar uniformly elliptic operator, and let $N > 1$. Then $\kappa(A) = 0$. Thus, for such operators, the condition $\nu(A) > 0$ is necessary and sufficient for invertibility of A.

The condition $\nu(A) > 0$ is satisfied if and only if the operator A has a trivial kernel and a closed range. Hence, if $A \in OP\mathfrak{A}^0_{1,0}$ is a scalar uniformly elliptic and Fredholm operator with $\kappa(A) = 0$, then A is invertible.

4.4.3 Operators with semi-almost periodic symbols

The class $\mathfrak{B}^0_{1,0}$ of the *semi-almost periodic symbols with respect to* x is defined as the closure in the topology of $S^0_{1,0}$ of the set of all functions of the form

$$a(x,\,\xi) = \sum_{j=1}^{J} c_j(x) b_j(x,\,\xi)$$

where $J \in \mathbb{N}$, $c_j \in AP^\infty(\mathbb{R}^N)$ and $b_j \in SO^0_{1,0} := S^0_{1,0} \cap SO^0_{0,0}$.

Theorem 4.4.8 *Let $N > 1$, and let $a \in \mathfrak{B}^0_{1,0}$. Then the operator $A := Op(a)$ is a Fredholm operator if and only if the following conditions are satisfied:*

(a) *A is uniformly elliptic, that is*

$$\lim_{R\to\infty} \inf_{x,\,\xi\in\mathbb{R}^N,\,|\xi|>R} |a(x,\,\xi)| > 0.$$

(b) *For each limit operator A_g of A which is defined by a sequence $g = (g_1,\, g_2) : \mathbb{N} \to \mathbb{Z}^N \times \mathbb{Z}^N$ with $g_2 \to \infty$, one has $\nu(A_g) > 0$.*

Proof. Let conditions (a) and (b) be satisfied. In the same way as in the proof of Theorem 4.4.2, we obtain that condition (a) implies the invertibility of all limit operators of A which correspond to sequences $g = (g_1,\, g_2)$ with $g_2 \to \infty$. Let now $g = (g_1,\, g_2)$ be a sequence with $g_1 \to \infty$ for which the limit operator A_g exists. Then, by the definition of the class $\mathfrak{B}^0_{1,0}$, this limit operator belongs to $OP\mathfrak{A}^0_{1,0}$ and, due to condition (a), the operator A_g is uniformly elliptic with $\kappa(A_g) = 0$ (since A is an operator with scalar-valued symbol). It follows from the last remark in the preceding section that A_g is invertible if the lower norm $\nu(A_g)$ is positive. Thus, conditions (a) and (b) provide us with the invertibility of all limit operators of A. By Theorem 4.3.16, A is a Fredholm operator.

Let, conversely, A be a Fredholm operator. Then, by Theorem 4.3.16 again, all limit operators of A are invertible. The invertibility of all limit operators with respect to sequences $g = (g_1, g_2)$ with $g_2 \to \infty$ yields the uniform ellipticity of A, that is condition (a), whereas the invertibility of all limit operators corresponding to sequences $g = (g_1, g_2)$ with $g_1 \to \infty$ evidently implies condition (b). □

4.4.4 Pseudodifferential operators of nonzero order

Let $a \in S^m_{0,0}$. By Theorem 4.1.18, the pseudodifferential operator $A := Op(a)$ acts as a bounded linear operator from $H^{s+m}(\mathbb{R}^N)$ into $H^s(\mathbb{R}^N)$ for every $s \in \mathbb{R}$. We are going to study the Fredholm properties of that operator by reducing it in a standard way to a pseudodifferential operator acting on $H^0(\mathbb{R}^N) = L^2(\mathbb{R}^N)$. For, let $\langle D \rangle^r$ refer to the pseudodifferential operator with symbol $(x, \xi) \mapsto (1 + |\xi|_2^2)^{r/2}$. The operator $\langle D \rangle^r$ is an isometry from $H^{s+r}(\mathbb{R}^N)$ onto $H^s(\mathbb{R}^N)$ for each real s. Thus,

$$A : H^{s+m}(\mathbb{R}^N) \to H^s(\mathbb{R}^N)$$

is a Fredholm operator if and only if

$$B := \langle D \rangle^s A \langle D \rangle^{-s-m} : L^2(\mathbb{R}^N) \to L^2(\mathbb{R}^N)$$

is a Fredholm operator. The operator B is a pseudodifferential operator in the class $OPS^0_{0,0}$ due to Proposition 4.1.7. Hence, Theorem 4.3.16 implies the following.

Theorem 4.4.9 *Let $a \in S^m_{0,0}$. Then the operator $A = Op(a) : H^{s+m}(\mathbb{R}^N) \to H^s(\mathbb{R}^N)$ is Fredholm operator if and only if all limit operators of the operator $B := \langle D \rangle^s A \langle D \rangle^{-s-m} : L^2(\mathbb{R}^N) \to L^2(\mathbb{R}^N)$ are invertible. In particular,*

$$\sigma_{ess}(A) = \cup_{B_h \in \sigma_{op}(B)} \sigma(B_h).$$

These conditions can be made more explicit for symbols which are slowly oscillating in the following sense.

Definition 4.4.10 *Let $m \in \mathbb{N}$. We say that the function a is in the class $SO^m_{0,0}$ if the function $(x, \xi) \mapsto a(x, \xi)\langle \xi \rangle^{-m}$ belongs to $SO^0_{0,0}$. Analogously, the double symbol a is said to be in $SO^m_{0,0,0}$ if the function $(x, y, \xi) \mapsto a(x, y, \xi)\langle \xi \rangle^{-m}$ belongs to $SO^0_{0,0,0}$.*

Proposition 4.4.11

(a) *If $A := Op(a) \in OPSO^{m_1}_{0,0}$ and $B := Op(b) \in OPSO^{m_2}_{0,0}$, then $AB \in OPSO^{m_1+m_2}_{0,0}$, and the formal symbol of AB is of the form* $\mathrm{sym}_{AB} = ab + t$ *with t satisfying*

$$\lim_{(x,\, \xi) \to \infty} t(x, \xi)\langle \xi \rangle^{-m_1-m_2} = 0. \tag{4.34}$$

(b) *Let* $A := Op_d(a) \in OPSO^m_{0,0,0}$. *Then* $A \in OPS^m_{0,0}$, *and the formal symbol of that operator is given by* $\mathrm{sym}_A(x,\xi) := a(x,x,\xi) + t(x,\xi)$ *where* t *is such that*

$$\lim_{(x,\,\xi)\to\infty} t(x,\xi)\langle\xi\rangle^{-m} = 0.$$

Proof. (a) By Proposition 4.1.3, the operator AB belongs to $OPSO^{m_1+m_2}_{0,0}$, and its formal symbol is given by

$$\mathrm{sym}_{AB}(x,\xi) = os \int\!\!\int_{\mathbb{R}^N} a(x,\xi+\eta)b(x+y,\xi)e^{-i\langle y,\eta\rangle}\,dy\,d\eta.$$

By Lagrange's formula, we have

$$a(x,\xi+\eta) = a(x,\xi) + \sum_{j=1}^N \eta_j \int_0^1 \partial_{\xi_j} a(x,\xi+\theta\eta)\,d\theta,$$

whence via Proposition 4.1.4

$$\mathrm{sym}_{AB}(x,\xi) = a(x,\xi)b(x,\xi) + t(x,\xi),$$

with

$$t(x,\xi) = \sum_{j=1}^N \int_0^1 L_j(x,\xi,\theta)\,d\theta$$

and

$$\begin{aligned} &L_j(x,\xi,\theta)\\ &\quad = os \int\!\!\int_{\mathbb{R}^N} \partial_{\xi_j} a(x,\xi+\theta\eta)(-i\partial_{x_j})b(x+y,\xi)e^{-i\langle y,\eta\rangle}\,dy\,d\eta\\ &\quad = os \int\!\!\int_{\mathbb{R}^N} \langle\eta\rangle^{-2k_2}\langle D_y\rangle^{2k_2}\\ &\qquad\times \left\{\langle y\rangle^{-2k_1}\langle D_\eta\rangle^{2k_1}\partial_{\xi_j} a(x,\xi+\theta\eta)(-i\partial_{x_j})b(x+y,\xi))\right\} e^{-i\langle y,\xi\rangle}\,dy\,d\eta \end{aligned}$$

for all k_1, k_2 with $2k_1 > N$ and $2k_2 > N+|m_1|$. Taking into account the elementary inequality

$$\langle\xi+\eta\rangle^l \le 2^{l/2}\langle\eta\rangle^{|l|}\langle\xi\rangle^l,$$

holding for arbitrary $l \in \mathbb{R}$, we obtain

$$L_j(x,\xi,\theta) \le C\langle\xi\rangle^{m_1+m_2} K_j(x,\xi,\theta)$$

where

$$\begin{aligned} &K_j(x,\xi,\theta)\\ &\quad = os \int\!\!\int_{\mathbb{R}^N} \langle\xi+\theta\eta\rangle^{-m_1}\langle\eta\rangle^{-2k_2+|m_1|}\\ &\qquad\times \left|\langle D_y\rangle^{2k_2}\langle y\rangle^{-2k_1}\langle D_\eta\rangle^{2k_1}\partial_{\xi_j} a(x,\xi+\theta\eta)(-i\partial_{x_j})b(x+y,\xi)\langle\xi\rangle^{-m_2})\right|\,dy\,d\eta. \end{aligned}$$

The latter integral converges uniformly with respect to $x, \xi \in \mathbb{R}^N$ and $\theta \in [0, 1]$. Hence, we can pass to the limit as $(x, \xi) \to \infty$ under this integral, which yields

$$\lim_{(x,\,\xi)\to\infty} \sup_{\theta\in[0,\,1]} K_j(x, \xi, \theta) = 0.$$

This implies (4.34). Assertion (b) can be checked in the same way. □

It follows from this proposition that if $A = Op(a) \in OPS^m_{0,0}$, then

$$B := \langle D\rangle^s A\langle D\rangle^{-(s+m)} = Op(a_m) + Op(t)$$

where

$$a_m(x, \xi) := a(x, \xi)\langle\xi\rangle^{-m} \quad \text{and} \quad \lim_{(x,\,\xi)\to\infty} t(x, \xi) = 0.$$

Thus, all limit operators B_g of B depend on the part a_m of the symbol of B only. Moreover, these limit operators are pseudodifferential operators $B_g = Op(b_g)$ which are invariant with respect to shifts (i.e., their symbols b_g depend on ξ only), or they are operators of multiplication (i.e., their symbols are only dependent on x). So we arrive at the following theorem.

Theorem 4.4.12

(a) *Consider the operator* $A = Op(a) \in OPSO^m_{0,0}$ *as acting from* $H^{s+m}(\mathbb{R}^N)$ *into* $H^s(\mathbb{R}^N)$. *Then all limit operators of* $B := \langle D\rangle^s A\langle D\rangle^{-(s+m)} : L^2(\mathbb{R}^N) \to L^2(\mathbb{R}^N)$ *are invertible if and only if*

$$\lim_{R\to\infty} \inf_{|x|+|\xi|\geq R} |a(x, \xi)|\langle\xi\rangle^{-m} > 0. \tag{4.35}$$

The condition (4.35) *is necessary and sufficient for the Fredholmness of A.*

(b) *Consider the operator* $A = Op_d(a) \in OPSO^m_{0,0,0}$ *as acting from* $H^{s+m}(\mathbb{R}^N)$ *into* $H^s(\mathbb{R}^N)$. *Then all limit operators of* $B := \langle D\rangle^s A\langle D\rangle^{-(s+m)} : L^2(\mathbb{R}^N) \to L^2(\mathbb{R}^N)$ *are invertible if and only if*

$$\lim_{R\to\infty} \inf_{|x|+|\xi|\geq R} |a(x, x, \xi)|\langle\xi\rangle^{-m} > 0. \tag{4.36}$$

Condition (4.36) *is necessary and sufficient for the Fredholmness of A.*

4.4.5 Differential operators

The results of the previous section apply to study the Fredholmness of differential operators on $\mathbb{R}^N$ by means of their limit operators. Let

$$P = \sum_{|\alpha|\leq m} a_\alpha D^\alpha$$

be a differential operator of order m with coefficients $a_\alpha \in C_b^\infty(\mathbb{R}^N)$. We consider this operator as acting from $H^{s+m}(\mathbb{R}^N)$ into $H^s(\mathbb{R}^N)$. The function

$$p_m : \mathbb{R}^N \times \mathbb{R}^N \to \mathbb{R}, \quad (x, \xi) \mapsto \sum_{|\alpha|=m} a_\alpha(x)\xi^\alpha$$

is called the *main symbol* of P, and the operator P is called *uniformly elliptic* if

$$\inf_{x\in\mathbb{R}^N} |p_m(x, \omega)| > 0 \quad \text{for all } \omega \in S^{N-1}.$$

Let $h : \mathbb{N} \to \mathbb{Z}^N$ be a sequence which tends to infinity. Then there exist a subsequence g of h and functions $a_\alpha^g \in C_b^\infty(\mathbb{R}^N)$ such that the functions $x \mapsto a_\alpha(x+g(k))$ converge to a_α^g in the topology of $C_b^\infty(\mathbb{R}^N)$ for every α. We set

$$P_g := \sum_{|\alpha|\leq m} a_\alpha^g D^\alpha,$$

consider P_g as an operator from $H^{s+m}(\mathbb{R}^N)$ into $H^s(\mathbb{R}^N)$ again, and denote by $\sigma_{op}^1(P)$ the set of all operators which arise in this way.

Theorem 4.4.13 *The differential operator $P : H^{s+m}(\mathbb{R}^N) \to H^s(\mathbb{R}^N)$ is Fredholm if and only if the following conditions are satisfied:*

(a) *All operators $P_g \in \sigma_{op}^1(P)$ are invertible.*

(b) *The operator P is uniformly elliptic.*

Proof. It follows from Theorem 4.4.9 that P is a Fredholm operator if and only if all limit operators of $\langle D\rangle^s P\langle D\rangle^{-s-m}$ are invertible on $L^2(\mathbb{R}^N)$.

Let $h = (h_1, h_2) : \mathbb{N} \to \mathbb{Z}^N \times \mathbb{Z}^N$ be a sequence such that $h_1 \to \infty$ but h_2 is bounded. Then there exists a subsequence $g = (g_1, g_2)$ of h such that, for every α, the functions $x \mapsto a_\alpha(x + g_1(k))$ converge to certain functions $a_\alpha^{g_1}$ in the topology of $C_b^\infty(\mathbb{R}^N)$ and that the sequence g_2 is constant, say $g_2(k) = \gamma_2 \in \mathbb{Z}^N$ for all k. In this case, it is easy to see that

$$\text{s-lim}_{k\to\infty} U_{g(k)}^* \langle D\rangle^s P\langle D\rangle^{-s-m} U_{g(k)} = E_{\gamma_2}^* \langle D\rangle^s P_{g_1} \langle D\rangle^{-s-m} E_{\gamma_2}$$

with $(E_\gamma u)(x) := e^{i\langle\gamma, x\rangle} u(x)$. Thus, the limit operators of $\langle D\rangle^s P\langle D\rangle^{-s-m}$ which are defined by sequences of this kind are invertible if and only if condition (a) holds.

Now consider limit operators of $\langle D\rangle^s P\langle D\rangle^{-s-m}$ which are defined by sequences $g = (g_1, g_2)$ such that $g_2 \to \infty$ and g_1 is constant, say $g_1(k) = \gamma_1 \in \mathbb{Z}^N$. Suppose for definiteness that g_2 tends to infinity into the direction of the infinitely distant point $\omega \in S^{N-1}$. Then

$$\text{s-lim}_{k\to\infty} E_{g_2(k)}^* \langle D\rangle^s P\langle D\rangle^{-s-m} E_{g_2(k)} = p_m(\,.\,, \omega)I$$

whence

$$\text{s-lim}_{k\to\infty} U^*_{g(k)} \langle D\rangle^s P \langle D\rangle^{-s-m} U_{g(k)} = \sum_{|\alpha|=m} a(\,.-\gamma_1)\omega^\alpha I.$$

Hence, all limit operators defined by these sequences are operators of multiplication by the functions

$$p_{m,\,g} : (x,\,\omega) \mapsto \sum_{|\alpha|=m} a(.-\gamma_1)\omega^\alpha.$$

Finally, if both g_1 and g_2 go to infinity, and if g_1 and g_2 are chosen such that the functions $x \mapsto a_\alpha(x + g_1(k))$ converge to certain functions $a_\alpha^{g_1}$ in the topology of $C^\infty(\mathbb{R}^N)$ and that g_2 tends to infinity into the direction of the infinitely distant point $\omega \in S^{N-1}$, then

$$\text{s-lim}_{k\to\infty} U^*_{g(k)} \langle D\rangle^s P \langle D\rangle^{-s-m} U_{g(k)} = \sum_{|\alpha|=m} a_\alpha^{g_1} \omega^\alpha I.$$

Thus, we get multiplication operators again, this time by the functions

$$p_{m,\,g} : (x,\,\omega) \mapsto \sum_{|\alpha|=m} a_\alpha^{g_1}(x)\omega^\alpha.$$

Evidently, if the operator is uniformly elliptic, then in all cases

$$\inf_{x\in\mathbb{R}^N} |p_{m,\,g}(x,\,\omega)| > 0.$$

Hence, the limit operators $Op(p_{m,\,g})I$ are invertible on $L^2(\mathbb{R}^N)$, and condition (b) implies the invertibility of all limit operators defined by sequences g with $g_2 \to \infty$. Conversely, choosing sequences $g = (g_1,\, g_2)$ with $g_1(k) = 0$ for all k and with g_2 tending to infinity into the direction of $\omega \in S^{N-1}$, we obtain that the invertibility of all associated limit operators implies condition (b). □

Corollary 4.4.14 *Let $P : H^m(\mathbb{R}^N) \to L^2(\mathbb{R}^N)$ be a uniformly elliptic differential operator of order m. Then*

$$\sigma_{ess}(P) = \cup_{P_g \in \sigma^1_{op}(P)} \sigma(P_g).$$

Proof. By Theorem 4.4.9, the operator $P - \lambda I : H^m(\mathbb{R}^N) \to L^2(\mathbb{R}^N)$ is Fredholm if and only if all limit operators

$$P_g - \lambda I : H^m(\mathbb{R}^N) \to L^2(\mathbb{R}^N), \qquad P_g \in \sigma^1_{op}(P),$$

are invertible and if $P - \lambda I$ is uniformly elliptic. Since the uniform ellipticity of a differential operator depends on its main symbol only, the uniform ellipticity of $P - \lambda I$ follows from the conditions of the corollary. □

We denote by $SO^\infty(\mathbb{R}^N)$ the class of the smooth slowly oscillating functions on $\mathbb{R}^N$, that is the class of all functions $a \in C_b^\infty(\mathbb{R}^N)$ with

$$\lim_{x\to\infty} \partial_{x_j} a(x) = 0 \quad \text{for all } j = 1, \ldots, N.$$

Let the coefficients a_α of the differential operator P belong to $SO^\infty(\mathbb{R}^N)$. Then all limit operators $P_g \in \sigma^1_{op}(P)$ are of the form

$$P_g = Op(p_g) = \sum_{|\alpha|\le m} a_\alpha^{g_1} D^\alpha$$

with *constant* coefficients $a_\alpha^{g_1}$. The operator P_g is invertible if and only if

$$\inf_{\xi\in\mathbb{R}^N} |p_g(\xi)|\langle\xi\rangle^{-m} = \inf_{\xi\in\mathbb{R}^N} \left|\sum_{|\alpha|\le m} a_\alpha^{g_1}\xi^\alpha\right| \langle\xi\rangle^{-m} > 0.$$

Hence, if P is a differential operator with smooth slowly oscillating coefficients, then

$$\sigma_{ess}(P) = \bigcup_{P_g\in\sigma^1_{op}(P)} \overline{\{p_g(\xi) : \xi \in \mathbb{R}^N\}}.$$

Remark. A differential operator P of order m can be considered as an unbounded operator on the Hilbert space $L^2(\mathbb{R}^N)$ with domain $H^m(\mathbb{R}^N)$. If P is uniformly elliptic, then P is a closed operator. An unbounded operator P is called a *Fredholm operator* if its range is closed in $L^2(\mathbb{R}^N)$ and if $\ker A$ and $\ker A^*$ are finite-dimensional spaces, and the *essential spectrum* $\sigma_{ess}(A)$ *of* A consists of all $\lambda \in \mathbb{C}$ for which $A - \lambda I$ is not a Fredholm operator.

It is well known that if P is uniformly elliptic, then P is a Fredholm operator in this sense (i.e., as an unbounded operator) if and only if $P : H^m(\mathbb{R}^N) \to L^2(\mathbb{R}^N)$ is a Fredholm operator in the common sense (i.e., as a bounded operator). Hence, if P is a uniformly elliptic differential operator, then

$$\sigma_{ess}(P) = \cup_{P_g\in\sigma^1_{op}(P)}\sigma(P_g),$$

where now both the essential spectrum on the left-hand side and the spectra on the right-hand side are understood in the unbounded operator sense. □

4.4.6 Schrödinger operators

Here we are going to specialize the results of the previous section to operators of the form

$$\mathbf{H} = \sum_{l,\,m=1}^{N} (i\partial_{x_l} + a_l I) g^{lm} (i\partial_{x_m} + a_m I) + wI$$

where g^{lm}, a_l and w are real-valued functions in $C_b^\infty(\mathbb{R}^N)$. This operator can be viewed of as the electro-magnetic Schrödinger operator on the Riemann space

$\mathbb{R}^N$ provided with the metric tensor $(g_{lm})_{l,\,m=1}^N$ which is the tensor inverse of $(g^{lm})_{l,\,m=1}^N$. Schrödinger operators of this form arise in multi-particle problems after separating the mass center of the system (see, for instance, [41], pp. 29–33 and [79], pp. 172–176).

Throughout this section, we will suppose that

$$\inf_{x\in\mathbb{R}^N,\eta\in S^{N-1}} \sum_{l,\,m=1}^N g_{lm}(x)\eta_l\eta_m > 0.$$

Let $h : \mathbb{N} \to \mathbb{Z}^N$ be a sequence which tends to infinity. Then there exists a subsequence k of h such that the functions

$$x \mapsto g^{lm}(x + k(n)), \quad x \mapsto a_l(x + k(n)) \quad \text{and} \quad x \mapsto w(x + k(n))$$

converge in the topology of $C_b^\infty(\mathbb{R}^N)$ to certain functions g_k^{lm}, a_l^k and w^k, respectively. In particular, these limit functions belong to $C_b^\infty(\mathbb{R}^N)$ again. If k is chosen in this way, then the limit operator $\mathbf{H}_k$ of $\mathbf{H}$ with respect to k exists, and

$$\mathbf{H}_k = \sum_{l,\,m=1}^N (i\partial_{x_l} + a_l^k I)g_k^{lm}(i\partial_{x_m} + a_m^k I) + w^k I.$$

We consider $\mathbf{H}$ as an unbounded operator on $L^2(\mathbb{R}^N)$ with domain $H^2(\mathbb{R}^N)$. Note that $\lambda \in \mathbb{C}$ is a point in the discrete spectrum of the unbounded operator $\mathbf{H}$ if and only if λ belongs to the discrete spectrum of the bounded operator $\mathbf{H} : H^2(\mathbb{R}^N) \to L^2(\mathbb{R}^N)$. Hence, the essential spectrum of $\mathbf{H}$, considered as an unbounded operator, coincides with the essential spectrum of the bounded operator $\mathbf{H} : H^2(\mathbb{R}^N) \to L^2(\mathbb{R}^N)$. With Corollary 4.4.14, we find

$$\sigma_{ess}(\mathbf{H}) = \bigcup_{\mathbf{H}_k\in\sigma_{op}^1(\mathbf{H})} \sigma(\mathbf{H}_k). \tag{4.37}$$

Here are a few instances where the structure of the limit operators is sufficiently simple such that their invertibility can be effectively checked.

Example A. Let the functions g^{lm}, a_l and w be in $SO^\infty(\mathbb{R}^N)$. Then each limit operator of $\mathbf{H}$ is a differential operator with constant coefficients, i.e.,

$$\mathbf{H}_k = \sum_{l,\,m=1}^N (i\partial_{x_l} + a_l^k I)g_k^{lm}(i\partial_{x_m} + a_m^k I) + w^k I.$$

with real numbers g_k^{lm}, a_l^k and w^k. Set $a^k := (a_1^k, \ldots, a_N^k)$ and $(E_\alpha u)(x) := e^{i\langle\alpha,\,x\rangle}u(x)$ for $\alpha \in \mathbb{R}^N$. Then

$$E_{a^k}\mathbf{H}_k E_{a^k}^{-1} = -\sum_{l,\,m=1}^N g_k^{lm}\partial_{x_l}\partial_{x_m} + w^k I.$$

Thus,

$$\sigma(\mathbf{H}_k) = \left\{ \sum_{l,m=1}^{N} g_k^{lm} \xi_l \xi_m + w^k : (\xi_1, \ldots, \xi_N) \in \mathbb{R}^N \right\} = [w^k, +\infty],$$

and the essential spectrum of $\mathbf{H}$ is

$$\sigma_{ess}(\mathbf{H}) = \bigcup [w^k, +\infty] = [m_w, +\infty]$$

where $m_w := \inf w^k = \liminf_{x \in \mathbb{R}^N} w(x)$. □

Example B. We let v_1, v_2 and v_{12} be C^∞-functions on $\mathbb{R}^3$ with

$$\lim_{y \to \infty} v_1(y) = \lim_{y \to \infty} v_2(y) = \lim_{y \to \infty} v_{12}(y) = 0,$$

define functions w_1, w_2, w_{12} on $\mathbb{R}^3 \times \mathbb{R}^3$ by

$$w_1(x) := v_1(x^{(1)}), \quad w_2(x) := v_2(x^{(2)}), \quad w_{12}(x) := v_{12}(x^{(1)} - x^{(2)})$$

where $x = (x^{(1)}, x^{(2)}) \in \mathbb{R}^3 \times \mathbb{R}^3$, and consider the Hamiltonian on $L^2(\mathbb{R}^3 \times \mathbb{R}^3)$,

$$\mathbf{H} := -\Delta_{x^{(1)}} - \Delta_{x^{(2)}} - w_1 I - w_2 I - w_{12} I.$$

Hamiltonians of this special structure arise in nuclear physics (but, usually, with non-smooth functions v_1, v_2 and v_{12}, which moreover will have singularities at 0; see, for instance, [44], p. 163, and [41], p. 29).

We will describe the essential spectrum of $\mathbf{H}$ by means of its limit operators. Let $h := (h_1, h_2) : \mathbb{N} \to \mathbb{Z}^3 \times \mathbb{Z}^3$ tend to infinity. After passing to suitable subsequences of h, if necessary, we have to distinguish between four cases.

[A] We have $h_1 \to \infty$, and h_2 is a constant sequence, say $h_2(k) = \gamma_2 \in \mathbb{Z}^N$ for all k. Then the limit operator of $\mathbf{H}$ with respect to h exists, and

$$(\mathbf{H}_h u)(x) = -(\Delta_{x^{(1)}} u)(x) - (\Delta_{x^{(2)}} u)(x) - w_2(x^{(2)} + \gamma_2) u(x).$$

The operator $\mathbf{H}_h$ is unitarily equivalent to the operator

$$\mathbf{H}^1 := -\Delta_{x^{(1)}} - \Delta_{x^{(2)}} - w_2 I.$$

[B] If $h_2 \to \infty$, and if $h_1(k) = \gamma_1 \in \mathbb{Z}^N$ for all k, then the limit operator of $\mathbf{H}$ with respect to h exists, and it is unitarily equivalent to the operator

$$\mathbf{H}^2 := -\Delta_{x^{(1)}} - \Delta_{x^{(2)}} - w_1 I.$$

[C] If both h_1 and h_2 tend to infinity, and if also $h_1 - h_2 \to \infty$, then the limit operator of $\mathbf{H}$ is equal to the Laplacian

$$\mathbf{H}^3 := -\Delta_{x^{(1)}} - \Delta_{x^{(2)}}.$$

[D] If, finally, h_1 and h_2 tend to infinity, and if the difference $h_1 - h_2$ is a constant sequence, then the limit operator of $\mathbf{H}$ with respect to h exists, and it is unitarily equivalent to the operator

$$\mathbf{H}^4 := -\Delta_{x^{(1)}} - \Delta_{x^{(2)}} - w_{12}I.$$

Let $j = 1, 2$. Applying the Fourier transform with respect to $x^{(j)}$, we obtain that the operator $\mathbf{H}^j$ is unitarily equivalent to the operator of multiplication by the operator-valued function

$$\widehat{\mathbf{H}}^j : \mathbb{R}^3 \to L(L^2(\mathbb{R}^3 \times \mathbb{R}^3)), \quad \xi \mapsto |\xi|^2 - \Delta_{x^{(3-j)}} - w_{3-j}I.$$

It is well known that the essential spectrum of the operator $A_j := -\Delta_{x^{(3-j)}} - w_{3-j}I$ is the interval $[0, \infty)$ and that its discrete spectrum consists of finitely many points in $(-\infty, 0)$. Let $\lambda_{\min}^{(j)} < 0$ be the minimal eigenvalue of A_j. Then, since $|\xi|^2$ varies over $[0, \infty)$, the spectrum of $\mathbf{H}^j$ is the interval $[\lambda_{\min}^{(j)}, \infty)$.

Now consider the operator $\mathbf{H}^4$. After a change of variables

$$y^{(1)} := x^{(1)} + x^{(2)}, \quad y^{(2)} := x^{(1)} - x^{(2)},$$

the operator $\mathbf{H}^4$ becomes

$$-2(\Delta_{y^{(1)}} + \Delta_{y^{(2)}}) - \hat{w}_{12}I$$

with $\hat{w}_{12}(y) := v_{12}(y^{(2)})$. The spectrum of this operator is the interval $[\lambda_{\min}^{(12)}, \infty)$ where $\lambda_{\min}^{(12)} < 0$ is the minimal eigenvalue of $-2\Delta_{y^{(2)}} - \hat{w}_{12}I$.

Summarizing, we get from (4.37) that

$$\sigma_{ess}(\mathbf{H}) = [\lambda_{\min}, \infty) \quad \text{where} \quad \lambda_{\min} := \min\{\lambda_{\min}^{(1)}, \lambda_{\min}^{(2)}, \lambda_{\min}^{(12)}\}.$$

4.4.7 Partial differential-difference operators

Consider differential-difference operators of the form

$$P := \sum_{|\alpha| \le m,\, j \le N} a_{\alpha j} D^\alpha V_{\beta_{\alpha j}}$$

where $(V_\beta u)(x) = u(x - \beta)$ for $\beta \in \mathbb{R}^N$ and where the coefficients $a_{\alpha j}$ belong to $SO^\infty(\mathbb{R}^N)$. The operator P is a pseudodifferential operator in the class $OPS_{0,0}^m$ with symbol

$$p(x, \xi) := \sum_{|\alpha| \le m,\, j \le N} a_{\alpha j}(x) \xi^\alpha e^{i\langle \beta_{\alpha j}, \xi \rangle}.$$

Hence, $P : H^m(\mathbb{R}^N) \to L^2(\mathbb{R}^N)$ is a Fredholm operator if and only if all limit operators of the operator $Q := P\langle D \rangle^{-m} : L^2(\mathbb{R}^N) \to L^2(\mathbb{R}^N)$ are invertible.

Let $h = (h_1, h_2) : \mathbb{N} \to \mathbb{Z}^N \times \mathbb{Z}^N$ be a sequence tending to infinity which defines a limit operator of Q. We distinguish between three cases for the sequence h.

[A] Let $h_1 \to \infty$, and let h_2 tend to infinity into the direction of the infinitely distant point $\eta \in S^{N-1}$. Then the limit operator of Q is a difference operator with constant coefficients of the form

$$Q_h := \sum_{|\alpha|=m,\, j\leq N} a^h_{\alpha j}\eta^\alpha V_{\beta_{\alpha j}},$$

i.e., with numbers $a^h_{\alpha j} \in \mathbb{C}$. It is evident that Q_h is invariant with respect to shifts, and this operator is invertible if and only if

$$\inf_{\xi\in\mathbb{R}^N} \left| \sum_{|\alpha|=m,\, j\leq N} a^h_{\alpha j}\eta^\alpha e^{i\langle\beta_{\alpha j},\,\xi\rangle} \right| > 0.$$

[B] Let $h_1 \to \infty$, and let h_2 be a constant sequence. Then the limit operator Q_h is unitarily equivalent to the pseudodifferential operator with symbol

$$q_h : \xi \mapsto \sum_{|\alpha|\leq m,\, j\leq N} a^h_{\alpha j}\frac{\xi^\alpha}{\langle\xi\rangle^m} e^{i\langle\beta_{\alpha j},\,\xi\rangle}.$$

Clearly, this operator is invertible if and only if

$$\inf\left\{|q_h(\xi)| : \xi \in \mathbb{R}^N\right\} > 0.$$

[C] Finally, let h_2 tend to infinity into the direction of the infinitely distant point $\eta \in S^{N-1}$, and let h_1 be a constant sequence. Then the limit operator Q_h is unitarily equivalent to the difference operator with variable coefficients,

$$\sum_{|\alpha|=m,\, j\leq N} a_{\alpha j}\eta^\alpha V_{\beta_{\alpha j}}.$$

Effective sufficient conditions for the invertibility of difference operators with variable coefficients can be found in the monographs [7, 8, 9].

4.5 Mellin pseudodifferential operators

In this section, we are going to summarize some facts on Fourier and Mellin pseudodifferential operators with analytic symbols on weighted Sobolev spaces. These results will be employed in the forthcoming section to study the Fredholmness of singular integral operators on Carleson curves.

4.5.1 Pseudodifferential operators with analytic symbols

Definition 4.5.1 *Let B be an open and convex domain in $\mathbb{R}^N$ which contains the origin. The class $S^m_{0,0,0}(B)$ consists of all functions*

$$p \in S^m_{0,0,0}, \qquad (x,\, y,\, \xi) \mapsto p(x,\, y,\, \xi)$$

which possess an analytic extension with respect to ξ *into the tube domain* $\mathbb{R}^N + iB$, *and for which*

$$\sup \sum_{|\alpha|\le l_1,\,|\beta|\le l_2,\,|\gamma|\le l_3} \langle\xi\rangle^{-m} |\partial_x^\beta \partial_y^\gamma \partial_\xi^\alpha p(x,\,y,\,\xi + i\eta)| < \infty \tag{4.38}$$

for all $l_1,\ l_2,\ l_3 \in \mathbb{N}$, *where the supremum is taken over all quadruples* $(x,\,y,\,\xi,\,\eta) \in \mathbb{R}^N \times \mathbb{R}^N \times \mathbb{R}^N \times B$.

We agree upon denoting the analytic continuation of $p \in S^m_{0,0,0}(B)$ by p again. As above, we associate with each double symbol $p \in S^m_{0,0,0}(B)$ a pseudodifferential operator $Op_d(p)$. The class of all pseudodifferential operators obtained in this way is denoted by $OPS^m_{0,0,0}(B)$.

Definition 4.5.2 *A* weight in the class $\Lambda(B)$ *is a function* $w : \mathbb{R}^N \to [0,\,\infty)$ *of the form* $x \mapsto e^{v(x)}$ *where* $v \in C^\infty(\mathbb{R}^N)$ *is subject to the conditions*

(a) $N_l(v) := \sup_{x\in\mathbb{R}^N} \sum_{|\beta|\le l} |\partial^\beta (\nabla v)(x)| < \infty$ *for all* $l \in \mathbb{N}$, *and*

(b) $(\nabla v)(x) \in B$ *for all* $x \in \mathbb{R}^N$.

Given a function v as in this definition, we define

$$g_v : \mathbb{R}^N \times \mathbb{R}^N, \quad (x,\,y) \mapsto \int_0^1 (\nabla v)(x - t(x-y))\,dt.$$

It is easy to check that g_v is a function in $C_b^\infty(\mathbb{R}^N \times \mathbb{R}^N)$ with values in B and that

$$\sum_{|\alpha|\le l_1,\,|\beta|\le l_2} |\partial_x^\alpha \partial_y^\beta g_v(x,\,y)| \le N_{l_1+l_2}(v).$$

The following theorem is a key result for the study of pseudodifferential operators in the class $OPS^m_{0,0,0}(B)$ on exponentially weighted Sobolev spaces.

Theorem 4.5.3 *If* $p \in S^m_{0,0,0}(B)$ *and* $w \in \Lambda(B)$, *then the operator* $wOp_d(p)w^{-1}I$ *belongs to* $OPS^m_{0,0,0}$, *and*

$$wOp_d(p)w^{-1}I = Op_d(p_v) \quad \text{with} \quad p_v(x,\,y,\,\xi) = p(x,\,y,\,\xi + ig_v(x,\,y)).$$

Proof. The inclusion $p_v \in S^m_{0,0,0}$ can be checked simply by differentiating. Let us show that $wOp_d(p)w^{-1}I = Op_d(p_v)$.

Let $u \in C_0^\infty(\mathbb{R}^N)$. By means of the Lagrange formula

$$v(x) - v(y) = \langle x - y,\, g_v(x,\,y)\rangle,$$

we have

$$\begin{aligned}&(wOp_d(p)w^{-1}u)(x)\\ &\quad = \lim_{\varepsilon\to 0} os \int\!\!\int_{\mathbb{R}^N} e^{-i\langle x-y,\,\xi - ig_v(x,\,y)\rangle - \varepsilon\xi^2} p(x,\,y,\,\xi) u(y)\,dy\,d\xi.\end{aligned} \tag{4.39}$$

For every $\varepsilon > 0$, the function under the integral sign in (4.39) has a compact support Ω with respect to y, and it is exponentially decreasing with respect to ξ as $\xi \to \infty$. Hence, the integral (4.39) can be written as

$$(wOp_d(p)w^{-1}u)(x) = \lim_{\varepsilon\to 0}(2\pi)^{-N}\int_\Omega J(\varepsilon, x, y)u(y)\,dy$$

where

$$J(\varepsilon, x, y) := \int_{\mathbb{R}^N} p(x, y, \xi)e^{-i\langle x-y, \xi - ig_v(x, y)\rangle - \varepsilon\xi^2}\,d\xi.$$

Substituting $\tau = \tau(\xi) = \xi - ig_v(x, y)$ in the latter integral, we obtain

$$J(\varepsilon, x, y) = \int_{\mathbb{R}^N - ig_v(x, y)} p(x, y, \tau + ig_v(x, y))e^{-i\langle x-y, \tau\rangle - \varepsilon(\tau + ig_v(x, y))^2}\,d\tau. \quad (4.40)$$

Notice that this change of variables is justified because $g_v(x, y) \in B$ for $x, y \in \mathbb{R}^N$. Since $(x, y, \zeta) \mapsto p(x, y, \zeta)$ is an analytic function with respect to $\zeta \in \mathbb{R}^N + iB$, and since $g_v(x, y) \in B$ for all $x, y \in \mathbb{R}^N$, we obtain that $(x, y, \tau) \mapsto p(x, y, \tau + ig_v(x, y))$ is an analytic function with respect to τ running through the layer

$$T_{x, y, \varepsilon} := \mathbb{R}^N + i\{-tg_v(x, y) : t \in (-\varepsilon, 1 + \varepsilon)\},$$

whenever $\varepsilon > 0$ is sufficiently small. Thus, the function under the integral sign in (4.40) is analytic with respect to $\tau \in T_{x, y, \varepsilon}$. Since this function is moreover exponentially decreasing with respect to $\operatorname{Im}\tau \in \operatorname{Im} T_{x, y, \varepsilon}$ as $\operatorname{Re}\tau \to \infty$, and since the layer $T_{x, y, \varepsilon}$ contains $\mathbb{R}^N$, the Cauchy-Poincaré theorem (see [185], p. 233) justifies to replace the integration along $\mathbb{R}^N - ig_v(x, y)$ in (4.40) by integration along $\mathbb{R}^N$. Hence,

$$J(\varepsilon, x, y) = \int_{\mathbb{R}^N} p(x, y, \tau + ig_v(x, y))e^{-i\langle x-y, \tau\rangle - \varepsilon(\tau + ig_v(x, y))^2}\,d\tau,$$

whence

$$\begin{aligned}(wOp_d(p)w^{-1}u)(x) &= \lim_{\varepsilon\to 0} os \int\int_{\mathbb{R}^N} e^{-i\langle x-y, \tau\rangle - \varepsilon(\tau + ig_v(x, y))^2} p(x, y, \tau + ig_v(x, y))u(y)\,dy\,d\tau \\ &= (2\pi)^{-N}\int_{\mathbb{R}^N}\int_{\mathbb{R}^N} e^{-i\langle x-y, \tau\rangle} p(x, y, \tau + ig_v(x, y))u(y)\,dy\,d\tau\end{aligned}$$

and $wOp(p)w^{-1}I = Op_d(p_v)$. □

Let $w \in \Lambda(B)$ and $s \in \mathbb{R}$. We say that a function u lies in the *weighted Sobolev space* $H^s_w(\mathbb{R}^N)$ if $wu \in H^s(\mathbb{R}^N)$. The norm of u in $H^s_w(\mathbb{R}^N)$ is defined as the norm of wu in $H^s(\mathbb{R}^N)$.

Proposition 4.5.4 *Let $P \in OPS^m_{0,0,0}(B)$ and $w \in \Lambda(B)$. Then P is a bounded operator from $H^{s+m}_w(\mathbb{R}^N)$ to $H^s_w(\mathbb{R}^N)$.*

The proof follows immediately from Proposition 4.1.10 and Theorem 4.1.18. The next result is a consequence of Theorem 4.4.9.

Theorem 4.5.5 *Let $p \in S^m_{0,0,0}(B)$ and $w = \exp v \in \Lambda(B)$. Then the operator $Op_d(p) : H^{s+m}_w(\mathbb{R}^N) \to H^s_w(\mathbb{R}^N)$ is Fredholm if and only if all limit operators of the operator*

$$\langle D\rangle^s Op_d(p_v)\langle D\rangle^{-s-m} : L^2(\mathbb{R}^N) \to L^2(\mathbb{R}^N)$$

are invertible.

The condition for the Fredholmness of $Op_d(p) : H^{s+m}_w(\mathbb{R}^N) \to H^s_w(\mathbb{R}^N)$ can be checked effectively if the symbol is slowly oscillating, and if also the weight function is slowly oscillating in the following sense.

Definition 4.5.6 *The weight function $w = \exp v \in \Lambda(B)$ is* slowly oscillating *and belongs to $S\Lambda(B)$ if*

$$\lim_{x\to\infty} (\partial_{x_i}\partial_{x_j} v)(x) = 0 \quad \textit{for all } i,\, j = 1,\, \ldots,\, N.$$

For example, if $B_a := \{x \in \mathbb{R}^N : |x|_2 < a\}$, then the weight functions $x \mapsto e^{a|x|_2}$ and $x \mapsto e^{a\langle x\rangle \sin(\log\langle x\rangle)}$ belong to $S\Lambda(B_{a\sqrt{2}})$.

Proposition 4.5.7 *Let $p \in SO^m_{0,0,0}(B) := SO^m_{0,0,0} \cap S^m_{0,0,0}(B)$ and $w = \exp v \in S\Lambda(B)$. Then*

$$wOp_d(p)w^{-1}I = Op(p_{[v]}) + Op(t) \tag{4.41}$$

where $p_{[v]}(x,\, \xi) := p(x,\, x,\, \xi + i(\nabla v)(x))$ and

$$\lim_{(x,\,\xi)\to\infty} t(x,\, \xi)\langle\xi\rangle^{-m} = 0.$$

Proof. From Theorem 4.5.3 we know that $wOp_d(p)w^{-1}I = Op_d(p_v)$. The function p_v belongs to $OPSO^m_{0,0,0}$ under the hypotheses of the proposition, as one easily checks by differentiating. Thus, by Proposition 4.4.11, $Op_d(p_v) = Op(\tilde{p}) + Op(t)$ with

$$\tilde{(p)}(x,\, \xi) := p_v(x,\, x,\, \xi) = p(x,\, x,\, \xi + ig_v(x,\, x)),$$

and with a function t such that

$$\lim_{(x,\,\xi)\to\infty} t(x,\, \xi)\langle\xi\rangle^{-m} = 0.$$

Since $g_v(x,\, x) = (\nabla v)(x)$, this is the assertion. □

Theorem 4.5.8 *Let $a \in SO^m_{0,0,0}(B)$ and $w = \exp v \in S\Lambda(B)$. Then the operator*

$$Op_d(a) : H^{s+m}_w(\mathbb{R}^N) \to H^s_w(\mathbb{R}^N)$$

is Fredholm if and only if all limit operators of the pseudodifferential operator with symbol $(x, \xi) \mapsto a(x, x, \xi + i(\nabla v)(x))\langle\xi\rangle^{-m}$ are invertible on $L^2(\mathbb{R}^N)$. Equivalently, $Op_d(a)$ is Fredholm if and only if

$$\lim_{R\to\infty} \inf_{|x|+|\xi|>R} |a(x, x, \xi + i(\nabla v)(x))|\langle\xi\rangle^{-m} > 0.$$

This is an evident consequence of the preceding proposition and Theorem 4.4.12.

4.5.2 Mellin pseudodifferential operators

Let $\mathbb{R}_+ := (0, \infty)$, and denote by $L^2(\mathbb{R}_+, d\mu)$ the Hilbert space with norm

$$\|u\|_{L^2(\mathbb{R}_+, d\mu)} := \left(\int_{\mathbb{R}_+} |u(r)|^2 \, d\mu(r)\right)^{1/2} \quad \text{with } \mu(r) = \frac{dr}{r}.$$

Note that $\mathbb{R}_+$ is a commutative locally compact group with respect to multiplication, and that $d\mu$ is just the Haar measure on this group. The Fourier transform associated with $(\mathbb{R}_+, d\mu)$ is called the *Mellin transform*. It acts via

$$M : L^2(\mathbb{R}_+, d\mu) \to L^2(\mathbb{R}), \quad (Mf)(\lambda) = \int_{\mathbb{R}_+} f(r) r^{-i\lambda} \frac{dr}{r},$$

and its inverse is given by

$$M^{-1} : L^2(\mathbb{R}) \to L^2(\mathbb{R}_+, d\mu), \quad (M^{-1}g)(r) = \frac{1}{2\pi} \int_{\mathbb{R}} g(\lambda) r^{i\lambda} \, d\lambda.$$

If $a \in L^\infty(\mathbb{R})$, then the operator $C_M(a) := M^{-1}aM$ is bounded on $L^2(\mathbb{R}_+, d\mu)$. It is referred to as the *Mellin convolution operator* with symbol a.

Definition 4.5.9 *An operator $A \in L(L^2(\mathbb{R}_+, d\mu))$ is called* locally invertible at the point 0 *if there exist an $\varepsilon > 0$ and operators L_ε, $R_\varepsilon \in L(L^2(\mathbb{R}_+, d\mu))$ such that*

$$L_\varepsilon A \chi_{(0,\varepsilon]} I = \chi_{(0,\varepsilon]} I \quad \text{and} \quad \chi_{(0,\varepsilon]} A R_\varepsilon = \chi_{(0,\varepsilon]} I$$

where $\chi_{(0,\varepsilon]}$ stands for the characteristic function of the interval $(0, \varepsilon]$. The set of all $\lambda \in \mathbb{C}$ for which the operator $A - \lambda I$ is not locally invertible at 0 is called the local spectrum *of A at 0. We denote it by $\sigma_0(A)$.*

Theorem 4.5.10 *If $a \in L^\infty(\mathbb{R})$ then*

$$\sigma_0(C_M(a)) = \sigma(C_M(a)) = R(a) \tag{4.42}$$

where $R(a)$ refers to the essential range of a, i.e., to the spectrum of a considered as an element of the C^*-algebra $L^\infty(\mathbb{R})$.

Proof. Let $A := C_M(a)$. It is evident that $\sigma_0(A) \subseteq \sigma(A) = R(a)$. To prove the inclusion $\sigma(A) \subseteq \sigma_0(A)$, assume that $A - \lambda I$ is locally invertible at 0 for some $\lambda \in \mathbb{C}$. Then there exists an $\varepsilon > 0$ and a constant $C > 0$ such that

$$\|(A - \lambda I)\chi_{(0,\,\varepsilon]} f\| \geq C\|\chi_{(0,\,\varepsilon]} f\| \quad \text{for each } f \in C_0^\infty(\mathbb{R}_+). \tag{4.43}$$

For $\delta > 0$, let

$$Z_\delta : L^2(\mathbb{R}_+, d\mu) \to L^2(\mathbb{R}_+, d\mu), \quad (Z_\delta u)(r) := u(r/\delta).$$

This operator is unitary on $L^2(\mathbb{R}_+, d\mu)$. Given $f \in C_0^\infty(\mathbb{R}_+)$, choose $\delta > 0$ small enough such that the support of $Z_\delta f$ lies in $(0, \varepsilon]$. Then (4.43) implies

$$\|(A - \lambda I)Z_\delta f\| \geq C\|Z_\delta f\|.$$

Since $(MZ_\delta u)(\lambda) = \delta^{i\lambda}(Mu)(\lambda)$, the Mellin convolution operator A commutes with Z_δ, and since Z_δ is a unitary operator,

$$\|(A - \lambda I)f\| \geq C\|f\| \quad \text{for all } f \in C_0^\infty(\mathbb{R}_+).$$

Analogously,

$$\|(A^* - \overline{\lambda} I)f\| \geq C\|f\| \quad \text{for all } f \in C_0^\infty(\mathbb{R}_+).$$

The latter two estimates imply the invertibility of $A - \lambda I$. Thus, $\lambda \notin \sigma(A)$. □

Next we are going to introduce function spaces $\mathcal{E}$ and $\mathcal{E}_d$, which are the analogues of the Hörmander classes $S^0_{0,0}$ and $S^0_{0,0,0}$, as well as classes of operators on $L^2(\mathbb{R}_+, d\mu)$ with symbols in $\mathcal{E}$ or $\mathcal{E}_d$ which correspond to the (Fourier) pseudodifferential operators. We say that a complex-valued function $a \in C^\infty(\mathbb{R}_+ \times \mathbb{R})$ belongs to $\mathcal{E}$ if

$$\sup_{(r,\,\lambda) \in \mathbb{R}_+ \times \mathbb{R}} |(r\partial_r)^\alpha \partial_\lambda^\beta a(r, \lambda)| < \infty$$

for all nonnegative integers α and β. For $a \in \mathcal{E}$, the operator $Op_M(a)$ defined at $u \in C_0^\infty(\mathbb{R}_+)$ by

$$(Op_M(a)u)(r) := (2\pi)^{-1} \int_{\mathbb{R}} \int_{\mathbb{R}_+} a(r, \lambda)\, (r\rho^{-1})^{i\lambda} u(\rho)\rho^{-1}\, d\rho\, d\lambda$$

is called the *Mellin pseudodifferential operator with symbol a*. The class of all operators of this kind is denoted by $OP\mathcal{E}$. Further, let $\mathcal{E}_d$ denote the set of all complex-valued functions $a \in C^\infty(\mathbb{R}_+ \times \mathbb{R}_+ \times \mathbb{R})$ which satisfy the conditions

$$\sup_{(r,\,\rho,\,\lambda) \in \mathbb{R}_+ \times \mathbb{R}_+ \times \mathbb{R}} |(r\partial_r)^\alpha (\rho\partial_\rho)^\beta \partial_\lambda^\gamma a(r, \rho, \lambda)| < \infty$$

for all non-negative integers α, β, γ. For $a \in \mathcal{E}_d$, the operator $Op_{M,d}(a)$ defined at $u \in C_0^\infty(\mathbb{R}_+)$ by

$$(Op_{M,d}(a)u)(r) := (2\pi)^{-1} \int_{\mathbb{R}} \int_{\mathbb{R}_+} a(r, \rho, \lambda)\, (r\rho^{-1})^{i\lambda} u(\rho)\rho^{-1}\, d\rho\, d\lambda$$

is called the *Mellin pseudodifferential operator with double symbol a*. The class of all these operators will be denoted by $OP\mathcal{E}_d$. That Mellin pseudodifferential operators are indeed the analogues of the (Fourier) pseudodifferential operators on $L^2(\mathbb{R})$ can be seen as follows. Let T be the operator

$$T : L^2(\mathbb{R}) \to L^2(\mathbb{R}_+, d\mu), \quad (Tu)(r) := u(\log r).$$

Then T is unitary, and

$$OP\mathcal{E} = T(OPS^0_{0,0})T^{-1} \quad \text{as well as} \quad OP\mathcal{E}_d = T(OPS^0_{0,0,0})T^{-1}.$$

Thus, the properties of the Mellin pseudodifferential operators follow straightforwardly from the corresponding properties of (Fourier) pseudodifferential operators on $L^2(\mathbb{R})$. Some of these properties are summarized in the following theorem.

Theorem 4.5.11

(a) *Let $a \in \mathcal{E}$. Then $Op_M(a)$ is bounded on $L^2(\mathbb{R}_+, d\mu)$, and there is a constant C such that*

$$\|Op_M(a)\| \le C \sum_{0\le\alpha,\,\beta\le 2} \sup_{(r,\,\lambda)\in\mathbb{R}_+\times\mathbb{R}} |(r\partial_r)^\alpha \partial_\lambda^\beta a(r,\,\lambda)|.$$

(b) *Let $a, b \in \mathcal{E}$. Then $Op_M(a)Op_M(b) \in OP\mathcal{E}$, and $Op_M(a)Op_M(b) = Op_M(c)$ where*

$$c(r,\,\lambda) = (2\pi)^{-1} \int\int_{\mathbb{R}_+\times\mathbb{R}} a(r,\,\lambda+\mu)b(r\rho,\,\mu)\rho^{-i\mu}\,d\rho\,d\mu.$$

(c) *Let $a \in \mathcal{E}$. Then the Hilbert space adjoint $Op_M(a)^*$ belongs to $OP\mathcal{E}$, and $Op_M(a)^* = Op_M(a^*)$ where*

$$a^*(r,\,\lambda) = (2\pi)^{-1} \int\int_{\mathbb{R}_+\times\mathbb{R}} \overline{a}(r\rho,\,\lambda+\mu)\rho^{-i\mu}\,d\rho\,d\mu.$$

(d) *Let $a \in \mathcal{E}_d$. Then the Mellin pseudodifferential operator $Op_{M,d}(a)$ with double symbol belongs to $OP\mathcal{E}$, and $Op_{M,d}(a) = Op_M(c)$ where $c \in \mathcal{E}$ is given by*

$$c(r,\,\lambda) = (2\pi)^{-1} \int\int_{\mathbb{R}_+\times\mathbb{R}} a(r,\,r\rho,\,\lambda+\mu)\rho^{-i\mu}d\rho\,d\mu.$$

The integrals in (b), (c) and (d) are understood as oscillatory integrals on $\mathbb{R}_+\times\mathbb{R}$, which are defined via the transformation $A \mapsto TAT^{-1}$ of oscillatory integrals on $\mathbb{R}\times\mathbb{R}$.

The next definition provides us with an adequate notion of slowly oscillating Mellin symbols.

Definition 4.5.12 *A symbol $a \in \mathcal{E}$ is called* slowly oscillating at the origin *if*

$$\lim_{r \to +0} \sup_{\lambda \in \mathbb{R}} |r\partial_r a(r, \lambda)| = 0.$$

Further, a double symbol $a \in \mathcal{E}_d$ is slowly oscillating at the origin *if, for every compact subset K of $\mathbb{R}_+$,*

$$\lim_{r \to +0} \sup_{(\rho, \lambda) \in K \times \mathbb{R}} |r\partial_r a(r, r\rho, \lambda)| = 0.$$

We denote the sets of all symbols in $\mathcal{E}$ and in $\mathcal{E}_d$ which are slowly oscillating at the origin by $\mathcal{E}^{SO}$ and $\mathcal{E}_d^{SO}$, respectively. The corresponding classes of Mellin pseudodifferential operators will be denoted by $OP\mathcal{E}^{SO}$ and $OP\mathcal{E}_d^{SO}$.

Proposition 4.5.13

(a) *Let $a, b \in \mathcal{E}^{SO}$. Then $Op_M(a)Op_M(b) = Op_M(ab + t)$ where $t \in \mathcal{E}$ and*

$$\lim_{r \to +0} \sup_{\lambda \in \mathbb{R}} |t(r, \lambda)| = 0. \tag{4.44}$$

(b) *Let $a \in \mathcal{E}_d^{SO}$ and $\tilde{a}(r, \lambda) := a(r, r, \lambda)$. Then $\tilde{a} \in \mathcal{E}^{SO}$ and $Op_{M,d}(a) = Op(\tilde{a} + t)$ with $t \in \mathcal{E}^{SO}$ satisfying* (4.44).

The proof is similar to that of Proposition 4.4.11. □

4.5.3 Mellin pseudodifferential operators with analytic symbols

Let $(c, d) \subset \mathbb{R}$ be an open interval which contains the origin. We say that $w = \exp v \in C^\infty(\mathbb{R}_+)$ is a weight function in the class $\mathcal{R}(c, d)$ if

- the function v satisfies

$$\sup_{r \in \mathbb{R}} \left|(rD_r)^k v(r)\right| < \infty \quad \text{for all } k \in \mathbb{N} \tag{4.45}$$

  where $D_r v = \frac{dv}{dr}$,

- the function $\kappa_w(r) := rv'(r)$ is such that

$$c < \inf_{r \in \mathbb{R}_+} \kappa_w(r) \leq \sup_{r \in \mathbb{R}_+} \kappa_w(r) < d, \tag{4.46}$$

- and if w is *slowly oscillating at the point* 0 in the sense that

$$\lim_{r \to +0} r\kappa_w'(r) = 0.$$

For $w \in \mathcal{R}(c, d)$, the weighted space $L^2(\mathbb{R}_+, w\,d\mu)$ consists of all functions u with

$$\|u\|_{L^2(\mathbb{R}_+,\, w\,d\mu)} = \|wu\|_{L^2(\mathbb{R}_+,\, d\mu)} < \infty.$$

Definition 4.5.14 *The class $\mathcal{E}(c, d)$ consists of all functions $(r, \lambda) \mapsto a(r, \lambda)$ in $C^\infty(\mathbb{R}_+ \times \mathbb{R})$ which admit an analytic extension with respect to λ into the strip $\Pi_{(c,\, d)} := \{\lambda \in \mathbb{C} : c < \operatorname{Im} \lambda < d\}$ such that*

$$\sup_{(r,\, \lambda) \in \mathbb{R}_+ \times \Pi_{(c,\, d)}} \|(r\partial_r)^\alpha \partial_\lambda^\beta a(r, \lambda)\| < \infty.$$

The class of Mellin pseudodifferential operators with symbols in $\mathcal{E}(c, d)$ will be denoted by $OP\mathcal{E}(c, d)$.

Here we used the same letter for a function a and for its analytic continuation. In the obvious way, one also introduces the classes $\mathcal{E}_d(c, d)$ of analytic double symbols as well as the classes $\mathcal{E}^{SO}(c, d)$ and $\mathcal{E}_d^{SO}(c, d)$ of slowly oscillating analytic symbols.

Proposition 4.5.15 *Let $a \in \mathcal{E}_d(c, d)$, and let w be a weight satisfying conditions* (4.45) *and* (4.46). *Then $wOp_{M,d}(a)w^{-1}I \in \mathcal{E}_d$, and $wOp_M(a)w^{-1}I = Op_{M,d}(a_w)$ where*

$$a_w(r, \rho, \lambda) := a(a, \rho, \lambda + i\theta_w(r, \rho)) \tag{4.47}$$

with

$$\theta_w : \mathbb{R}_+ \times \mathbb{R}_+ \to \mathbb{R}, \quad (r, \rho) \mapsto \int_0^1 \kappa_w(r^{1-\tau}\rho^\tau)\, d\tau$$

Notice that condition (4.46) implies that $\theta_w(r, \rho) \in (c, d)$ for all $r, \rho \in \mathbb{R}_+$.

Corollary 4.5.16 *Let $a \in \mathcal{E}_d(c, d)$, and let w be a weight satisfying conditions* (4.45) *and* (4.46). *Then the operator $Op_{M,d}(a)$ is bounded on $L^2(\mathbb{R}_+, w\, d\mu)$.*

Proposition 4.5.17 *Let $a \in \mathcal{E}_d^{SO}(c, d)$ and $w \in \mathcal{R}(c, d)$. Then*

$$wOp_{M,d}(a)w^{-1}I = Op_M(a_{[w]} + t)$$

where $a_{[w]} \in \mathcal{E}$ is given by $a_{[w]}(r, \lambda) := a(r, r, \lambda + irv'(r))$, and where $t \in \mathcal{E}$ satisfies

$$\lim_{r \to +0} \sup_{\lambda \in \mathbb{R}} |t(r, \lambda)| = 0.$$

The Propositions 4.5.15 and 4.5.17 follow from Theorem 4.5.3 and Proposition 4.5.7 by the change of variables $x = \log r$.

4.5.4 Local invertibility of Mellin pseudodifferential operators

Recall that, for $\alpha > 0$, the (multiplicative) shift operator $Z_\alpha : L^2(\mathbb{R}_+, d\mu) \to L^2(\mathbb{R}^+, d\mu)$ is defined by $(Z_\alpha u)(r) := u(r/\alpha)$. Clearly, Z_α is unitary, and $Z_\alpha^{-1} = Z_{1/\alpha}$. Further, for $\beta \in \mathbb{R}$, set

$$F_\beta : L^2(\mathbb{R}_+, d\mu) \to L^2(\mathbb{R}^+, d\mu), \quad r \mapsto r^{i\beta} u(r)$$

and, for $\delta := (\alpha, \beta) \in \mathbb{R}_+ \times \mathbb{R})$, let $\mathcal{U}_\delta := Z_\alpha F_\beta$.

A sequence $h = (h_1, h_2) : \mathbb{N} \to \mathbb{R}_+ \times \mathbb{Z}$ is called *admissible* if $h_1 = \exp g$ with a sequence $g : \mathbb{N} \to \mathbb{Z}$, and if $h_1(n) \to 0$ as $n \to \infty$.

Definition 4.5.18 *Let $A \in L(L^2(\mathbb{R}_+, d\mu))$, and let h be an admissible sequence. The operator $A_h \in L(L^2(\mathbb{R}_+, d\mu))$ is called a* limit operator *of A with respect to h if*

$$\mathcal{U}_{h(n)}^{-1} A \mathcal{U}_{h(n)} \to A_h \qquad * \text{-strongly as } n \to \infty.$$

We denote by $\sigma_{op}^0(A)$ the set of all limit operators of A defined by admissible sequences.

Theorem 4.5.19 *A Mellin pseudodifferential operator $A \in OP\mathcal{E}$ is locally invertible at 0 if and only if all limit operators of A defined by admissible sequences are invertible. In particular,*

$$\sigma_0(A) = \bigcup_{A_h \in \sigma_{op}^0(A)} \sigma(A_h).$$

Proof. If $A \in OP\mathcal{E}$, then $T^{-1}AT \in OPS_{0,0}^0$, and A is locally invertible at the point 0 if and only if the operator $T^{-1}AT$ is locally invertible at $-\infty$. Moreover, if $h = (\exp h_1, h_2)$, then

$$T^{-1}\mathcal{U}_{h(n)}T = U_{(h_1(n),\, h_2(n))} \quad \text{for all } n \in \mathbb{N},$$

and $h_1(n) \to -\infty$ as $n \to \infty$. Hence, the assertion follows from Theorem 4.3.17.□

Corollary 4.5.20 *Let $A := Op_M(a) \in OP\mathcal{E}_d(c, d)$, and let W be a weight which satisfies the conditions* (4.45) *and* (4.46). *Then $A : L^2(\mathbb{R}_+, w\, d\mu) \to L^2(\mathbb{R}_+, w\, d\mu)$ is locally invertible at the point 0 if and only if all limit operators of the operator $B := Op_M(a_w) : L^2(\mathbb{R}_+, d\mu) \to L^2(\mathbb{R}_+, d\mu)$ with respect to admissible sequences are invertible. Here, the symbol a_w is given by* (4.47). *In particular,*

$$\sigma_0(A) = \bigcup_{B_h \in \sigma_{op}^0(B)} \sigma(B_h).$$

It can be seen as in the proof of Theorem 4.3.15 that the limit operators of the Mellin pseudodifferential operator $Op_M(a) \in OP\mathcal{E}$ with respect to the admissible sequence $h = (h_1, h_2)$ can be obtained in the following way. One has $\mathcal{U}_{h(n)}^{-1} Op_M(a) \mathcal{U}_{h(n)} = Op_M(a_n)$ where

$$a_n(r, \lambda) := a(h_1(n)r, \lambda + h_2(n)), \qquad n \in \mathbb{N}. \tag{4.48}$$

The sequence $(a_n)_{n\in\mathbb{N}}$ possesses a subsequence $(a_{r(n)})_{n\in\mathbb{N}}$ which converges in the topology of $C_b^\infty(\mathbb{R}_+ \times \mathbb{R})$. We denote its limit by a_g where g is the admissible sequence $h \circ r$. Then the function a_g belongs to $\mathcal{E}$, the limit operator of $Op_M(a)$ with respect to g exists, and $Op_M(a)_g = Op_M(a_g)$.

The limit operators get a simpler form for Mellin pseudodifferential operators with slowly oscillating symbol. For, let $a \in \mathcal{E}^{SO}$, and let h be an admissible sequence, such that the sequence (a_n) with a_n as in (4.48) converges in $C_b^\infty(\mathbb{R}_+ \times \mathbb{R})$

to a function a_h. Then one gets as in the proof of Proposition 4.4.1 that the function a_h does not depend on the variable r.

In the very same way, one also obtains the limit operators of Mellin pseudodifferential operators with double symbols $Op_{M,d}(a) \in OP\mathcal{E}_d$. In this case, one chooses a subsequence of the sequence $(a_n)_{n\in\mathbb{N}}$ which converges in $C_b^\infty(\mathbb{R}_+ \times \mathbb{R}_+ \times \mathbb{R})$ to a certain function a_g where now

$$a_n(r, \rho, \lambda) := a(h_1(n)r, h_1(n)\rho, \lambda + h_2(n)), \qquad n \in \mathbb{N}.$$

If $a \in \mathcal{E}_d^{SO}$ is a slowly oscillating double symbol, and if h is an admissible sequence such that the functions a_n converge to a limit function a_h, then this limit function does not depend on r and ρ. Consequently, the limit operator $Op_{M,d}(a)_h = Op_{M,d}(a_h)$ of A is a Mellin convolution,

$$(A_h u)(r) = (C_M(a_h)u)(r) = (2\pi)^{-1} \int_{\mathbb{R}} a_h(\lambda)\tilde{u}(\lambda) r^{i\lambda}\, d\lambda,$$

where $\tilde{u}$ is the Mellin transform of $u \in C_0^\infty(\mathbb{R}_+)$. In this case, it follows from Theorem 4.5.10 that the operator $A_h : L^2(\mathbb{R}_+, d\mu) \to L^2(\mathbb{R}_+, d\mu)$ is invertible if and only if

$$\inf_{\lambda\in\mathbb{R}} |a_h(\lambda)| > 0, \tag{4.49}$$

and that

$$\sigma(A_h) = \sigma_0(A_h) = \operatorname{clos}\{a_h(\lambda) : \lambda \in \mathbb{R}\}.$$

With these remarks, the following theorem can be proved as Theorem 4.4.2.

Theorem 4.5.21 *The operator $A = Op_{M,d}(a) \in \mathcal{E}_d^{SO}$, considered as an operator on $L^2(\mathbb{R}_+, d\mu)$, is locally invertible at the point 0 if and only if, for every limit operator $A_h = C_M(a_h) \in \sigma_{op}^0(A)$, the condition (4.49) holds. Equivalently, A is locally invertible at 0 if*

$$\lim_{\varepsilon\to 0} \inf_{(r,\,\lambda)\in(0,\,\varepsilon)\times\mathbb{R}} |a(r, r, \lambda)| > 0.$$

In particular,

$$\sigma_0(A) = \bigcup_{C_M(a_h)\in\sigma_{op}^0(A)} \operatorname{clos}\{a_h(\lambda) : \lambda \subset \mathbb{R}\}.$$

Corollary 4.5.22 *Let $A = Op_{M,d}(a) \in \mathcal{E}_d^{SO}(c, d)$ and $w \in \mathcal{R}(c, d)$. Then the operator A, considered as an operator on $L^2(\mathbb{R}_+, w\, d\mu)$, is locally invertible at the point 0 if and only if all limit operators of the operator $B := Op_{M,d}(a_w) \in L(L^2(\mathbb{R}_+, d\mu))$ with respect to admissible sequences are invertible. (Recall the definition of a_w in (4.47).) This condition is equivalent to*

$$\lim_{\varepsilon\to 0} \inf_{(r,\,\lambda)\in(0,\,\varepsilon)\times\mathbb{R}} |a(r, r, \lambda + i\theta_w(r))| > 0.$$

In particular,

$$\sigma_0(A) = \bigcup_{C_M(b_h)\in\sigma_{op}^0(B)} \operatorname{clos}\{b_h(\lambda) : \lambda \in \mathbb{R}\}.$$

4.6 Singular integrals over Carleson curves with Muckenhoupt weights

In this section, we will consider singular integral operators S_Γ of Cauchy type on spaces $L^p(\Gamma, w)$ over curves Γ possessing the Carleson property and with weight functions w satisfying the Muckenhoupt condition. These are the general conditions under which the operator S_Γ is bounded. In particular, we will determine the local essential spectra of the singular integral operator S_Γ by identifying the local representatives of S_Γ with certain Mellin pseudodifferential operators, which allows us to have recourse to the results presented in Section 4.5. For the sake of simplicity, we confine ourselves to the Hilbert space case $p = 2$.

4.6.1 Carleson curves and Muckenhoupt weights

Let Γ be an oriented rectifiable simple arc in the complex plane and set $\Gamma(t, \varepsilon) := \{\tau \in \Gamma : |\tau - t| < \varepsilon\}$ for $\varepsilon > 0$ and $t \in \Gamma$. Given $p \in (1, \infty)$ and a measurable function $w : \Gamma \to [0, \infty]$ such that $w^{-1}(\{0, \infty\})$ has Lebesgue measure zero, consider the weighted Lebesgue space $L^p(\Gamma, w)$ provided with norm

$$\|f\|_{L^p(\Gamma, w)} := \left(\int_\Gamma |f(\tau)|^p w(\tau)^p \, |d\tau|\right)^{1/p}.$$

Let $f \in L^1(\Gamma)$. Then the *Cauchy singular integral* $S_\Gamma f$,

$$(S_\Gamma f)(t) := \lim_{\varepsilon \to 0} \frac{1}{\pi i} \int_{\Gamma \setminus \Gamma(t, \varepsilon)} \frac{f(\tau)}{\tau - t} \, d\tau$$

exists for almost all $t \in \Gamma$. As a result of a long development culminating in the work by Hunt, Muckenhoupt, Wheeden [78] and David [42], it became clear that S_Γ is a well-defined and bounded operator on $L^p(\Gamma, w)$ if and only if

$$\sup_{\varepsilon > 0} \sup_{t \in \Gamma} \frac{1}{\varepsilon} \left(\int_{\Gamma(t, \varepsilon)} w(\tau)^p \, |d\tau|\right)^{1/p} \left(\int_{\Gamma(t, \varepsilon)} w(\tau)^{-q} |d\tau|\right)^{1/q} < \infty \tag{4.50}$$

where $1/p + 1/q = 1$ (see also [51] and [23]). We write A_p for the set of all pairs (Γ, w) satisfying (4.50). By Hölder's inequality, (4.50) implies that

$$\sup_{\varepsilon > 0} \sup_{t \in \Gamma} |\Gamma(t, \varepsilon)|/\varepsilon < \infty \tag{4.51}$$

where $|\Gamma(t, \varepsilon)|$ stands for the Lebesgue (length) measure of $\Gamma(t, \varepsilon)$. Conditions (4.50) and (4.51) are frequently referred to as the Muckenhoupt and the Carleson condition, respectively, and weights w and curves Γ satisfying these conditions are called *Muckenhoupt weights* and *Carleson curves*.

4.6.2 Logarithmic spirals and power weights

For $\delta \in \mathbb{R}$, the curve

$$G_\delta := \{re^{i\delta \log r} : r > 0\} \tag{4.52}$$

is called a *logarithmic spiral.* It is easily seen that logarithmic spirals are Carleson curves. Further, given $\gamma \in \mathbb{R}$, we call the function

$$u_\gamma : G_\delta \to (0, \infty), \quad \tau \mapsto |\tau|^\gamma \tag{4.53}$$

a *power weight.* Thus, if $\tau = re^{i\delta \log r}$, then

$$u_\gamma(\tau) = r^\gamma = e^{\gamma \log r}. \tag{4.54}$$

It turns out that u_γ is a Muckenhoupt weight on G_δ if and only if $-1/2 < \gamma < 1/2$ (see, e.g., Theorem 2.2 in [23]). Hence, under this condition, S_{G_δ} acts as a linear and bounded operator on $L^2(G_\delta, u_\gamma)$.

Let $\delta \in \mathbb{R}$ and $\gamma \in (-1/2, 1/2)$. The mapping $\Phi_{\delta,\gamma}$ defined by

$$(\Phi_{\delta,\gamma} f)(r) := |1 + i\delta|^{1/2} r^{\gamma+1/2} f(re^{i\delta \log r}), \quad r \in \mathbb{R}_+,$$

is an isometric isomorphism from $L^2(G_\delta, u_\gamma)$ onto $L^2(\mathbb{R}_+, d\mu)$ where $d\mu$ again refers to the Haar measure of the multiplicative group $\mathbb{R}_+$. A straightforward computation yields

$$(\Phi_{\delta,\gamma} S_{G_\delta} \Phi_{\delta,\gamma}^{-1} g)(r) = \int_{G_\delta} k\left(\frac{r}{\rho}\right) g(\rho)\, \frac{d\rho}{\rho}, \quad r \in \mathbb{R}_+,$$

for all $g \in C_0^\infty(\mathbb{R}_+)$, where

$$k(\rho) := \frac{1 + i\delta}{\pi i} \frac{\rho^{\gamma+1/2}}{1 - \rho^{1+i\delta}}. \tag{4.55}$$

To determine the Mellin transform of k, we need the following technical lemma.

Lemma 4.6.1 *Let $0 < \operatorname{Im} \lambda < \operatorname{Re} \xi$. Then*

$$\frac{\xi}{\pi i} \int_0^\infty (1 - t^\xi)^{-1} t^{-i\lambda} \, \frac{dt}{t} = \coth \frac{\pi\lambda}{\xi}, \tag{4.56}$$

where the integral is understood in the sense of the Cauchy principle value.

Proof. For real (hence, positive) ξ, a change of variables yields

$$\frac{\xi}{\pi i} \int_0^\infty (1 - t^\xi)^{-1} t^{-i\lambda} \, \frac{dt}{t} = \frac{1}{\pi i} \int_0^\infty (1 - s)^{-1} s^{-i\lambda/\xi} \, \frac{ds}{s}.$$

On the other hand, adding the Formulas 3.238.1 and 3.238.2 in [70] gives

$$\frac{1}{\pi i} \int_0^\infty (1 - x)^{-1} x^{y-1} \, dx = \coth(y\pi i)$$

whenever $0 < \operatorname{Re} y < 1$. Thus, if $\xi > 0$ and $0 < \operatorname{Im}\lambda < \xi$, then the identity (4.56) holds. Clearly, this identity makes sense and is still valid by analytic continuation if $\operatorname{Re}\xi > 0$ and $\operatorname{Im}\lambda \in (0, \operatorname{Re}\xi)$. □

Proposition 4.6.2 *Let $\delta \in \mathbb{R}$ and $\gamma \in (-1/2, 1/2)$, and let k be given by (4.55). Then the Mellin transform of k equals*

$$(Mk)(\lambda) = \coth\left(\pi\frac{\lambda + i(\gamma+1/2)}{1+i\delta}\right) =: \Sigma(\delta, \gamma, \lambda), \quad \lambda \in \mathbb{R}.$$

Hence, the transformed singular integral $\Phi_{\delta,\gamma} S_{G_\delta} \Phi_{\delta,\gamma}^{-1}$ is the Mellin convolution operator $C_M(\Sigma_{\delta,\gamma})$ with symbol $\Sigma_{\delta,\gamma}(\lambda) := \Sigma(\delta, \gamma, \lambda)$.

Proof. By the previous lemma, we have

$$(Mk)(\lambda) = \frac{1+i\delta}{\pi i}\int_0^\infty \frac{\rho^{\gamma+1/2-i\lambda}}{1-\rho^{1+i\delta}}\frac{d\rho}{\rho} = \coth\left(\pi\frac{\lambda+i(\gamma+1/2)}{1+i\delta}\right)$$

for all $\lambda \in \mathbb{R}$. □

To determine the essential range of the function $\Sigma_{\delta,\gamma}$, notice that $\coth(\pi z) = m_{-1,1}(e^{2\pi z})$ where $m_{-1,1}$ is the Möbius transform $\zeta \mapsto (\zeta+1)/(\zeta-1)$. Now, if λ runs through $\mathbb{R}$, then the point $(\lambda + i(\gamma+1/2))/(1+i\delta)$ traverses the straight line

$$\{x + iy \in \mathbb{C} : y = -\delta x + (\gamma + 1/2)\}. \tag{4.57}$$

The function $z \mapsto e^{2\pi z}$ maps this line to the logarithmic spiral $e^{2\pi i(\gamma+1/2)}G_{-\delta}$, which on its hand is mapped by the Möbius transform $m_{-1,1}$ to a curve $\mathcal{S}_{\delta,\gamma}$ joining -1 to 1 which might be called a *logarithmic double spiral.* (The points $\lambda = \pm\infty$ are mapped to ± 1.) Thus, the essential range of $\Sigma_{\delta,\gamma}$ is

$$\mathcal{S}_{\delta,\gamma} = m_{-1,1}(e^{2\pi i(\gamma+1/2)}G_{-\delta}) \cup \{-1, 1\}. \tag{4.58}$$

An operator $A \in L(L^2(G_\delta, u_\gamma))$ is called *locally invertible at the origin* if there are an $\varepsilon > 0$, a function $\varphi \in C(G_\delta)$ which is identically 1 on $\{re^{i\delta\log r} : r \in (0, \varepsilon)\}$, and operators $D_l, D_r \in L(L^2(G_\delta, u_\gamma))$ such that

$$\varphi A D_r = \varphi I \quad \text{and} \quad D_l A \varphi I = \varphi I.$$

The local spectrum $\sigma_0(A)$ is the set of all $\lambda \in \mathbb{C}$ such that $A - \lambda I$ is not locally invertible at the origin. Clearly, $\sigma_0(A) = \sigma_0(\Phi_{\delta,\gamma} A \Phi_{\delta,\gamma}^{-1})$. Thus, from what has been said above and from Theorem 4.5.10, we get the following.

Theorem 4.6.3 *Let $\delta \in \mathbb{R}$ and $\gamma \in (-1/2, 1/2)$, and consider the singular integral operator S_{G_δ} as acting on $L^2(G_\delta, u_\gamma)$. Then*

$$\sigma(S_{G_\delta}) = \sigma_0(S_{G_\delta}) = \mathcal{S}_{\delta,\gamma}.$$

Notice that the condition $\gamma \in (-1/2, 1/2)$ guarantees that the line (4.57) intersects the imaginary axis strictly between 0 and i. Therefore, the singularities $i\mathbb{Z}$ of $z \mapsto \coth(\pi z)$ are not met.

4.6.3 Curves and weights with slowly oscillating data

Here we are going to introduce a class of Carleson curves and a class of Muckenhoupt weights which behave locally as logarithmic spirals and as power weights, respectively.

Slowly oscillating functions. Fix $s > 0$. A function $a \in C^\infty(0, s)$ is said to be *slowly oscillating* (at the origin) if

$$\sup_{r\in(0,\,s)} |(rD_r)^j a(r)| < \infty \quad \text{for all } j \geq 0 \tag{4.59}$$

and if

$$\lim_{r\to 0} |ra'(r)| = 0. \tag{4.60}$$

Note that (4.59) for $j = 0$ is the statement that a is bounded, and that (4.59) and (4.60) imply that actually

$$\lim_{r\to 0} |(rD_r)^j a(r)| = 0 \quad \text{for all } j \geq 1.$$

For example, if $f \in C_b^\infty(\mathbb{R})$, then the function $r \mapsto f(\log(-\log r))$ is slowly oscillating.

Slowly oscillating curves. Let Γ be a bounded oriented simple arc with starting point t. We say that the curve Γ is *slowly oscillating* (at t) if there are an $s > 0$ and a real-valued function θ in $C^\infty(0, s)$ such that the function $r \mapsto r\theta'(r)$ is slowly oscillating at the origin and Γ can be parametrized as

$$\Gamma = \{t + re^{i\theta(r)} : r \in (0,\, s)\}. \tag{4.61}$$

Recall that the slow oscillation property of the function $r \mapsto r\theta'(r)$ means that

$$\sup_{r\in(0,\,s)} |(rD_r)^j \theta(r)| < \infty \quad \text{for all } j \geq 1 \tag{4.62}$$

and that

$$\lim_{r\to 0} |(rD_r)^2 \theta(r)| = 0, \tag{4.63}$$

but that $\theta(r)$ may be unbounded as $r \to 0$. Since

$$|d\tau| = \sqrt{1 + (r\theta'(r))^2}\, dr,$$

the case $j = 1$ in condition (4.62) ensures that Γ is a Carleson curve.

For example, if $g \in C^\infty(\mathbb{R})$, then

$$\theta(r) := g(\log(-\log r)) \log r, \qquad r \in (0,\, 1)$$

satisfies (4.62) and (4.63). In case g is constant, say $g(x) = \delta \in \mathbb{R}$ for all $x \in (0,\, \infty)$, the curve (4.61) is the starting piece of the logarithmic spiral G_δ introduced in Section 4.6.2.

Slowly oscillating weights. Let Γ be a slowly oscillating curve with starting point t. A function $w : \Gamma \to [0, \infty]$ is called a *slowly oscillating weight* (at the point t) if there are an $s > 0$ and a function $v \in C^\infty(0, s)$ such that

$$w(t + re^{i\theta(r)}) = e^{v(r)} \quad \text{for } r \in (0, s) \tag{4.64}$$

and the function $r \mapsto rv'(r)$ is slowly oscillating at the origin:

$$\sup_{r\in(0,\, s)} |(rD_r)^j v(r)| < \infty \quad \text{for all } j \geq 1 \tag{4.65}$$

and

$$\lim_{r\to 0} |(rD_r)^2 v(r)| = 0. \tag{4.66}$$

One can show (Lemma 4.2 in [22] or Theorem 2.36 in [23]) that the Muckenhoupt condition (4.50) with $p = q = 2$ is satisfied if and only if

$$-1/2 < \liminf_{r\to 0} rv'(r) \leq \limsup_{r\to 0} rv'(r) < 1/2. \tag{4.67}$$

For example, if $h \in C^\infty(\mathbb{R})$ is bounded together with all its derivatives, then the function

$$v : (0, 1) \to \mathbb{R}, \quad r \mapsto h(\log(-\log r)) \log r$$

obviously satisfies (4.65) and (4.66), whereas (4.67) is equivalent to the inequalities

$$-1/2 < \liminf_{x\to+\infty} (h(x) + h'(x)) \leq \limsup_{x\to+\infty} (h(x) + h'(x)) < 1/2.$$

Note that in case h is a constant function, say $h(x) = \gamma$ for all $x \in \mathbb{R}$, the weight (4.64) is the power weight $r \mapsto r^\gamma$.

We denote by A_2^{SO} the set of all pairs $(\Gamma, w) \in A_2$ where Γ is a slowly oscillating curve and w is a slowly oscillating weight.

Slowly oscillating coefficients. Let $(\Gamma, w) \in A_2^{SO}$. A function $c : \Gamma \to \mathbb{C}$ is called *slowly oscillating* (at the point t) if there are an $s > 0$ and a slowly oscillating function $d \in C^\infty(0, s)$ such that

$$c(t + re^{i\theta(r)}) = d(r) \quad \text{for } r \in (0, s).$$

The class of all these functions will be denoted by $SO^\infty(\Gamma)$.

4.6.4 Local representatives and local spectra of singular integral operators

Let $(\Gamma, w) \in A_2^{SO}$. In particular we assume that Γ and w are given via the representations (4.61) and (4.64), and that condition (4.67) holds. The starting point of Γ is denoted by t again. The goal of this section is to represent the singular integral operator $S_\Gamma : L^2(\Gamma, w) \to L^2(\Gamma, w)$ locally as a Mellin pseudodifferential operator.

The map

$$\Phi_t : L^2(\Gamma, w) \to L^2((0, s), d\mu), \quad (\Phi_t f)(r) := e^{v(r)} r^{1/2} f(t + re^{i\theta(r)})$$

is a bijective isometry. We write it as $\Phi_t = \Phi_{t,1} \circ \Phi_{t,2}$ where

$$(\Phi_{t,1}\psi)(r) := e^{v(r)}\psi(r) \quad \text{and} \quad (\Phi_{t,2} f)(r) := r^{1/2} f(t + re^{i\theta(r)}).$$

Given $\varepsilon \in (0, s/2)$, let φ_ε a function in $C^\infty([0, \infty))$ which takes its values in $[0, 1]$, is equal to 1 on $[0, \varepsilon]$ and vanishes outside $[0, 2\varepsilon]$. Further, we associate the weight function $w_+(r) := e^{v(r)}$ on $\mathbb{R}_+$ with the weight w on Γ.

Proposition 4.6.4 *Let $(\Gamma, w) \in A_2^{SO}$ be as above, and consider the operator S_Γ as acting on $L^2(\Gamma, w)$. Then the operator*

$$\varphi_\varepsilon \Phi_{t,2} S_\Gamma \Phi_{t,2}^{-1} \varphi_\varepsilon I : L^2(\mathbb{R}_+, w_+\, d\mu) \to L^2(\mathbb{R}_+, w_+\, d\mu)$$

is the Mellin pseudodifferential operator $Op_{M,d}(b)$ with the double symbol

$$b(r, \rho, \lambda) := \varphi_\varepsilon(r)\varphi_\varepsilon(\rho) \frac{1 + i\rho\theta'(\rho)}{1 + iM_\theta(r, \rho)} \coth \frac{\lambda + i/2}{1 + iM_\theta(r, \rho)}$$

where

$$M_\theta(r, \rho) := \int_0^1 r^{1-\tau} \rho^\tau \theta'(r^{1-\tau}\rho^\tau)\, dx.$$

Proof. Let $f \in C_0^\infty(\mathbb{R}_+)$. A straightforward computations yields

$$(\varphi_\varepsilon \Phi_{t,2} S_\Gamma \Phi_{t,2}^{-1} \varphi_\varepsilon f)(r) = \int_{\mathbb{R}_+} l(r, \rho) f(\rho)\, d\mu(\rho)$$

with

$$l(r, \rho) := \varphi_\varepsilon(r)\varphi_\varepsilon(\rho) \frac{1}{\pi i} \frac{(1 + i\rho\theta'(\rho))\,(r/\rho)^{1/2}}{1 - (r/\rho)e^{i(\theta(r) - \theta(\rho))}}.$$

Applying the Lagrange formula

$$\theta(r) - \theta(\rho) = M_\theta(r, \rho)(\log r - \log \rho),$$

we obtain

$$l(r, \rho) = \varphi_\varepsilon(r)\varphi_\varepsilon(\rho) \frac{1}{\pi i} \frac{(1 + i\rho\theta'(\rho))\,(r/\rho)^{1/2}}{1 - (r/\rho)^{1 + iM_\theta(r, \rho)}}.$$

If $\rho > 0$, $0 < \eta < 1$ and $\operatorname{Re} \xi > 1$, then

$$\frac{1}{\pi i} \frac{\xi \rho^\eta}{1 - \rho^\xi} = \frac{1}{2\pi} \int_{\mathbb{R}} \coth(\pi \frac{\lambda + i\eta}{\xi}) \rho^{i\lambda}\, d\lambda$$

by Lemma 4.6.1. Consequently,

$$l(r, \rho) = \varphi_\varepsilon(r)\varphi_\varepsilon(\rho)\frac{1}{2\pi}\frac{1+i\rho\theta'(\rho)}{1+iM_\theta(r, \rho)}\int_{\mathbb{R}} \coth\left(\pi\frac{\lambda+i/2}{1+iM_\theta(r, \rho)}\right)\rho^{i\lambda}\,d\lambda.$$

This implies the assertion. □

Proposition 4.6.5 *Let* $(\Gamma, w) \in A_2^{SO}$ *be as above, and consider the operator* S_Γ *as acting on* $L^2(\Gamma, w)$. *Then the operator*

$$\Phi_{t,1}\varphi_\varepsilon\Phi_{t,2}S_\Gamma\Phi_{t,2}^{-1}\varphi_\varepsilon\Phi_{t,1}^{-1} : L^2(\mathbb{R}_+, d\mu) \to L^2(\mathbb{R}_+, d\mu) \tag{4.68}$$

is the Mellin pseudodifferential operator $Op_M(a+q)$ *with*

$$a(r, \lambda) := \varphi_\varepsilon^2(r)\coth\pi\frac{\lambda+i(rv'(r))+1/2}{1+ir\theta'(r)}$$

and

$$\lim_{r\to 0}\sup_{\lambda\in\mathbb{R}}|q(r, \lambda)| = 0.$$

Proof. The condition (4.67) on the weight implies that there is an r_0 such that $-1/2 < rv'(r) < 1/2$ for $r \in (0, r_0)$. Since the behavior of the weight w outside an open neighborhood of the origin does not influence the operator (4.68) if ε is small enough, we may a priori assume that $w \in \mathcal{R}(-1/2, 1/2)$. (Recall the definition of the class $\mathcal{R}(c, d)$ of slowly oscillating weights given before Definition 4.5.14.) The double symbol

$$(r, \rho, \lambda) \mapsto \coth\pi\frac{\lambda+i/2}{1+irM_\theta(r, \rho)}$$

is an analytic function of λ in the domain $\{\lambda \in \mathbb{C} : -1/2 < \operatorname{Im}\lambda < 1/2\}$, and it satisfies the conditions of Proposition 4.5.17 on this domain. So the assertion follows from this proposition. □

Remark. The functions given on $(0, \varepsilon)$ by

$$r \mapsto r\theta'(r) =: \kappa_\theta(r) \quad \text{and} \quad r \mapsto rv'(r) := \kappa_v(r)$$

are local characteristics of the curve Γ and the weight w at the end-point t of Γ. If $\Gamma = G_\delta = \{z \in \mathbb{C} : z = re^{i\delta\log r}, r \in (0, s)\}$ is a piece of a logarithmic spiral, and if $w(r) = r^\alpha$ with $|\alpha| < 1/2$ is a power weight, then $\kappa_\theta(r) = \delta$ and $\kappa_v(r) = \alpha$, and the operator (4.68) is a pure Mellin convolution operator as described in Proposition (4.6.2). □

After these preparations, we turn over to more involved singular integral operators. These are constituted by the operator S_Γ of singular integration, the identity operator, and certain multiplication operators. More precisely, we assume that the curve Γ is slowly oscillating at the origin, that the weight w is in $\mathbb{R}(-1/2, 1/2)$, and that a and b are functions from $SO^\infty(\Gamma)$, and we consider the singular integral operator

$$A_\Gamma := aI + bS_\Gamma : L^2(\Gamma, w) \to L^2(\Gamma, w).$$

We will apply the above representations to describe the local spectrum at the origin of the operator

$$(A_\Gamma)_{t,\varepsilon} := \Phi_{t,1}\varphi_\varepsilon\Phi_{t,2}A_\Gamma\Phi_{t,2}^{-1}\varphi_\varepsilon\Phi_{t,1}^{-1} := L^2(\mathbb{R}_+, w\,d\mu) \to L^2(\mathbb{R}_+, w\,d\mu).$$

Proposition 4.6.6 *Under these assumptions on* Γ, w *and* a, b,

$$(A_\Gamma)_{t,\varepsilon} = Op_M(a_{t,\varepsilon}) \in OP\mathcal{E}^{SO}$$

where

$$a_{t,\varepsilon}(r, \lambda) := \varphi_\varepsilon^2(r)\left(a(r) + b(r)\coth\pi\frac{\lambda + i(rv'(r) + 1/2)}{1 + ir\theta'(r)} + q(r, \lambda)\right)$$

with

$$\lim_{r\to 0}\sup_{\lambda\in\mathbb{R}}|q(r, \lambda)| = \lim_{\lambda\to\infty}\sup_{r\in(0,\, s)}|q(r, \lambda)| = 0.$$

Indeed, this is a direct consequence of Proposition 4.6.5. □

To describe the local spectrum at t of the operator A_Γ via Theorem 4.5.21, we need the limit operators of $(A_\Gamma)_{t,\varepsilon}$ at the origin. Thus, let $h = (h_1, h_2)$ be an admissible sequence which defines a limit operator of $(A_\Gamma)_{t,\varepsilon}$. In case $h_2(n) \to \pm\infty$ as $n \to \infty$, the functions defined by

$$a^n(r, \lambda) := a_{t,\varepsilon}(h_1(n)r, \lambda + h_2(n))$$

converge in the topology of $C_b^\infty(\mathbb{R}_+ \times \mathbb{R})$ to $a_h \pm b_h \in \mathbb{C}$, where

$$a_h = \lim_{n\to\infty} a(h_1(n)) \quad \text{and} \quad b_h = \lim_{n\to\infty} b(h_1(n))$$

are the partial limits of a and b defined by the sequence $h_1 \to 0$. In case h is an admissible sequence for which h_2 is a constant sequence, the limit operator of $(A_\Gamma)_{t,\varepsilon}$ defined by this sequence is unitarily equivalent to the Mellin convolution operator $C_M(a_{t,\varepsilon,h})$ with

$$a_{t,\varepsilon,h}(r, \lambda) := a_h + b_h\coth\pi\frac{\lambda + i(\kappa_{v,\,h} + 1/2)}{1 + i\kappa_{\theta,\,h}}$$

where a_h, b_h, $\kappa_{w,\,h}$ and $\kappa_{\theta,\,h}$ refer to the partial limits of a, b, κ_v and κ_θ defined by the sequence h_1 tending to 0. Note that the symbol $a_{t,\varepsilon,h}$ is independent on the variable r. Further define

$$\mathcal{S}_{\delta,\,\gamma} := \left\{\zeta \in \mathbb{C} : \zeta = \coth\pi\frac{\lambda + i(\gamma + 1/2)}{1 + i\delta} \text{ with } \lambda \in [-\infty, +\infty]\right\}.$$

Then Theorem 4.5.21 implies the following.

Theorem 4.6.7 *Under the assumptions made above, the singular integral operator* $A_\Gamma : L^2(\Gamma, w) \to L^2(\Gamma, w)$ *is locally invertible at the point* t *if and only if*

$$\lim_{\varepsilon \to 0} \inf_{(r, \lambda) \in (0, \varepsilon) \times \mathbb{R}} \left| a(r) + b(r) \coth \pi \frac{\lambda + i(rv'(r) + 1/2)}{1 + ir\theta'(r)} \right| > 0.$$

The local spectrum of A_Γ *at* t *is*

$$\sigma_t(A_\Gamma) = \bigcup_h (a_h + b_h \mathcal{S}_{\kappa_{\theta,h}, \kappa_{v,h}})$$

where the union is taken over all admissible sequences $h = (h_1, h_2)$ *for which the partial limits*

$$a_h := \lim_{n \to \infty} a(h_1(n)), \qquad b_h := \lim_{n \to \infty} b(h_1(n)),$$

$$\kappa_{v,h} := \lim_{n \to \infty} \kappa_v(h_1(n)), \qquad \kappa_{\theta,h} := \lim_{n \to \infty} \kappa_\theta(h_1(n))$$

exist.

Proposition 4.6.2 and the representation of the operator (4.68) derived in Proposition 4.6.5 show that the local spectrum of the operator $A_\Gamma : L^2(\Gamma. w) \to L^2(\Gamma, w)$ is the union of the spectra of the singular integral operators with constant coefficients

$$A_{G_{\kappa_{\theta,h}}} := a_h I + b_h S_{G_{\kappa_{\theta,h}}} : L^2(G_{\kappa_{\theta,h}}, r^{\kappa_{v,h}}) \to L^2(G_{\kappa_{\theta,h}}, r^{\kappa_{v,h}})$$

acting on $L^2(G_{\kappa_{\theta,h}}, r^{\kappa_{v,h}})$ over the logarithmic spiral $G_{\kappa_{\theta,h}}$ and with the power weight $r^{\kappa_{v,h}}$. In particular, the local spectrum of the operator $S_\Gamma : L^2(\Gamma, w) \to L^2(\Gamma, w)$ is the union of the double logarithmic spirals

$$\mathcal{S}_{\kappa_{\theta,h}, \kappa_{v,h}} := M_{-1,1}(e^{2\pi i(\kappa_{w,h}+1/2)} G_{-\kappa_{vh}}) \cup \{-1, 1\}.$$

4.6.5 Singular integral operators on composed curves

The results of the previous section can be extended to singular integral operators on composed curves without cusps. The point is that these curves behave locally as a union of a finite number of semi-axes (a *star*), and singular integral operators on stars can be transformed into Mellin pseudodifferential operators with matrix-valued symbols.

Let $\Gamma := \Gamma_1 \cup \cdots \cup \Gamma_K$ where the curves

$$\Gamma_k := \left\{ t + re^{i(\theta_0(r) + \theta_k(r))} : r \in (0, s) \right\}$$

are oriented simple arcs with common point t. The point t is also called a *node* of the curve Γ. Further, let the weight function $w : \Gamma \to [0, \infty]$ be given by

$$w(r) = e^{v(r)}, \quad \tau \in \Gamma.$$

Moreover, we assume that there are numbers

$$0 \leq m_1 < M_1 < m_2 < M_2 < \cdots < m_K < M_K < 2\pi$$

such that $m_k < \theta_k(r) < M_k$ for all $k = 1, \ldots, K$, and that the functions $r \mapsto r\theta_k'(r)$ for $k = 0, \ldots, K$ as well as the function $r \mapsto rv'(r)$ are in $C^\infty(0, s)$, that these functions are slowly oscillating at the origin, and that (4.67) is satisfied. Finally, we write $SO^\infty(\Gamma)$ for the class of all functions $a_\Gamma : \Gamma \to \mathbb{C}$ which are slowly oscillating in the sense that the restrictions $a_\Gamma|_{\Gamma_k}$ belong to $SO^\infty(\Gamma_k)$ for every k. Equivalently, the functions

$$a_k : r \mapsto a_\Gamma(t + re^{i(\theta_0(r)+\theta_k(r))})$$

are in $C^\infty(0, s)$ and slowly oscillating at the origin. Let

$$a(r) := \operatorname{diag}(a_1(r), \ldots, a_K(r)). \tag{4.69}$$

As above we introduce the map

$$\Phi_t : L^2(\Gamma, w) \to \oplus_{j=1}^K L^2((0, s), d\mu), \quad f \mapsto (f_1, \ldots, f_K)$$

with

$$f_k(r) = r^{1/2} e^{v(r)} f(t + re^{i(\theta_0(r)+\theta_k(r))})$$

for $k = 1, \ldots, K$. Clearly, Φ_t is an isometric bijection. For each bounded linear operator $A : L^2(\Gamma, w) \to L^2(\Gamma, w)$, we set

$$A_{t,\varepsilon} := \varphi_\varepsilon \Phi_t A \Phi_t^{-1} \varphi_\varepsilon I.$$

Finally, given real parameters α, β, we introduce the matrix-function

$$\mathbb{C} \setminus i\mathbb{Z} \to \mathbb{C}^{K \times K}, \quad \lambda \mapsto \nu(\alpha, \beta, \lambda) = (\nu_{jk}(\alpha, \beta, \lambda))_{j,k=1}^K$$

where

$$\nu_{jk}(\alpha, \beta, \lambda) = \begin{cases} e^{(\beta-\alpha-\pi)\lambda}/\sinh(\pi\lambda) & \text{if } k > j, \\ e^{(\beta-\alpha+\pi)\lambda}/\sinh(\pi\lambda) & \text{if } k < j, \\ \coth(\pi\lambda) & \text{if } k = j. \end{cases}$$

Proposition 4.6.8 *Let the curve Γ and the weight w satisfy the above conditions in a neighborhood of the point $t \in \Gamma$. Then $(S_\Gamma)_{t,\varepsilon}$ is a matrix Mellin pseudodifferential operator with components in $OP\mathcal{E}^{SO}$. More precisely,*

$$(S_\Gamma)_{t,\varepsilon} = Op_M(\varphi_\varepsilon^2 s + q)$$

where

$$s(r, \lambda) = (s_{jk}(r, \lambda))_{j,k=1}^K, \qquad q(r, \lambda) = (q_{jk}(r, \lambda))_{j,k=1}^K$$

with

$$s_{jk}(r, \lambda) = \epsilon_k \nu_{jk}\left(\theta_j(r), \theta_k(r), \frac{\lambda + i(rv'(r) + 1/2)}{1 + ir\theta_0'(r)}\right)$$

and with $\epsilon_k = 1$ *if the curve* Γ_k *is oriented away from the point* t *and* $\epsilon_k = -1$ *in the opposite case. Finally,*

$$\lim_{r\to 0} \sup_{\lambda\in\mathbb{R}} |q_{jk}(r, \lambda)| = 0, \qquad \lim_{\lambda\to\infty} \sup_{r\in(0,\, s)} |q_{jk}(r, \lambda)| = 0.$$

The proof of Proposition 4.6.8 proceeds similar to the proofs of Propositions 4.6.4 and 4.6.5. □

Corollary 4.6.9 *Let* $A_\Gamma := a_\Gamma I + b_\Gamma S_\Gamma : L^2(\Gamma, w) \to L^2(\Gamma, w)$*, where the curve* Γ *and the weight* w *satisfy the above-mentioned conditions, and where the coefficients* a_Γ *and* b_Γ *are slowly oscillating in a neighborhood of* $t \in \Gamma$*. Then*

$$(A_\Gamma)_{t,\,\varepsilon} = Op_M(\varphi_\varepsilon^2(a + bs) + q)$$

where a *and* b *are the diagonal matrices given by* (4.69) *and with* s *and* q *as in the preceding proposition.*

Employing the obvious generalization of Theorem 4.5.21 to the case of vector-valued functions, we obtain the following result.

Theorem 4.6.10 *Let* Γ, w, a_Γ *and* b_Γ *be as in Corollary* 4.6.9. *Then the operator* $A_\Gamma = a_\Gamma I + b_\Gamma S_\Gamma : L^2(\Gamma, w) \to L^2(\Gamma, w)$ *is locally invertible at the point* $t \in \Gamma$ *if and only if*

$$\lim_{\varepsilon\to 0} \inf_{(r,\,\lambda)\in(0,\,\varepsilon)\times\mathbb{R}} |\det(a(r) + b(r)s(r, \lambda))| > 0. \tag{4.70}$$

The local spectrum of A_Γ *at the point* t *is given by*

$$\sigma_t(A_\Gamma) = \bigcup_h \{\zeta \in \mathbb{C} : \zeta = sp\,(a_h + b_h(r)s_h(\lambda)),\ \lambda \in [-\infty, +\infty]\} \tag{4.71}$$

where $sp\,c$ *refers to the spectrum of the matrix* c*. The union in* (4.71) *is taken over all admissible sequences* $h = (h_1, 0)$ *for which the limits*

$$\lim_{n\to\infty} a(h_1(n)) =: a_h, \qquad \lim_{n\to\infty} b(h_1(n)) =: b_h$$

exist and for which also the limit

$$\lim_{n\to\infty} \sigma(h_1(n), \lambda) = \sigma_h(\lambda)$$

exists uniformly with respect to $\lambda \in [-R, R]$ *for every* $R > 0$.

Let us finally consider a composed curve Γ which is the union of a finite number of simple smooth curves, and let $t_1, \ldots, t_L$ refer to the nodes of Γ. We suppose that each node t_l has a neighborhood U_l which contains no other node of Γ and such that $\Gamma \cap U_l$ is a curve of the form considered above.

We consider the singular integral operator

$$A_\Gamma = a_\Gamma I + b_\Gamma S_\Gamma : L^2(\Gamma, w) \to L^2(\Gamma, w)$$

where the coefficients a_Γ, b_Γ and the weight w are C^∞-functions on $\Gamma \setminus \{t_1, \ldots, t_L\}$ which moreover satisfy the above-given conditions on each of the neighborhoods U_l. The operator A_Γ is bounded, which follows from the general theory of singular integral operators on Carleson curves (see, for instance, [23]). From Allan's local principle (Theorem 2.3.16) it becomes obvious that

$$\sigma_{ess}(A_\Gamma) = \bigcup_{t \in \Gamma} \sigma_t(A_\Gamma). \tag{4.72}$$

It is further well known (and not hard to see) that if $t \in \Gamma$ is not a node, then

$$\sigma_t(A_\Gamma) = \{a(t) + b(t),\, a(t) - b(t)\}. \tag{4.73}$$

Thus, equality (4.72) together with (4.71) and (4.73) gives a complete description of the essential spectrum of a singular integral operator with slowly oscillating coefficients on $L^2(\Gamma, w)$ if (Γ, w) belongs locally (on a neighborhood of each point $t \in \Gamma$) to A_2^{SO}.

4.7 Comments and references

Comprehensive introductions into the world of pseudodifferential operators are [76, 90, 175, 181]. Our presentation of the material in Section 4.1 follows [127].

The bi-discretization method proposed in Section 4.2, the construction of the Wiener algebra of operators on $L^2(\mathbb{R}^N)$, and the basic Fredholm results for operators in the Wiener algebra (hence, for pseudodifferential operators in $OPS^0_{0,0}$) from Section 4.3, together with the applications to concrete operators presented in Section 4.4, are taken from [134]. Similar operator algebras of Wiener type were considered by Sjöstrand [170, 171] and Boulkhemair [36].

One should mention that the standard approach to Fredholmness of pseudodifferential operators makes use of composition formulas (see, for instance, [164, 175, 76, 127]). This approach does not work for operators in $OPS^m_{0,0}$. So new tools are needed, and it is our hope that we convinced the readers of that the limit operators method is a quite promising tool.

The Fredholm theory of pseudodifferential operators in $OPS^m_{1,0}$ with symbols which are slowly oscillating with respect to the spatial variable x has been considered by Grushin [71].

For elliptic differential operators with almost-periodic symbols with respect to x, conditions for the invertibility are given in Shubin [162, 163], Fedosov and Shubin [54] and Coburn, Moyer and Singer [39]. These conditions are based upon the concept of the almost periodic index.

Pseudodifferential operators with analytic symbols on exponentially weighted spaces are considered in [122], whereas [128] deals with a more general class of pseudodifferential operators on such spaces. The results concerning Mellin pseudodifferential on weighted spaces are taken from [121] and [26].

In Section 4.6 we follow the paper [27] by Böttcher, Karlovich and one of the authors. In the case of nice curves Γ and nice weights w, the spectrum of the operator S_Γ of singular integration along Γ, considered on the Lebesgue space $L^p(\Gamma, w)$ with power weight w has been known since the sixties from the work of Widom, Gohberg and Krupnik [65, 186]. This spectrum had been determined for nice curves Γ and arbitrary Muckenhoupt weights w by Spitkovsky [173] in the early nineties, and only very recently, the spectrum of S_Γ was completely identified by Böttcher, Karlovich [24] and Bishop, Böttcher, Karlovich and Spitkovsky [17] for general (composed) Carleson curves Γ and general Muckenhoupt weights w (also see [23]).

The approach of [24] and [17] is based on Wiener-Hopf factorization and on the use of sub-multiplicative functions. An entirely different approach, making use of Mellin convolutions and Mellin pseudodifferential operators, was elaborated in [150, 169] and [119, 123, 125]. In [25, 26], it is shown how Mellin techniques can be applied to study the spectrum of S_Γ on $L^p(\Gamma, w)$ for large classes of Carleson curves and Muckenhoupt weights.

Chapter 5

Pseudodifference Operators

The aim of this chapter is to study pseudodifference operators on weighted l^p-spaces over $\mathbb{Z}^N$. Pseudodifference operators can be viewed of as the discrete analogs of the standard pseudodifferential operators on $\mathbb{R}^N$ (see, for instance, the monographs [175, 164, 174, 76]). Notice also that, under some special conditions on the symbol a, the dual operator to a pseudodifference operator with respect to the discrete Fourier transform is a pseudodifferential operator on the torus $\mathbb{T}^N$. Operators of this kind have been studied in [2, 3, 4, 5, 154].

Particular attention will be paid to the following questions:

- Fredholm properties and essential spectrum of pseudodifference operators on general exponentially weighted spaces,
- Phragmen-Lindelöf type theorems on the exponential decay at infinity of solutions to pseudodifference equations,
- the description of the essential spectrum of discrete Schrödinger operators and to the decay of their eigenfunctions at infinity.

5.1 Pseudodifference operators

This section deals with some general aspects of pseudodifference operators. This class includes difference operators as well as their inverses and inverses modulo compact operators, if the latter exist. We formulate here the main theorems on the calculus of pseudodifference operators, the boundedness in l^p-spaces, and the inverse closedness of the Frechét algebra of the pseudodifference operators. These results are analogs of the well-known results for pseudodifferential operators as mentioned in Chapter 4 (see [175, 164, 174, 127]).

For w a positive function on $\mathbb{Z}^N$ and $1 \leq p \leq \infty$, we write $l^p_w(\mathbb{Z}^N)$ for the Banach space of all functions f on $\mathbb{Z}^N$ such that $wf \in l^p(\mathbb{Z}^N)$ and

$$\|f\|_{p,w} := \|wf\|_p.$$

We further denote by $S(\mathbb{Z}^N)$ the Fréchet space of all functions u on $\mathbb{Z}^N$ such that

$$[u]_k := \sup_{x \in \mathbb{Z}^N} \langle x \rangle^k |u(x)| < \infty, \quad \text{where} \quad \langle x \rangle := (1 + |x|^2)^{1/2}$$

for each non-negative integer k. The space $S(\mathbb{Z}^N)$ is densely embedded into each of the spaces $l^p(\mathbb{Z}^N)$ with $1 \le p < \infty$. We denote by $S'(\mathbb{Z}^N)$ the corresponding space of tempered distributions on $S(\mathbb{Z}^N)$. Finally, we will need the space $c_{00}(\mathbb{Z}^N)$ of all function on $\mathbb{Z}^N$ which have a finite support. This space lies densely both in $S(\mathbb{Z}^N)$ and in all spaces $l^p(\mathbb{Z}^N)$ with $1 \le p < \infty$. The *discrete Fourier transform* $\hat{u}$ of $u \in S(\mathbb{Z}^N)$ is defined by

$$\hat{u}(t) := (Fu)(t) = \sum_{x \in \mathbb{Z}^N} u(x) t^{-x}, \quad t \in \mathbb{T}^N.$$

The operator F of discrete Fourier transform provides us with a topological isomorphism

$$F : S(\mathbb{Z}^N) \to C^\infty(\mathbb{T}^N),$$

the inverse of which is given by

$$(F^{-1} f)(x) = \int_{\mathbb{T}^N} f(t) t^x \, d\mu(t), \quad x \in \mathbb{Z}^N$$

where

$$\mu(t) := (2\pi i)^{-N} \frac{dt_1}{t_1} \cdots \frac{dt_N}{t_N}$$

is the Haar measure on the N-dimensional torus $\mathbb{T}^N$. It is well known that the discrete Fourier transform extends by continuity to an (almost) unitary operator $F : l^2(\mathbb{Z}^N) \to L^2(\mathbb{T}^N)$ such that Parseval's identity

$$\|\hat{u}\|_{L^2(\mathbb{T}^N)} = (2\pi)^{N/2} \|u\|_{l^2(\mathbb{Z}^N)}$$

holds.

Definition 5.1.1 *Let a be a complex-valued function defined on $\mathbb{Z}^N \times \mathbb{T}^N$ such that*

$$|a|_k := \sup_{(x,\,t) \in \mathbb{Z}^N \times \mathbb{T}^N,\, \alpha \in \mathbb{N}^N : |\alpha|_\infty \le k} |\partial_t^\alpha a(x,\, t)| < \infty \tag{5.1}$$

for each non-negative integer k. The class $\mathcal{S}$ of all functions enjoying this property is provided with the topology defined by the family of semi-norms $a \mapsto |a|_k$ with $k \ge 0$. To each function $a \in \mathcal{S}$, there is associated a pseudodifference operator *$Op(a)$, which maps a function $u \in S(\mathbb{Z}^N)$ into the function $Op(a)u$ on $\mathbb{Z}^N$, given by*

$$(Op(a)u)(x) := \int_{\mathbb{T}^N} a(x,\, t) \hat{u}(t) t^x \, d\mu(t), \quad x \in \mathbb{Z}^N. \tag{5.2}$$

The class of all pseudodifference operators with symbol $a \in \mathcal{S}$ will be denoted by $OP\mathcal{S}$.

In case

$$a(x, t) = \sum_{|\alpha|_\infty \le m} a_\alpha(x) t^{-\alpha}$$

is a polynomial with respect to $t \in \mathbb{T}^N$, the pseudodifferential operator $Op(a)$ acts as the difference operator

$$u \mapsto \sum_{|\alpha|_\infty \le m} a_\alpha V_\alpha u, \tag{5.3}$$

where the V_α are shift operators,

$$(V_\alpha u)(x) := u(x - \alpha) \quad \text{for } \alpha \in \mathbb{Z}^N.$$

Among the most important and best-studied classes of difference operators are those with $m = 1$, which are also called *Jacobi operators.* An archetypal example of a Jacobi operator is the discrete Schrödinger operator,

$$\mathbf{H} := \sum_{j=1}^{N} (V_{e_j} + V_{e_j}^{-1}) + aI, \tag{5.4}$$

where the $e_j := (0, \ldots, 0, 1, 0, \ldots, 0)$ (with the 1 standing at the jth place) are unit vectors and where aI is the operator of multiplication by the function $a \in l^\infty(\mathbb{Z}^N)$.

Below we will give, partially without proofs, some results concerning the calculus of pseudodifference operators. These results are completely parallel to the well-known results for usual pseudodifferential operators on $\mathbb{R}^N$, as they are cited in Section 4.1 and as they can be found in [76, 127, 164, 174, 175], for instance.

Proposition 5.1.2 *Let $a \in \mathcal{S}$ and $u \in S(\mathbb{Z}^N)$. Then $Op(a)u$ is in $S(\mathbb{Z}^N)$ again, and the operator $Op(a) : S(\mathbb{Z}^N) \to S(\mathbb{Z}^N)$ is bounded.*

By the definition of the discrete Fourier transform,

$$(Op(a)u)(x) = \int_{\mathbb{T}^N} a(x, t) \sum_{y \in \mathbb{Z}^N} u(y) t^{x-y} \, d\mu(t), \quad u \in S(\mathbb{Z}^N),$$

which suggests to write the operator $Op(a)$ as

$$(Op(a)u)(x) = \sum_{y \in \mathbb{Z}^N} \int_{\mathbb{T}^N} a(x, t) u(y) t^{x-y} \, d\mu(t). \tag{5.5}$$

This makes perfectly sense if the right side part of (5.5) is interpreted as follows, which is an analogue of an oscillatory integral, namely

$$(Op(a)u)(x) = \sum_{y \in \mathbb{Z}^N} \int_{\mathbb{T}^N} \langle D_t \rangle^{2k} a(x, t) \langle y \rangle^{-2k} u(x + y) t^{-y} \, d\mu(t) \tag{5.6}$$

(cf. Section 4.1 and [127, 175]). Here we choose $2k > N$ and set

$$\langle D_t \rangle^2 := I - \sum_{j=1}^{N} (t_j \frac{\partial}{\partial t_j})^2.$$

One can show that the right-hand side of (5.6) does not depend on k if $2k > N$. The representation (5.6) allows us to define $Op(a)u$ also for functions u in $l^\infty(\mathbb{Z}^N)$.

Proposition 5.1.3 *If $a \in \mathcal{S}$, then the operator $Op(a) : l^\infty(\mathbb{Z}^N) \to l^\infty(\mathbb{Z}^N)$ defined by* (5.6) *is bounded.*

Proposition 5.1.4 *Let $a, b \in \mathcal{S}$. Then $Op(a)Op(b) \in OP\mathcal{S}$, and $Op(a)Op(b) = Op(c)$ with*

$$c(x, t) := \sum_{y \in \mathbb{Z}^N} \int_{\mathbb{T}^N} a(x, t\tau) b(x + y, t) \tau^{-y} \, d\mu(\tau),$$

with the right-hand side being understood in the oscillatory sense.

Given an operator A on $S(\mathbb{Z}^N)$, we define its *formal adjoint* as the operator A^* which satisfies

$$\langle Au, v \rangle_{l^2(\mathbb{Z}^N)} = \langle u, A^* v \rangle_{l^2(\mathbb{Z}^N)} \quad \text{for all } u, v \in S(\mathbb{Z}^N).$$

Proposition 5.1.5 *If $a \in \mathcal{S}$, then the formal adjoint $Op(a)^*$ belongs to $OP\mathcal{S}$ again, and $Op(a)^* = Op(a^*)$ with*

$$a^*(x, t) := \sum_{y \in \mathbb{Z}^N} \int_{\mathbb{T}^N} \overline{a(x + y, t\tau)} \tau^{-y} \, d\mu(\tau).$$

Corollary 5.1.6 *Operators in $OP\mathcal{S}$ are bounded on $S'(\mathbb{Z}^N)$.*

Indeed, one defines the action of $Op(a) \in OP\mathcal{S}$ at $u \in S'(\mathbb{Z}^N)$ by

$$(Op(a)u)(v) := u(Op(a)^* v), \qquad v \in S(\mathbb{Z}^N),$$

where $Op(a)^*$ is the formal adjoint of $Op(a)$. This definition makes sense since $Op(a)^*$ is a bounded operator on $S(\mathbb{Z}^N)$ by the preceding proposition. Moreover, the boundedness of $Op(a)^*$ on $S(\mathbb{Z}^N)$ implies the boundedness of $Op(a)$ on $S'(\mathbb{Z}^N)$. □

Next we present the discrete analogs of pseudodifferential operators with double symbols. A *double symbol* is a function a on $\mathbb{Z}^N \times \mathbb{Z}^N \times \mathbb{T}^N$ which satisfies the condition

$$|a|_k := \sup_{(x, y, t) \in \mathbb{Z}^N \times \mathbb{Z}^N \times \mathbb{T}^N, \, \alpha \in \mathbb{N}^N : |\alpha| \le k} |\partial_t^\alpha a(x, y, t)| < \infty \tag{5.7}$$

for each non-negative integer k. We let $\mathcal{S}_d$ refer to the set of all double symbols. Each function $a \in \mathcal{S}_d$ induces an operator $Op_d(a)$, called the *pseudodifference operator with double symbol* a, via

$$\begin{aligned}(Op_d(a)u)(x) &:= \int_{\mathbb{T}^N} \sum_{y \in \mathbb{Z}^N} a(x, y, t)u(y)t^{x-y}\, d\mu(t) \\ &= \sum_{y \in \mathbb{Z}^N} \int_{\mathbb{T}^N} a(x, y, t)u(y)t^{x-y}\, d\mu(t) \qquad (5.8)\end{aligned}$$

where $u \in S(\mathbb{Z}^N)$, and where the right-hand side of (5.8) is understood in the oscillatory sense. We denote the class of all pseudodifference operators with double symbol by $OP\mathcal{S}_d$.

Proposition 5.1.7 *Let $a \in \mathcal{S}_d$. Then $Op_d(a) \in OP\mathcal{S}$ and $Op_d(a) = Op(a^\sharp)$ with*

$$a^\sharp(x, t) = \sum_{y \in \mathbb{Z}^N} \int_{\mathbb{T}^N} a(x, x + y, t\tau)\tau^{-y}\, d\mu(\tau), \qquad (5.9)$$

where the right-hand side is understood in the oscillatory sense.

The following can be considered as the discrete version of the Calderon-Vaillancourt theorem for pseudodifferential operators.

Theorem 5.1.8 *If $a \in \mathcal{S}$, then the pseudodifference operator $Op(a)$ is bounded on $l^p(\mathbb{Z}^N)$ for all $1 \le p \le \infty$, and*

$$\|Op(a)\|_{L(l^p(\mathbb{Z}^N))} \le C|a|_{2k}$$

whenever $2k > N$ and with a constant $C > 0$ independent of A (but depending on k).

Thus, one has an obvious difference between pseudodifferential operators on $\mathbb{R}^N$, which are bounded on $L^p(\mathbb{R}^N)$ only when $1 < p < \infty$ (see, for instance, [174], Chapter VII), and pseudodifference operators on $l^p(\mathbb{Z}^N)$, which are bounded for all $1 \le p \le \infty$.

Proof. Write $a \in \mathcal{S}$ as

$$a(x, t) = \sum_{\alpha \in \mathbb{Z}^N} a_\alpha(x)t^{-\alpha}$$

where

$$a_\alpha(x) := \int_{\mathbb{T}^N} a(x, t)t^\alpha\, d\mu(t). \qquad (5.10)$$

Consequently,

$$Op(a) = \sum_{\alpha \in \mathbb{Z}^N} a_\alpha V_\alpha. \qquad (5.11)$$

Integrating by parts in (5.10), we obtain the estimate

$$|a_\alpha(x)| \le |\langle D_t\rangle^{2k} a(x, t)| \langle\alpha\rangle^{-2k} \le C|a|_{2k}\langle\alpha\rangle^{-2k},$$

whence

$$\|a_\alpha\|_\infty \le C|a|_{2k}\langle\alpha\rangle^{-2k}$$

for all $k \ge 0$. In case $2k > N$, this implies

$$\sum_{\alpha\in\mathbb{Z}^N} \|a_\alpha\|_\infty \le C|a|_{2k}$$

with a constant C depending on k only. Since $\|V_\alpha\|_{L(l^p(\mathbb{Z}^N))} = 1$ for all $p \in [1, \infty]$, we conclude from (5.11) that

$$\|Op(a)\|_{L(l^p(\mathbb{Z}^N))} \le \sum_{\alpha\in\mathbb{Z}^N} \|a_\alpha\|_\infty \le C|a|_{2k}$$

for each k greater than $2N$. □

Corollary 5.1.9 *If $a \in S_d$, then the pseudodifference operator with double symbol a is bounded on $l^p(\mathbb{Z}^N)$ for every $1 \le p \le \infty$, and*

$$\|Op_d(a)\|_{L(l^p(\mathbb{Z}^N))} \le C|a|_{4k}$$

whenever $2k > N$ and with a constant $C > 0$ independent of A (but depending on k).

Proof. From Proposition 5.1.7 we conclude that $Op_d(a)$ is the pseudodifference operator $Op(a^\sharp)$ where

$$a^\sharp(x, t) := \sum_{y\in\mathbb{Z}^N} \langle y\rangle^{-2k} \int_{\mathbb{T}^N} \left(\langle D_\tau\rangle^{2k} a(x, x+y, t\tau)\right) \tau^{-y}\, d\mu(\tau)$$

with a k greater than $N/2$. Thus,

$$|a^\sharp(x, t)| \le C|a|_{2k}$$

with a constant $C > 0$ independent of a. In the same way, one gets

$$|\partial_t^\alpha a^\sharp(x, t)| \le C|a|_{2k+|\alpha|},$$

which implies that $|a^\sharp|_{2k} \le C|a|_{4k}$. Hence, if we choose $k > N/2$, then Theorem 5.1.8 implies the boundedness of $Op_d(a)$ on each of the spaces $l^p(\mathbb{Z}^N)$ with $p \in [1, \infty]$ as well as the desired estimate. □

Given a linear operator $B : S(\mathbb{Z}^N) \to S'(\mathbb{Z}^N)$ and an integer $j \in \{1, \dots, N\}$, we let x_jI stand for the operator of multiplication by the function

$$x = (x_1, \dots, x_N) \mapsto x_j,$$

and we define

$$L_j B := -(ix_j B - Bix_j I).$$

Further, for every multi-index $\alpha := (\alpha_1, \ldots, \alpha_N)$, we set $B^{(\alpha)} := L_1^{\alpha_1} \cdots L_N^{\alpha_N} B$. The following theorem establishes an equivalent characterization of pseudodifference operators. It is the analogue of a well-known result by Beals ([14]) holding for pseudodifferential operators.

Theorem 5.1.10 *A linear operator $B : \mathcal{S}(\mathbb{Z}^N) \to \mathcal{S}'(\mathbb{Z}^N)$ belongs to $OP\mathcal{S}$ if and only if, for every multi-index α, the operator $B^{(\alpha)}$ has a continuous extension to a bounded operator from $l^2(\mathbb{Z}^N)$ into itself.*

Corollary 5.1.11 *If the operator $A \in OP\mathcal{S}$ is invertible when considered as an operator on $l^2(\mathbb{Z}^N)$, then A^{-1} belongs to $OP\mathcal{S}$ again.*

Proof. For $j \in \{1, \ldots, N\}$, one has $L_j A^{-1} = -A^{-1}(L_j A)A^{-1}$ and, thus,

$$\|L_j A^{-1}\| \leq \|A^{-1}\|^2 \|L_j A\| =: C_j.$$

Iterated application of this estimate gives, for each multi-index α, a constant C_α such that

$$\|L_1^{\alpha_1} \cdots L_N^{\alpha_N} A^{-1}\| \leq C_\alpha.$$

By (the analogue of) Beals' Theorem 5.1.10, the operator A^{-1} belongs to $OP\mathcal{S}$.□

5.2 Fredholmness of pseudodifference operators

In this section, we are going to show that the class of pseudodifference operators under consideration is included in the Wiener algebra $\mathcal{W}$ of band-dominated operators introduced in Section 2.5, and we apply Theorem 2.5.7 to establish Fredholm criteria and to describe the essential spectrum of pseudodifference operators in terms of their limit operators.

The following proposition can be derived from Theorem 5.1.8 in the very same way as its analogue for pseudodifferential operators on $\mathbb{R}^N$ (Corollary 4.3.3).

Proposition 5.2.1 $OP\mathcal{S} \subset \mathcal{W}$.

Let us mention that in the case at hand the Wiener algebra $\mathcal{W}$ coincides with the rich Wiener algebra $\mathcal{W}^\$$. Thus, by Proposition 5.1.7, both $OP\mathcal{S}$ and $OP\mathcal{S}_d$ are contained in the rich Wiener algebra. Thus, as consequences of Theorem 2.5.7 and its Corollary 2.5.8, we have the following results which we formulate for pseudodifference operators in $OP\mathcal{S}$ only. In order to indicate the dependence on the underlying Banach space, we denote the essential spectrum of an operator $A \in L(X)$ by $\sigma_X^{ess}(A)$ and we write $\sigma_X(A)$ for the usual spectrum of A on X.

Theorem 5.2.2 *The following assertions are equivalent for an operator $A \in OPS$:*

(a) *There is a $p \in [1, \infty]$ such that A is Fredholm on $l^p(\mathbb{Z}^N)$.*

(b) *A is a Fredholm operator on $l^p(\mathbb{Z}^N)$ for each $p \in [1, \infty]$.*

(c) *There is a $p \in [1, \infty]$ such that all limit operators of A are invertible on $l^p(\mathbb{Z}^N)$.*

(d) *All limit operators of A are invertible on $l^p(\mathbb{Z}^N)$ for each $p \in [1, \infty]$, and the norms of their inverses are uniformly bounded.*

Theorem 5.2.3 *If $A \in OPS$, then the essential spectrum $\sigma^{ess}_{l^p(\mathbb{Z}^N)}(A)$ is independent of $p \in [1, \infty]$, and*

$$\sigma^{ess}_{l^p(\mathbb{Z}^N)} = \bigcup_{A_h \in \sigma_{op}(A)} \sigma_{l^r(\mathbb{Z}^N)}(A_h)$$

for each choice of $r \in [1, \infty]$.

The application of these results requires some knowledge about limit operators of pseudodifference operators.

Proposition 5.2.4 *Let $a \in \mathcal{S}$. Then each sequence $h : \mathbb{N} \to \mathbb{Z}^N$ which tends to infinity possesses a subsequence g such that there exists a function $a_g \in \mathcal{S}$ with*

$$\lim_{n\to\infty} \sup_{t\in\mathbb{T}^N} |(\partial_t^\beta a)(x + g(n), t) - (\partial_t^\beta a_g)(x, t)| = 0$$

at each point $x \in \mathbb{Z}^N$ and for each multi-index β.

Proof. Write $a \in \mathcal{S}$ as

$$a(x, t) = \sum_{\alpha\in\mathbb{Z}^N} a_\alpha(x) t^{-\alpha},$$

where the series is convergent absolutely and uniformly with respect to $(x, t) \in \mathbb{Z}^N \times \mathbb{T}^N$ together with all derivatives with respect to $t \in \mathbb{T}^N$. By a Cantor diagonal argument, one can easily check that the sequence h has a subsequence g such that

$$a_\alpha(x + g(n)) \to a_\alpha^g(x) \quad \text{as } n \to \infty \tag{5.12}$$

for each point $x \in \mathbb{Z}^N$ and each multi-index α. Moreover, the inequality (5.13) holds. Since $a \in \mathcal{S}$, one has

$$\|a_\alpha\|_\infty \le C|a|_{2k}\langle x\rangle^{-2k}$$

for arbitrary $k \in \mathbb{N}$. Hence,

$$\|a_\alpha^g\|_\infty \le C|a|_{2k}\langle x\rangle^{-2k}.$$

These estimates together with (5.12) imply the assertion. □

The analogue of Proposition 5.2.4 for double symbols reads as follows. Its proof is similar to that of the preceding proposition.

Proposition 5.2.5 *Let $a \in \mathcal{S}_d$. Then each sequence $h : \mathbb{N} \to \mathbb{Z}^N$ which tends to infinity possesses a subsequence g such that there exists a function $a_g \in \mathcal{S}_d$ with*

$$\lim_{n\to\infty} \sup_{t\in\mathbb{T}^N} |(\partial_t^\beta a)(x + g(n), y + g(n), t) - (\partial_t^\beta a_g)(x, y, t)| = 0$$

at each point $(x, y) \in \mathbb{Z}^N \times \mathbb{Z}^N$ and for each multi-index β.

In the settings of the preceding propositions, we call a_g the *limit function* of a with respect to g.

Theorem 5.2.6 *Let $a \in \mathcal{S}$ ($a \in \mathcal{S}_d$), and let $a_h \in \mathcal{S}$ ($a_h \in \mathcal{S}_d$) be the limit function of a with respect to a sequence $h : \mathbb{N} \to \mathbb{Z}^N$ tending to infinity. Then $Op(a_h)$ ($Op_d(a_h)$) is the limit operator of the pseudodifference operator $Op(a)$ ($Op_d(a)$) with respect to this sequence.*

Proof. If $A \in OP\mathcal{S}$ is of the form $\sum_\alpha a_\alpha V_\alpha$ with $\sum_\alpha \|a_\alpha\|_\infty < \infty$, and if h is a sequence tending to infinity, then

$$(V_{-h(n)} A V_{h(n)} u)(x) = \sum_{\alpha\in\mathbb{Z}^N} a_\alpha(x + h(n))(V_\alpha u)(x).$$

A Cantor diagonal argument yields that the sequence h has a subsequence g such that, for each $\alpha \in \mathbb{Z}^N$, there exists a function $a_\alpha^g \in l^\infty(\mathbb{Z}^N)$ with

$$\|a_\alpha^g\|_\infty \le \|a_\alpha\|_\infty \tag{5.13}$$

and

$$\lim_{n\to\infty} a_\alpha(x + g(n)) = a_\alpha^g(x),$$

for every $x \in \mathbb{Z}^N$. Hence, for each $1 \le p \le \infty$, the limit operator A_g of A with respect to g exists and that

$$A_g = \sum_{\alpha\in\mathbb{Z}^N} a_\alpha^g V_\alpha.$$

This settles the assertion in case $a \in \mathcal{S}$.

If $a \in \mathcal{S}_d$, then, by Proposition 5.1.7, $Op_d(a) = Op(a^\sharp)$ where

$$a^\sharp(x, t) = \sum_{y\in\mathbb{Z}^N} \int_{\mathbb{T}^N} a(x, x + y, \tau t)\tau^{-y}\, d\mu(\tau).$$

Thus,

$$a^\sharp(x + h(n), t) = \sum_{y\in\mathbb{Z}^N} \int_{\mathbb{T}^N} a(x + h(n), x + h(n) + y, \tau t)\tau^{-y}\, d\mu(\tau), \tag{5.14}$$

with the right-hand side understood in the oscillatory sense. By means of Proposition 5.2.5, one can pass to the limit as $n \to \infty$ on the right-hand side of (5.14), which yields that the function $a_h^\sharp$,

$$a_h^\sharp(x, t) := \sum_{y \in \mathbb{Z}^N} \int_{\mathbb{T}^N} a_h(x, x+y, \tau t)\tau^{-y} \, d\mu(\tau),$$

is the limit function of $a^\sharp$ with respect to h in the sense of Proposition 5.2.4. Hence, $Op(a_h^\sharp) = Op_d(a_h)$ is the limit operator of $Op_d(a)$ with respect to h. □

Thus, all limit operators of operators in $OP\mathcal{S}$ and $OP\mathcal{S}_d$ belong $OP\mathcal{S}$ and $OP\mathcal{S}_d$, respectively, and their symbols are the limit functions defined in Propositions 5.2.4 and 5.2.5.

5.3 Fredholm properties of pseudodifference operators on weighted spaces

Here we are going to consider pseudodifference operators with symbols which are analytic with respect to the second variable. We prove their boundedness in exponentially weighted spaces, with weights being compatible with the domains of analyticity of the symbols, and we consider again Fredholm properties and essential spectrum.

5.3.1 Boundedness on weighted spaces

We start with introducing an appropriate class of pseudodifference operators which are distinguished by analyticity properties of their symbols, and with defining a corresponding class of exponentially weighted functions. For $r > 1$, set

$$\mathbb{K}_r := \{z \in \mathbb{C} : 1/r < |t| < r\},$$

and for an N-tuple $r = (r_1, \ldots, r_N)$ of real numbers $r_j > 1$, we consider the multi-dimensional annular domain in $\mathcal{S}^N$,

$$\mathbb{K}_r^N := \mathbb{K}_{r_1} \times \cdots \times \mathbb{K}_{r_N}.$$

The N-tuple r and the domain $\mathbb{K}_r^N$ will be fixed throughout this section.

Definition 5.3.1 *Let $\mathcal{S}(\mathbb{K}_r^N)$ denote the set of all functions $a : \mathbb{Z}^N \times \mathbb{K}_r^N$, $(x, t) \mapsto a(x, t)$, which are analytic with respect to the variable t in the domain $\mathbb{K}_r^N$ and which satisfy the estimates*

$$|a|_k := \sum_{\alpha \in \mathbb{N}^N : |\alpha| \le k} \sup_{x \in \mathbb{Z}^N,\, t \in \mathbb{K}_r^N} |(\partial_t^\alpha a)(x, t)| < \infty$$

for each non-negative integer k. Further, each function $a \in \mathcal{S}(\mathbb{K}_r^N)$ determines a pseudodifference operator $Op(a)$ on $c_{00}(\mathbb{Z}^N)$ as in (5.2). We denote the class of all pseudodifference operators with symbol in $\mathcal{S}(\mathbb{K}_r^N)$ by $OP\mathcal{S}(\mathbb{K}_r^N)$.

Evidently, $OPS(\mathbb{K}_r^N) \subset OPS$. For the following definition, we agree upon denoting a function on $\mathbb{R}^N$ and its restriction onto $\mathbb{Z}^N$ by the same letter.

Definition 5.3.2 *Let $W(\mathbb{K}_r^N)$ denote the class of exponential weights $w = \exp v$, where v is the restriction onto $\mathbb{Z}^N$ of a function $v \in C^1(\mathbb{R}^N)$ such that, for each point $x \in \mathbb{R}^N$ and each $j \in \{1, \ldots, N\}$,*

$$-\log r_j < \frac{\partial v}{\partial x_j}(x) < \log r_j. \tag{5.15}$$

Theorem 5.3.3 *If $a \in \mathcal{S}(\mathbb{K}_r^N)$ and $w \in W(\mathbb{K}_r^N)$, then the operator $wOp(a)w^{-1}I$ (which is previously defined on $c_{00}(\mathbb{Z}^N)$) belongs to the class OPS_d. Moreover, $wOp(a)w^{-1}I = Op_d(\tilde{a})$ with*

$$\tilde{a}(x, y, t) = a(x, e^{-\theta(x, y)} \cdot t)$$

where $s \cdot t$ stands for the N-tuple $(s_1t_1, \ldots, s_Nt_N)$ whenever $s, t \in \mathbb{C}^N$, and where

$$\theta(x, y) := \int_0^1 (\nabla v)((1-t)x + ty)dt.$$

Proof. Let $u \in c_{00}(\mathbb{Z}^N)$. Then $wAw^{-1}u$ can be represented in the form

$$(wAw^{-1}u)(x) = (2\pi)^{-N} \sum_{y \in \mathbb{Z}^N} u(y) \int_{[-\pi,\,\pi]^N} a(x, e^{i\xi}) e^{i\langle x-y,\, \xi - i\theta(x,\,y)\rangle}\, d\xi. \tag{5.16}$$

Notice that the convexity of the rectangle $Q := \prod_{j=1}^N (-\log a_j, \log a_j)$ implies that $\theta(x, y) \in Q$ for arbitrary points $x, y \in \mathbb{Z}^N$. Changing the variable $\eta := \xi - i\theta(x, y)$ in (5.16), we find

$$(wAw^{-1}u)(x) = (2\pi)^{-N} \sum_{y \in \mathbb{Z}^N} u(y) \int_{[-\pi,\,\pi]^N - i\theta(x,\,y)} a(x, e^{-\theta(x,y)} \cdot e^{i\eta}) e^{i\langle x-y,\,\eta\rangle}\, d\eta.$$

Setting $t = e^{i\eta} = (e^{i\eta_1}, \ldots, e^{i\eta_N})$, we further obtain

$$(wAw^{-1}u)(x) = \sum_{y \in \mathbb{Z}^N} \int_{e^{\theta(x,\,y)} \cdot \mathbb{T}^N} a(x, e^{-\theta(x,y)} \cdot t) u(y) t^{x-y} d\mu(t) \tag{5.17}$$

where

$$e^{\theta(x,\,y)} \cdot \mathbb{T}^N = \{z \in \mathbb{C}^N : |z_j| = e^{\theta_j(x,\,y)} \text{ for } j = 1, \ldots, N\}.$$

Note that $t \mapsto a(x, e^{-\theta(x,y)} \cdot t)$ is an analytic function of the complex variable $t \in \mathbb{C}^N$ in the domain

$$e^{\theta(x,\,y)} \cdot \mathbb{K}_r^N = \{z \in \mathbb{C}^N : \frac{e^{\theta_j(x,\,y)}}{r_j} < |z_j| < e^{\theta_j(x,\,y)} r_j \text{ for } j = 1, \ldots, N\}.$$

We claim that the domain $e^{\theta(x,y)} \cdot \mathbb{K}_r^N$ contains both the deformed torus $e^{\theta(x,y)} \cdot \mathbb{T}^N$ and the torus $\mathbb{T}^N$. Indeed, the inclusion $e^{\theta(x,y)} \cdot \mathbb{T}^N \subset e^{\theta(x,y)} \cdot \mathbb{K}_r^N$ is evident. Further, the estimates (5.15) imply that

$$\frac{e^{\theta_j(x,y)}}{r_j} < 1 < e^{\theta_j(x,y)} r_j \quad \text{for } j = 1, \ldots, N,$$

whence the second inclusion. Thus, the Cauchy-Poincaré theorem ([185], p. 233) entitles us to replace the integration along the deformed torus $e^{\theta(x,y)} \cdot \mathbb{T}^N$ by integration along $\mathbb{T}^N$ in (5.17). So we arrive at

$$(wAw^{-1}u)(x) = \sum_{y \in \mathbb{Z}^N} \int_{\mathbb{T}^N} a(x,\, e^{-\theta(x,y)} \cdot t) t^{x-y} u(y)\, d\mu(t) = Op_d(\tilde{a}).$$

Finally, it follows from Definition 5.3.1 that $\tilde{a}$ is a double symbol in $\mathcal{S}_d$. □

Corollary 5.3.4 *Let $a \in \mathcal{S}(\mathbb{K}_r^N)$ and $w \in W(\mathbb{K}_r^N)$. Then the operator $Op(a)$ is bounded on each of the spaces $l_w^p(\mathbb{Z}^N)$ with $p \in [1, \infty]$.*

Indeed, an operator $A : l_w^p(\mathbb{Z}^N) \to l_w^p(\mathbb{Z}^N)$ is bounded if and only if the operator $wAw^{-1}I : l^p(\mathbb{Z}^N) \to l^p(\mathbb{Z}^N)$ is bounded. Since Theorem 5.3.3 states that the operator $wOp(a)w^{-1}I$ belongs to $OP\mathcal{S}_d$, the assertion follows from Proposition 5.1.7 and Theorem 5.1.8. □

5.3.2 Fredholmness on weighted spaces

Here we consider the Fredholmness of operators in $OP\mathcal{S}(\mathbb{K}_r^N)$ on weighted spaces $l_w^p(\mathbb{Z}^N)$. It is evident that an operator $A : l_w^p(\mathbb{Z}^N) \to l_w^p(\mathbb{Z}^N)$ is Fredholm operator if and only if the operator $wAw^{-1}I : l^p(\mathbb{Z}^N) \to l^p(\mathbb{Z}^N)$ is Fredholm. Now, if $A = Op(a) \in OP\mathcal{S}(\mathbb{K}_r^N)$, then, by Theorem 5.3.3, the operator $wAw^{-1}I$ belongs to $OP\mathcal{S}_d$, and $wAw^{-1}I = Op_d(\tilde{a})$ with the double symbol $\tilde{a}$ defined as in Theorem 5.3.3. Thus, and due to Propositions 5.1.7 and 5.2.1, the Theorems 5.2.2 and 5.2.3 apply to the problem of Fredholmness of $Op(a)$ on $l_w^p(\mathbb{Z}^N)$, and they yield the following.

Theorem 5.3.5 *Let $a \in OP\mathcal{S}(\mathbb{K}_r^N)$ and $w \in W(\mathbb{K}_r^N)$, and let $\tilde{a}$ be as in Theorem 5.3.3. Then the following assertions are equivalent.*

(a) *There is a $p \in [1, \infty]$ such that the operator $Op(a)$ is Fredholm on $l_w^p(\mathbb{Z}^N)$.*

(b) *The operator $Op(a)$ is Fredholm on each of the spaces $l_w^p(\mathbb{Z}^N)$ with $1 \le p \le \infty$.*

(c) *There is an $r \in [1, \infty]$ such that all limit operators of $Op(\tilde{a})$ are invertible on $l^r(\mathbb{Z}^N)$.*

(d) *All limit operators of $Op(\tilde{a})$ are invertible on each of the spaces $l^p(\mathbb{Z}^N)$ with $1 \le p \le \infty$, and the norms of their inverses are uniformly bounded.*

Theorem 5.3.6 *Let the notations and hypotheses be as in the preceding theorem. Then* $\sigma^{ess}_{l^p_w(\mathbb{Z}^N)}(A)$ *does not depend on the choice of* $p \in [1, \infty]$, *and*

$$\sigma^{ess}_{l^p_w(\mathbb{Z}^N)}(A) = \bigcup_{Op(\tilde{a}_h) \in \sigma_{op}(Op(\tilde{a}))} \sigma_{l^r(\mathbb{Z}^N)}(Op(\tilde{a}_h))$$

for arbitrarily chosen $r \in [1, \infty]$.

5.3.3 The Phragmen-Lindelöf principle

The following result can be viewed of as an analogue of the well-known Phragmen-Lindelöf principle, which states that the solutions to certain equations are better than expected (see, for example, [96], page 284).

Theorem 5.3.7 *Let* $a \in \mathcal{S}(\mathbb{K}^N_r)$, *and let* w *be a weight function in* $W(\mathbb{K}^N_r)$ *with* $\lim_{x\to\infty} w(x) = +\infty$. *Suppose further that, for each* $\mu \in [-1, 1]$, *all limit operators of* $w^\mu Op(a) w^{-\mu} I$ *are invertible on* $l^r(\mathbb{Z}^N)$ *for some* $r \in [1, \infty]$.

If $1 < p < \infty$, *and if* u *is a solution to the equation*

$$Au = f \qquad \text{with } f \in l^p_w(\mathbb{Z}^N) \tag{5.18}$$

(which a priori belongs to the space $l^p_{w^{-1}}(\mathbb{Z}^N)$*), then* $u \in l^p_w(\mathbb{Z}^N)$.

Proof. For $w \in W(\mathbb{K}^N_r)$, it is immediate from Definition 5.3.2 that $w^\mu \in W(\mathbb{K}^N_r)$ for every $\mu \in [-1, 1]$. From Theorem 5.3.3 we know that $w^\mu Op(a) w^{-\mu} I = Op_d(\tilde{a}_\mu)$, where

$$\tilde{a}_\mu(x, y, t) := a(x, e^{-\mu\theta(x, y)} \cdot t).$$

Since all limit operators of $Op_d(\tilde{a}_\mu)$ are invertible for each $\mu \in [-1, 1]$ by hypothesis, Theorem 5.3.5 implies the Fredholmness of $w^\mu Op(a) w^{-\mu} I$ on each of the spaces $l^p(\mathbb{Z}^N)$ with $1 < p < \infty$.

Moreover, it follows from Corollary 5.1.9, that the family

$$w^\mu Op(a) w^{-\mu} I \,.\, l^r(\mathbb{Z}^N) \to l^r(\mathbb{Z}^N) \tag{5.19}$$

depends continuously on the parameter $\mu \in [-1, 1]$. Hence, the Fredholm index of the operators (5.19) is independent of the choice of μ. Consequently, also the operator

$$Op(a) : l^p_{w^\mu}(\mathbb{Z}^N) \to l^p_{w^\mu}(\mathbb{Z}^N)$$

is a Fredholm operator for each $\mu \in [-1, 1]$, and its Fredholm index does not depend on μ.

Now we can suppose without loss of generality that $w(x) > 1$ for all $x \in \mathbb{Z}^N$. Then $w^\mu \le w$. Hence, the space $l^p_w(\mathbb{Z}^N)$ is densely embedded into $l^p_{w^\mu}(\mathbb{Z}^N)$ for all $p \in (1, \infty)$ and all $\mu \in [-1, 1]$. Then it follows from [59], p. 308, that equation (5.18) has the same solutions in all spaces $l^p_{w^\mu}(\mathbb{Z}^N)$ with $\mu \in [-1, 1]$. This implies the assertion. □

5.4 Slowly oscillating pseudodifference operators

This section is devoted to pseudodifference operators with slowly oscillating symbols with respect to the spatial variable $x \in \mathbb{Z}^N$, considered as acting on spaces with slowly oscillating weights. For operators in this class, the limit operators are invariant with respect to shift, which allows us to give an effective description of the essential spectrum of such operators. Other results of this section are a Fredholm index formula for pseudodifference operators on $l^p_w(\mathbb{Z})$, and a Phragmen-Lindelöf type theorem on the exponential decay of the solutions to pseudodifference equations.

5.4.1 Fredholmness on l^p-spaces

Here we are going to introduce a class of symbols a for which the limit operators of the pseudodifference operator $Op(a)$ get a quite simple form, which will allow us to derive explicit and effective Fredholm criteria. We start with recalling the definition of a slowly oscillating function from Section 2.4.1 and with defining corresponding classes of slowly oscillating symbols.

Definition 5.4.1

(a) *A function* $f \in l^\infty(\mathbb{Z}^N)$ *is called* slowly oscillating *if*

$$\lim_{z \to \infty} |f(x+z) - f(z)| = 0$$

for each $x \in \mathbb{Z}^N$. *The class of all slowly oscillating functions is denoted by* SO.

(b) *A symbol* $a \in \mathcal{S}$ *is* slowly oscillating *if*

$$\lim_{z \to \infty} \sup_{t \in \mathbb{T}^N} |a(x+z, t) - a(z, t)| = 0$$

for each $x \in \mathbb{Z}^N$. *The class of all slowly oscillating symbols is denoted by* $\mathcal{SO}$.

(c) *A double symbol* $a \in \mathcal{S}_d$ *is* slowly oscillating *if*

$$\lim_{z \to \infty} \sup_{t \in \mathbb{T}^N} |a(x+z, y+z, t) - a(z, z, t)| = 0$$

for each pair $(x, y) \in \mathbb{Z}^N \times \mathbb{Z}^N$. *We denote the class of all slowly oscillating double symbols by* $\mathcal{SO}_d$.

(d) *We write* $\mathcal{C}_0$ *for the subset of* $\mathcal{SO}$ *consisting of all symbols* a *with*

$$\lim_{x \to \infty} \sup_{t \in \mathbb{T}^N} |(\partial_t^\alpha a)(x, t)| = 0$$

for each multi-index α.

The classes of associated pseudodifference operators will be denoted by $OP\mathcal{SO}$, $OP\mathcal{SO}_d$, and $OP\mathcal{C}_0$. It is easy to check that if a is a double symbol in $\mathcal{SO}_d$, then the function $a^\sharp$ defined by (5.9) is a symbol in $\mathcal{SO}$.

Proposition 5.4.2

(a) *If* $a, b \in \mathcal{SO}$, *then*

$$Op(a)Op(b) - Op(ab) \in OP\mathcal{C}_0.$$

(b) *Let* $a \in \mathcal{SO}_d$ *and* $b(x, \xi) := a(x, x, \xi)$. *Then*

$$Op_d(a) - Op(b) \in OP\mathcal{C}_0.$$

(c) *If* $a \in \mathcal{SO}$, *then*

$$Op(a)^* - Op(a^*) \in OP\mathcal{C}_0.$$

Proof. To prove assertion (a), recall from Proposition 5.1.4 that $Op(a)Op(b) = Op(c)$ with

$$c(x, t) := \sum_{y \in \mathbb{Z}^N} \int_{\mathbb{T}^N} a(x, t\tau) b(x + y, t) \tau^{-y} \, d\mu(\tau).$$

Hence,

$$c(x, t) = \left(\sum_{y \in \mathbb{Z}^N} \int_{\mathbb{T}^N} a(x, t\tau) \tau^{-y} \, d\mu(\tau) \right) b(x, t) + r(x, t) \tag{5.20}$$

where

$$r(x, t) := \sum_{y \in \mathbb{Z}^N} \int_{\mathbb{T}^N} a(x, t\tau)(b(x + y, t) - b(x, t)) \tau^{-y} \, d\mu(\tau).$$

The first term on the right-hand side of (5.20) is nothing but $a(x, t)b(x, t)$. To estimate the second term, write $r(x, t)$ as

$$\sum_{y \in \mathbb{Z}^N} \langle y \rangle^{-2(k+1)} \int_{\mathbb{T}^N} \langle D_\tau \rangle^{2(k+1)} a(x, t\tau)(b(x + y, t) - b(x, t)) \tau^{-y} \, d\mu(\tau)$$

where k is an arbitrary non-negative integer. Choose $2k + 1 > N$. Then

$$\sup_{t \in \mathbb{T}^N} |r(x, t)| \le C |a|_{2(k+1)} \sup_{t \in \mathbb{T}^N, \, y \in \mathbb{Z}^N} \frac{|b(x + y, t) - b(x, t)|}{\langle y \rangle}.$$

The right-hand side of this estimate tends to 0 as $x \to \infty$. In the same way, one can show that

$$\lim_{x \to \infty} \sup_{t \in \mathbb{T}^N} |(\partial_t^\alpha r)(x, t)| = 0 \quad \text{for each } \alpha \in \mathbb{N}^N.$$

This proves assertion (a), and assertions (b) and (c) can be checked similarly. □

Proposition 5.4.3 *Let* $a \in \mathcal{SO}$. *Then each sequence* $h : \mathbb{N} \to \mathbb{Z}^N$ *which tends to infinity possesses a subsequence* g *such that the limit functions* a_g *of* a *exists (in the sense of Proposition* 5.2.4*), and this limit function is constant with respect to the variable* $x \in \mathbb{Z}^N$.

Proof. In virtue of Proposition 5.2.4, we only have to check the independence of $a_g(x, t)$ from x. By the definition of the limit function,

$$\lim_{n\to\infty} \sup_{t\in\mathbb{T}^N} |a(x+g(n), t) - a_g(x, t)| = 0 \quad \text{for each } x \in \mathbb{Z}^N.$$

Together with the definition of the class $\mathcal{SO}$, this implies that

$$\begin{aligned}\sup_{t\in\mathbb{T}^N} &|a_g(x, t) - a_g(y, t)| \\ &\le \lim_{n\to\infty} \sup_{t\in\mathbb{T}^N} |a_g(x, t) - a(x+g(n), t)| + \lim_{n\to\infty} \sup_{t\in\mathbb{T}^N} |a(y+g(n), t) - a_g(y, t)| \\ &\quad + \lim_{n\to\infty} \sup_{t\in\mathbb{T}^N} |a(x+g(n), t) - a(y+g(n), t)| = 0\end{aligned}$$

for arbitrary $x, y \in \mathbb{Z}$, whence the assertion. □

Consequently, if the symbol a is slowly oscillating, then all limit operators of $Op(a)$ are of the form $Op(a_h)$ where the limit function a_h depends on the variable $t \in \mathbb{T}^N$ only. So we can identify a_h with a function in $C^\infty(\mathbb{T}^N)$, and the pseudodifference operator $Op(a_h)$ acts as an operator of discrete convolution on $\mathbb{Z}^N$, i.e.,

$$(Op(a_h)u)(x) = \sum_{y\in\mathbb{Z}^N} \check{a}_h(x-y)u(y)$$

where

$$\check{a}_h(x) := \int_{\mathbb{T}^N} a_h(t)t^x \, d\mu(t).$$

Note that $\check{a}_h$ is a function in $S(\mathbb{Z}^N) \subset \mathcal{W}$. Hence, the spectrum of $Op(a_h)$, considered as an operator on $l^p(\mathbb{Z}^N)$, does not depend on $p \in [1, \infty]$, and

$$\sigma_{l^p(\mathbb{Z}^N)}(Op(a_h)) = \{a_h(t) : t \in \mathbb{T}^N\}.$$

So we obtain the following as a corollary of Theorem 5.2.3.

Theorem 5.4.4 *If $a \in \mathcal{SO}$, then, for each $p \in [1, \infty]$,*

$$\sigma^{ess}_{l^p(\mathbb{Z}^N)}(Op(a)) = \bigcup_{Op(a_h)\in\sigma_{op}(Op(a))} \{a_h(t) : t \in \mathbb{T}^N\}.$$

In particular, the essential spectrum of $Op(a)$ is independent of p.

Theorem 5.4.5 *Let $a \in \mathcal{SO}$ and $p \in [1, \infty]$. Then $Op(a) : l^p(\mathbb{Z}^N) \to l^p(\mathbb{Z}^N)$ is a Fredholm operator if and only if*

$$\lim_{R\to\infty} \inf_{|x|>R,\, t\in\mathbb{T}^N} |a(x, t)| > 0. \tag{5.21}$$

Proof. Let the condition (5.21) be satisfied, and let h be a sequence which tends to infinity, and for which the limit function a_h of a exists (in the sense of Proposition 5.2.4). Then, in particular,

$$a(x + h(n), t) \to a_h(t) \quad \text{as } n \to \infty$$

for all $t \in \mathbb{T}^N$, and (5.21) clearly implies that

$$\inf_{t \in \mathbb{T}^N} |a_h(t)| > 0.$$

Hence, the limit operator $Op(a)_h = Op(a_h)$ is invertible. Since h has been chosen arbitrarily, the operator $Op(a)$ is Fredholm on $l^p(\mathbb{Z}^N)$ by Theorem 5.2.2.

Conversely, let $Op(a)$ be a Fredholm operator on $l^p(\mathbb{Z}^N)$, but assume that condition (5.21) does not hold. Then there exists a sequence $\mathbb{N} \to \mathbb{Z}^N \times \mathbb{T}^N$, $n \mapsto (h(n), t(n))$ with $h(n) \to \infty$ and $t(n) \to t_0 \in \mathbb{T}^N$ such that

$$\lim_{n \to \infty} a(h(n), t(n)) \to 0.$$

We can moreover choose the sequence h such that it defines a limit function a_h of a (otherwise we pass to a suitable subsequence). Thus, we have

$$a(h(n), t) \to a_h(t) \quad \text{for all } t \in \mathbb{T}^N$$

whence, in particular, $a_h(t_0) = 0$. On the other hand, the Fredholmness of $Op(a)$ implies via Theorem 5.2.2, that the limit operator $Op(a_h)$ of $Op(a)$ is invertible. This means that $\inf_{t \in \mathbb{T}^N} |a_h(t)| > 0$, which contradicts $a_h(t_0) = 0$. □

Proposition 5.4.6 *Let $a \in SO$ be a slowly oscillating symbol which satisfies condition (5.21), and let $p \in [1, \infty]$. If $u \in l^p(\mathbb{Z}^N)$ and $Op(a)u \in S(\mathbb{Z}^N)$, then $u \in S(\mathbb{Z}^N)$.*

Proof. Choose $R > 0$ such that

$$\inf_{|x|_\infty > R,\, t \in \mathbb{T}^N} |a(x, t)| > 0,$$

and let ψ_R be the function defined by $\psi_R(x) = 1$ if $|x|_\infty > R$ and $\psi_R(x) = 0$ if $|x|_\infty \le R$. Further, set $b(x, t) := 1/a(x, t)$ if $|x|_\infty > R$ and $b(x, t) := 1$ if $|x|_\infty \le R$, such that we can formally write $b(x, t) = \psi_R(x)/a(x, t)$ for all $(x, t) \in \mathbb{Z}^N \times \mathbb{T}^N$. Then, by Proposition 5.4.2 (a),

$$Op(b)Op(a)\psi_R I = (Op(\psi_R) + Op(r))\psi_R I = (I + Op(r)\psi_R I)\psi_R I$$

with $r \in \mathcal{C}_0$. From Theorem 5.1.8 we conclude that $Op(r)\psi_R I$, considered as an operator on $l^p(\mathbb{Z}^N)$, has a small norm if only R is chosen large enough. Hence, the operator $I + Op(r)\psi_R I$ is invertible on $l^p(\mathbb{Z}^N)$, and from Corollary 5.1.11 we infer that $(I + Op(r)\psi_R I)^{-1}$ lies in OPS. Thus, the operator

$$C := (I + Op(r)\psi_R I)^{-1} Op(b)$$

belongs to OPS, and

$$C = Op(a)\psi_R u = \psi_R u \qquad \text{for each } u \in l^p(\mathbb{Z}^N).$$

Consequently, for all $u \in l^p(\mathbb{Z}^N)$,

$$\psi_R u = COp(a)u - COp(a)(1 - \psi_R)u.$$

If now $Op(a)u \in S(\mathbb{Z}^N)$, then the right-hand side of this equality belongs to $S(\mathbb{Z}^N)$, too. Indeed, the operator C is in OPS. So it maps $S(\mathbb{Z}^N)$ into $S(\mathbb{Z}^N)$ by Proposition 5.1.2, whence $COp(a)u \in S(\mathbb{Z}^N)$. Analogously, since $(1-\psi_R)u$ belongs to $S(\mathbb{Z}^N)$, and since $Op(a)$ and C are in OPS, we also have $COp(a)(1 - \psi_R)u \in S(\mathbb{Z}^N)$. Thus, $\psi_R u$ is in $S(\mathbb{Z}^N)$ for some $R > 0$. Of course, this implies that u is in $S(\mathbb{Z}^N)$. □

Corollary 5.4.7 *Let $a \in SO$, and let $Op(a)$ be a Fredholm operator on $l^p(\mathbb{Z}^N)$ for some $p \in (1, \infty)$. Then $Op(a)$ is Fredholm on each of the spaces $l^p(\mathbb{Z}^N)$, and its index does not depend on $p \in (1, \infty)$.*

Proof. The only fact which needs a proof is that the Fredholm index of $Op(a)$ is independent of p. Indeed, the preceding proposition shows that each solution $u \in l^p(\mathbb{Z}^N)$ to the equation $Op(a)u = 0$ belongs to $S(\mathbb{Z}^N)$. Hence, the kernel of $Op(a)$ is independent of p. Analogously, the kernel of $Op(a)^* = Op(a^*)$ does not depend on p. Since both the kernel and the cokernel dimension of $Op(a)$ are independent of p, we get the assertion. □

5.4.2 Fredholmness on weighted spaces, Phragmen-Lindelöf principle

Here we are going to consider pseudodifference operators on weighted l^p-spaces. The appropriate class of weight functions is introduced in the following definition.

Definition 5.4.8 *The weight $w = e^v \in W(\mathbb{K}_r^N)$ is called* slowly oscillating *if the functions $\frac{\partial v}{\partial x_j}$ are slowly oscillating for every $j = 1, \ldots, N$.*

Here are two examples of slowly oscillating weights. In the first one, the weight function tends to infinity as $x \to \infty$, whereas it has no limit as $x \to \infty$ in Example B.

Example A. For $v(x) := a|x|$, one has

$$\frac{\partial v}{\partial x_j}(x) = a\frac{x_j}{|x|} \quad \text{for } j = 1, \ldots, N \text{ and } x \neq 0.$$

Thus, if $r = (R, \ldots, R)$ and $0 < a < \log R$, then $w \in W(\mathbb{K}_r^N)$, and the weight w is slowly oscillating. □

Example B. Let $v(x) := a|x| \sin \log |x|$ for $x \neq 0$. Then

$$\frac{\partial v}{\partial x_j}(x) = a \frac{x_j}{|x|}(\sin \log |x| + \cos \log |x|) \quad \text{for } j = 1, \ldots, N \text{ and } x \neq 0.$$

It easy to check that the functions $\frac{\partial v}{\partial x_j}$ are slowly oscillating, and that

$$\left| \frac{\partial v}{\partial x_j}(x) \right| < \sqrt{2} a$$

for $j = 1, \ldots, N$. Let R and r be as in Example A, and let $a \leq \frac{1}{\sqrt{2}} \log R$. Then $w \in W(\mathbb{K}_r^N)$, and w is slowly oscillating. □

In order to obtain Fredholm criteria for the difference operator $Op(a)$ with symbol $a \in \mathcal{S}(\mathbb{K}_r^N) \cap \mathcal{SO}$ on the space $l_w^p(\mathbb{Z}^N)$ with slowly oscillating weight $w \in W(\mathbb{K}_r^N)$, we need to understand the structure of the limit operators of the operator $wOp(a)w^{-1}I$. It is easy to see that

$$wOp(a)w^{-1}I = Op_d(\tilde{a})$$

where $\tilde{a}(x, y, t) := a(x, e^{-\theta(x,y)} \cdot t)$ is now a double symbol in $\mathcal{SO}_d$ (recall the definition of θ from Theorem 5.3.3).

Let $h : \mathbb{N} \to \mathbb{Z}^N$ be a sequence tending to infinity such that the limit function $a_h \in \mathcal{S}(\mathbb{K}_r^N)$ of a exists. In particular,

$$\lim_{n\to\infty} a(x + h(n), t) = \lim_{n\to\infty} a(h(n), t) = a_h(t). \tag{5.22}$$

Moreover, we choose the sequence h such that the limits

$$\lim_{n\to\infty} \frac{\partial v}{\partial x_j}(h(n)) =: \theta_j^h \tag{5.23}$$

exist for each $j = 1, \ldots, N$. Since the functions $\frac{\partial v}{\partial x_j}$ are slowly oscillating, we get

$$\begin{aligned} \lim_{n\to\infty} \theta_j(x + h(n), y + h(n)) &= \lim_{n\to\infty} \int_0^1 \frac{\partial v}{\partial x_j}((1-t)x + ty + h(n))\, dt \\ &= \lim_{n\to\infty} \frac{\partial v}{\partial x_j}(h(n)) = \theta_j^h. \end{aligned}$$

Thus, if the sequence h satisfies (5.22) and (5.23), then

$$\lim_{n\to\infty} \tilde{a}(x + h(n), y + h(n), t) = a_h(e^{-\theta^h} \cdot t) \quad \text{with} \quad \theta^h := (\theta_1^h, \ldots, \theta_N^h).$$

Notice that the function

$$a_h^\dagger : \mathbb{Z}^N \times \mathbb{T}^N \to \mathbb{C}, \quad (x, t) \mapsto a_h(e^{-\theta^h} \cdot t) \tag{5.24}$$

(which is constant with respect to x) belongs to $\mathcal{S}$. So we arrive at the following proposition.

Proposition 5.4.9 *Let $a \in \mathcal{S}(\mathbb{K}_r^N) \cap \mathcal{SO}$, and let $w \in W(\mathbb{K}_r^N)$ be a slowly oscillating weight. Then all limit operators of $wOp(a)w^{-1}I$ are of the form $Op(a_h^\dagger)$ where $a_h^\dagger \in \mathcal{S}$ is defined by* (5.24).

Theorems 5.4.4, 5.4.5 and 5.3.7 specialize to the present context as follows.

Corollary 5.4.10 *Under the notations and hypotheses of Proposition* 5.4.9,

$$\sigma^{ess}_{l^p_w(\mathbb{Z}^N)}(Op(a)) = \cup_h \{a_h(e^{-\theta^h} \cdot t) : t \in \mathbb{T}^N\}$$

for each $p \in [1, \infty]$, where the union is taken over all sequences h for which the limits (5.22) *and* (5.23) *exist.*

Theorem 5.4.11 *If the hypotheses of Proposition* 5.4.9 *are satisfied, then the operator $Op(a) : l^p_w(\mathbb{Z}^N) \to l^p_w(\mathbb{Z}^N)$ is Fredholm if and only if*

$$\lim_{R\to\infty} \inf_{|x|>R,\, t\in\mathbb{T}^N} |\tilde{a}(x,\, t)| > 0$$

where $\tilde{a}(x,\, y,\, t) := a(x,\, e^{-\theta(x,\, y)} \cdot t)$.

Proof. The operator $Op(a) : l^p_w(\mathbb{Z}^N) \to l^p_w(\mathbb{Z}^N)$ is Fredholm if and only if the operator $wOp(a)w^{-1}I = Op_d(\tilde{a}) : l^p(\mathbb{Z}^N) \to l^p(\mathbb{Z}^N)$ is Fredholm (Theorem 5.3.3), where the double symbol $\tilde{a}$ belongs to $\mathcal{SO}_d$. Hence, by Proposition 5.4.2 (b),

$$Op_d(\tilde{a}) - Op(a^\circ) \in OPC_0$$

with

$$a^\circ(x,\, t) = \tilde{a}(x,\, x,\, t) = a(x,\, e^{-(\nabla v)(x)} \cdot t).$$

Thus, the assertion follows from Theorem 5.4.5. □

Theorem 5.4.12 (Phragmen-Lindelöf principle). *Let $a \in \mathcal{S}(\mathbb{K}_r^N) \cap \mathcal{SO}$, $p \in (1, \infty)$, and let $w \in W(\mathbb{K}_r^N)$ be a slowly oscillating weight with $\lim_{x\to\infty} w(x) = +\infty$. Let further $u \in l^p_{w^{-1}}(\mathbb{Z}^N)$ be a solution to the equation*

$$Au = f, \qquad f \in l^p_w(\mathbb{Z}^N).$$

If the condition

$$\lim_{R\to\infty} \inf_{|x|>R,\, t\in\mathbb{K}_r^N} |a(x,\, t)| > 0$$

is satisfied, then u belongs to $l^p_w(\mathbb{Z}^N)$ again.

Indeed, this is a direct consequence of Theorem 5.3.7.

5.4.3 Fredholm index for operators in $OPSO$

Here we are going to calculate the Fredholm index of operators in $OPSO$ in terms of indices of their limit operators. In this subsection, we consider only operators on sequence spaces over $\mathbb{Z}$, i.e., we let $N = 1$.

Given an operator $A \in \mathcal{W}$, we denote the subsets of the operator spectrum of A which consist of all limit operators with respect to sequences h tending to $+\infty$ and ∞ by $\sigma_+(A)$ and $\sigma_-(A)$, respectively. Obviously,

$$\sigma_{op}(A) = \sigma_+(A) \cup \sigma_-(A).$$

We further write $\mathbb{Z}_+$ and $\mathbb{Z}_-$ for the non-negative and negative integers, and we let P_+ and P_- denote the orthogonal projections from $l^2(\mathbb{Z})$ onto $l^2(\mathbb{Z}_+)$ and $l^2(\mathbb{Z}_-)$, which we identify with closed subspaces of $l^2(\mathbb{Z})$ in the natural way.

If $A \in \mathcal{W}$ is a Fredholm operator (considered as an operator on $l^2(\mathbb{Z})$), then all limit operator are invertible, and the limit operators A_h of A belong to $\mathcal{W}$ again. The simple observation

$$A_h - (P_+A_hP_+ + P_-)(P_+ + P_-A_hP_-) \in K(l^2(\mathbb{Z}))$$

shows that then $P_+A_hP_+$ and $P_-A_hP_-$ are Fredholm operators on $l^2(\mathbb{Z}_+)$ and $l^2(\mathbb{Z}_-)$, respectively.

The following theorem is proved in [136].

Theorem 5.4.13 *Let $A \in W$ be a Fredholm operator on $l^2(\mathbb{Z})$. Then*

(a) *the Fredholm index of $P_+A_hP_+$ is the same for all limit operators $A_h \in \sigma_+(A)$, and the Fredholm index of $P_-A_hP_-$ is the same for all limit operators $A_h \in \sigma_-(A)$.*

(b) *if $A_+ \in \sigma_+(A)$ and $A_- \in \sigma_-(A)$ are chosen arbitrarily, then*

$$\operatorname{ind} A = \operatorname{ind}(P_+A_+P_+) + \operatorname{ind}(P_-A_-P_-).$$

This result takes an especially simple form for operators in $OPSO$ (with $N = 1$).

Let $a : \mathbb{T} \to \mathbb{C}$ be a continuous function which has no zeros on $\mathbb{T}$. Then there is a real-valued function $b : \mathbb{T} \to \mathbb{C}$, which is continuous on the punctured circle $\mathbb{T} \setminus \{1\}$, for which $a = |a|e^{2\pi ib}$. The increment of b as the result of a counter-clockwise circuit around $\mathbb{T}$ is an integer, which only depends on a (and not on the particular choice of b). This integer is called the *winding number* of a. We denote it by $\text{wind}_{\mathbb{T}}a$.

Theorem 5.4.14 *Let $Op(a) \in OPSO$, and let all limit operators of $Op(a)$ be invertible on $l^p(\mathbb{Z})$ for some $p \in [1, \infty]$. Then $Op(a)$ is a Fredholm operator on $l^p(\mathbb{Z})$ for each $p \in [1, \infty]$, and its Fredholm index* ind $Op(a)$ *does not depend on p. Moreover,*

$$\operatorname{ind} Op(a) = -\text{wind}_{\mathbb{T}}a_+ + \text{wind}_{\mathbb{T}}a_-, \tag{5.25}$$

where a_+ and a_- are limit functions of a defined by sequences h_+ and h_- tending to $+\infty$ and $-\infty$, respectively, i.e.,

$$a_+(t) = \lim_{n\to\infty} a(h_+(n), t), \qquad a_-(t) = \lim_{n\to\infty} a(h_-(n), t).$$

In particular, the Fredholm index of $Op(a)$ is independent of the choice of the sequences h_+ and h_-.

Proof. The independence of $\operatorname{ind} Op(a)$ of p follows from Corollary 5.4.7. Let $Op(a)_{h_+}$ and $Op(a)_{h_-}$ be the limit operators defined by sequences h_+ and h_-. Then

$$Op(a)_{h_+} = Op(a_+) \quad \text{and} \quad Op(a)_{h_-} = Op(a_-).$$

Applying Theorem 5.4.13 and the well-known Gohberg-Krein formula for the Fredholm index of the operators $P_\pm Op(a_\pm) P_\pm$ (see, for instance, [59]), we obtain formula (5.25). □

The index formula has to be modified for operators acting on weighted spaces. For, we define the winding number $\operatorname{wind}_{\gamma\mathbb{T}}\, a$ of a function $a : \gamma\mathbb{T} \to \mathbb{C}$ where $\gamma > 0$ in complete analogy with the case $\gamma = 1$.

Theorem 5.4.15 *Let $a \in \mathcal{SO} \cap \mathcal{S}(\mathbb{K}_r^1)$, and let $w = e^v \in W(\mathbb{K}_r^1)$ be a slowly oscillating weight. Let further $Op(a) : l_w^p(\mathbb{Z}) \to l_w^p(\mathbb{Z})$ be a Fredholm operator for some $p \in [1, \infty]$. Then $Op(a)$ is a Fredholm operator on $l_w^p(\mathbb{Z})$ for every $p \in [1, \infty]$, and its Fredholm index does not depend on p. Moreover,*

$$\operatorname{ind} Op(a) = -\operatorname{wind}_{\exp(\theta_+)\mathbb{T}} a_+ + \operatorname{wind}_{\exp(\theta_-)\mathbb{T}} a_-,$$

where the functions $a_\pm$ are defined as in the preceding theorem, and where

$$\theta_\pm := \lim_{n\to\infty} \frac{dv}{dt}(h_\pm(n)).$$

5.5 Almost periodic pseudodifference operators

Here we will see that, for pseudodifference operators with almost periodic symbol, their essential spectrum coincides with their common spectrum for every $p \in [1, \infty]$.

Recall that function a in $l^\infty(\mathbb{Z}^N)$ is *almost periodic* if the set of all functions

$$\mathbb{Z}^N \to \mathbb{C}, \quad x \mapsto a(x+k) \tag{5.26}$$

with $k \in \mathbb{Z}^N$ is precompact in $l^\infty(\mathbb{Z}^N)$. This means that each sequence of functions of the form (5.26) possesses a subsequence which converges in the norm of $l^\infty(\mathbb{Z}^N)$. We let $\mathcal{AS}$ stand for the class of all symbols $a \in \mathcal{S}$ such that $x \mapsto a(x, t)$ is an almost periodic function on $\mathbb{Z}^N$ for each $t \in \mathbb{T}^N$.

It turns out that if $a \in \mathcal{AS}$ and if $h : \mathbb{N} \to \mathbb{Z}^N$ is a sequence which tends to infinity, then there are a subsequence g of h and a symbol $a_g \in \mathcal{AS}$ such that

$$\lim_{n\to\infty} \sup_{x\in\mathbb{Z}^N,\, t\in\mathbb{T}^N} |\partial_t^\alpha (a(x+g(n),\, t) - a_g(x,\, t))| = 0$$

for each multi-index α. Thus, we have the following special cases of Proposition 2.6.1 and Theorem 2.6.2.

Proposition 5.5.1 *Let $a \in \mathcal{AS}$, and let $h : \mathbb{N} \to \mathbb{Z}^N$ be a sequence which tends to infinity. Then there are a subsequence g of h and a symbol $a_g \in \mathcal{AS}$ such that*

$$\lim_{n\to\infty} \|V_{-g(n)} A V_{g(n)} - Op(a_g)\|_{L(l^p(\mathbb{Z}^N))} = 0 \tag{5.27}$$

for all $p \in [1,\, \infty]$.

Theorem 5.5.2 *The following assertions are equivalent for an operator $Op(a) \in OP\mathcal{AS}$:*

(a) *$Op(a)$ is Fredholm on $l^p(\mathbb{Z}^N)$ for some $p \in [1,\, \infty]$.*

(b) *$Op(a)$ is invertible on $l^p(\mathbb{Z}^N)$ for each $p \in [1,\, \infty]$.*

Corollary 5.5.3 *If $Op(a) \in OP\mathcal{AS}$, then*

$$\sigma^{ess}_{l^p(\mathbb{Z}^N)}(Op(a)) = \sigma_{l^p(\mathbb{Z}^N)}(Op(a)).$$

5.6 Periodic pseudodifference operators

The topic of this section are periodic pseudodifference operators. We show that a scalar periodic pseudodifference operator can be considered as a matrix pseudodifference operator which is invariant with respect to shifts. This identification allows us to describe the spectrum of such operators in an explicit form.

Let $g = (g_1,\, \ldots,\, g_N) \in \mathbb{N}^N$. We say that a function $a \in l^\infty(\mathbb{Z}^N)$ is *g-periodic* if

$$a(x+g) = a(x_1+g_1,\, \ldots,\, x_N+g_N) = a(x) \qquad \text{for each } x \in \mathbb{Z}^N.$$

We denote the class of all g-periodic functions by $\mathcal{P}_g(\mathbb{Z}^N)$. Further, we write $\mathcal{PS}_g$ for the set of all symbols $a \in \mathcal{S}$ which are g-periodic with respect to x, i.e.,

$$a(x+g,\, t) = a(x,\, t) \quad \text{for all } (x,\, t) \in \mathbb{Z}^N \times \mathbb{T}^N.$$

The class of associated pseudodifference operators is denoted by $OP\mathcal{PS}_g$.

Let $Op(a) \in OP\mathcal{PS}_g$ and $u \in S(\mathbb{Z}^N)$. Then $Op(a)u$ can be represented as

$$(Op(a)u)(x) = \sum_{\alpha\in\mathbb{Z}^N} a_\alpha(x)(V_\alpha u)(x), \qquad x \in \mathbb{Z}^N, \tag{5.28}$$

where the $a_\alpha \in l^\infty(\mathbb{Z}^N)$ are g-periodic. We are going to show that the operator $Op(a) : l^p(\mathbb{Z}^N) \to l^p(\mathbb{Z}^N)$ is similar to an operator of multiplication by a matrix-valued function $\mathcal{A} : \mathbb{T}^N \to L(l^p(\mathbb{Z}^N, \mathbb{C}^M))$ where $M := g_1 \cdots g_N$. Here, $l^p(\mathbb{Z}^N, \mathbb{C}^M)$ stands for the Banach space consisting of all vector-valued functions $u = (u_1, \ldots, u_M)$ where $u_j \in l^p(\mathbb{Z}^N)$ for each j between 1 and M. We provide this space with the norm

$$\|u\|^p_{l^p(\mathbb{Z}^N, \mathbb{C}^M)} := \sum_{j=1}^{M} \sum_{x \in \mathbb{Z}^N} |u_j(x)|^p$$

if $p \in [1, \infty)$, and

$$\|u\|_{l^\infty(\mathbb{Z}^N, \mathbb{C}^M)} := \sup_{1 \le j \le M} \sup_{x \in \mathbb{Z}^N} |u_j(x)|.$$

Alternatively, one can consider $l^p(\mathbb{Z}^N, \mathbb{C}^M)$ with $p < \infty$ as the space of functions $u : \mathbb{Z}^N \to \mathbb{C}^M$ which satisfy

$$\|u\|^p_{l^p(\mathbb{Z}^N, \mathbb{C}^M)} := \sum_{x \in \mathbb{Z}^N} \|u(x)\|^p_{\mathbb{C}^M} < \infty$$

where we have to provide $\mathbb{C}^M$ with the corresponding p-norm. The case $p = \infty$ can be treated similarly. In what follows, we will switch freely between both perspectives of $l^p(\mathbb{Z}^N, \mathbb{C}^M)$.

5.6.1 The one-dimensional case

We start with the case when $N = 1$. Given a positive integer M, we consider the mapping

$$T_M : l^p(\mathbb{Z}) \to l^p(\mathbb{Z}, \mathbb{C}^M), \quad u \mapsto (v_1, \ldots, v_M)$$

where $v_j(y) := u(My + j - 1)$ for $y \in \mathbb{Z}$ and $j \in \{1, \ldots, M\}$. Evidently, T_M is an isometry for each $p \in [1, \infty]$.

Let us look at how certain operators $A \in L(l^p(\mathbb{Z}))$ are transformed under the mapping $A \mapsto T_M A T_M^{-1}$. If g is a positive integer and $M = g$ (recall that $N = 1$), and if a is a g-periodic function, then the operator aI of multiplication by a is transformed into the constant diagonal matrix $\operatorname{diag}(a(1), \ldots, a(M))$, which acts on $l^p(\mathbb{Z}, \mathbb{C}^M)$ as

$$\operatorname{diag}(a(1), \ldots, a(M)) : (u_1, \ldots, u_M) \mapsto (a(1)u_1, \ldots, a(M)u_M).$$

Next we consider the operator $T_M V_{-1} T_M^{-1}$, where $(V_{-1}u)(x) := u(x+1)$. Let $u \in l^p(\mathbb{Z})$ and $T_M u = (v_1, \ldots, v_M)$. Then

$$(T_M V_{-1} u)(y) = (v_2(y), v_3(y), \ldots, v_M(y), (V_{-1} v_1)(y)) \quad \text{for } y \in \mathbb{Z},$$

i.e., the action of the operator $T_M V_{-1} T_M^{-1} : l^p(\mathbb{Z}, \mathbb{C}^M) \to l^p(\mathbb{Z}, \mathbb{C}^M)$ is described by the $M \times M$ matrix

$$T_M V_{-1} T_M^{-1} = \begin{pmatrix} 0 & 1 & 0 & \dots & 0 & 0 \\ 0 & 0 & 1 & \dots & 0 & 0 \\ 0 & 0 & 0 & \dots & 0 & 0 \\ \vdots & \vdots & \vdots & \ddots & \vdots & \vdots \\ 0 & 0 & 0 & \dots & 0 & 1 \\ V_{-1} & 0 & 0 & \dots & 0 & 0 \end{pmatrix} =: \Lambda.$$

Consequently, if $Op(a)$ is the operator (5.28) with g-periodic coefficients a_α, then the operator $T_M Op(a) T_M^{-1}$ acts on $l^p(\mathbb{Z}, \mathbb{C}^M)$ as

$$T_M Op(a) T_M^{-1} = \sum_{\alpha \in \mathbb{Z}} \operatorname{diag}(a_\alpha(1), \dots, a_\alpha(M)) \Lambda^{-\alpha}.$$

Let now $p = 2$. Then the discrete Fourier transform maps the operator $\Lambda_M : l^2(\mathbb{Z}, \mathbb{C}^M) \to l^2(\mathbb{Z}, \mathbb{C}^M)$ to the operator of multiplication by the matrix-valued function

$$\mathbb{T} \to L(\mathbb{C}^M), \quad t \mapsto \hat{\Lambda}(t) := \begin{pmatrix} 0 & 1 & 0 & \dots & 0 & 0 \\ 0 & 0 & 1 & \dots & 0 & 0 \\ 0 & 0 & 0 & \dots & 0 & 0 \\ \vdots & \vdots & \vdots & \ddots & \vdots & \vdots \\ 0 & 0 & 0 & \dots & 0 & 1 \\ t & 0 & 0 & \dots & 0 & 0 \end{pmatrix}$$

which we consider as an multiplication operator acting on $L^2(\mathbb{T}, \mathbb{C}^M)$ where $\mathbb{C}^M$ is provided with the Euklidean norm.

Thus, if we write an operator $A \in \mathcal{OPPS}_g$ in the form (5.28), then it becomes unitarily equivalent to the operator of multiplication by the matrix-valued C^∞-function

$$\mathbb{T} \to L(\mathbb{C}^M), \quad t \mapsto \mathcal{A}(t) := \sum_{\alpha \in \mathbb{Z}} \operatorname{diag}(a_\alpha(1), \dots, a_\alpha(M)) \hat{\Lambda}^{-\alpha}(t), \tag{5.29}$$

thought of as acting on $L^2(\mathbb{T}, \mathbb{C}^M)$. This yields the following theorem.

Theorem 5.6.1 *Let $N = 1$, and let $Op(a) \in \mathcal{OPPS}_g$ be given by (5.28) with associated symbol $\mathcal{A}$ defined by (5.29). Then the operator $Op(a)$ is invertible on $l^2(\mathbb{Z})$ (hence, on all spaces $l^p(\mathbb{Z})$ with $p \in [1, \infty]$) if and only if*

$$\det \mathcal{A}(t) \neq 0 \qquad \text{at each } t \in \mathbb{T}.$$

Moreover,

$$\sigma^{ess}_{l^p(\mathbb{Z})}(Op(a)) = \sigma_{l^p(\mathbb{Z})}(Op(a)) = \bigcup_{t \in \mathbb{T}} \{\lambda \in \mathbb{C} : \det(\mathcal{A}(t) - \lambda I) = 0\}.$$

Remark. The matrix $\mathcal{A}(t)$ has M eigenvalues $\lambda_1(t), \ldots, \lambda_M(t)$ (counted with respect to their multiplicity) which can be numbered in such a way that the $t \mapsto \lambda_j(t)$ become continuous for each j ([82], Chapter II, Theorem 5.2). Let

$$\Gamma_j := \{\lambda_j(t) : t \in \mathbb{T}\}.$$

The continuity of the function λ_j implies that each Γ_j is a compact connected curve in $\mathbb{C}$. Hence, under the hypotheses of the preceding theorem, the spectrum of $Op(a)$ is a finite union of curves:

$$\sigma^{ess}_{l^p(\mathbb{Z})}(Op(a)) = \sigma_{l^p(\mathbb{Z})}(Op(a)) = \bigcup_{j=1}^{M} \Gamma_j.$$

5.6.2 The multi-dimensional case

Let now $g = (g_1, \ldots, g_N)$ and $M = g_1 \cdots g_N$ with $N \geq 1$. We write

$$\mathbb{C}^M = \mathbb{C}^{g_1} \otimes \cdots \otimes \mathbb{C}^{g_N} = \otimes_{j=1}^N \mathbb{C}^{g_j}.$$

If $e_{j,\,k}$ denotes the unit vector of length k having a 1 at the jth place and zeros at the other places, then the set

$$\{e_{j_1,\,g_1} \otimes \cdots \otimes e_{j_N,\,g_N}\}_{j_1=1,\,\ldots,\,j_N=1}^{g_1,\,\ldots,\,g_N} \tag{5.30}$$

forms a basis in $\mathbb{C}^M$. We order its elements lexicographically. Further, we consider the mapping

$$T_g : l^p(\mathbb{Z}^N) \to l^p(\mathbb{Z}^N, \mathbb{C}^M)$$

defined by

$$(T_g u)(y_1, \ldots, y_N) := (u(g_1 y_1 + j_1 - 1, \ldots, g_N y_N + j_N - 1))_{j_1=1,\,\ldots,\,j_N=1}^{g_1,\,\ldots,\,g_N}$$

which again acts as an isometry for each $p \in [1, \infty]$. As in the one-dimensional case, we obtain

$$T_g(V_{\alpha_1,\,\ldots,\,\alpha_N})T_g^{-1} = \Lambda_1^{-\alpha_1} \otimes \cdots \otimes \Lambda_N^{-\alpha_N}$$

where Λ_j is the $g_j \times g_j$-matrix operator

$$\Lambda_j := \begin{pmatrix} 0 & 1 & 0 & \ldots & 0 & 0 \\ 0 & 0 & 1 & \ldots & 0 & 0 \\ 0 & 0 & 0 & \ldots & 0 & 0 \\ \vdots & \vdots & \vdots & \ddots & \vdots & \vdots \\ 0 & 0 & 0 & \ldots & 0 & 1 \\ V_{e_{j,\,N}}^{-1} & 0 & 0 & \ldots & 0 & 0 \end{pmatrix}.$$

For $t \in \mathbb{T}$ we further denote by $\hat{\Lambda}_j(t)$ the $g_j \times g_j$-matrix

$$\hat{\Lambda}_j(t) := \begin{pmatrix} 0 & 1 & 0 & \dots & 0 & 0 \\ 0 & 0 & 1 & \dots & 0 & 0 \\ 0 & 0 & 0 & \dots & 0 & 0 \\ \vdots & \vdots & \vdots & \ddots & \vdots & \vdots \\ 0 & 0 & 0 & \dots & 0 & 1 \\ t & 0 & 0 & \dots & 0 & 0 \end{pmatrix}$$

If $p = 2$, then we see as in case $N = 1$ that the operator $Op(a) \in OP\mathcal{PS}_g$, given by (5.28) and acting on $l^2(\mathbb{Z}^N)$, is unitarily equivalent to the operator of multiplication by the matrix-valued function $\mathcal{A} : \mathbb{T}^N \to \mathbb{C}^M$ acting on $L^2(\mathbb{T}^N, \mathbb{C}^M)$, where

$$\mathcal{A}(t) := \sum_{\alpha \in \mathbb{Z}^N} \operatorname{diag}\left(a_\alpha(j_1 - 1, \dots, j_N - 1)\right)_{j_1=1, \dots, j_N=1}^{g_1, \dots, g_N} \hat{\Lambda}_1^{-\alpha_1}(t_1) \otimes \dots \otimes \hat{\Lambda}_N^{-\alpha_N}(t_N).$$

The notation $\operatorname{diag}\left(a_\alpha(j_1 - 1, \dots, j_N - 1)\right)_{j_1=1, \dots, j_N=1}^{g_1, \dots, g_N}$ means a diagonal matrix with lexicographically ordered elements on its main diagonal. With these notations, we have the following generalization of Theorem 5.6.1.

Theorem 5.6.2 *Let the operator $Op(a) \in OP\mathcal{PS}_g$ be given by* (5.28) *with associated symbol $\mathcal{A}$ defined as above. Then the operator $Op(a)$ is invertible on $l^2(\mathbb{Z}^N)$ (hence, on all spaces $l^p(\mathbb{Z}^N)$ with $p \in [1, \infty]$) if and only if*

$$\det \mathcal{A}(t) \neq 0 \qquad \textit{at each } t \in \mathbb{T}^N.$$

Moreover,

$$\sigma^{ess}_{l^p(\mathbb{Z}^N)}(Op(a)) = \sigma_{l^p(\mathbb{Z}^N)}(Op(a)) = \bigcup_{t \in \mathbb{T}^N} \{\lambda \in \mathbb{C} : \det(\mathcal{A}(t) - \lambda I) = 0\}.$$

Again, the spectrum of $Op(a)$ is a finite union of connected compact curves.

5.7 Semi-periodic pseudodifference operators

Here we consider so-called semi-periodic pseudodifference operators which can be thought of as certain perturbations of periodic operators. Limit operators of semi-periodic operators are operators with periodic coefficients. This property makes it possible to derive effective conditions for semi-periodic operators to be Fredholm on weighted spaces, to describe their essential spectrum, and to calculate their Fredholm index (in dimension one).

5.7.1 Fredholmness on unweighted spaces

Let again $g = (g_1, \dots, g_N) \in \mathbb{N}^N$. A function $a \in l^\infty(\mathbb{Z}^N)$ is called *g-semi-periodic* if it is the limit in $l^\infty(\mathbb{Z}^N)$ of functions of the form

$$\sum_{j=1}^{m} b_j c_j \quad \text{with } b_j \in SO(\mathbb{Z}^N) \text{ and } c_j \in \mathcal{P}_g(\mathbb{Z}^N). \tag{5.31}$$

We denote the set of all g-semi-periodic functions by $SP_g(\mathbb{Z}^N)$. Further, we let $\mathcal{SPS}_g$ stand for the set of all symbols $a \in \mathcal{S}$ for which the functions $x \mapsto a(x, t)$ belong to $SP_g(\mathbb{Z}^N)$ for each point $t \in \mathbb{T}^N$. The corresponding class of pseudodifference operators is denoted by $OP\mathcal{SPS}_g$.

Our treatment of the Fredholm properties of pseudodifference operators with symbol in $\mathcal{SPS}_g$ is based on the following proposition.

Proposition 5.7.1 *Let $a \in \mathcal{SPS}_g$, and consider the operator $Op(a)$ as acting on $l^p(\mathbb{Z}^N)$ where $1 \le p \le \infty$. Then all limit operators of $Op(a)$ belong to $OP\mathcal{PS}_g$.*

Proof. Let $a \in SP(\mathbb{Z}^N)$, and let $h : \mathbb{N} \to \mathbb{Z}^N$ be a sequence tending to infinity such that the limit $a_h(x) := \lim_{n\to\infty} a(x + h(n))$ exists for each point $x \in \mathbb{Z}^N$. We claim that the function a_h is g-periodic.

It is certainly sufficient to prove this claim for functions a of the form (5.31) (since $\mathcal{P}_g(\mathbb{Z}^N)$ is closed) and, hence, for functions $a = bc$ with $b \in SO(\mathbb{Z}^N)$ and $c \in \mathcal{P}_g(\mathbb{Z}^N)$ (since $\mathcal{P}_g(\mathbb{Z}^N)$ is a linear space). In this case, we can find a subsequence k of h, a complex number b_k and a function $c_k \in \mathcal{P}_g(\mathbb{Z}^N)$ such that $b(x+k(n)) \to b_k$ and $c(x + k(n)) \to c_k(x)$ for each $x \in \mathbb{Z}^N$ (compare Proposition 5.4.3). Hence,

$$a_h(x) = \lim_{n\to\infty} a(x + k(n)) = \lim_{n\to\infty} b(x + k(n))c(x + k(n)) = b_k c_k(x),$$

whence $a_h \in \mathcal{P}_g(\mathbb{Z}^N)$.

Let now the operator $A \in OP\mathcal{SPS}_g$ have the representation

$$A = \sum_{\alpha\in\mathbb{Z}^N} a_\alpha V_\alpha \quad \text{with } a_\alpha \in SP_g(\mathbb{Z}^N).$$

Then

$$V_{-h(n)}AV_{h(n)} = \sum_{\alpha\in\mathbb{Z}^N} (V_{-h(n)}a_\alpha V_{h(n)})V_\alpha$$

and, if the sequence h defines a limit operator of A, we have

$$(V_{-h(n)}a_\alpha V_{h(n)})(x) = a_\alpha(x + h(n)) \to a_{\alpha,h}(x) \quad \text{for each } x \in \mathbb{Z}^N$$

with certain g-periodic functions $a_{\alpha,h}$, as we have seen above. Consequently, $A_h = \sum_{\alpha\in\mathbb{Z}^N} a_{\alpha,h} V_\alpha \in OP\mathcal{PS}_g$. □

To the limit operator $A_h = \sum_{\alpha\in\mathbb{Z}^N} a_{\alpha,h} V_\alpha$ of $A = \sum_{\alpha\in\mathbb{Z}^N} a_\alpha V_\alpha$, we associate the matrix-valued function

$$\mathcal{A}_h(t) := \sum_{\alpha\in\mathbb{Z}^N} \operatorname{diag}\left(a_{\alpha,h}(j_1-1, \ldots, j_N-1)\right)_{j_1=1, \ldots, j_N=1}^{g_1, \ldots, g_N} \hat{\Lambda}_1^{-\alpha_1}(t_1)\otimes\cdots\otimes\hat{\Lambda}_N^{-\alpha_N}(t_N)$$

where the notations are as in Section 5.6.2.

Theorem 5.7.2 *Let $a \in \mathcal{SPS}_g$, and consider the operator $Op(a)$ as acting on $l^p(\mathbb{Z}^N)$ where $1 \leq p \leq \infty$. Then the operator $Op(a)$ is Fredholm if and only if*

$$\det \mathcal{A}_h(t) \neq 0 \quad \textit{at each point } t \in \mathbb{T}^N$$

for every limit operator $Op(a)_h$ of $Op(a)$. Moreover,

$$\sigma^{ess}_{l^p(\mathbb{Z}^N)} Op(a) = \bigcup \{\lambda \in \mathbb{C} : \det(\mathcal{A}_h(t) - \lambda I) = 0\}$$

where the union is taken over all limit operators of $Op(a)$ and over all $t \in \mathbb{T}^N$.

The proof follows immediately from Theorems 5.2.2 and 5.6.2 and from the remark following Theorem 5.6.1.

The image of operators in $OP\mathcal{SPS}_g$ under the transformation $A \mapsto T_g A T_g^{-1}$ is described in the following proposition.

Proposition 5.7.3 *If $a \in \mathcal{SPS}_g$, then $T_g Op(a) T_g^{-1} : l^p(\mathbb{Z}^N, \mathbb{C}^M) \to l^p(\mathbb{Z}^N, \mathbb{C}^M)$ with $M := g_1 \cdots g_N$ is a matrix operator with entries in $OP\mathcal{SO}$.*

Proof. Let $a \in SP(\mathbb{Z}^N)$. We are going to show that the operator $T_g a T_g^{-1}$ has a representation as an $M \times M$-matrix-valued multiplication operator

$$(T_g a T_g^{-1})(y) = \operatorname{diag}(a_{j_1 j_2 \cdots j_N}(y))^{g_1, \dots, g_N}_{j_1=1, \dots, j_N=1} \tag{5.32}$$

with slowly oscillating entries $a_{j_1 j_2 \cdots j_N}$.

Again, it is sufficient to prove this for $a = bc$ where $b \in SO(\mathbb{Z}^N)$ and $c \in \mathcal{P}_g(\mathbb{Z}^N)$. In this case we have

$$\begin{aligned}(T_g bc T_g^{-1})(y) &= (T_g b T_g^{-1})(y) \cdot (T_g c T_g^{-1})(y) \\ &= \operatorname{diag}(b(g_1 y_1 + j_1 - 1, \dots, g_N y_N + j_N - 1))^{g_1, \dots, g_N}_{j_1=1, \dots, j_N=1} \\ &\qquad \cdot \operatorname{diag}(c(j_1 - 1, \dots, j_N - 1))^{g_1, \dots, g_N}_{j_1=1, \dots, j_N=1}.\end{aligned}$$

Since the functions

$$b_{j_1 \cdots j_N} : (y_1, \dots, y_N) \mapsto b(g_1 y_1 + j_1 - 1, \dots, g_N y_N + j_N - 1)$$

are slowly oscillating, and since the $c_{j_1 \cdots j_N} := c(j_1 - 1, \dots, j_N - 1)$ are constants, we obtain the assertion of the proposition for the operator $T_g bc T_g^{-1}$ and, thus, for all operators $T_g a T_g^{-1}$ with $a \in SP(\mathbb{Z}^N)$.

The operators $T_g V_\alpha T_g^{-1}$ have already been described in Section 5.6.2. These two partial results combine to give the proof of the assertion for general operators in $OP\mathcal{SPS}_g$. □

The preceding proposition shows in particular that if $A = \sum_{\alpha \in \mathbb{Z}^N} a^\alpha V_\alpha$ with $a^\alpha \in OP\mathcal{SPS}_g$, then the operator $T_g A T_g^{-1}$ is the pseudodifference operator acting on the space $l^p(\mathbb{Z}^N, \mathbb{C}^M)$ and having the matrix-valued symbol

$$\mathcal{A}(y, t) := \sum_{\alpha \in \mathbb{Z}^N} \operatorname{diag}(a^\alpha_{j_1 \cdots j_N}(y))^{g_1, \dots, g_N}_{j_1=1, \dots, j_N=1} \hat{\Lambda}_1^{-\alpha_1}(t_1) \otimes \cdots \otimes \hat{\Lambda}_N^{-\alpha_N}(t_N)$$

with slowly oscillating functions $a^\alpha_{j_1 \cdots j_N}$.

The following theorem is a corollary of Theorem 5.4.5.

Theorem 5.7.4 *An operator $A \in OPSPS_g$ is Fredholm on $l^p(\mathbb{Z}^N)$ with $1 \le p \le \infty$ if and only if*

$$\lim_{R\to\infty} \inf_{|y|>R,\, t\in\mathbb{T}^N} |\det \mathcal{A}(y, t)| > 0$$

where $\mathcal{A}$ is the matrix-symbol of the pseudodifference operator $T_g A T_g^{-1}$.

5.7.2 Fredholmness on weighted spaces

The methods developed in Section 5.3 apply to derive the following versions of Theorems 5.7.2 and 5.7.4, holding for pseudodifference operators with semi-periodic symbol which act on spaces with slowly oscillating weight. The notations are as in Section 5.3.

Theorem 5.7.5 *Let $Op(a)$ be a pseudodifference operator with symbol in $SPS_g \cap S(\mathbb{K}_r^N)$, let $p \in [1, \infty]$, and let $w = e^v \in W(\mathbb{K}_r^N)$ be a slowly oscillating weight. Then $Op(a)$ is a Fredholm operator on $l^p_w(\mathbb{Z}^N)$ if and only if*

$$\det \mathcal{A}_h(e^{-\theta_h} \cdot t) \neq 0$$

for every $t \in \mathbb{T}^N$ and every limit operator A_h of $Op(a)$. Here, the function $\mathcal{A}_h$ is defined as before Theorem 5.7.2, and $\theta_h := \lim_{n\to\infty}(\nabla v)(h(n))$. In particular,

$$\sigma^{ess}_{l^p_w(\mathbb{Z}^N)}(Op(a)) = \bigcup \{\lambda \in \mathbb{C} : \det(\mathcal{A}_h(e^{-\theta_h} \cdot t) - \lambda I) = 0\},$$

where the union is taken over all limit operators of $Op(a)$ and over all $t \in \mathbb{T}^N$.

Theorem 5.7.6 *Let $a \in SPS_g \cap S(\mathbb{K}_r^N)$, let $p \in [1, \infty]$, and let $w = e^v \in W(\mathbb{K}_r^N)$ be a slowly oscillating weight. Then the operator $Op(a)$ is Fredholm on $l^p_w(\mathbb{Z}^N)$ if and only if*

$$\lim_{R\to\infty} \inf_{|y|>R,\, t\in\mathbb{T}^N} |\det \mathcal{A}(y, e^{-(\nabla v)(y)} \cdot t)| > 0.$$

Theorem 5.7.7 (Phragmen-Lindelöf principle.) *Suppose that $a \in SPS_g \cap S(\mathbb{K}_r^N)$, that $p \in (1, \infty)$, and that $w \in W(\mathbb{K}_r^N)$ is a slowly oscillating weight with $\lim_{x\to\infty} w(x) = +\infty$. Further, let u be a solution to the equation $Op(a)u = f$ with right-hand side $f \in l^p_w(\mathbb{Z}^N)$, which a priori lies in $l^p_{w^{-1}}(\mathbb{Z}^N)$. If*

$$\lim_{R\to\infty} \inf_{|y|>R,\, t\in\mathbb{K}_r^N} |\det \mathcal{A}(y, t)| > 0,$$

then $u \in l^p_w(\mathbb{Z}^N)$.

5.7.3 Fredholm index

The assertions on Fredholm indices of operators in $\mathcal{OPSO}$ derived in Section 5.4.3 can be modified in an evident way to give results on indices of operators in $\mathcal{OPSPS}_g$. The point is that operators in $\mathcal{OPSPS}_g$ are unitarily equivalent to matrix operators with components in $\mathcal{OPSO}$, as we have seen in Proposition 5.7.3. We formulate the resulting Fredholm index theorem for operators on weighted spaces only.

Theorem 5.7.8 *Let $a \in \mathcal{SPS}_g \cap S(\mathbb{K}_r^N)$, $p \in [1, \infty]$, and let the weight $w = e^v \in W(\mathbb{K}_r^N)$ be slowly oscillating. If the operator $Op(a) : l^p_w(\mathbb{Z}) \to l^p_w(\mathbb{Z})$ is Fredholm for at least one $p \in [1, \infty]$, then it is Fredholm for each p, and its Fredholm index does not depend on p. Moreover,*

$$\operatorname{ind} Op(a) = -\operatorname{wind}_{e^{\theta^+}\mathbb{T}} \det \mathcal{A}^+ + \operatorname{wind}_{e^{\theta^-}\mathbb{T}} \det \mathcal{A}^-$$

where $\mathcal{A}^+$ and $\mathcal{A}^-$ are the limit functions of the matrix symbol $\mathcal{A}$ defined before Theorem 5.7.4 *with respect to certain sequences h_+ and h_- tending to $+\infty$ and $-\infty$,*

$$\mathcal{A}^+(t) := \lim_{n\to\infty} \mathcal{A}(h_+(n), t), \quad \mathcal{A}^-(t) := \lim_{n\to\infty} \mathcal{A}(h_-(n), t) \quad \text{for } t \in \mathbb{T},$$

and where the $\theta^\pm$ are as in Theorem 5.4.15. *In particular, the index of $Op(a)$ does not depend on the special choice of the sequences $h_\pm$.*

5.8 Discrete Schrödinger operators

Here we are going to apply the results of the previous sections to the discrete analogue of the Schrödinger operator $-\Delta + aI$, viz. to the operator $\mathbf{H}$ defined by

$$\mathbf{H}u := \mathbf{L}u + au \tag{5.33}$$

where $\mathbf{L}$ is the discrete Laplace operator

$$\mathbf{L} := \sum_{j=1}^{N} (V_{e_j} + V_{-e_j})$$

and where the potential a is a function in $l^\infty(\mathbb{Z}^N)$. We consider $\mathbf{L}$ and $\mathbf{H}$ as bounded operators on the Banach space $l^p(\mathbb{Z}^N)$ with $p \in [1, \infty]$. In particular, we will present estimates of the decay at infinity of the eigenfunctions of Schrödinger operators with slowly oscillating potential.

Given $a \in l^\infty(\mathbb{Z}^N)$, we let $\operatorname{Lim}(a)$ denote the set of all limit functions of a, i.e., the set of all functions $a_h \in l^\infty(\mathbb{Z}^N)$ for which there exists a sequence $h : \mathbb{N} \to \mathbb{Z}^N$ tending to infinity such that

$$a_h(x) = \lim_{n\to\infty} a(x + h(n)) \quad \text{for each } x \in \mathbb{Z}^N.$$

Then, clearly,

$$\sigma_{op}(\mathbf{H}) = \{\mathbf{L} + a_h I : a_h \in \mathrm{Lim}\,(a)\}.$$

Thus, the following is a corollary of Theorems 5.2.2 and 5.2.3 and of Proposition 5.2.1.

Theorem 5.8.1 *The essential spectrum of the operator* $\mathbf{H} : l^p(\mathbb{Z}^N) \to l^p(\mathbb{Z}^N)$ *does not depend on* $p \in [1, \infty]$, *and*

$$\sigma^{ess}_{l^p(\mathbb{Z}^N)}(\mathbf{H}) = \bigcup_{a_h \in \mathrm{Lim}\,(a)} \sigma_{l^p(\mathbb{Z}^N)}(\mathbf{L} + a_h I).$$

This theorem reduces the calculation of the essential spectra of discrete Schrödinger operators to the calculation of the common spectra of their limit operators. In many cases, these limit operators are of a more simple structure than the original operators, which allows one to determine their spectra. Some particular instances where this calculation can be performed explicitly are examined in the following sections, where we let $p = 2$.

5.8.1 Slowly oscillating potentials

If the potential a belongs to $SO(\mathbb{Z}^N)$, then all limit operators of $\mathbf{H} = \mathbf{L} + aI$ are of the form $\mathbf{H} = \mathbf{L} + a_h I$, and each limit function $a_h \in \mathrm{Lim}\,(a)$ is constant, i.e., $a_h = \lim_{n\to\infty} a(h(n))$ for some sequence $h \to \infty$. So we can identify $\mathrm{Lim}\,(a)$ with a closed subset of $\mathbb{C}$. In that sense, we have the following.

Theorem 5.8.2 *If* $a \in SO(\mathbb{Z}^N)$, *then*

$$\sigma^{ess}_{l^2(\mathbb{Z}^N)}(\mathbf{H}) = \mathrm{Lim}\,(a) + [-2N, 2N].$$

Proof. The operator $\mathbf{H} = \mathbf{L} + a_h I$, considered as an operator on $l^2(\mathbb{Z}^N)$, is unitarily equivalent (by means of the discrete Fourier transform) to the multiplication operator

$$\hat{\mathbf{H}}_h := \mathcal{L}I + a_h I : L^2(\mathbb{T}^N) \to L^2(\mathbb{T}^N)$$

where $\mathcal{L}$ refers to the function

$$\mathcal{L} : \mathbb{T}^N \to \mathbb{R}, \quad t \mapsto \sum_{j=1}^{N} (t_j + t_j^{-1}).$$

Since

$$\sigma(\hat{\mathbf{H}}_h) = [-2N, 2N] + a_h,$$

we get the assertion via Theorem 5.8.1. □

Corollary 5.8.3 *Let $N > 1$, let the potential $a \in SO(\mathbb{Z}^N)$ be a real-valued function, and set*

$$m := \liminf_{x \to \infty} a(x), \qquad M := \limsup_{x \to \infty} a(x).$$

Then

$$\sigma^{ess}_{l^2(\mathbb{Z}^N)}(\mathbf{H}) = [m - 2N,\, M + 2N].$$

Moreover, all points in the spectrum of $\mathbf{H}$ *which lie outside the segment* $[m - 2N,\, M + 2N]$ *are isolated eigenvalues, and their only possible cluster points are* $m - 2N$ *and* $M + 2N$. *Finally, all eigenfunctions of* $\mathbf{H}$ *belong to* $S(\mathbb{Z}^N)$.

Proof. Since the potential a is real, all partial limits of a are real, and since a is slowly oscillating, the set of partial limits of a is connected. Indeed, this set is just the range of the restriction of the Gelfand transform of a onto the fiber $M^\infty(SO(\mathbb{Z}^N))$, i.e., it is the image of a compact and connected (by Theorem 2.4.7) set under a continuous mapping. So we conclude from Theorem 5.8.2 that

$$\sigma^{ess}_{l^2(\mathbb{Z}^N)}(\mathbf{H}) = [m,\, M] + [-2N,\, 2N] = [m - 2N,\, M + 2N].$$

Since the operator $\mathbf{H}$ is self-adjoint on $l^2(\mathbb{Z}^N)$, the assertions concerning the location of the eigenvalues of $\mathbf{H}$ are immediate consequences of the Gohberg-Sigal theorem (which can be found, for instance, in [60]). Finally, the eigenfunctions of $\mathbf{H}$ belong to $S(\mathbb{Z}^N)$ due to Proposition 5.4.6. □

Obvious modifications yield an analogous result for the case $N = 1$ where one has to consider the fibers of $M(SO(\mathbb{Z}))$ over $-\infty$ and $+\infty$.

5.8.2 Exponential decay of eigenfunctions

Here we apply the Phragmen-Lindelöf principle from Section 5.4.2 to Schrödinger operators $\mathbf{H} = \mathbf{L} + aI$ with slowly oscillating and real-valued potential a. We will consider two situations:

(i) $\liminf_{x \to \infty} a(x) =: a_+ > 2N$, and

(ii) $\limsup_{x \to \infty} a(x) =: a_- < -2N$.

Note that the symbol h of the difference operator $\mathbf{H}$ is

$$h : \mathbb{Z}^N \times \mathbb{T}^N \to \mathbb{R}, \quad (x,\, t) \mapsto \mathcal{L}(t) + a(x).$$

For $\varepsilon > 0$ sufficiently small, we set

$$Q^+_\varepsilon := \left\{ \eta \in \mathbb{R}^N : |\eta|_\infty < \log\left(\frac{a_+ - \varepsilon}{2N} + \sqrt{\left(\frac{a_+ - \varepsilon}{2N}\right)^2 - 1} \right) \right\}$$

and

$$Q^-_\varepsilon := \left\{ \eta \in \mathbb{R}^N : |\eta|_\infty < \log\left(\frac{|a_-| - \varepsilon}{2N} + \sqrt{\left(\frac{|a_-| - \varepsilon}{2N}\right)^2 - 1} \right) \right\}.$$

Since, as one easily checks,

$$\operatorname{Re} h(x, e^{i(\xi+i\eta)}) = 2\sum_{j=1}^{N} \cos\xi_j \cosh\eta_j + a(x),$$

we have in case (i)

$$\lim_{R\to\infty} \inf_{|x|\geq R,\, \xi\in\mathbb{R}^N,\, \eta\in Q_\varepsilon^+} \operatorname{Re} h(x, e^{i(\xi+i\eta)}) \geq \inf_{\eta\in Q_\varepsilon^+} \left(a_+ - 2\sum_{j=1}^{N} \cosh\eta_j \right) \geq \varepsilon,$$

whereas in case (ii)

$$\lim_{R\to\infty} \inf_{|x|\geq R,\, \xi\in\mathbb{R}^N,\, \eta\in Q_\varepsilon^-} |\operatorname{Re} h(x, e^{i(\xi+i\eta)})| \geq \inf_{\eta\in Q_\varepsilon^-} \left(|a_-| - 2\sum_{j=1}^{N} \cosh\eta_j \right) \geq \varepsilon.$$

Thus, the conditions of Theorem 5.4.12 are satisfied for operators $\mathbf{L} + aI$ with potential a subject to condition (i) on the domain Q_ε^+, and they hold for potentials subject to condition (ii) on Q_ε^-. As a consequence of that theorem, we obtain the following.

Theorem 5.8.4 *Let $w = e^v$ be a weight function with $\lim_{x\to\infty} w(x) = +\infty$ and $\nabla v \in SO(\mathbb{Z}^N)$. Further we suppose that if a is a potential satisfying condition (i), then*

$$\left|\frac{\partial v}{\partial x_j}(x)\right| \leq \log\left(\frac{a_+ - \varepsilon}{2N} + \sqrt{\left(\frac{a_+ - \varepsilon}{2N}\right)^2 - 1} \right),$$

and if the potential a satisfies condition (ii), then we suppose

$$\left|\frac{\partial v}{\partial x_j}(x)\right| \leq \log\left(\frac{|a_-| - \varepsilon}{2N} + \sqrt{\left(\frac{|a_-| - \varepsilon}{2N}\right)^2 - 1} \right)$$

for all $j = 1, \ldots, N$, for all sufficiently large $x \in \mathbb{R}^N$ and for all sufficiently small $\varepsilon > 0$. Let u be a solution to the equation

$$(\mathbf{L} + aI)u = f, \quad f \in l^2_w(\mathbb{Z}^N)$$

which a priori belongs to the space $l^2_{w^{-1}}(\mathbb{Z}^N)$. Then u lies in $l^2_w(\mathbb{Z}^N)$.

Corollary 5.8.5 *Let a be a potential satisfying one of the conditions (i) and (ii). If $u_\lambda \in l^2(\mathbb{Z}^N)$ is an eigenfunction of $\mathbf{L} + aI$ which corresponds to an eigenvalue $\lambda < a_+ - 2N$, then $u_\lambda \in l^2_{w_\varepsilon}(\mathbb{Z}^N)$ for sufficiently small $\varepsilon > 0$, with the weight w_ε given by*

$$w_\varepsilon(x) = \left(\frac{\lambda - a_+ - \varepsilon}{2N} + \sqrt{\left(\frac{\lambda - a_+ - \varepsilon}{2N}\right)^2 - 1} \right)^{|x|_2^2}$$

where $|x|_2$ refers to the Euklidean norm of $x \in \mathbb{R}^N$. Similarly, if u_λ is an eigenfunction corresponding to an eigenvalue $\lambda > a_- + 2N$, then $u_\lambda \in l^2_{w_\varepsilon}(\mathbb{Z}^N)$ for sufficiently small ε, where now

$$w_\varepsilon(x) = \left(\frac{a_- - \lambda - \varepsilon}{2N} + \sqrt{\left(\frac{a_- - \lambda - \varepsilon}{2N}\right)^2 - 1} \right)^{|x|_2^2}.$$

This corollary can be viewed of as the discrete analogue of the well-known Agmon estimates for the decay of the eigenfunctions of the Schrödinger operator $-\Delta + aI$, compare [1].

5.8.3 Semi-periodic Schrödinger operators

Finally, we consider Schrödinger operators with real-valued semi-periodic potentials $a \in SP_g(\mathbb{Z}^N)$. In this case, all all limit operators of $\mathbf{H} = \mathbf{L} + aI$ have the form $\mathbf{H}_h := \mathbf{L} + a_h I$ where a_h is a g-periodic function, that is

$$a_h(x) = \lim_{n \to \infty} a(x + h(n)) \quad \text{for } x \in \mathbb{Z}^N$$

for some sequence $h \to \infty$.

For simplicity, we will treat the case $N = 1$ only, i.e., g is a positive integer. As we have seen in Section 5.6.1, the limit operator $\mathbf{H}_h : l^2(\mathbb{Z}) \to l^2(\mathbb{Z})$ is unitarily equivalent to the operator $\hat{\mathbf{H}}_h : L^2(\mathbb{T}, \mathbb{C}^g) \to L^2(\mathbb{T}, \mathbb{C}^g)$ of multiplication by the Hermitian matrix

$$\mathcal{H}_h(t) := \begin{pmatrix} a_h(0) & 1 & 0 & \dots & 0 & t \\ 1 & a_h(1) & 1 & \dots & 0 & 0 \\ 0 & 1 & a_h(2) & \dots & 0 & 0 \\ \vdots & \vdots & \vdots & \ddots & \vdots & \vdots \\ 0 & 0 & 0 & \dots & a_h(g-2) & 1 \\ t & 0 & 0 & \dots & 1 & a_h(g-1) \end{pmatrix}, \quad t \in \mathbb{T}.$$

Let $\lambda_j^h(t)$, $j = 1, \dots, g$, refer to the eigenvalues of the matrix $\mathcal{H}_h(t)$. We enumerate these eigenvalues such that the functions $t \mapsto \lambda_j^h(t)$ become continuous for each j. Since all eigenvalues of $\mathcal{H}_h(t)$ are real, the (compact and connected) curves

$$\Gamma_j^h := \{\lambda_j^h(t) : t \in \mathbb{T}\}$$

become compact intervals in $\mathbb{R}$. From Theorem 5.8.1 we further infer that

$$\sigma^{ess}_{l^2(\mathbb{Z})}(\mathbf{H}) = \bigcup_{a_h \in \operatorname{Lim}(a)} \bigcup_{j=1}^{g} \Gamma_j^h.$$

Thus, the spectrum of a discrete Schrödinger operator $\mathbf{H}$ with semi-periodic potential is an infinite union of closed intervals, which are labeled by the limit operators of $\mathcal{H}$.

5.9 Comments and references

Special classes of pseudodifference operators have intensively been studied for about a century. In the case where the symbol $a : \mathbb{Z}^N \times \mathbb{T}^N, (x, t) \mapsto a(x, t)$ is independent of x, we are in the context of discrete convolution operators. The monographs [56] and [59] as well as the classical papers [63, 64] by Gohberg and Krein contain interesting historical notes. In particular, it has been known for a very long time that the spectral properties of convolution operators depend on the exponent $p \in [1, \infty]$ and on the weight. For continuous symbols, there is no dependence on p, but the presence of an exponential weight may drastically change the spectrum (incidentally, this phenomenon was employed in [28] to study the invertibility of convolution operators in unweighted Lebesgue spaces over the quarter-plane). One of our results is that this behavior of discrete convolution operators carries over to large classes of pseudodifference operators.

We remark that the spectra of convolution operators with piecewise continuous symbols depend on the space in much more a sensitive way than in the case of continuous symbols: in this situation already the exponent $p \in (1, \infty)$ influences the shape of the spectrum. For details of this fascinating topic (which goes far beyond the scope of this text: for example, a Laurent operator on $l^2(\mathbb{Z})$ is band-dominated if and only if its generating function is *continuous*) we refer to the books [30, 50, 65, 150, 23] and the references therein.

In the case of singular integral operators on L^p-spaces with general Muckenhoupt weight, this effect has been observed by Spitkovski [173] for the first time. See also [23, 27, 26] for the further development of the story, and [29] for analogous effects for discrete Wiener-Hopf operators with discontinuous symbols.

The derivation of the Fredholm index formula for pseudodifference operators on $l^p_w(\mathbb{Z})$ in Section 5.4 follows the paper [136] by two of the authors with John Roe, and the Phragmen-Lindelöf theorem on the exponential decay of the solutions to pseudodifference equations is a discrete analogue of a result for pseudodifferential equations on $\mathbb{R}^N$ which has been established by one of the authors in [128]).

Notice that the approach presented in Section 5.6 is completely different from the approach proposed in [179], where the spectral theory of one-dimensional periodic Jacobi operators (the so-called Floquet theory) is studied. There is an extensive bibliography devoted to partial differential equations with periodic coefficients from which we only cite the well-known books [143] (Section XIII.16) and [89].

Discrete Schrödinger operators of the form (5.4) appear in the theory of the spin waves [142], the description of random walks on $\mathbb{Z}^N$ [104], the propagation of waves in crystals [179], the theory of nonlinear integrable lattices [179, 45], etc. There is an extensive bibliography devoted to one-dimensional Jacobi operators, see for instance the monographs [179, 45, 80] and references therein.

The presentation of this section follows [135].

Chapter 6

Finite Sections of Band-dominated Operators

Let A be a band-dominated operator on the Banach space E^∞ introduced in Section 2.1.1. The practical solution of an operator equation $Ax = y$ with given $y \in E^\infty$ usually requires the use of a suitable discretization method which allows one to compute approximate solutions to $Ax = y$ numerically. For discretization by a *projection method* one replaces the equation $Ax = y$ by the sequence of the approximate equations

$$R_n A R_n x_n = R_n y, \quad n \geq 1, \tag{6.1}$$

where the R_n are projection operators which converge in a suitable sense (strongly or $\mathcal{P}$-strongly) to the identity operator on E^∞. We will examine a concrete projection method where the projections are specified as follows: We let Ω be a compact subset of $\mathbb{R}^N$ which has the point $0 \in \mathbb{R}^N$ in its interior. The characteristic function of Ω is the function χ_Ω which is 1 for $x \in \Omega$ and 0 for $x \notin \Omega$. Then we denote by R_n the operator of multiplication by the restriction of the function $\chi_{n\Omega}$ onto $\mathbb{Z}^N$. Clearly, every R_n is a projection operator on E^∞, and the condition $0 \in \operatorname{int} \Omega$ guarantees the strong convergence of the sequence (P_n) to the identity operator on every space E, whereas this sequence converges $\mathcal{P}$-strongly to I on $l^\infty(\mathbb{Z}^N, X)$. If the projections are specified in this way, we call (6.1) the *finite section method* for the approximate solution of $Ax = y$.

Our assumptions guarantee that $\mathcal{R} := (R_n)$ is an approximate identity which is equivalent to the approximate identity $\mathcal{P}$ specified in Section 2.1.1. Since equivalent approximate identities determine identical notions of compactness, Fredholmness and convergence, we can assume for simplicity and without loss of generality that $\mathcal{R} = \mathcal{P}$. Thus, the equations (6.1) become

$$P_n A P_n x_n = P_n y, \quad n \geq 1, \tag{6.2}$$

in what follows.

The crucial question is whether the sequence $(P_n A P_n)$ is *stable*, i.e., whether the operators $P_n A P_n : \operatorname{Im} P_n \to \operatorname{Im} P_n$ are invertible for sufficiently large n and whether the norms of their inverses are uniformly bounded. The stability of (P_nAP_n) together with the $\mathcal{P}$-strong convergence of the P_n and with the invertibility of A imply the applicability of the finite section method in the sense that, for every $y \in E$ and n large enough, the equations (6.2) possess unique solutions x_n which converge to a solution of $Ax = y$ as $n \to \infty$. If the norm convergence on $l^\infty(\mathbb{Z}^N, X)$ is replaced by a weaker one, then an analogous result holds for all spaces E^∞.

The formulation of a stability criterion is in Section 6.1. Its proof is surprisingly simple and rests on an application of the results on the Fredholmness of band-dominated operators. In Section 6.2, we will consider some special choices of Ω and make the stability conditions more explicit. After this, we turn our attention to the problem of the approximate determination of spectra of band-dominated operators in the third section. In particular, we will demonstrate how our stability results apply to establish the convergence of the eigenvalues or pseudo-eigenvalues of the matrices P_nAP_n. The concluding forth section is devoted to a discussion of the fractality properties of the finite section method for band-dominated operators. Fractality is a property of an approximation sequence which guarantees that certain limiting processes behave more uniformly than expected (see [73, 146, 151]). For example, the condition numbers of a stable approximation method (A_n) are bounded:

$$\limsup \|A_n\| \, \|A_n^{-1}\| < \infty,$$

but if the sequence (A_n) is fractal, then this limes superior is actually a limes. It turns out that, in general, the finite section method for band-dominated operators is non-fractal in an essential manner. So it is one of our goals to specify classes of band-dominated operators for which fractality of the finite section method can be guaranteed.

6.1 Stability of the finite section method

In this section, we are going to summarize some general facts on approximation methods and to explain the relationship between the stability of the finite section method for a band-dominated operator on $\mathbb{Z}^N$ and the Fredholmness of an associated band-dominated operator on $\mathbb{Z}^{N+1}$.

6.1.1 Approximation sequences

Let E be a Banach space, and let $\mathcal{P} = (P_n)$ be an approximate identity on E the elements of which are projection operators. In order to solve the operator equation $Ax = y$ for the operator $A \in L(E, \mathcal{P})$ approximately, we choose a sequence (A_n) of operators $A_n : \operatorname{Im} P_n \to \operatorname{Im} P_n$ and replace the equation $Ax = y$ by the sequence of the equations

$$A_n x_n = P_n y, \quad n = 1, 2, 3, \ldots \tag{6.3}$$

the solutions x_n of which are sought in $\operatorname{Im} P_n$. Observe that $A_nP_n = P_nA_nP_n$ belongs to $K(E, \mathcal{P})$ and, hence, to $L(E, \mathcal{P})$ for every operator $A_n \in L(\operatorname{Im} P_n)$ and for every n.

Of course, one has to assume some consistency between the operator A and its approximations A_n. At least one has to fix the sense in which the A_n approximate A. The weakest requirement one usually imposes is that the sequence $(A_n)_{n=1}^{\infty}$ is an *approximation method* for A in the following sense:

Definition 6.1.1 *A sequence (A_n) of operators $A_n \in L(\operatorname{Im} P_n)$ is an* approximation method *for $A \in L(E, \mathcal{P})$ if the operators $A_n P_n$ converge $\mathcal{P}$-strongly to A as $n \to \infty$.*

Being an approximation method says nothing about the solvability of the equations (6.3) and about the relations between possible solutions x_n of (6.3) and a possible solution x of $Ax = y$. What one is interested in is that the equations (6.3) possess unique solutions x_n for every $n \geq n_0$ and every right-hand side $y \in E$ and that these solutions converge in the norm of E to a solution of $Ax = y$. Equivalently, the operators A_n should be invertible for sufficiently large n, and their inverses $A_n^{-1} P_n$ should converge strongly. We summarize these requirements in the following definition.

Definition 6.1.2

(a) *The approximation method (A_n) for A is* applicable *if there exists a number n_0 such that the operators A_n are invertible for every $n \geq n_0$ and if their inverses A_n^{-1} converge strongly.*

(b) *The approximation method (A_n) for A is* $\mathcal{P}$-applicable *if there exists a number n_0 such that the operators A_n are invertible for every $n \geq n_0$ and if their inverses A_n^{-1} converge $\mathcal{P}$-strongly.*

(c) *A sequence (A_n) of operators $A_n \in L(\operatorname{Im} P_n)$ is* stable *if there exists a number n_0 such that the operators A_n are invertible for every $n \geq n_0$ and if the norms of their inverses are uniformly bounded:*

$$\sup_{n \geq n_0} \| A_n^{-1} P_n \|_{L(E)} < \infty.$$

The following basic result connects these notions. Recall the Definition 1.1.18 of the class $\mathcal{F}(E, \mathcal{P})$.

Theorem 6.1.3 *Let (P_n) be a uniform approximate identity consisting of projection operators, and let (A_n) with $A_n \in L(\operatorname{Im} P_n)$ be an approximation method for the operator $A \in L(E, \mathcal{P})$. Further suppose that $(A_n P_n) \in \mathcal{F}(E, \mathcal{P})$. Then this method is $\mathcal{P}$-applicable if and only if the operator A is invertible and the sequence (A_n) is stable.*

Proof. Let A be invertible and let (A_n) be stable. Then, for all sufficiently large n and all $P_k \in \mathcal{P}$,

$$\begin{aligned} \|(A_n^{-1} P_n - A^{-1}) P_k\| &\leq \|(A_n^{-1} P_n - P_n A^{-1}) P_k\| + \|(P_n - I) A^{-1} P_k\| \\ &\leq \|A_n^{-1} P_n\| \, \|(A - A_n P_n) A^{-1} P_k\| + \|(P_n - I) A^{-1} P_k\| \end{aligned}$$

and, similarly,

$$\|P_k (A_n^{-1} P_n - A^{-1})\| \leq \|A_n^{-1} P_n\| \, \|P_k A^{-1} (A - A_n P_n)\| + \|P_k A^{-1} (P_n - I)\|.$$

These estimates imply the $\mathcal{P}$-strong convergence of $A_n^{-1}P_n$ to A^{-1} and, thus, the $\mathcal{P}$-applicability of the method (A_n).

Let, conversely, (A_n) be a $\mathcal{P}$-applicable approximation method for A. Thus, the operators A_n are invertible for large n, and the sequence $(A_n^{-1}P_n)$ of their inverses is $\mathcal{P}$-strongly convergent. Then Proposition 1.1.17 (a) yields the stability of (A_n).

To verify the invertibility of A, recall that the definition of an applicable method implies the solvability of the equation $Au = f$ for every right-hand side f. Hence, A is surjective. For the injectivity of A observe that, for all $x \in E$, all k and m, and for all sufficiently large n,

$$\begin{aligned}
\|P_k(I - A_n^{-1}P_nA)x\| &= \|P_k(P_n - A_n^{-1}P_nA)x\| \\
&= \|P_kA_n^{-1}P_n(A_nP_n - A)x\| \\
&\le \|P_kA_n^{-1}P_nP_m(A_nP_n - A)x\| + \|P_kA_n^{-1}P_nQ_m(A_nP_n - A)x\| \\
&\le \|P_kA_n^{-1}P_n\|\,\|P_m(A_nP_n - A)x\| + \|P_kA_n^{-1}P_nQ_m\|\,\|(A_nP_n - A)x\| \\
&\le C(\|P_m(A_nP_n - A)\| + \|P_kA_n^{-1}P_nQ_m\|).
\end{aligned}$$

For fixed k and given $\varepsilon > 0$, choose m such that $\|P_kA_n^{-1}P_nQ_m\| < \varepsilon/2$ uniformly with respect to n (which is possible due to the assumption $(A_nP_n) \in \mathcal{F}(E, \mathcal{P})$ and to the inverse closedness of $\mathcal{F}(E, \mathcal{P})$ by Theorem 1.1.19). Then choose n_0 such that $\|P_m(A_nP_n - A)\| < \varepsilon/2$ for all $n \ge n_0$. This shows that

$$\|P_k(I - A_n^{-1}P_nA)x\| \to 0 \quad \text{as } n \to \infty$$

for every $k \in \mathbb{N}$ and $x \in E$. If, in particular, x is in the kernel of A, then we obtain $P_kx = 0$ for every k whence $x = 0$ because $\mathcal{P}$ is an approximate identity. □

6.1.2 Stability vs. invertibility

Theorem 6.1.3 indicates that a main problem in numerical analysis is to study the stability of a given sequence of approximation matrices. We will see now that this stability problem can be realized as an invertibility problem in a suitably chosen Banach algebra. For this goal, it will be convenient to consider approximation operators which act on all of E rather than on $\operatorname{Im} P_n$. Further, we will admit two-sided approximation sequences (i.e., sequences on $\mathbb{Z}$) rather than the usual one-sided sequences (on $\mathbb{N}$). For example, we will rewrite the equations (6.2) as

$$(P_nAP_n + Q_n)X_n = y \quad \text{where} \quad Q_n := I - P_n. \tag{6.4}$$

Clearly, the solutions X_n are related to the solutions x_n of (6.2) by $x_n = P_nX_n$ and $X_n := x_n + Q_ny$. It is also evident that the sequence (P_nAP_n) is stable if and only if the sequence $(P_nAP_n + Q_n)$ has this property. Moreover, if we set $P_n := 0$ or $P_n := P_{-n}$ for $n < 0$, then the stability of $(P_nAP_n+Q_n)_{n\ge0}$ becomes equivalent to the stability of the sequence $(P_nAP_n + Q_n)_{n\in\mathbb{Z}}$.

By $\mathcal{F}$ we denote the set of all bounded sequences $(A_n)_{n\in\mathbb{Z}}$ of operators $A_n \in L(E)$. Provided with elementwisely defined operations and the supremum norm (see Section 1.1.5), this set becomes a Banach algebra, and the set $\mathcal{G}$ of all sequences $(G_n) \in \mathcal{F}$ with $\|G_n\| \to 0$ as $n \to \infty$ becomes a closed ideal of $\mathcal{F}$. Let furthermore $\mathcal{F}_c$ refer to the set of all sequences $(A_n) \in \mathcal{F}$ with the properties that all operators A_n belong to $L(E, \mathcal{P})$ and that all differences $A_m - A_n$ are $\mathcal{P}$-compact, and abbreviate $\mathcal{F}_c \cap \mathcal{G}$ to $\mathcal{G}_c$. It is elementary to check that $\mathcal{F}_c$ is a closed and inverse closed subalgebra of $\mathcal{F}$ and that $\mathcal{G}_c$ is a closed ideal of $\mathcal{F}_c$. If (G_n) is a sequence in $\mathcal{G}_c$ then, necessarily, every operator G_n is $\mathcal{P}$-compact. Indeed, since $(G_n) \in \mathcal{F}_c$, all operators G_n have the same essential norm $\|G_n\|_{ess} := \|G_n + K(E^\infty, \mathcal{P})\|$. Thus, if the essential norm of one of these operators is positive, say $c > 0$, then $\|G_n\| \geq \|G_n\|_{ess} = c > 0$ for all n, which is in contradiction with $\|G_n\| \to 0$.

Lemma 6.1.4

(a) *A sequence* $\mathbf{A} \in \mathcal{F}$ *is stable if and only of the coset* $\mathbf{A} + \mathcal{G}$ *is invertible in the quotient algebra* $\mathcal{F}/\mathcal{G}$.

(b) *Let* $\mathcal{P}$ *be a uniform approximate identity. Then a sequence* $\mathbf{A} \in \mathcal{F}_c$ *is stable if and only of the coset* $\mathbf{A} + \mathcal{G}_c$ *is invertible in the quotient algebra* $\mathcal{F}_c/\mathcal{G}_c$.

Proof. (a) Let $\mathbf{A} = (A_n)_{n\in\mathbb{Z}}$ be a stable sequence. Thus, the operators A_n are invertible for $|n| \geq n_0$, and the norms of their inverses are uniformly bounded. We set $B_n := A_n^{-1}$ for $|n| \geq n_0$ and choose B_n arbitrarily for $|n| < n_0$. Then the sequence (B_n) belongs to $\mathcal{F}$, and it is an inverse of (A_n) modulo $\mathcal{G}$.

Let, conversely, the coset $(A_n) + \mathcal{G}$ be invertible in $\mathcal{F}/\mathcal{G}$. Then there are sequences (B_n) in $\mathcal{F}$ and (G_n) and (H_n) in $\mathcal{G}$ such that $A_n B_n = I + G_n$ and $B_n A_n = I + H_n$ for all $n \in \mathbb{Z}$. If $|n|$ is large enough, then $\|G_n\| < 1/2$ and $\|H_n\| < 1/2$. Thus, for these n, the operators $I + G_n$ and $I + H_n$ are invertible, and the norms of their inverses are uniformly bounded by 2. Hence, $A_n B_n (I + G_n)^{-1} = I$ and $(I + H_n)^{-1} B_n A_n = I$, and the norms of $B_n (I + G_n)^{-1} = (I + H_n)^{-1} B_n = A_n^{-1}$ are uniformly bounded. Consequently, (A_n) is a stable sequence.
(b) If (A_n) is a stable sequence in $\mathcal{F}_c$, then we choose B_n as above for $|n| \geq n_0$ and set $B_n = A_{n_0}^{-1}$ for $|n| < n_0$. Then all operators B_n belong to $L(E, \mathcal{P})$ by Theorem 1.1.9, and

$$B_m - B_n = A_m^{-1}(A_n - A_m)A_n^{-1} \in K(E, \mathcal{P}).$$

Consequently, the sequence (B_n) belongs to $\mathcal{F}_c$, and it is an inverse of (A_n) modulo $\mathcal{G}_c$. The reverse implication follows from assertion (a). □

6.1.3 Stability vs. Fredholmness

Beginning with this section, let again E^∞ refer to one of the sequence spaces $l^p(\mathbb{Z}^N, X)$ with $1 \leq p \leq \infty$ or $c_0(\mathbb{Z}^N, X)$. In order to indicate the dependence of E^∞ on the dimension N, we will sometimes write E_N^∞ in place of E^∞. Let further $\mathcal{P} = (P_n)$ be an approximate identity on E^∞ which satisfies

$$P_m P_n = P_n P_m = P_m \quad \text{whenever } m \leq n.$$

In particular, all P_n are projection operators. Our goal is to establish a fairly general criterion for the stability of approximation methods for operators which act on one of the spaces E_N^∞. This criterion relates the stability of a sequence (A_n) of operators on E_N^∞ with the $\mathcal{P}'$-Fredholmness of a related operator which lives on E_{N+1}^∞. Here, $\mathcal{P}'$ is an approximate identity on E_{N+1}^∞ which is related with $\mathcal{P}$ and which will be defined later.

We agree upon writing every vector $x \in \mathbb{Z}^{N+1}$ as $x = (x', x_{N+1}) \in \mathbb{Z}^N \times \mathbb{Z}$. The Banach space E_{N+1}^∞ can be viewed as the direct sum over all $m \in \mathbb{Z}$ of its closed subspaces

$$E_{N+1,m}^\infty := \{f \in E_{N+1}^\infty : f(x) = 0 \text{ if } x_{N+1} \neq m\}.$$

Further, for every $m \in \mathbb{Z}$, consider the operator of restriction

$$R_m : E_{N+1}^\infty \to E_N^\infty, \quad (R_m f)(x') := f(x', m)$$

as well as the operator of embedding

$$S_m : E_N^\infty \to E_{N+1}^\infty, \quad (S_m)(x', n) := \begin{cases} f(x') & \text{if } n = m \\ 0 & \text{if } n \neq m. \end{cases}$$

The operator $T_m := S_m R_m$ is the projection (in case E is a Hilbert space the orthogonal projection) of E_{N+1}^∞ onto $E_{N+1,m}^\infty$ parallel to the direct sum of all $E_{N+1,r}^\infty$ with $r \neq m$, whereas $R_m S_m$ acts on E_N^∞ as the identity operator for every m. In particular, each of the spaces $E_{N+1,m}^\infty$ is isometrically isomorphic to E_N^∞.

We say that an operator $A \in L(E_{N+1}^\infty)$ is *reduced by the family* $(T_m)_{m\in\mathbb{Z}}$ if

$$AT_m = T_m A \quad \text{for all } m \in \mathbb{Z}.$$

The class of all of these operators will be denoted by $L_{\text{red}}(E_{N+1}^\infty)$. Clearly, this class is a closed, unital, and inverse closed subalgebra of $L(E_{N+1}^\infty)$. To every sequence $\mathbf{A} := (A_n) \in \mathcal{F}$, we finally associate an operator $\text{Op}\,(\mathbf{A})$ on E_{N+1}^∞ by

$$(\text{Op}\,(\mathbf{A})f)(x', m) := (A_m R_m f)(x').$$

Further, if $\mathbf{A}$ is the constant sequence $(A) \in \mathcal{F}$, then we agree upon writing $\text{Op}(A)$ in place of $\text{Op}(\mathbf{A})$.

Lemma 6.1.5 *The mapping* Op *is an isometric isomorphism from $\mathcal{F}$ onto the algebra $L_{\text{red}}(E_{N+1}^\infty)$.*

Proof. Let $\mathbf{A} := (A_n) \in \mathcal{F}$. Then, for every $f \in E_{N+1}^\infty$ and every $m \in \mathbb{Z}$,

$$\begin{aligned} (\text{Op}\,(\mathbf{A})T_r f)(x', m) &= (A_m R_m T_r f)(x') \\ &= \begin{cases} (A_r R_r f)(x') & \text{if } m = r \\ 0 & \text{if } m \neq r \end{cases} \;=\; (T_r \text{Op}\,(\mathbf{A})f)(x', m), \end{aligned}$$

implying that (T_r) reduces the operator $\mathrm{Op}\,(\mathbf{A})$. Conversely, if the operator A is reduced by (T_m), then $\mathbf{A} := (R_n A S_n)$ is a sequence in $\mathcal{F}$ with

$$(\mathrm{Op}\,(\mathbf{A})f)(x', m) = (R_m A S_m R_m f)(x') = (R_m A T_m f)(x')$$
$$= (R_m T_m A f)(x') = (S_m A f)(x') = (Af)(x', m).$$

Thus, the mapping $\mathbf{A} \mapsto \mathrm{Op}(\mathbf{A})$ is onto. To see that this mapping is an isometry notice that

$$\|\mathrm{Op}\,(\mathbf{A})f\|^p = \sum_m \sum_{x'} \|(\mathrm{Op}\,(\mathbf{A}))(x', m)\|^p = \sum_m \sum_{x'} \|(A_m R_m f)(x')\|^p$$
$$\sum_m \|A_m R_m f\|^p \le \sup_m \|A_m\|^p \sum_m \|R_m f\|^p = \|\mathbf{A}\|^p \|f\|^p$$

whence the estimate $\|\mathrm{Op}\,(\mathbf{A})\| \le \|\mathbf{A}\|$ in case $1 \le p < \infty$. In case of l^∞ or c_0, this estimate follows similarly. Finally, given $\varepsilon > 0$, choose m such that $\|A_m\| \ge \|\mathbf{A}\| - \varepsilon$, and choose a unit vector $x \in E_N^\infty$ such that $\|A_m x\| \ge \|A_m\| - \varepsilon$. Then $\|S_m x\| = 1$ and

$$\|\mathrm{Op}\,(\mathbf{A}) S_m x\| = \|A_m x\| \ge \|\mathbf{A}\| - 2\varepsilon,$$

which gives the reverse estimate. □

The next result settles the connection between stability and Fredholmness for sequences in the algebra $\mathcal{F}_c$. For every non-negative integer n, define

$$P'_n := \sum_{k=-n}^{n} S_k P_n R_k \quad \text{and} \quad Q'_n := I - P'_n.$$

The family $\mathcal{P}' := (P'_n)_{n \ge 0}$ forms an approximate identity on E_{N+1}^∞. Evidently, one has $P'_n T_m = T_m P'_n$ for all choices of m and n. Thus, $P'_n \in L_{\mathrm{red}}(E_{N+1}^\infty)$ for every n and $T_m \in L(E_{N+1}^\infty, \mathcal{P}')$ for every m.

Theorem 6.1.6

(a) *If* $\mathbf{A}$ *is a sequence in the algebra* $\mathcal{F}_c$, *then* $\mathrm{Op}(\mathbf{A})$ *belongs to* $L(E_{N+1}^\infty, \mathcal{P}')$.

(b) *A sequence* $\mathbf{G} \in \mathcal{F}_c$ *is in* $\mathcal{G}_c$ *if and only if* $\mathrm{Op}(\mathbf{G})$ *is* $\mathcal{P}'$*-compact.*

(c) *Let* $\mathcal{P}$ *be a uniform approximate identity on* E_N^∞. *Then a sequence* $\mathbf{A} \in \mathcal{F}_c$ *is stable if and only if the operator* $\mathrm{Op}(\mathbf{A})$ *is* $\mathcal{P}'$*-Fredholm.*

Proof. (a) Let $\mathbf{A} = (A_n) \in \mathcal{F}_c$. It is easy to check that, for every fixed $k \ge 0$ and every $n \ge k$,

$$P'_k \mathrm{Op}(\mathbf{A}) Q'_n = \sum_{|r| \le k} S_r P_k A_r Q_n R_r.$$

The right-hand side of this equality tends to zero as $n \to \infty$ since $A_r \in L(E_N^\infty, \mathcal{P})$. Similarly one gets $\|Q'_n \mathrm{Op}(\mathbf{A}) P'_k\| \to 0$. Thus, $\mathrm{Op}(\mathbf{A})$ is in $L(E_{N+1}^\infty, \mathcal{P}')$.

(b) Let $\mathbf{G} = (G_n)$ be a sequence in $\mathcal{G}_c$. Given $\varepsilon > 0$, choose n_0 such that $\|G_n\| < \varepsilon$ for all $|n| > n_0$, and choose $m \geq n_0$ such that

$$\|G_n - P_m G_n P_m\| < \varepsilon \quad \text{for all } |n| \geq n_0.$$

Further, let $H_n := P_m$ if $|n| \geq n_0$ and $H_n := 0$ for $|n| > n_0$. With $\mathbf{H} := (H_n)$, one then has $\|\mathbf{G} - \mathbf{HGH}\| \leq \varepsilon$ and, thus,

$$\|\mathrm{Op}(\mathbf{G}) - \mathrm{Op}(\mathbf{H})\mathrm{Op}(\mathbf{G})\mathrm{Op}(\mathbf{H})\| \leq \varepsilon.$$

The operator $\mathrm{Op}(\mathbf{H})$ belongs to $K(E^\infty_{N+1})$. Indeed,

$$\mathrm{Op}(\mathbf{H}) = \sum_{|n| \leq n_0} S_n P_m R_n = \sum_{|n| \leq n_0} T_n P'_m,$$

and $T_n P'_m$ belongs to $K(E^\infty_{N+1})$ since $T_m \in L(E^\infty_{N+1})$. Since furthermore $\mathrm{Op}(\mathbf{G}) \in L(E^\infty_{N+1})$ by assertion (a), we conclude that $\mathrm{Op}(\mathbf{H})\mathrm{Op}(\mathbf{G})\mathrm{Op}(\mathbf{H})$ is $\mathcal{P}'$-compact and, hence, $\mathrm{Op}(\mathbf{G})$ can be approximated by $\mathcal{P}'$-compact operators as closely as desired.

For the reverse implication, let $\mathbf{G} = (G_n) \in \mathcal{F}_c$ and let $\mathrm{Op}(\mathbf{G})$ be $\mathcal{P}'$-compact. We claim that

$$\|G_n\| = \|R_n \mathrm{Op}(\mathbf{G}) S_n\| \to 0 \quad \text{as } n \to \infty.$$

Since

$$\|R_n \mathrm{Op}(\mathbf{G}) S_n\| = \|R_n T_n \mathrm{Op}(\mathbf{G}) S_n\| \leq \|T_n \mathrm{Op}(\mathbf{G})\|,$$

it is sufficient to show that

$$\|T_n K\| \to 0 \quad \text{for every } K \in K(E^\infty_{N+1}, \mathcal{P}'). \tag{6.5}$$

Given $\varepsilon > 0$, choose m such that $\|Q'_m K\| < \varepsilon$. Then, for all $n > m$,

$$\|T_n K\| \leq \|T_n P'_m K\| + \|T_n Q'_m K\| \leq \|Q'_m K\| < \varepsilon,$$

whence the assertion.

(c) If $\mathbf{A} \in \mathcal{F}_c$ is a stable sequence then, by Lemma 6.1.4 (b), there are sequences $\mathbf{B} \in \mathcal{F}_c$ and $\mathbf{G}, \mathbf{H} \in \mathcal{G}_c$ such that $\mathbf{AB} = \mathbf{I} + \mathbf{G}$ and $\mathbf{BA} = \mathbf{I} + \mathbf{H}$ where $\mathbf{I}$ refers to the constant sequence (I). Applying the isomorphism Op to these equalities and taking into account that $\mathrm{Op}(\mathbf{A})$ as well as $\mathrm{Op}(\mathbf{B})$ belong to $L(E^\infty_{N+1}, \mathcal{P}')$ and that $\mathrm{Op}(\mathbf{G})$ as well as $\mathrm{Op}(\mathbf{H})$ are $\mathcal{P}'$-compact by the previous parts of the present theorem, we get the $\mathcal{P}'$-Fredholmness of $\mathrm{Op}(\mathbf{A})$.

Let, conversely, $\mathbf{A} = (A_n) \in \mathcal{F}_c$ and let $\mathrm{Op}(\mathbf{A})$ be a $\mathcal{P}'$-Fredholm operator. Then there are an operator $B \in L(E^\infty_{N+1}, \mathcal{P}')$ as well as $\mathcal{P}'$-compact operators G and H such that $\mathrm{Op}(\mathbf{A})B = I + G$ and $B\mathrm{Op}(\mathbf{A}) = I + H$. Multiplying the first of these equalities by R_n from the left- and by S_n from the right-hand side and taking into account that $\mathrm{Op}(\mathbf{A})$ is reduced by $(T_n)_{n\in\mathbb{Z}}$, we obtain

$$\begin{aligned} R_n \mathrm{Op}(\mathbf{A}) B S_n &= R_n T_n \mathrm{Op}(\mathbf{A}) B S_n = R_n \mathrm{Op}(\mathbf{A}) T_n B S_n \\ &= R_n \mathrm{Op}(\mathbf{A}) S_n R_n B S_n = A_n R_n B S_n = I + R_n G S_n. \end{aligned}$$

The sequence (R_nGS_n) converges to 0 in the operator norm as $n \to \infty$, which follows from $R_nG = R_nT_nG$ and from (6.5). Similarly, $R_nBS_nA_n = I + R_nHS_n$ with the sequence (R_nHS_n) tending to zero. Since the sequence (R_nBS_n) is bounded, Lemma 6.1.4 (a) implies the stability of the sequence (A_n). □

So we are left with the problem of the $\mathcal{P}'$-Fredholmness of the operator $\mathrm{Op}(\mathbf{A})$. It turns out that if $\mathbf{A} = (A_n)$ is the sequence of the approximation operators in (6.4), i.e., if $A_n = P_nAP_n + Q_n$ with a band-dominated operator A on E_N^∞, then $\mathrm{Op}(\mathbf{A})$ is a band-dominated operator on E_{N+1}^∞. More generally, we let $\mathcal{B}$ (resp. $\mathcal{B}^\$$) stand for the smallest closed subalgebra of the algebra $\mathcal{F}$ which contains all constant sequences (A) where A is a band-dominated operator (resp. a rich band-dominated operator) on E_N^∞ and which contains the sequence (P_n) with $P_n = 0$ for $n < 0$.

Lemma 6.1.7 *If* $\mathbf{A} \in \mathcal{B}$, *then* $\mathbf{A} \in \mathcal{F}_c$, *and* $\mathrm{Op}(\mathbf{A})$ *is a band-dominated operator on* E_{N+1}^∞. *If* $\mathbf{A} \in \mathcal{B}^\$$, *then* $\mathrm{Op}(\mathbf{A})$ *is a rich band-dominated operator.*

Proof. The algebra $\mathcal{B}$ can be characterized as the smallest closed subalgebra of $\mathcal{F}$ which contains all constant sequences (V_m) and (aI) with $m \in \mathbb{Z}^N$ and $a \in l^\infty(\mathbb{Z}^N, L(X))$ as well as the sequence (P_n). It suffices to check the assertions of the lemma for these generating sequences in place of $\mathbf{A}$. The first assertion is evident. It is further clear that $\mathrm{Op}(aI)$ and $\mathrm{Op}(P_n)$ are multiplication operators on E_{N+1}^∞ and that $\mathrm{Op}(V_m)$ is the shift operator $V_{(m,0)}$. Hence, in any case, $\mathrm{Op}(\mathbf{A})$ is band-dominated. It is finally evident that the operator $\mathrm{Op}(P_n)$ possesses a rich operator spectrum and that every operator $\mathrm{Op}(A)$ is rich whenever A is rich. □

In particular, this result offers the possibility to study the stability of the sequence by means of the Fredholm criterion in Theorem 2.2.1. This theorem has been proved for rich band-dominated operators acting on spaces E only. Thus, we will restrict our attention to these spaces in what follows, and we will write E_N in place if E_N^∞ in accordance with our earlier notations. For these spaces, Theorem 2.2.1 in combination with Theorem 6.1.6 (c) states that a sequence $\mathbf{A} \in \mathcal{B}^\$$ is stable if and only if all limit operators of the operator $\mathrm{Op}(\mathbf{A})$ are invertible and if the norms of their inverses are uniformly bounded.

Since the entries of the sequence $\mathbf{A}$ are operators acting on E_N, one would prefer a criterion for the stability of $\mathbf{A}$ in terms of operators on E_N, not on E_{N+1}. The next lemma shows that the stability criterion for sequences in $\mathcal{B}^\$$ can be formulated in the desired way.

Lemma 6.1.8 *Every limit operator* A_h *of an operator* $A \in L_{\mathrm{red}}(E_{N+1})$ *belongs to* $L_{\mathrm{red}}(E_{N+1})$, *too. In particular,* A_h *is invertible if and only if all operators* $R_mA_hS_m$ *with* $m \in \mathbb{Z}$ *are invertible in* E_N *and if the norms of their inverses are uniformly bounded.*

Proof. The only thing which requires a proof is that A_h is reduced by the family (T_m) if A is so.

Indeed, for every $k \in \mathbb{Z}$ and $n \in \mathbb{N}$, one has

$$\begin{aligned} T_k V_{-h(n)} A V_{h(n)} &= T_k V_{-(h(n)', h_{N+1}(n))} A V_{(h(n)', h_{N+1}(n))} \\ &= V_{-(h(n)', h_{N+1}(n))} T_{k+h_{N+1}(n)} A V_{(h(n)', h_{N+1}(n))} \\ &= V_{-(h(n)', h_{N+1}(n))} A T_{k+h_{N+1}(n)} V_{(h(n)', h_{N+1}(n))} \\ &= V_{-(h(n)', h_{N+1}(n))} A V_{(h(n)', h_{N+1}(n))} T_k = V_{-h(n)} A V_{h(n)} T_k. \end{aligned}$$

Letting n go to infinity yields $T_k A_h = A_h T_k$. □

For every sequence $\mathbf{A} \in \mathcal{F}$, let $\sigma_{stab}(\mathbf{A})$ stand for the set of all operators $R_m B S_m$ where $B \in \sigma_{op}(\mathrm{Op}(\mathbf{A}))$ and $m \in \mathbb{Z}$. Summarizing the previous results, we arrive at the following stability criterion for sequences in $\mathcal{B}^\$$.

Theorem 6.1.9 *A sequence $\mathbf{A} \in \mathcal{B}^\$$ is stable if and only if all operators in $\sigma_{stab}(\mathbf{A})$ are invertible and if the norms of their inverses are uniformly bounded.*

6.2 Finite sections of band-dominated operators on $\mathbb{Z}^1$ and $\mathbb{Z}^2$

Our next goal is to compute the stability spectrum $\sigma_{stab}(\mathbf{A})$ for the finite section sequences $\mathbf{A} = (P_n A P_n + Q_n)$ with

$$P_n := \begin{cases} \chi_{n\Omega} I|_{\mathbb{Z}^N} & \text{if} \quad n \geq 0 \\ 0 & \text{if} \quad n < 0 \end{cases} \qquad \text{and} \quad Q_n := I - P_n$$

in the cases when A is band-dominated on E_1 and $\Omega = [-1, 1]$, and when A is band-dominated on E_2 and Ω is a convex and compact polygon with 0 as its interior point and with vertices in $\mathbb{Z}^2$. We start with describing the limit operators of the operators $\mathrm{Op}(A)$ where A is band-dominated on E_N and N is arbitrary.

Proposition 6.2.1 *For every operator $A \in L(E_N)$,*

$$\sigma_{stab}(\mathrm{Op}(A)) = \sigma_{op}(A) \cup \{V_{-c} A V_c : c \in \mathbb{Z}^N\}. \tag{6.6}$$

Proof. Again, we write each sequence $h : \mathbb{N} \to \mathbb{Z}^{N+1}$ as $h = (h', h_{N+1}) : \mathbb{N} \to \mathbb{Z}^N \times \mathbb{Z}$. Then

$$\begin{aligned} & V_{-h(m)} \,\mathrm{Op}(A)\, V_{h(m)} \\ &= V_{-(h'(m),0)} V_{-(0,\, h_{N+1}(m))} \,\mathrm{Op}(A)\, V_{(0,\, h_{N+1}(m))} V_{(h'(m),0)} \\ &= V_{-(h'(m),0)} \,\mathrm{Op}(A)\, V_{(h'(m),0)} = \mathrm{Op}(V_{-h'(m)} A V_{h'(m)}). \end{aligned}$$

Let now h be a sequence which tends to infinity and for which the limit operator $\mathrm{Op}(A)_h$ exists. Then the sequence h' can be bounded or unbounded. If it is bounded, then there is a subsequence η of $\mathbb{N}$ such that $h' \circ \eta$ is a constant sequence, (c), say. For the sequence $h \circ \eta$ one has

$$\mathrm{Op}(A)_h = \mathrm{Op}(A)_{h \circ \eta} = \mathrm{Op}(V_{-(h' \circ \eta)(m)} A V_{(h' \circ \eta)(m)}) = \mathrm{Op}(V_{-c} A V_c).$$

If h' is unbounded, then there is a subsequence η of $\mathbb{N}$ such that the limit operator $A_{h'\circ\eta}$ exists, and consequently

$$Op(A)_h = Op(A)_{h\circ\eta} = Op(A_{h'\circ\eta}).$$

Thus, the operators in $\sigma_{stab}(\mathrm{Op}(A))$ are necessarily limit operators of A or operators of the form $V_{-c}AV_c$ with $c \in \mathbb{Z}^N$. Conversely, if h' is a sequence such that the limit operator $A_{h'}$ exists, then $h(m) := (h'(m), 0)$ defines a sequence which tends to infinity and for which the limit operator of $\mathrm{Op}(A)$ exists and satisfies $\mathrm{Op}(A)_h = \mathrm{Op}(A_{h'})$. Finally, if $c \in \mathbb{Z}^N$, then $h(m) := (c, m)$ defines a sequence which tends to infinity and for which the limit operator $\mathrm{Op}(A)_h$ exists and coincides with $\mathrm{Op}(V_{-c}AV_c)$. This shows the identity (6.6). □

6.2.1 Band-dominated operators on $\mathbb{Z}^1$: the general case

Let $N = 1$, define P_n as above with respect to $\Omega = [-1, 1]$, and abbreviate the sequence (P_n) by $\mathbf{P}$. Then $\mathrm{Op}(\mathbf{P})$ is the operator of multiplication by the characteristic function of the cone $\{(x_1, x_2) \in \mathbb{R}^2 : x_2 \geq |x_1|\}$. This operator can be written as the product $\hat{\chi}_{\mathbb{H}_1} \hat{\chi}_{\mathbb{H}_{-1}}$ where $\mathbb{H}_1$ and $\mathbb{H}_{-1}$ refer to the half-planes

$$\mathbb{H}_1 := \{(x_1, x_2) : x_2 \geq x_1\} \quad \text{and} \quad \mathbb{H}_{-1} := \{(x_1, x_2) : x_2 \geq -x_1\}.$$

The results of Section 2.6.3 yield the following description of the local operator spectra of the operator $\mathrm{Op}(\mathbf{P})$. For, write $\eta \in S^1$ as $\eta = \exp(i\varphi)$ with $\varphi \in [0, 2\pi)$ and, for $m \in \mathbb{Z}$, let U_m stand for the shift operator $V_{(0,m)}$ on $\mathbb{Z}^2$. Then

$$\sigma_\eta(\mathrm{Op}(\mathcal{P})) = \begin{cases} \{0, I, U_{-m}\, \chi_{H_1} I|_{\mathbb{Z}^2}\, U_m : m \in \mathbb{Z}\} & \text{if } \varphi = \pi/4 \\ \{I\} & \text{if } \varphi \in (\pi/4, 3\pi/4) \\ \{0, I, U_{-m}\, \chi_{H_{-1}} I|_{\mathbb{Z}^2}\, U_m : m \in \mathbb{Z}\} & \text{if } \varphi = 3\pi/4 \\ \{0\} & \text{if } \varphi \notin [\pi/4, 3\pi/4]. \end{cases}$$

Let, finally, $Q : E_1 \to E_1$ stand for the projection operator which sends the sequence $(\ldots, x_{-2}, x_{-1}, x_0, x_1, \ldots)$ into $(\ldots, 0, 0, x_0, x_1, \ldots)$, and set $Q := I - P$. Then

$$\hat{\chi}_{\mathbb{H}_{-1}} I = \mathrm{Op}((V_{-n}PV_n)_{n\in\mathbb{Z}}) \quad \text{and} \quad \hat{\chi}_{\mathbb{H}_1} I = \mathrm{Op}((V_{-n}QV_n)_{n\in\mathbb{Z}}).$$

Theorem 6.2.2 *Let A be a band-dominated operator on E_1 with rich spectrum, and let $\Omega = [-1, 1]$. Then the finite section method $(P_nAP_n + Q_n)$ is stable if and only if the operator A is invertible, if the operators*

$$PA_sP + Q \tag{6.7}$$

are invertible for all limit operators $A_s \in \sigma_{-1}(A)$, if the operators

$$QA_sQ + P \tag{6.8}$$

are invertible for all limit operators $A_s \in \sigma_1(A)$, and if the norms of the inverses of the operators in (6.7) and (6.8) are uniformly bounded.

Proof. If the sequence $\mathbf{A} := (P_n A P_n + Q_n)$ is stable, then the operator A is invertible by Theorem 6.1.3. We have to check that the operators in (6.7) are invertible, too.

Let $s : \mathbb{N} \to \mathbb{Z}$ be a sequence which tends to infinity in the direction of $-1 \in S^0$ and for which the limit operator A_s exists. Then $h(n) := (s(n), -s(n))$ defines a sequence in $\mathbb{Z}^2$ which tends to infinity in the direction of $\exp(i\varphi)$ with $\varphi = 3\pi/4$, and the limit operators of $\mathrm{Op}(A)$ and $\mathrm{Op}(\mathbf{P})$ with respect to h exist and are equal to

$$\mathrm{Op}(A_s) \quad \text{and} \quad \hat{\chi}_{\mathbb{H}_{-1}} I = \mathrm{Op}((V_{-n} P V_n)_{n \in \mathbb{Z}}),$$

respectively. Thus, $\sigma_{stab}(\mathbf{A})$ contains all operators

$$V_{-n} P V_n A_s V_{-n} P V_n + V_{-n} Q V_n$$

with $n \geq 0$. In particular, it contains the operator $PA_sP + Q$ and, thus, all operators (6.7). The uniform boundedness of the inverses of these operators is a consequence of Theorem 6.1.9. The proof for the operators in (6.8) is analogous.

Now suppose that the operator A as well as all operators in (6.7) and (6.8) are invertible and that the norms of their inverses are uniformly bounded. By Theorem 6.1.9, we have to show that all operators in $\sigma_{stab}(\mathbf{A})$ are invertible and that the norms of their inverses are uniformly bounded.

If $h \in \mathbb{Z}^2$ is a sequence which tends to infinity such that the limit operator $Op(\mathbf{P})_h$ exists and is equal to 0, and if g is a subsequence of h such that the limit operator $Op(A)_g$ exists, then $\mathrm{Op}(\mathbf{A})_g = I$, and nothing is to prove. If $\mathrm{Op}(\mathbf{P})_h$ exists and is equal to I, and if g is as above, then either

$$\mathrm{Op}(\mathbf{A})_g = \mathrm{Op}(A_s) \quad \text{or} \quad \mathrm{Op}(\mathbf{A})_g = \mathrm{Op}(V_{-c} A V_c)$$

with a sequence $s : \mathbb{N} \to \mathbb{Z}$ tending to infinity or with a constant $c \in \mathbb{Z}$ (Proposition 6.2.1). In both cases, the invertibility of $\mathrm{Op}(\mathbf{A})_g$ follows from the invertibility of A, and one has $\|(\mathrm{Op}(\mathbf{A})_g)^{-1}\| \leq \|A^{-1}\|$.

It remains to study sequences h which tend to infinity in the direction $\exp(i\varphi) \in S^1$ with $\varphi = \pi/4$ and $\varphi = 3\pi/4$ and for which $\mathrm{Op}(\mathbf{P})_h$ exists and is not equal to 0 or I. Let, for example, $\varphi = 3\pi/4$, and let g be a subsequence of h such that the limit operator $\mathrm{Op}(A)_g$ exists. Then the first component of $g = (g_1, g_2)$ is a sequence in $\mathbb{Z}$ which tends to $-\infty$ (i.e., to infinity in the direction of $-1 \in S^0$) and for which the limit operator A_s exists. Thus, there is an $m \in \mathbb{Z}$ such that

$$\begin{aligned}
\mathrm{Op}(\mathbf{A})_g &= U_{-m}\, \hat{\chi}_{\mathbb{H}_{-1}}\, U_m \mathrm{Op}(A_s) U_{-m}\, \hat{\chi}_{\mathbb{H}_{-1}} I|_{\mathbb{Z}^2}\, U_m + U_{-m}\, (1 - \chi_{H_{-1}})\, U_m \\
&= U_{-m} \mathrm{Op}((V_{-n} P V_n)_{n \in \mathbb{Z}}) U_m \mathrm{Op}(A_s) U_{-m} \mathrm{Op}((V_{-n} P V_n)_{n \in \mathbb{Z}}) U_m \\
&\quad + U_{-m} \mathrm{Op}((V_{-n} Q V_n)_{n \in \mathbb{Z}}) U_m \\
&= \mathrm{Op}((V_{-n-m} P V_{n+m})_{n \in \mathbb{Z}}) \mathrm{Op}(A_s) \mathrm{Op}((V_{-n-m} P V_{n+m})_{n \in \mathbb{Z}}) \\
&\quad + \mathrm{Op}((V_{-n-m} Q V_{n+m})_{n \in \mathbb{Z}}) \\
&= \mathrm{Op}((V_{-n-m} (P V_{n+m} A_s V_{-n-m} P + Q) V_{n+m})_{n \in \mathbb{Z}}).
\end{aligned}$$

Clearly, this operator is invertible if and only if all operators

$$PV_{n+m}A_sV_{-n-m}P + Q \quad \text{with} \quad m,\, n \in \mathbb{Z}$$

are invertible and if the norms of their inverses are uniformly bounded. But the uniform invertibility of these inverses is an obvious consequence of the invertibility of all operators in (6.7) and of the shift invariance of local operator spectra (see Section 2.3.2). □

6.2.2 Band-dominated operators on $\mathbb{Z}^1$: slowly oscillating coefficients

The results of the previous section take their simplest and most satisfying form for band-dominated operators with slowly oscillating coefficients. To get this, we will have to employ the index formula derived in Theorem 2.7.1. So we let $p = 2$ in this section.

First we have to fix some notations. The projections P, Q and P_n are as in the previous subsection. Besides these projections, we consider the operators

$$R_n : l^2(\mathbb{N}) \to l^2(\mathbb{N}), \quad R_n = PP_nP.$$

Recall further that a function $a \in l^\infty(\mathbb{Z})$ is slowly oscillating if

$$\lim_{m\to-\infty}(a_{m+1} - a_m) = 0 \quad \text{and} \quad \lim_{m\to+\infty}(a_{m+1} - a_m) = 0,$$

and that we call a function $a \in l^\infty(\mathbb{N})$ slowly oscillating if it satisfies the second of these conditions. Finally, if A is a band-dominated operator on $l^2(\mathbb{Z})$, then we call the operator PAP, considered as acting on $\operatorname{Im} P$, a band-dominated operator on $l^2(\mathbb{N})$.

A simple necessary condition for the stability of the finite section method of the operator A is the invertibility of A (Theorem 6.1.3). Our first result says that the invertibility of A is also sufficient for the stability of the finite section method if A is a band operator with slowly oscillating coefficients on $l^2(\mathbb{N})$.

Theorem 6.2.3 *Let $A \in L(l^2(\mathbb{N}))$ be a band operator with slowly oscillating coefficients. Then the finite section method (R_nAR_n) is stable if and only if the operator A is invertible.*

This result is well known in case of band operators on $l^2(\mathbb{N})$ with constant coefficients, which are called Toeplitz operators. Recall in this connection that every function $a \in C(\mathbb{T})$ induces a bounded *Laurent operator* $L(a)$ on $l^2(\mathbb{Z})$ by $(L(a)x)_n := \sum_{k\in\mathbb{Z}} a_{n-k}x_k$ where a_n refers to the nth Fourier coefficient of a. The compression $PL(a)P$ of $L(a)$ onto $l^2(\mathbb{N})$ is the *Toeplitz operator* $T(a)$, and these operators have a band structure if and only if a is a trigonometric polynomial. Overviews on Toeplitz and Laurent operators with continuous generating functions, including their Fredholmness, invertibility, and stability of the finite section method, can be found in [30, 32, 59], for example.

One cannot expect that Theorem 6.2.3 remains valid for band operators with slowly oscillating coefficients on $l^2(\mathbb{Z})$ – it is even wrong for band Laurent operators. In fact, the finite section method $(P_nL(a)P_n)$ for the *Laurent operator* $L(a)$ is stable if and only if the *Toeplitz operator* $T(a)$ is invertible. (Notice that the invertibility of $T(a)$ implies that of $L(a)$, but if $a(t) := t^{-1}$, then the Laurent operator $L(a)$ is invertible, whereas the Toeplitz operator $T(a)$ has a non-trivial kernel.)

To establish the result for band operators with slowly oscillating coefficients on $l^2(\mathbb{Z})$, we will have recourse to the notions of the plus-index and the minus-index of a Fredholm band-dominated operator introduced in Section 2.7.

Theorem 6.2.4 *Let $A \in L(l^2(\mathbb{Z}))$ be a band operator with slowly oscillating coefficients. Then the finite section method (P_nAP_n) is stable if and only if the operator A is invertible and if the plus-index of A is zero.*

If these conditions are satisfied, then, by (2.136), the minus-index of A is zero, too. Notice also that for operators with constant coefficients, i.e., for band Laurent operators $A = L(a)$, the Fredholmness of $L(a)$ implies the Fredholmness of the Toeplitz operator $T(a)$, and that the plus-index of A is just the common Fredholm index of $T(a)$. Since Fredholm Toeplitz operators with index zero are invertible (this is Coburn's theorem, see [59], Chapter 1, Theorem 3.1, for the case of polynomial generating functions and [30], Theorem 2.38 and Corollary 2.40, for the general case), we rediscover the classical result for the finite section method for band Laurent operators from Theorem 6.2.4.

Proof of Theorem 6.2.3. Let $A \in L(l^2(\mathbb{N}))$ be a band operator with slowly oscillating coefficients. If the finite section method (R_nAR_n) is stable, then A is invertible, as we have already remarked. Let, conversely, A be an invertible operator. We identify A with the operator $PAP + Q$ acting on $l^2(\mathbb{Z})$. Clearly, this operator is invertible, too. Hence, all limit operators of $PAP + Q$ are invertible by Theorem 2.2.1. It is easy to check that the part $\sigma_{-1}(PAP + Q)$ of the operator spectrum of $PAP + Q$ consists of the identity operator only. Let A_h be a limit operator in $\sigma_1(PAP + Q)$. Since the coefficients of A (hence, the coefficients of $PAP + Q$) are slowly oscillating, the operator A_h is shift invariant (Proposition 2.4.1). Thus, there is a trigonometric polynomial a_h such that $A_h = L(a_h)$. Further, an elementary calculation shows that $JQA_hQJ = PJL(a_h)JP$ is the Toeplitz operator $T(\widetilde{a_h})$ where $\tilde{a}(t) := a(1/t)$ for a function a on the unit circle.

Since the operator A is invertible, the plus-index of $PAP + Q$ is zero. By Theorem 2.7.1, the plus- and minus-indices of each limit operator of $PAP + Q$ are zero, too. In particular, the index of $QA_hQ + P$ (which is the minus-index of A_h) is zero. This implies that the index of $JQA_hQJ = T(\widetilde{a_h})$ is zero, whence the invertibility of $T(\widetilde{a_h})$ via Coburn's theorem, stating that the kernel or the cokernel of a non-zero Toeplitz operator are trivial ([30], Theorem 2.38). By Theorem 6.2.2, the finite section method (R_nAR_n) is stable. □

Proof of Theorem 6.2.4. Let now $A \in L(l^2(\mathbb{Z}))$ be a band operator with slowly oscillating coefficients. If the finite section method (P_nAP_n) is stable, then A is invertible. Let A_h be a limit operator of A which lies in $\sigma_1(A)$. Then $QA_hQ + P$ is invertible by Theorem 6.2.2, whence

$$0 = \text{ind}\,(QA_hQ + P) = \text{ind}_-(A_h) = \text{ind}_-(A)$$

by Theorem 2.7.1. Since A is invertible, this implies that $\text{ind}_+(A) = 0$.

Let, conversely, A be invertible and $\text{ind}_+(A) = 0$, and let A_h be a limit operator in $\sigma_{\pm 1}(A)$. Then also $\text{ind}_-(A) = 0$, and we get as above that

$$0 = \text{ind}_\pm(A) = \text{ind}_\pm(A_h),$$

whence

$$\text{ind}\,(PA_hP + Q) = \text{ind}\,(QA_hQ + P) = 0.$$

Since the limit operators A_h are again Laurent operators $L(a_h)$ with some trigonometric polynomial a_h, we conclude via Coburn's theorem again, that the operators $PA_hP + Q$ and $QA_hQ + P$ are invertible. This implies the stability of the finite section method (P_nAP_n) via Theorem 6.2.2. □

We conclude this section with two results which can be proved in the same vein. The first one concerns compactly perturbed band operators.

Theorem 6.2.5

(a) *Let $A \in L(l^2(\mathbb{N}))$ be a band operator with slowly oscillating coefficients, and let $K \in L(l^2(\mathbb{N}))$ be compact. Then the finite section method $(R_n(A+K)R_n)$ is stable if and only if the operator $A + K$ is invertible.*

(b) *Let $A \in L(l^2(\mathbb{Z}))$ be a band operator with slowly oscillating coefficients, and let $K \in L(l^2(\mathbb{Z}))$ be compact. Then the finite section method $(P_n(A+K)P_n)$ is stable if and only if the operator $A + K$ is invertible and if the plus-index of A is zero.*

The proof is as above. One only has to take into account Proposition 1.2.6 (b) and the obvious fact that A and $A + K$ have the same plus-index.

One subtlety should be mentioned. Every compact operator is a band dominated operator with slowly oscillating coefficients (in fact, its coefficients have limit zero at $\pm\infty$). Thus, if the compact operator K is a band operator itself, then Theorem 6.2.5 says nothing new when compared with Theorems 6.2.3 and 6.2.4.

It is still an open question whether Theorems 6.2.3, 6.2.4, and 6.2.5 remain true for general band-dominated operators A with slowly oscillating coefficients. We only have the following result which holds for band-dominated operators in the discrete Wiener algebra $\mathcal{W}$ introduced in Section 2.5. The point is that a band-dominated operator in the Wiener algebra is Fredholm if and only if each of its limit operators is invertible (without requiring the uniform boundedness of the norms of their inverses, see Theorem 2.5.7). So, the proofs of Theorems 6.2.3 and 6.2.4 can be modified in an evident manner to yield the following.

Theorem 6.2.6

(a) *Let $A \in L(l^2(\mathbb{N}))$ be an operator in the Wiener algebra with slowly oscillating coefficients, and let $K \in L(l^2(\mathbb{N}))$ be compact. Then the finite section method $(R_n(A+K)R_n)$ is stable if and only if the operator $A+K$ is invertible.*

(b) *Let $A \in L(l^2(\mathbb{Z}))$ be an operator in the Wiener algebra with slowly oscillating coefficients, and let $K \in L(l^2(\mathbb{Z}))$ be compact. Then the finite section method $(P_n(A+K)P_n)$ is stable if and only if the operator $A+K$ is invertible and if the plus-index of A is zero.*

6.2.3 Band-dominated operators on $\mathbb{Z}^2$

Here we will study the stability of the finite section method $(P_nAP_n+Q_n)$ in case when A is a band-dominated operator with rich spectrum on E_2 and when Ω is a convex and compact polygon with vertices in $\mathbb{Z}^2$ and with 0 as its interior point.

Let $u_1, \ldots, u_k \in \mathbb{Z}^2$ denote the vertices of Ω, define $u_{k+1} := u_1$ and $u_0 := u_k$, and abbreviate the open segment $\{x \in \mathbb{R}^2 : x = (1-t)u_j + tu_{j+1} \text{ with } 0 < t < 1\}$ to (u_j, u_{j+1}). For $j = 1, \ldots, k$, let $H_j \subseteq \mathbb{R}^2$ refer to the half-plane which is bounded by the straight line through u_j and u_{j+1} and which contains the point 0, and let $K_j \subseteq \mathbb{R}^2$ refer to the angle with vertex at u_j which is bounded by the straight lines through u_j and u_{j-1} resp. through u_j and u_{j+1} and which contains the point 0. Further, set $H_j^0 := H_j - u_j$ and $K_j^0 := K_j - u_j$ (= the algebraic difference).

Besides these notations for subsets of the plane $\mathbb{R}^2$ we need some notations for their spatial counterparts. If Ω is as above, then the operator $\text{Op}(\mathbf{P})$ (where $\mathbf{P} = (P_n)$ again) turns out to be the characteristic function of a pyramid in $\mathbb{R}^3$ with its vertex at the origin. The faces of this pyramid are in correspondence with the edges of Ω. We let $\mathbb{H}_j$ stand for the half-space of $\mathbb{R}^3$ which is bounded by the plane through the face which corresponds to (u_j, u_{j+1}), and which contains the pyramid. Similarly, the edges of the pyramid are in correspondence with the vertices of Ω; the edge corresponding to u_j will be denoted by G_j.

Since the operator $\text{Op}(\mathbf{P})$ is the projection onto a cone, we can refer to the results of Section 2.6.3 in order to get the following description of the local operator spectra of this operator. Given a point $x \in \Omega$, let η_x refer to the intersection of the ray $\{(tx, t) : t \geq 0\} \subseteq \mathbb{R}^3$ with the sphere S^2. Then one has to distinguish the following cases.

- If $\eta \in S^2$ is not of the form η_x for some $x \in \Omega$, then $\sigma_\eta(\text{Op}(\mathbf{P})) = \{0\}$.
- If x is in the interior of Ω, then $\sigma_{\eta_x}(\text{Op}(\mathbf{P})) = \{I\}$.
- If $x \in (u_j, u_{j+1})$, then $\sigma_{\eta_x}(\text{Op}(\mathbf{P}))$ consists of the operators 0, I and of the operators of multiplication by the characteristic functions of the half-planes $U_{-m}\hat{\chi}_{\mathbb{H}_j}U_m$ with $m \in \mathbb{Z}$.

- If $x = u_j$, then $\sigma_{\eta_x}(\mathrm{Op}(\mathbf{P}))$ consists of the operators 0 and I, the characteristic functions of the half-planes

$$U_{-k}\hat{\chi}_{\mathbb{H}_{j-1}}U_k \quad \text{and} \quad U_{-m}\hat{\chi}_{\mathbb{H}_j}U_m \quad \text{with } k,\, m \in \mathbb{Z},$$

and of the characteristic functions of the dihedrals

$$U_{-k}\hat{\chi}_{\mathbb{H}_{j-1}}U_kU_{-m}\hat{\chi}_{\mathbb{H}_j}U_m \quad \text{with } k,\, m \in \mathbb{Z}.$$

Observe further that

$$\hat{\chi}_{\mathbb{H}_j}I = \mathrm{Op}((V_{-n}\Pi_jV_n)_{n\in\mathbb{Z}}) \tag{6.9}$$

where Π_j stands for the projection $\hat{\chi}_{H_j^0}I$. Finally, given a point x in the boundary of Ω, the ray $\{tx : t \geq 0\} \subseteq \mathbb{R}^2$ intersects the sphere S^1 in a point which we denote by μ_x.

Theorem 6.2.7 *Let A be a band-dominated operator on E_2 with rich operator spectrum, and let Ω a convex and compact polygon with 0 as its interior point and with vertices $u_1, \ldots, u_k \in \mathbb{Z}^2$. Then the finite section method $(P_nAP_n + Q_n)$ is stable if and only if the operator A is invertible, if the operators*

$$\Pi_jA_s\Pi_j + (I - \Pi_j) \tag{6.10}$$

are invertible for all limit operators $A_s \in \sigma_{\mu_x}(A)$ and for all $x \in (u_j,\, u_{j+1})$, if the operators

$$\Pi_j\Pi_{j-1}A_s\Pi_{j-1}\Pi_j + (I - \Pi_j\Pi_{j-1}) \tag{6.11}$$

are invertible for all limit operators $A_s \in \sigma_{\mu_x}(A)$ and for all $x = u_j$, and if the norms of the inverses of the operators in (6.10) *and* (6.11) *are uniformly bounded.*

Proof. Most parts of the proof run completely parallel to the proof of Theorem 6.2.2. The only essential difference is that now, at points $x = u_j$, we have (besides the trivial operators 0 and I) two kinds of limit operators of $\mathrm{Op}(\mathbf{P})$, namely half-plane projections and dihedral projections. For instance, one has now to verify that, for every fixed vertex $x = u_j$ of Ω, the invertibility of all operators in (6.11) implies the invertibility of the operators

$$U_{-m}\hat{\chi}_{\mathbb{H}_j}U_m\mathrm{Op}(A_s)U_{-m}\hat{\chi}_{\mathbb{H}_j}U_m + U_{-m}(1 - \hat{\chi}_{\mathbb{H}_j})U_m$$

for every m and every limit operator $A_s \in \sigma_{\mu_x}(A)$ as well as the uniform boundedness of the inverses of these operators. Due to (6.9), this is equivalent to the invertibility of all operators

$$V_{-m-n}\Pi_jV_{m+n}A_sV_{-m-n}\Pi_jV_{m+n} + V_{-m-n}(I - \Pi_j)V_{m+n}.$$

Employing, as in the proof of preceding theorem, the shift invariance of the local spectra, we see that this is further equivalent to the invertibility of the operator

$$\Pi_jA_s\Pi_j + (I - \Pi_j). \tag{6.12}$$

That this operator is invertible for every operator $A_s \in \sigma_{\mu_x}(A)$ can be seen as follows: Choose a sequence $h \in \mathbb{Z}^2$ such that

$$V_{-h(n)}\Pi_{j-1}V_{h(n)} \to I \quad \mathcal{P}\text{-strongly as } n \to \infty,$$

but

$$V_{-h(n)}\Pi_j V_{h(n)} = \Pi_j \quad \text{for all } n.$$

Since every shifted limit operator of A is a limit operator of A again, the uniform invertibility of all operators in (6.11) implies that the operators

$$\Pi_j\Pi_{j-1}V_{h(n)}A_sV_{-h(n)}\Pi_{j-1}\Pi_j + (I - \Pi_j\Pi_{j-1})$$

are invertible for all n and that the norms of their inverses are uniformly bounded. Then also the operators

$$\Pi_j V_{-h(n)}\Pi_{j-1}V_{h(n)}A_sV_{-h(n)}\Pi_{j-1}V_{h(n)}\Pi_j + (I - \Pi_j V_{-h(n)}\Pi_{j-1}V_{h(n)}) \tag{6.13}$$

are invertible for all n, and the norms of their inverses are uniformly bounded. Letting n go to infinity, the operators in (6.13) converge $\mathcal{P}$-strongly to the operator (6.12). Since the adjoints of the operators (6.13) converge $\mathcal{P}$-strongly, too, the invertibility of the operator (6.12) follows. Finally, the Banach-Steinhaus theorem entails that $\|(\Pi_jA_s\Pi_j + (I - \Pi_j))^{-1}\|$ is bounded from above by the same constant as the norms of the inverses of the operators (6.13). □

It is evident that an analogous result holds for the finite section method for band-dominated operators on E_N with $N > 2$ if the corresponding projections P_n are defined with respect to convex polyhedra having 0 as interior point and vertices in $\mathbb{Z}^N$.

6.2.4 Finite sections of convolution type operators

We conclude this section by an application of the results of the previous subsections to the convergence of the finite section method for convolution type operators. If we restrict ourselves to finite sections of a very special shape such as squares and cubes with integer vertices, then the discretizations of the finite sections of an operator A yield exactly the finite sections of the discretization of A. Hence, the results from Section 6.2.3 can be applied directly.

Let $A \in L(L^p(\mathbb{R}^2))$ be an operator which belongs to the algebra $\mathcal{B}^\$$ introduced in Definition 3.1.5. For m a positive integer, let Ω_m denote the square $[-m, m]^2$. We consider the finite sections

$$\chi_{\Omega_m}A\chi_{\Omega_m}I : L^p(\Omega_m) \to L^p(\Omega_m) \tag{6.14}$$

of A. Since $G\chi_{\Omega_m}IG^{-1} = \hat{\chi}_{\Omega_m}I$ (with the discretization operator G defined by (3.2)), it is evident that the sequence of the operators (6.14) is stable if and only if the sequence of the operators

$$\hat{\chi}_{\Omega_m}A_G\hat{\chi}_{\Omega_m}I : l^p(\hat{\Omega}_m, L^p(I_0)) \to l^p(\hat{\Omega}_m, L^p(I_0))$$

with $\hat{\Omega}_m := \Omega_m \cap \mathbb{Z}^2$, $I_0 := [0, 1)^2$ and $A_G := GAG^{-1}$ is stable. The operator A_G is a rich band-dominated operator on $l^p(\hat{\Omega}_m, L^p(I_0))$ by Proposition 3.1.6. Thus, Theorem 6.2.7 applies to the finite sections of A_G, and it yields the following result. The notations are as in Section 6.2.3.

Theorem 6.2.8 *Let $A \in \mathcal{B}_p^\$$ over $\mathbb{R}^2$. Then the sequence $(\chi_{\Omega_m} A \chi_{\Omega_m} I)_{m\geq 1}$ of the finite sections of A is stable if and only if the operator $A : L^p(\mathbb{R}^2) \to L^p(\mathbb{R}^2)$ is invertible, if the operators*

$$\chi_{H_j^0} A_h \chi_{H_j^0} I : L^p(H_j^0) \to L^p(H_j^0)$$

with $A_h \in \sigma_{\eta_x}(A)$ are uniformly invertible for each non-vertex point $x \in (u_j, u_{j+1})$ in the boundary of Ω_1, and if the operators

$$\chi_{H_j^0 \cap H_{j-1}^0} A_h \chi_{H_j^0 \cap H_{j-1}^0} I : L^p(H_j^0 \cap H_{j-1}^0) \to L^p(H_j^0 \cap H_{j-1}^0)$$

with $A_h \in \sigma_{\eta_x}(A)$ are uniformly invertible for each vertex $x = u_j$ of Ω_1.

6.3 Spectral approximation

Throughout this section, we will work in Hilbert spaces only, i.e., we let $p = 2$ and $X = H$ a Hilbert space. Then, accordingly, the projections P_n are self-adjoint.

Our concern in this section is relations between the spectrum of a band-dominated operator A and the spectra of its approximations $A_n = P_n A P_n$ obtained by the finite section method. We will pay particular attention to the asymptotic behavior of the spectra of the operators A_n. As a rule, one will observe that these results are satisfactory only in case of self-adjoint or, at least, normal operators A_n. In general, the connection between the spectrum of an operator A and the spectra of its approximations A_n proves to be quite loose. For example, the spectrum of the shift operator V_1 on $l^2(\mathbb{Z})$ is the unit circle $\mathbb{T}$, whereas each of its finite sections has the spectrum $\{0\}$.

This observation suggests to look for other spectral quantities which behave more robustly and which exhibit much better convergence properties than common spectra. We will discuss two of these quantities, namely pseudospectra and numerical ranges.

The results on the convergence of spectra, pseudospectra and numerical ranges of the approximation operators A_n will be obtained as special cases of some general theorems on spectral approximation. The most convenient way to formulate these general theorems makes use of the language of C^*-algebras. Thus, to apply these results to a concrete approximation sequence (A_n) requires not only a precise knowledge on stability properties of the single sequence (A_n) itself, but for a whole C^*-algebra of sequences which contains (A_n). Moreover, it will prove to be advantageous to work with C^*-algebra homomorphisms instead of limit operators. Recall in this connection the notion of a (weakly) sufficient family of homomorphisms introduced in Section 2.2.2. This notion allows us to reformulate the result of Theorem 6.1.9 in an appropriate way.

Indeed, given a sequence $\mathbf{A} = (A_n)$ in $\mathcal{B}^\$$, we let $C^*(\mathbf{A})$ stand for the smallest closed subalgebra of $\mathcal{F}$ which contains the sequences $\mathbf{A}$, $\mathbf{A}^* := (A_n^*)$ and $\mathbf{I} := (I)$. Let $h \in \mathcal{H}_{\mathrm{Op}(\mathbf{A})}$, the set of all sequences h for which the limit operator $Op(\mathbf{A})_h$ exists. Then the limit operator $(\mathrm{Op}(\mathbf{B}))_h$ exists for every sequence $\mathbf{B} \in C^*(\mathbf{A})$, and every limit operator of $\mathrm{Op}(\mathbf{B})_h$ arises in this way. Thus, the mapping $\mathbf{B} \mapsto R_m(\mathrm{Op}(\mathbf{B}))_h S_m$ is a *-homomorphism from $C^*(\mathbf{A})$ into $L(E)$ for each $m \in \mathbb{Z}$. Since the ideal $C^*(\mathbf{A}) \cap \mathcal{G}$ lies in the kernel of that homomorphism, the mapping

$$W_{h,m} : C^*(\mathbf{A})/(C^*(\mathbf{A}) \cap \mathcal{G}) \to L(E), \quad \mathbf{B} + C^*(\mathbf{A}) \cap \mathcal{G} \mapsto R_m(\mathrm{Op}(\mathbf{B}))_h S_m$$

is well defined for every sequence $h \in \mathcal{H}_{\mathrm{Op}(\mathbf{A})}$ and every $m \in \mathbb{Z}$. The following is an immediate consequence of Theorem 6.1.9.

Theorem 6.3.1 *Let $\mathbf{A} \in \mathcal{B}^\$$. Then the family of C^*-homomorphisms $\{W_{h,m}\}$ with $h \in \mathcal{H}_{\mathrm{Op}(\mathbf{A})}$ and $m \in \mathbb{Z}$ is weakly sufficient for the algebra $C^*(\mathbf{A})/(C^*(\mathbf{A}) \cap \mathcal{G})$.*

6.3.1 Weakly sufficient families and spectra

We will see now how certain spectral quantities of $b \in \mathcal{B}$ can be expressed by the corresponding spectral quantities of the $W_t(b)$, provided that $\{W_t\}$ is a weakly sufficient or a sufficient family of homomorphisms. It will be convenient to use the following notation. Given a family $(M_t)_{t\in T}$ of subsets of $\mathbb{C}$, we set

$$\sup_{t\in T} M_t := \mathrm{clos}\,(\cup_{t\in T} M_t).$$

Further, we call $\sup_{t\in T} M_t$ the *maximum* of the family (M_t) if $\cup_{t\in T} M_t$ is closed.

Spectra. As usual, we let $\sigma_{\mathcal{B}}(b) := \{\lambda \in \mathbb{C} : b - \lambda e \text{ is not invertible in } \mathcal{B}\}$.

Theorem 6.3.2 *Let $\mathcal{B}$, $\mathcal{B}_t$ and W_t be as in Definition 2.2.6. If $\{W_t\}_{t\in T}$ is a weakly sufficient family for $\mathcal{B}$, then*

$$\sigma_{\mathcal{B}}(b) = \sup_{t\in T} \sigma_{\mathcal{B}_t}(W_t(b)) \tag{6.15}$$

for all normal elements b of $\mathcal{B}$. If the family $\{W_t\}_{t\in T}$ is sufficient, then the supremum in (6.15) is a maximum, and the assertion holds for every $b \in \mathcal{B}$.

Proof. The inclusion $\supseteq$ in (6.15) is trivial and holds also in the context of general Banach algebras. It remains to show that, for every normal element $b \in \mathcal{B}$,

$$\lambda \notin \sup_{t\in T} \sigma_{\mathcal{B}_t}(W_t(b)) \quad \Rightarrow \quad \lambda \notin \sigma_{\mathcal{B}}(b).$$

Without loss, let $\lambda = 0$. Then, since $\sup_{t\in T} \sigma_{\mathcal{B}_t}(W_t(b))$ is compact, there is a closed disk with center 0 and with positive radius r which has no points in common with

$\sup_{t\in T}\sigma_{\mathcal{B}_t}(W_t(b))$. Thus, $W_t(b)$ is invertible for every $t \in T$, and since $(W_t(b))^{-1}$ is normal, we get

$$\begin{aligned}\|(W_t(b))^{-1}\| &= \rho((W_t(b))^{-1}) \\ &= \sup\{\lambda \in \mathbb{C} : \lambda \in \sigma((W_t(b))^{-1})\} \\ &= \inf\{\lambda \in \mathbb{C} : \lambda \in \sigma(W_t(b))\}^{-1} < 1/r.\end{aligned}$$

The weak sufficiency of $\{W_t\}$ implies the invertibility of b. The assertion for sufficient families follows immediately from the definitions. □

Pseudospectra. Let $\varepsilon > 0$. The *ε-pseudospectrum* of $b \in \mathcal{B}$ is the set

$$\sigma^\varepsilon(b) := \{\lambda \in \mathbb{C} : b - \lambda e \text{ is not invertible or } \|(b-\lambda e)^{-1}\| \geq 1/\varepsilon\}.$$

Pseudospectra are non-empty and compact, and for the pseudospectral radius of an element $b \in \mathcal{B}$ one has

$$\max\{|\lambda| : \lambda \in \sigma^\varepsilon(b)\} \leq \|b\| + \varepsilon.$$

Theorem 6.3.3 *Let $\mathcal{B}$, $\mathcal{B}_t$ and W_t be as in Definition* 2.2.6, *and let $\varepsilon > 0$. If $\{W_t\}_{t\in T}$ is a weakly sufficient family for $\mathcal{B}$, then*

$$\sigma^\varepsilon(b) = \sup_{t\in T} \sigma^\varepsilon(W_t(b)) \quad \textit{for every } b \in \mathcal{B}. \tag{6.16}$$

If the family $\{W_t\}_{t\in T}$ is sufficient, then the supremum in (6.16) *is a maximum.*

Again, one inclusion holds in a more general context, and we formulate it separately.

Lemma 6.3.4 *Let $\mathcal{B}$ and $\mathcal{C}$ be unital Banach algebras and $W : \mathcal{B} \to \mathcal{C}$ be a unital and contractive homomorphism. Then*

$$\sigma^\varepsilon(W(b)) \subseteq \sigma^\varepsilon(b) \quad \textit{for every } b \in \mathcal{B}.$$

Proof. Let $\lambda \in \sigma^\varepsilon(W(b))$. If $W(b) - \lambda e = W(b - \lambda e)$ is not invertible, then $b - \lambda e$ is not invertible. Hence, $\lambda \in \sigma^\varepsilon(b)$ in this case. Let now $W(b - \lambda e)$ be invertible and $\|(W(b-\lambda e)^{-1}\| \geq 1/\varepsilon$. If $b - \lambda e$ is not invertible, then we have $\lambda \in \sigma^\varepsilon(b)$ again. If $b - \lambda e$ is invertible, then $(W(b-\lambda e))^{-1} = W((b-\lambda e)^{-1})$, whence $\|W((b-\lambda e)^{-1})\| \geq 1/\varepsilon$. Since W is a contraction, this shows that $\|(b-\lambda e)^{-1}\| \geq 1/\varepsilon$, i.e., $\lambda \in \sigma^\varepsilon(b)$. □

The proof of Theorem 6.3.3 exploits the following result by Daniluk stating that the maximum principle (which does not hold for operator-valued analytic functions in general) holds for the resolvent function $z \mapsto (a - ze)^{-1}$.

Theorem 6.3.5 *Let $\mathcal{B}$ be a C^*-algebra with identity e, and let $a \in \mathcal{B}$ be such that $a - ze$ is invertible for all z in some open subset U of the complex plane. If $\|(a - ze)^{-1}\| \leq C$ for all $z \in U$, then $\|(a - ze)^{-1}\| < C$ for all $z \in U$.*

A proof is in [32], Theorem 3.14 and also in [73], Theorem 3.32.

Proof of Theorem 6.3.3. From the preceding lemma we conclude that

$$\sigma^\varepsilon(W_t(b)) \subseteq \sigma^\varepsilon(b) \quad \text{for every } b \in \mathcal{B} \text{ and } t \in T.$$

Since pseudospectra are closed, this implies

$$\sup_{t\in T} \sigma^\varepsilon(W_t(b)) \subseteq \sigma^\varepsilon(b) \quad \text{for every } b \in \mathcal{B}$$

and for every family $\{W_t\}$ of *-homomorphisms. For the reverse inclusion, let $\{W_t\}$ be a weakly sufficient family of *-homomorphisms, and let $\lambda \in \sigma^\varepsilon(b)$. If there is a $t \in T$ such that $\lambda \in \sigma^\varepsilon(W_t(b))$, then nothing is to prove. So let us assume that all elements $W_t(b - \lambda e)$ are invertible and that $\|(W_t(b - \lambda e))^{-1}\| < 1/\varepsilon$. Then

$$\sup_{t\in T} \|(W_t(b - \lambda e))^{-1}\| \leq 1/\varepsilon.$$

Since $\{W_t\}$ is a weakly sufficient family, the element $b - \lambda e$ is invertible, and from Theorem 2.2.7 we conclude that $\|(b-\lambda e)^{-1}\| \leq 1/\varepsilon$. Since $\lambda \in \sigma^\varepsilon(b)$ by hypothesis, this implies $\|(b - \lambda e)^{-1}\| = 1/\varepsilon$.

In every open neighborhood U of λ, there is a $\tilde{\lambda}$ such that $\|(b-\tilde{\lambda}e)^{-1}\| > 1/\varepsilon$. Indeed, otherwise we would have $\|(b - \tilde{\lambda}e)^{-1}\| \leq 1/\varepsilon$ for all $\tilde{\lambda} \in U$ whence, via Theorem 6.3.5, $\|(b - \tilde{\lambda}e)^{-1}\| < 1/\varepsilon$ for all $\tilde{\lambda} \in U$ including $\tilde{\lambda} = \lambda$. Thus, for $k \in \mathbb{N}$ being sufficiently large, there are $\lambda_k \in \mathbb{C}$ such that

$$|\lambda - \lambda_k| < 1/k \quad \text{and} \quad \|(b - \lambda_k e)^{-1}\| \geq \frac{1}{\varepsilon - 2/k}.$$

Further, again by Theorem 2.2.7, there are $t_k \in T$ such that

$$\|W_{t_k}(b - \lambda_k e))^{-1}\| = \|(W_{t_k}(b) - \lambda_k e_{t_k})^{-1}\| \geq \frac{1}{\varepsilon - 1/k}.$$

Since $\frac{1}{\varepsilon-1/k} > \frac{1}{\varepsilon}$, we have $\lambda_k \in \sigma^\varepsilon(W_{t_k}(b))$, and since $\lambda_k \to \lambda$ as $k \to \infty$, we get $\lambda \in \sup_{t\in T} \sigma^\varepsilon(W_t(b))$. Thus, (6.16) is verified.

In case of a sufficient family of homomorphisms we still have to show that

$$\sigma^\varepsilon(b) \subseteq \bigcup_{t\in T} \sigma^\varepsilon(W_t(b))$$

for every $b \in \mathcal{B}$. If $\lambda \notin \cup_{t\in T} \sigma^\varepsilon(W_t(b))$, then $W_t(b - \lambda e)$ is invertible and $\|W_t(b - \lambda e)^{-1}\| < 1/\varepsilon$ for every $t \in T$. Hence, $b - \lambda e$ is invertible, and

$$\|(b - \lambda e)^{-1}\| = \sup_{t\in T} \|W_t(b - \lambda e)^{-1}\| < 1/\varepsilon$$

because the supremum is attained by the second assertion of Theorem 2.2.7. Thus, $\lambda \notin \sigma^\varepsilon(b)$. □

Numerical ranges. Let $\mathcal{B}$ be a Banach algebra with identity e and $S(\mathcal{B})$ its *state space*, i.e., the set of all $f \in \mathcal{B}^*$ with $f(e) = 1$ and $\|f\| = 1$. The *numerical range* of $b \in \mathcal{B}$ is the set

$$N(b) := \{f(b) : f \in S(\mathcal{B})\}.$$

Numerical ranges are non-empty, compact and convex subsets of $\mathbb{C}$. For a bounded linear operator A on a Hilbert space H, one also considers its *spatial numerical range*

$$SN(A) := \{\langle Ax, x\rangle : x \in H, \|x\| = 1\}.$$

Let $A \in L(H)$. Then the spatial numerical range $SN_H(A)$ (where A is considered as an operator on H) and the numerical range $N_{L(H)}(A)$ (where A is considered as an element of the C^*-algebra $L(H)$) are related by

$$N_{L(H)}(A) = \operatorname{clos} SN_H(A).$$

Finally, if $\mathcal{J}$ is a closed ideal of the C^*- algebra $\mathcal{B}$, then

$$N(a) = \cap_{j \in \mathcal{J}} N(a + j) \quad \text{for every } a \in \mathcal{B}. \tag{6.17}$$

These and further properties of numerical ranges can be found in [34, 35].

Theorem 6.3.6 *Let $\mathcal{B}$, $\mathcal{B}_t$ and W_t be as in Definition 2.2.6. If $\{W_t\}_{t\in T}$ is a weakly sufficient family for $\mathcal{B}$, then*

$$N(b) = \operatorname{conv} \sup_{t \in T} N(W_t(b)) \quad \textit{for every } b \in \mathcal{B}. \tag{6.18}$$

One of the inclusions in (6.18) holds in the more general context of Banach algebras.

Lemma 6.3.7 *Let $\mathcal{B}$ and $\mathcal{C}$ be unital Banach algebras and $W : \mathcal{B} \to \mathcal{C}$ be a unital and contractive homomorphism. Then*

$$N(W(b)) \subseteq N(b) \quad \textit{for every } b \in \mathcal{B}.$$

Proof. Let $\lambda \in N(W(b))$, and let f be a state of $\mathcal{C}$ with $f(W(b)) = \lambda$. Since W is unital and contractive, one has $(f \circ W)(e) = 1$ and $\|f \circ W\| \leq 1$. Thus, $f \circ W$ is a state of $\mathcal{B}$, whence $\lambda \in N(b)$. □

Proof of Theorem 6.3.6. From Lemma 6.3.7 we infer that

$$\cup_{t \in T} N(W_t(b)) \subseteq N(b).$$

Since $N(b)$ is a closed and convex set, this implies the inclusion $\supseteq$ in (6.18). For the reverse inclusion, we think of each $\mathcal{B}_t$ as a C^*-algebra of bounded linear operators on some Hilbert space H_t (which is possible by the GNS-construction). Let $H := \oplus H_t$ refer to the orthogonal sum of the Hilbert spaces H_t, $t \in T$, and write W for the mapping from $\mathcal{B}$ into $L(H)$ which associates with every $b \in \mathcal{B}$ the operator

$$(x_t)_{t\in T} \mapsto (W_t(b)x_t)_{t\in T}.$$

This mapping is an isometry from $\mathcal{B}$ onto the C^*-subalgebra $W(\mathcal{B})$ of $L(H)$. Thus, $N_{\mathcal{B}}(b) = N_{W(\mathcal{B})}(W(b))$. It is further a simple consequence of the Hahn-Banach Theorem that $N_{W(\mathcal{B})}(W(b)) = N_{L(H)}(W(b))$, which implies that

$$N_{\mathcal{B}}(b) = N_{W(\mathcal{B})}(W(b)) = N_{L(H)}(W(b)) = \operatorname{clos} SN_H(W(b)).$$

Thus, given $\lambda \in N_{\mathcal{B}}(b)$ and $\varepsilon > 0$, there is a vector $(x_t)_{t\in T} \in H$ with norm 1 such that

$$|\lambda - \langle (W_t(b)x_t), (x_t)\rangle_H| = \left|\lambda - \sum_{t\in T} \langle W_t(b)x_t, x_t\rangle_{H_t}\right| < \varepsilon.$$

Let $\mathbb{M}$ denote the (at most countable) set of all $t \in T$ with $x_t \neq 0$ and set $y_t := x_t/\|x_t\|$ for $t \in \mathbb{M}$. Then,

$$\left|\lambda - \sum_{t\in\mathbb{M}} \langle W_t(b)x_t, x_t\rangle_{H_t}\right| = \left|\lambda - \sum_{t\in\mathbb{M}} \|x_k\|^2 \langle W_t(b)y_t, y_t\rangle_{H_t}\right| < \varepsilon.$$

Since $\|x_t\|^2 \geq 0$ and $\sum_{t\in\mathbb{M}} \|x_t\|^2 = \|(x_t)\|_H^2 = 1$, this shows that λ can be approximated by convex linear combinations of points $\langle W_t(b)y_t, y_t\rangle \in \cup_{t\in T} SN_{H_t}(W_t(b))$ as closely as desired. Hence,

$$\lambda \in \operatorname{clos}\operatorname{conv} \cup_{t\in T} SN_{H_t}(W_t(b)) \subseteq \operatorname{clos}\operatorname{conv} \cup_{t\in T} N_{L(H_t)}(W_t(b)),$$

which gives

$$\lambda \in \operatorname{clos}\operatorname{conv} \cup_{t\in T} N_{W_t(\mathcal{B})}(W_t(b)). \tag{6.19}$$

Since $\operatorname{clos}\operatorname{conv} M = \operatorname{conv}\operatorname{clos} M$ for every bounded subset M of the complex plane, (6.19) is just the assertion. □

6.3.2 Interlude: Spectra of band-dominated operators on Hilbert spaces

We will see now how the results of the preceding subsections apply to describe several spectral quantities of rich band-dominated operators. Let $E = l^2(\mathbb{Z}^N, H)$ with a Hilbert space H, and let A be a rich band-dominated operator on E. By $C^*(A)$ we denote the smallest closed subalgebra of $L(E)$ which contains the operators A and A^*, the identity operator I, and the ideal $K(E, \mathcal{P})$. Further, let $\mathcal{H}_A$ stand for the set of all sequences $h \in \mathcal{H}$ for which the limit operator A_h exists. Then $C^*(A)$ is a separable C^*-subalgebra of $L^\$(E, \mathcal{P})$, the limit operator B_h exists for every operator $B \in C^*(A)$ and every sequence $h \in \mathcal{H}_A$, and the mapping $B \mapsto B_h$ is a symmetric algebra homomorphism for every $h \in \mathcal{H}_A$. Finally,

$$\sigma_{op}(B) = \{B_h : h \in \mathcal{H}_A\} \quad \text{for every } B \in C^*(A)$$

(compare Proposition 1.3.3). Since the limit operator of $K \in K(E, \mathcal{P})$ exists with respect to every sequence in $\mathcal{H}$, the mappings

$$W_h : C^*(A)/K(E, \mathcal{P}) \to L(E), \quad B + K(E, \mathcal{P}) \mapsto B_h$$

are well defined algebra homomorphisms. Thus, the family $\{W_h\}_{h \in \mathcal{H}_A}$ is weakly sufficient for the algebra $C^*(A)/K(E, \mathcal{P})$, and it is sufficient if A is in the discrete Wiener algebra or if all coefficients of A are slowly oscillating.

Theorem 6.3.8

(a) *If $B \in C^*(A)$ is normal, then*

$$\sigma_{ess}(B) = \sigma_{C^*(A)/K(E, \mathcal{P})}(B + K(E, \mathcal{P})) = \sup_{h \in \mathcal{H}_A} \sigma(B_h).$$

If the family $\{W_h\}$ is sufficient, then the supremum is a maximum, and the assertion holds for every $B \in C^*(A)$.

(b) *Let $\varepsilon > 0$ and $B \in C^*(A)$. Then*

$$\sigma^{\varepsilon}_{C^*(A)/K(E, \mathcal{P})}(B + K(E, \mathcal{P})) = \sup_{h \in \mathcal{H}_A} \sigma^{\varepsilon}(B_h),$$

and the supremum is a maximum if the family $\{W_h\}$ is sufficient.

(c) *For every $B \in C^*(A)$,*

$$N_{C^*(A)/K(E, \mathcal{P})}(B + K(E, \mathcal{P})) = \text{conv} \sup_{h \in \mathcal{H}_A} N(B_h).$$

6.3.3 Asymptotic behavior of norms

The following result relates the asymptotic behavior of the norms of the approximation operators A_n with the norm of the coset of the sequence (A_n) modulo the ideal $\mathcal{G}$.

Proposition 6.3.9 *For all sequences $(A_n) \in \mathcal{F}$,*

$$\|(A_n) + \mathcal{G}\|_{\mathcal{F}/\mathcal{G}} = \limsup_{n \to \infty} \|A_n\|.$$

Proof. Let $(A_n) \in \mathcal{F}$. Then, for every sequence $(G_n) \in \mathcal{G}$,

$$\limsup \|A_n\| = \limsup \|A_n + G_n\| \leq \sup \|A_n + G_n\| = \|(A_n) + (G_n)\|_{\mathcal{F}},$$

whence the estimate $\limsup \|A_n\| \leq \|(A_n) + \mathcal{G}\|$. For the reverse inequality, let $\varepsilon > 0$, and choose n_0 such that $\|A_n\| \leq \limsup_{n \to \infty} \|A_n\| + \varepsilon$ for all $n \geq n_0$. The sequence (G_n) with

$$G_n := \begin{cases} -A_n & \text{if } n < n_0 \\ 0 & \text{if } n \geq n_0 \end{cases}$$

belongs to $\mathcal{G}$, and

$$\begin{aligned}\|(A_n)+\mathcal{G}\| \leq \|(A_n)+(G_n)\| &= \|(0,\ldots,0,A_{n_0},A_{n_0+1},\ldots)\|\\ &= \sup_{n\geq n_0}\|A_n\| \leq \limsup\|A_n\|+\varepsilon.\end{aligned}$$

Letting ε go to zero yields the desired result. □

Combining this result with Theorems 2.2.7 and 6.1.9 we get:

Theorem 6.3.10 *Let* $\mathbf{A}=(A_n)\in\mathcal{B}^{\$}$. *Then*

$$\limsup\|A_n\| = \|\mathbf{A}+\mathcal{G}\| = \sup\{\|A_h\| : A_h\in\sigma_{stab}(\mathbf{A})\}.$$

Proof. The first equality follows from Proposition 6.3.9. The second one is a consequence of Theorems 2.2.7 and 6.1.9 which follows since the operators in $\sigma_{stab}(\mathbf{A})$ are just the operators of the form $W_{h,m}(\mathbf{A})$ with $h\in\mathcal{H}_{\mathrm{Op}(\mathbf{A})}$ and $m\in\mathbb{Z}$. □

6.3.4 Asymptotic behavior of spectra

Let $(M_n)_{n=1}^{\infty}$ be a sequence of subsets of the complex plane. The *limes superior* or *partial limiting set* $\limsup M_n$ consists of all points $m\in\mathbb{C}$ which are a partial limit of a sequence (m_n) of points $m_n\in M_n$. Equivalently,

$$\limsup M_n = \cap_k \mathrm{clos}\,(\cup_{n\geq k}M_n). \tag{6.20}$$

We will see now how the limes superior of the spectrum of the A_n as well of certain generalized spectra of the A_n can be expressed in terms of the coset $(A_n)+\mathcal{G}$ and, hence, in terms of the stability spectrum of the sequence (A_n).

Spectra. Let $(A_n)\in\mathcal{F}$. It turns out that $\limsup\sigma(A_n)$ is related to some kind of stability which might be called 'spectral' stability. The sequence (A_n) is called *spectrally stable* if the operators A_n are invertible for all sufficiently large n and if the spectral radii $\rho(A_n^{-1})$ of their inverses are uniformly bounded. Clearly, every stable sequence is also spectrally stable.

Theorem 6.3.11 *Let* $(A_n)\in\mathcal{F}$ *and* $\lambda\in\mathbb{C}$. *Then* $\lambda\in\limsup\sigma(A_n)$ *if and only if the sequence* $(A_n-\lambda I)$ *is not spectrally stable.*

Proof. Let $(A_n-\lambda I)$ be a spectrally stable sequence, i.e.,

$$\sup_{n\geq n_0}\rho((A_n-\lambda I)^{-1}) =: m<\infty \quad \text{for a certain } n_0.$$

Then, for all $n\geq n_0$,

$$m\geq\sup\{|t| : t\in\sigma((A_n-\lambda I)^{-1})\} = (\inf\{|t| : t\in\sigma(A_n-\lambda I)\})-1,$$

whence

$$1/m \le \inf\{|t| : t \in \sigma(A_n) - \lambda\} = \inf\{|t-\lambda| : t \in \sigma(A_n)\} = \mathrm{dist}\,(\lambda, \sigma(A_n)).$$

Hence, λ cannot belong to $\limsup \sigma(A_n)$.

For the reverse direction suppose $(A_n - \lambda I)$ is a sequence which fails to be spectrally stable. Then either there is an infinite subsequence $(A_{n_k} - \lambda I)$ which consists of non-invertible elements only, or all operators $A_n - \lambda I$ with sufficiently large n are invertible, but $\rho((A_{n_k} - \lambda I)^{-1}) \to \infty$ as $k \to \infty$ for some subsequence.

In the first case one has $\lambda \in \sigma(A_{n_k})$ for every k and, thus, $\lambda \in \limsup \sigma(A_n)$. In the second case, there are numbers $t_{n_k} \in \sigma(A_{n_k})$ such that $|t_{n_k} - \lambda|^{-1} \to \infty$ resp. $|t_{n_k} - \lambda| \to 0$ as $k \to \infty$ which also implies that $\lambda \in \limsup \sigma(A_n)$. □

Thus, in order to determine the limiting set $\limsup \sigma(A_n)$ one has to investigate the spectral stability of the sequences $(A_n - \lambda I)$. This problem proves to be much more involved than the investigation of the common stability. These difficulties disappear if we restrict our attention to sequences for which stability and spectral stability coincide.

Corollary 6.3.12 *Let $(A_n) \in \mathcal{F}$ be a sequence of normal operators. Then*

$$\limsup \sigma(A_n) = \sigma_{\mathcal{F}/\mathcal{G}}((A_n) + \mathcal{G}).$$

Proof. The spectral radius and the norm of a normal element coincide. Hence, the sequence $(A_n - \lambda I)$ is spectrally stable if and only if it is stable. The stability of $(A_n - \lambda I)$ is equivalent to the invertibility of the coset $(A_n + \lambda I) + \mathcal{G}$ by Lemma 6.1.4. □

These results lead to the following theorem in a similar way as for Theorem 6.3.10.

Theorem 6.3.13 *Let $\mathbf{A} = (A_n) \in \mathcal{B}^\$$ be a sequence of normal operators. Then*

$$\limsup \sigma(A_n) = \sigma(\mathbf{A} + \mathcal{G}) = \sup \sigma(A_h)$$

where the supremum is taken over all operators $A_h \in \sigma_{stab}(\mathbf{A})$.

Pseudospectra. Here are the analogous results for pseudospectra. It is remarkable that these results hold without assuming that the operators A_n are normal.

Theorem 6.3.14 *Let $(A_n) \in \mathcal{F}$ and $\varepsilon > 0$. Then*

$$\limsup \sigma^\varepsilon(A_n) = \sigma^\varepsilon_{\mathcal{F}/\mathcal{G}}((A_n) + \mathcal{G}).$$

Proof. Let $\lambda \in \sigma^\varepsilon_{\mathcal{F}/\mathcal{G}}((A_n) + \mathcal{G})$. Then either $(A_n - \lambda I) + \mathcal{G}$ is not invertible, or the inverse of this coset exists, but $\|((A_n - \lambda I) + \mathcal{G})^{-1}\| \ge 1/\varepsilon$.

In the first case, the sequence $(A_n - \lambda I)$ fails to be stable, and the existence of an infinite subsequence (n_k) of $\mathbb{N}$ such that $\lambda \in \sigma^\varepsilon(A_{n_k})$ for each k follows easily. In particular, $\lambda \in \limsup \sigma^\varepsilon(A_n)$ in this case.

In the second case, we infer from Theorem 6.3.5 that, in every open neighborhood U of λ, there is a λ_0 such that $\|((A_n - \lambda_0 I) + \mathcal{G})^{-1}\| > 1/\varepsilon$. (Otherwise we would have $\|((A_n - \lambda_0 I) + \mathcal{G})^{-1}\| \le 1/\varepsilon$ for all $\lambda_0 \in U$ which implies via Theorem 6.3.5 that $\|((A_n - \lambda_0 I) + \mathcal{G})^{-1}\| < 1/\varepsilon$ for all $\lambda_0 \in U$ including $\lambda_0 = \lambda$.)

Thus, for all sufficiently large k, there are numbers λ_k with $\lambda_k \to \lambda$ as $k \to \infty$ such that

$$\|((A_n - \lambda_k I) + \mathcal{G})^{-1}\| \ge 1/(\varepsilon - 1/k).$$

By Proposition 6.3.9, this is equivalent to the inequality

$$\limsup_{n\to\infty} \|(A_n - \lambda_k I)^{-1}\| \ge 1/(\varepsilon - 1/k)$$

(with the invertibility of $A_n - \lambda_k I$ for all sufficiently large n being a consequence of a Neumann series argument). Since $1/\varepsilon < 1/(\varepsilon - 1/k)$, there are numbers n_k tending to infinity as $k \to \infty$ such that

$$\|(A_{n_k} - \lambda_k I)^{-1}\| \ge 1/\varepsilon \quad \text{for all } k.$$

In other words, $\lambda_k \in \sigma^\varepsilon(A_{n_k})$ for each sufficiently large k, which implies that $\lambda = \lim \lambda_k \in \limsup \sigma^\varepsilon(A_n)$.

For the reverse inclusion, let $\lambda \in \limsup \sigma^\varepsilon(A_n)$, but assume contrary to what we want that $\lambda \notin \sigma^\varepsilon_{\mathcal{F}/\mathcal{G}}((A_n) + \mathcal{G})$. Thus, $(A_n - \lambda I) + \mathcal{G}$ is invertible in $\mathcal{F}/\mathcal{G}$, and $\|((A_n - \lambda I) + \mathcal{G})^{-1}\| = 1/\varepsilon - 2\delta < 1/\varepsilon$ with some $\delta > 0$. Then the operators $A_n - \lambda I$ are invertible for sufficiently large n and, by Proposition 6.3.9,

$$\limsup_{n\to\infty} \|(A_n - \lambda I)^{-1}\| = 1/\varepsilon - 2\delta.$$

This shows that $\|(A_n - \lambda I)^{-1}\| < 1/\varepsilon - \delta$ for n sufficiently large, say for $n \ge n_0$. If $n \ge n_0$ and $|\lambda - \mu| < \varepsilon\delta(1/\varepsilon - \delta)^{-1}$ then a Neumann's series argument yields

$$\begin{aligned} \|(A_n - \mu I)^{-1}\| &\le \frac{\|(A_n - \lambda I)^{-1}\|}{1 - |\lambda - \mu|\,\|(A_n - \lambda I)^{-1}\|} \\ &< \frac{1/\varepsilon - \delta}{1 - \varepsilon\delta(1/\varepsilon - \delta)^{-1}(1/\varepsilon - \delta)} = \frac{1}{\varepsilon}, \end{aligned}$$

whence $\mu \notin \sigma^\varepsilon(A_n)$ for all μ in a certain open neighborhood of λ and for all sufficiently large n. But then λ cannot belong to the limit superior of the ε-pseudospectra $\sigma^\varepsilon(A_n)$, which is a contradiction. □

Theorem 6.3.15 *Let* $\mathbf{A} = (A_n) \in \mathcal{B}^\$$ *and* $\varepsilon > 0$. *Then*

$$\limsup \sigma^\varepsilon(A_n) = \sigma^\varepsilon(\mathbf{A} + \mathcal{G}) = \sup \sigma^\varepsilon(A_h)$$

where the supremum is taken over all operators $A_h \in \sigma_{stab}(\mathbf{A})$.

Indeed, this is the outcome of a combination of the preceding theorem with Theorems 6.3.1 and 6.3.3.

Numerical ranges. Here is a result of the same flavor holding for numerical ranges.

Theorem 6.3.16 *Let* $(A_n) \in \mathcal{F}$. *Then*

$$\operatorname{conv} \limsup_{n\to\infty} N(A_n) = N_{\mathcal{F}/\mathcal{G}}((A_n) + \mathcal{G}).$$

Observe that the limes superior of a sequence of convex sets need not be convex again, which explains the conv operator on the left-hand side. For the proof of Theorem 6.3.16 we need one more auxiliary result, the proof of which is left as an exercise.

Lemma 6.3.17 *Let* (M_k) *be a monotonically decreasing sequence of compact subsets of* $\mathbb{C}$, *i.e.,* $M_k \supseteq M_{k+1}$ *for all* k. *Then* $\operatorname{conv} \cap_k M_k = \cap_k \operatorname{conv} M_k$.

Proof of Theorem 6.3.16. We know from identity (6.17) that

$$N_{\mathcal{F}/\mathcal{G}}((A_n) + \mathcal{G}) = \cap_{(G_n)\in\mathcal{G}}\, N_{\mathcal{F}}((A_n) + (G_n))$$

whence, in combination with Theorem 6.3.6,

$$N_{\mathcal{F}/\mathcal{G}}((A_n) + \mathcal{G}) = \cap_{(G_n)\in\mathcal{G}} \operatorname{conv} \operatorname{clos} \cup_n N(A_n + G_n). \tag{6.21}$$

For every $k \in \mathbb{N}$, we choose an $m_k \in N(A_n)$ and define a sequence $(G_n^{(k)}) \in \mathcal{G}$ by

$$G_n^{(k)} = \begin{cases} -A_n + m_k I & \text{if} \quad n \le k-1 \\ 0 & \text{if} \quad n \ge k. \end{cases}$$

Then

$$N(A_n + G_n^{(k)}) = \begin{cases} \{m_k\} \subseteq N(A_n) & \text{if} \quad n \le k-1 \\ N(A_n) & \text{if} \quad n \ge k \end{cases}$$

and, hence,

$$N_{\mathcal{F}/\mathcal{G}}((A_n) + \mathcal{G}) \subseteq \cap_{k=1}^{\infty} \operatorname{conv} \operatorname{clos} \cup_{n\ge k} N(A_n).$$

Applying Lemma 6.3.17 to the sets $M_k := \operatorname{clos} \cup_{n\ge k} N(A_n)$ we get

$$N_{\mathcal{F}/\mathcal{G}}((A_n) + \mathcal{G}) \subseteq \operatorname{conv}\, \cap_{k=1}^{\infty} \operatorname{clos} \cup_{n\ge k} N(A_n).$$

The set on the right-hand side coincides with $\operatorname{conv} \limsup N(A_n)$ by (6.20), which proves the inclusion

$$N_{\mathcal{F}/\mathcal{G}}((A_n) + \mathcal{G}) \subseteq \operatorname{conv} \limsup_{n\to\infty} N(A_n).$$

For the reverse inclusion, let $m \in \limsup N(A_n)$. Then there exist a subsequence $(n_k) \subseteq \mathbb{N}$ tending to infinity as $k \to \infty$ as well as points $m_k \in N(a_{n_k})$ which converge to m as $k \to \infty$. For every k, we choose a state ϕ_k of $L(E)$ with $\phi_k(A_{n_k}) = m_k$.

Let $\mathcal{L}$ stand for the linear subspace of $\mathcal{F}$ consisting of all sequences $(\alpha A_n + \beta I + G_n)$ where $\alpha, \beta \in \mathbb{C}$ and where (g_n) runs through the ideal $\mathcal{G}$. For $(B_n) \in \mathcal{L}$, the limit $\lim_{k\to\infty} \phi_k(B_{n_k})$ exists. Indeed, it is sufficient to check the existence of that limit for the generating sequences of $\mathcal{L}$. For these sequences, one has $\phi_k(A_{n_k}) = m_k \to m$, $\phi_k(I) = 1 \to 1$, and

$$|\phi_k(G_{n_k})| \le \|\phi_k\| \, \|G_{n_k}\| = \|G_{n_k}\| \to 0.$$

Thus, there is a linear functional ϕ on $\mathcal{L}$ defined by

$$\phi((b_n)) := \lim_{k\to\infty} \phi_k(B_{n_k})$$

which maps the sequences $(A_n), (I)$ and (G_n) into m, 1 and 0, respectively, and which is continuous with norm 1:

$$|\phi((B_n))| = |\lim \phi_k(B_{n_k})| \le \sup_k |\phi_k(B_{n_k})| \le \sup_k \|B_{n_k}\| \le \|(B_n)\|_{\mathcal{F}}.$$

By means of the Hahn-Banach theorem, one can extend ϕ to a linear functional with norm 1 on all of $\mathcal{F}$, and we denote this extension by ϕ again. Since $\|\phi\| = \phi((I)) = 1$, this extension is a state on $\mathcal{F}$. Further, the ideal $\mathcal{G}$ lies in the kernel of ϕ by construction, so one can define a functional ψ on $\mathcal{F}/\mathcal{G}$ by

$$\psi : \mathcal{F}/\mathcal{G} \to \mathbb{C}, \quad (C_n) + \mathcal{G} \mapsto \phi((C_n)).$$

Clearly, $\psi((I) + \mathcal{G}) = 1$ and $\psi((A_n) + \mathcal{G}) = m$, and ψ is continuous with norm 1. Thus, ψ is a state of $\mathcal{F}/\mathcal{G}$ which implies that $m \in N_{\mathcal{F}/\mathcal{G}}((A_n) + \mathcal{G})$ and, consequently,

$$\limsup N(A_n) \subseteq N_{\mathcal{F}/\mathcal{G}}((A_n) + \mathcal{G}).$$

Finally, algebraic numerical ranges are convex. Thus,

$$\text{conv} \limsup N(A_n) \subseteq N_{\mathcal{F}/\mathcal{G}}((A_n) + \mathcal{G}),$$

which verifies the second half of the theorem. □

Theorem 6.3.18 *Let* $\mathbf{A} = (A_n) \in \mathcal{B}^{\$}$. *Then*

$$\text{conv} \limsup N(A_n) = N_{\mathcal{F}/\mathcal{G}}(\mathbf{A} + \mathcal{G}) = \sup N(A_h)$$

where the supremum is taken over all operators $A_h \in \sigma_{stab}(\mathbf{A})$.

This follows from the previous theorem and Theorems 6.3.1 and 6.3.6.

6.4 Fractality of approximation methods

Here we consider a class of approximation sequences (A_n) which, roughly speaking, own the following property: the sequence (A_n) can be reconstructed, modulo a sequence tending to zero, from each of its infinite subsequences. Due to this self-similarity, we call these sequences *fractal.* We will see that fractality of a sequence is a property which makes several approximation processes uniform: e.g., the norms $\|A_n\|$ form a *convergent* sequence if (A_n) is fractal, and also the sequences $(\sigma(A_n))$, $(\sigma^\varepsilon(A_n))$ and $(N(A_n))$ of the spectra, pseudospectra, and of the numerical ranges are convergent with respect to the Hausdorff metric.

6.4.1 Fractal approximation sequences

Throughout this section, let E be a Hilbert space, let $\mathcal{F}$ stand for the C^*-algebra of all bounded sequences $(A_n)_{n\in\mathbb{N}}$ with $A_n \in L(E)$, and write $\mathcal{G}$ for the closed ideal of $\mathcal{F}$ which consists of all sequences with $\lim \|A_n\| = 0$. (The restriction onto one-sided infinite approximation sequences is only for simplicity.)

Given a strongly monotonically increasing sequence $\eta : \mathbb{N} \to \mathbb{N}$, let R_η stand for the restriction mapping

$$R_\eta : \mathcal{F} \to \mathcal{F}, \quad (A_n)_{n\in\mathbb{N}} \mapsto (A_{\eta(n)})_{n\in\mathbb{N}}.$$

The mapping R_η is a *-homomorphism from $\mathcal{F}$ onto $\mathcal{F}$ which maps $\mathcal{G}$ onto $\mathcal{G}$. Further, given a C^*-subalgebra $\mathcal{A}$ of $\mathcal{F}$, we let $\mathcal{A}_\eta$ refer to the image of $\mathcal{A}$ under R_η. Clearly, $\mathcal{A}_\eta$ is a C^*-subalgebra of $\mathcal{F}$.

Definition 6.4.1 *Let $\mathcal{A}$ be a C^*-subalgebra of $\mathcal{F}$.*

(a) *A *-homomorphism W of $\mathcal{A}$ into a C^*-algebra $\mathcal{B}$ is* fractal *if, for every strongly monotonically increasing sequence $\eta : \mathbb{N} \to \mathbb{N}$, there is a *-homomorphism $W_\eta : \mathcal{A}_\eta \to \mathcal{B}$ such that $W = W_\eta R_\eta$.*

(b) *The algebra $\mathcal{A}$ is* fractal, *if the canonical homomorphism $\pi : \mathcal{A} \to \mathcal{A}/(\mathcal{A} \cap \mathcal{G})$ is fractal.*

(c) *A sequence $(A_n) \in \mathcal{F}$ is* fractal *if the smallest C^*-subalgebra of $\mathcal{F}$ which contains (A_n) is fractal.*

There are several criteria for fractality. The proof of the following two results can be found in [73], Theorem 1.66 and Corollary 1.67.

Theorem 6.4.2 *A C^*-subalgebra $\mathcal{A}$ of $\mathcal{F}$ is fractal if and only if the following implication holds for every sequence $\mathbf{A} = (A_n) \in \mathcal{A}$ and for every strongly monotonically sequence $\eta : \mathbb{N} \to \mathbb{N}$:*

$$R_\eta(\mathbf{A}) \in \mathcal{G} \Rightarrow \mathbf{A} \in \mathcal{A} \cap \mathcal{G}.$$

Theorem 6.4.3 *Let $\mathcal{A}$ be a C^*-subalgebra of $\mathcal{F}$. Then the following assertions are equivalent:*

(a) *$\mathcal{A}$ is fractal.*

(b) *Every C^*-subalgebra of $\mathcal{A}$ is fractal.*

(c) *Every sequence in $\mathcal{A}$ is fractal.*

Theorem 6.4.4 *A unital C^*-subalgebra $\mathcal{A}$ of $\mathcal{F}$ is fractal if and only if there exists a family $(W_t)_{t\in T}$ of unital and fractal *-homomorphisms from $\mathcal{A}$ into certain C^*-algebras $\mathcal{B}_t$ such that the following equivalence holds for every sequence $\mathbf{A} \in \mathcal{A}$: The coset $\mathbf{A} + \mathcal{A} \cap \mathcal{G}$ is invertible in $\mathcal{A}/(\mathcal{A} \cap \mathcal{G})$ if and only if $W_t(\mathbf{A})$ is invertible in $\mathcal{B}_t$ for every $t \in T$ and if the norms $\|W_t(\mathbf{A})^{-1}\|$ are uniformly bounded with respect to $t \in T$.*

It is not hard to see that $\mathcal{A} \cap \mathcal{G} \subseteq \ker W$ for every fractal homomorphism $W : \mathcal{A} \to \mathcal{B}$ (see, e.g., Lemma 1.64 in [73]). Thus, under the assumptions of the theorem, the mappings

$$W'_t : \mathcal{A}/(\mathcal{A} \cap \mathcal{G}) \to \mathcal{B}_t, \quad \mathbf{A} + \mathcal{A} \cap \mathcal{G} \mapsto W_t(\mathbf{A})$$

are well defined, and the family $(W'_t)_{t \in T}$ forms a weakly sufficient family of homomorphisms for $\mathcal{A}/(\mathcal{A} \cap \mathcal{G})$.

We prepare the proof of Theorem 6.4.4 by a simple lemma.

Lemma 6.4.5 *Let $\mathcal{A}$ be a C^*-subalgebra of $\mathcal{F}$, and let $\mathcal{B}$ be a C^*-algebra. If $W : \mathcal{A} \to \mathcal{B}$ is a fractal *-homomorphism, then $\mathcal{A} \cap \mathcal{G}$ lies in the kernel of W.*

Proof. Let $(G_n) \in \mathcal{A} \cap \mathcal{G}$. Given $\varepsilon > 0$, there is an n_0 such that $\|G_n\| \leq \varepsilon$ for $n \geq n_0$. Consider the sequence $\eta(n) := n + n_0$. Since W is fractal,

$$W(G_n) = (W_\eta R_\eta)(G_n) = W_\eta(G_{\eta(n)}) = W_\eta(G_{n+n_0})$$

and, consequently,

$$\|W(G_n)\| = \|W_\eta(G_{n+n_0})\| \leq \|W_\eta\| \, \|(G_{n+n_0})\|_{\mathcal{F}} \leq \varepsilon.$$

Letting ε go to zero we get the assertion. □

Proof of Theorem 6.4.4. If $\mathcal{A}$ is fractal, then the canonical homomorphism $\pi : \mathcal{A} \to \mathcal{A}/(\mathcal{A} \cap \mathcal{G})$ provides a 'family' (π) with the desired properties.

Let, conversely, $(W_t)_{t \in T}$ be a family of unital and fractal *-homomorphisms which is subject to the conditions of the theorem. Given a strongly monotonically increasing sequence η, define the operator S_η by

$$S_\eta : \mathcal{A}/(\mathcal{A} \cap \mathcal{G}) \to \mathcal{A}_\eta/(\mathcal{A} \cap \mathcal{G})_\eta, \quad \mathbf{A} + \mathcal{A} \cap \mathcal{G} \mapsto R_\eta(\mathbf{A}) + (\mathcal{A} \cap \mathcal{G})_\eta.$$

One easily checks that this definition is correct and that S_η is a *-homomorphism from $\mathcal{A}/(\mathcal{A} \cap \mathcal{G})$ onto $\mathcal{A}_\eta/(\mathcal{A} \cap \mathcal{G})_\eta$. We claim that S_η is an isometry.

In every C^*-algebra $\mathcal{B}$ with identity, one has

$$\|b\|^2 = \|b^* b\| = \rho(b^* b) \quad \text{for each } b \in \mathcal{B}.$$

Hence, in order to show that S_η is an isometry, it is sufficient to show that S_η preserves invertibility, i.e., that if $S_\eta(\mathbf{A} + \mathcal{A} \cap \mathcal{G}) = R_\eta(\mathbf{A}) + (\mathcal{A} \cap \mathcal{G})_\eta$ is invertible in $\mathcal{A}_\eta/(\mathcal{A} \cap \mathcal{G})_\eta$ for a sequence $\mathbf{A} \in \mathcal{A}$, then $\mathbf{A} + \mathcal{A} \cap \mathcal{G}$ is invertible in $\mathcal{A}/(\mathcal{A} \cap \mathcal{G})$. For this goal, let $\mathbf{A}$ be a sequence in $\mathcal{A}$ for which $R_\eta(\mathbf{A}) + (\mathcal{A} \cap \mathcal{G})_\eta$ is invertible. Then there are a sequence $\mathbf{B} \in \mathcal{A}$ as well as sequences $\mathbf{G}$ and $\mathbf{H}$ in $\mathcal{A} \cap \mathcal{G}$ such that

$$R_\eta(\mathbf{A}) R_\eta(\mathbf{B}) = R_\eta(\mathbf{I}) + R_\eta(\mathbf{G}), \quad R_\eta(\mathbf{B}) R_\eta(\mathbf{A}) = R_\eta(\mathbf{I}) + R_\eta(\mathbf{H}) \tag{6.22}$$

with $\mathbf{I} := (I)_{n \in \mathbb{N}}$. The homomorphisms W_t are fractal by assumption, i.e., there are homomorphisms $W_{t,\eta}$ such that $W_t = W_{t,\eta} R_\eta$. Applying $W_{t,\eta}$ to the equalities (6.22) we get

$$W_t(\mathbf{A}) W_t(\mathbf{B}) = W_t(\mathbf{I}) + W_t(\mathbf{G}), \quad W_t(\mathbf{B}) W_t(\mathbf{A}) = W_t(\mathbf{I}) + W_t(\mathbf{H}).$$

Since $W_t(\mathbf{I})$ is the identity element I_t of $\mathcal{B}_t$, and since $W_t(\mathbf{G}) = W_t(\mathbf{H}) = 0$ by Lemma 6.4.5, we get

$$W_t(\mathbf{A})W_t(\mathbf{B}) = W_t(\mathbf{B})W_t(\mathbf{A}) = I_t \quad \text{for all } t \in T.$$

Hence, all elements $W_t(a_n)$ are invertible, and from

$$\|W_t(\mathbf{A})^{-1}\| = \|W_t(\mathbf{B})\| \leq \|\mathbf{B}\|_{\mathcal{F}}$$

we conclude that the norms of their inverses are uniformly bounded. Thus, by hypothesis, the coset $\mathbf{A} + \mathcal{A} \cap \mathcal{G}$ is invertible, whence the isometry of S_η.

Let finally Π_η denote the canonical homomorphism from $\mathcal{A}_\eta$ onto the quotient algebra $\mathcal{A}_\eta/(\mathcal{A}\cap\mathcal{G})_\eta$. Then $S_\eta^{-1}\Pi_\eta R_\eta = \pi$, i.e., π is fractal with $\pi_\eta := S_\eta^{-1}\Pi_\eta$, and we are done. □

6.4.2 Fractality and norms

Here we consider once more the asymptotic behavior of the norms $\|A_n\|$ for a sequence $(A_n) \in \mathcal{F}$, but now in the presence of fractality of (A_n).

Theorem 6.4.6 *Let $\mathcal{A}$ be a fractal C^*-subalgebra of $\mathcal{F}$, and let $(A_n) \in \mathcal{A}$.*

(a) *For every strongly monotonically increasing sequence $\eta : \mathbb{N} \to \mathbb{N}$,*

$$\|(A_n) + \mathcal{G}\|_{\mathcal{F}/\mathcal{G}} = \|(a_{\eta(n)}) + \mathcal{G}\|_{\mathcal{F}/\mathcal{G}}.$$

(b) *The limit* $\lim \|A_n\|$ *exists and is equal to* $\|(A_n) + \mathcal{G}\|$.

Proof. (a) Let $\mathbf{A} = (A_n) \in \mathcal{A}$ and $\mathbf{G} = (G_n) \in \mathcal{A}\cap\mathcal{G}$, and write π for the canonical homomorphism from $\mathcal{A}$ onto $\mathcal{A}/(\mathcal{A} \cap \mathcal{G})$. Then

$$\begin{aligned}
\|\mathbf{A} + \mathcal{G}\|_{\mathcal{F}/\mathcal{G}} &= \|\mathbf{A} + \mathcal{A} \cap \mathcal{G}\|_{\mathcal{A}/(\mathcal{A}\cap\mathcal{G})} \\
&= \|\pi(\mathbf{A} + \mathbf{G})\|_{\mathcal{A}/(\mathcal{A}\cap\mathcal{G})} && \text{(definition of } \pi) \\
&= \|\pi_\eta R_\eta(\mathbf{A} + \mathbf{G})\|_{\mathcal{A}/(\mathcal{A}\cap\mathcal{G})} && \text{(fractality of } \pi) \\
&\leq \|(A_{\eta(n)} + G_{\eta(n)})_{n\in\mathbb{N}}\|_{\mathcal{F}},
\end{aligned}$$

the first identity being a consequence of the third isomorphy theorem for C^*-algebras. Taking the infimum over all $\mathbf{G} \in \mathcal{A} \cap \mathcal{G}$, and taking into account that $\mathcal{A}_\eta \cap \mathcal{G}_\eta = (\mathcal{A} \cap \mathcal{G})_\eta$, we obtain

$$\begin{aligned}
\|(A_n) + \mathcal{G}\|_{\mathcal{F}/\mathcal{G}} &\leq \|(A_{\eta(n)})_{n\in\mathbb{N}} + (\mathcal{A} \cap \mathcal{G})_\eta\|_{\mathcal{A}_\eta/(\mathcal{A}\cap\mathcal{G})_\eta} \\
&= \|(A_{\eta(n)})_{n\in\mathbb{N}} + \mathcal{A}_\eta \cap \mathcal{G}_\eta\|_{\mathcal{A}_\eta/(\mathcal{A}_\eta\cap\mathcal{G}_\eta)}.
\end{aligned}$$

A further application of the third isomorphy theorem gives

$$\|(A_n) + \mathcal{G}\|_{\mathcal{F}/\mathcal{G}} \leq \|(A_{\eta(n)})_{n\in\mathbb{N}} + \mathcal{G}_\eta\|_{\mathcal{F}_\eta/\mathcal{G}_\eta}. \tag{6.23}$$

On the other hand, by Proposition 6.3.9 (= the lim sup-formula for the norm in $\mathcal{F}/\mathcal{G}$) we have

$$\|(A_{\eta(n)}) + \mathcal{G}_\eta\|_{\mathcal{F}_\eta/\mathcal{G}_\eta} = \limsup \|A_{\eta(n)}\| \le \limsup \|A_n\| = \|(A_n) + \mathcal{G}\|_{\mathcal{F}/\mathcal{G}}$$

which together with (6.23) proves assertion (a).

(b) Let $(A_n) \in \mathcal{A}$ be a sequence with $\liminf \|A_n\| < \limsup \|A_n\|$. Then there is a strongly monotonically sequence $\eta : \mathbb{N} \to \mathbb{N}$ such that the limit $\lim \|A_{\eta(n)}\|$ exists and is equal to $\liminf \|A_n\|$. By part (a) of this theorem and by Proposition 6.3.9 again, we have

$$\begin{aligned} \limsup \|A_n\| &= \|(A_n) + \mathcal{G}\|_{\mathcal{F}/\mathcal{G}} = \|(A_{\eta(n)}) + \mathcal{G}\|_{\mathcal{F}_\eta/\mathcal{G}_\eta} \\ &= \limsup \|A_{\eta(n)}\| = \lim \|A_{\eta(n)}\| = \liminf \|A_n\| \end{aligned}$$

which is a contradiction. □

For the following corollary, recall that the *condition number* $\operatorname{cond} b$ of an invertible element b of a unital normed algebra is defined as $\|b\| \, \|b^{-1}\|$.

Corollary 6.4.7 *Let $\mathcal{A}$ be a fractal C^*-subalgebra of $\mathcal{F}$, and let $(A_n) \in \mathcal{A}$ be a stable sequence. Then the limit $\lim_{n\to\infty} \operatorname{cond} A_n$ exists, and it is equal to $\|(A_n) + \mathcal{G}\| \, \|((A_n) + \mathcal{G})^{-1}\|$.*

6.4.3 Fractality and spectra

Let $(M_n)_{n=1}^\infty$ be a sequence of subsets of the complex plane. The *limes inferior* or *uniform limiting set* $\liminf M_n$ consists of all points $m \in \mathbb{C}$ which are the limit of a sequence (m_n) of points $m_n \in M_n$. It is well known that the limes superior and the limes inferior of a sequence (M_n) of non-empty and compact subsets of $\mathbb{C}$ coincide if and only if the sequence (M_n) converges with respect to the Hausdorff metric, and that

$$\limsup M_n = \liminf M_n = \text{h-lim}\, M_n$$

in this case. Recall that the *Hausdorff distance* of the non-empty and compact subsets A and B of $\mathbb{C}$ is defined by

$$h(A, B) := \max\{\max_{a\in A} \operatorname{dist}(a,\, B),\, \max_{b\in B} \operatorname{dist}(b,\, A)\}$$

where $\operatorname{dist}(a, B) = \min_{b\in B} |a - b|$. We denote limits with respect to this metric by h-lim. For details we refer to Hausdorff's monograph [74] and to Section 3.1 of [73].

Spectra. The consequences of fractality for the asymptotic behaviour of spectra are as follows.

Theorem 6.4.8 *Let $\mathcal{A}$ be a fractal C^*-subalgebra of $\mathcal{F}$ which contains the identity sequence (I), and let $\mathbf{A} = (A_n) \in \mathcal{A}$. Then*

(a) *the sequence $\mathbf{A}$ is stable if and only if it possesses a stable subsequence.*

(b) *if $\mathbf{A}$ is normal, then $(\sigma(A_n))$ converges with respect to the Hausdorff metric.*

Proof. (a) Let $\eta : \mathbb{N} \to \mathbb{N}$ be a strongly monotonically increasing sequence for which sequence $(A_{\eta(n)}) = R_\eta(\mathbf{A})$ is stable. Clearly, η can be chosen in such a way that all operators $A_{\eta(n)}$ are invertible. Then, by the inverse closedness of $\mathcal{A}_\eta$ in $\mathcal{F}$, there is a sequence $\mathbf{B} \in \mathcal{A}$ such that

$$R_\eta(\mathbf{A})\, R_\eta(\mathbf{B}) = R_\eta(\mathbf{B})\, R_\eta(\mathbf{A}) = R_\eta(\mathbf{I}). \tag{6.24}$$

The canonical homomorphism $\pi : \mathcal{A} \to \mathcal{A}/(\mathcal{A} \cap \mathcal{G})$ is fractal by hypothesis, i.e., $\pi = \pi_\eta R_\eta$ with a certain homomorphism π_η. Applying π_η to (6.24) one gets the invertibility of $\pi(\mathbf{A}) = \mathbf{A} + \mathcal{A} \cap \mathcal{G}$ in $\mathcal{A}/(\mathcal{A} \cap \mathcal{G})$, whence the stability of $\mathbf{A}$.
(b) Assume, there is a $\lambda \in \limsup \sigma(A_n) \setminus \liminf \sigma(A_n)$. Then one finds a $\delta > 0$ as well as a strongly monotonically increasing sequence $\eta : \mathbb{N} \to \mathbb{N}$ such that $\operatorname{dist}(\lambda, \sigma(A_{\eta(n)})) \geq \delta$ for all n. The operators $A_{\eta(n)}$ are normal. So, this estimate guarantees that the sequence $(A_{\eta(n)} - \lambda I)$ is stable (with the norms of the inverses of the operators $A_{\eta(n)} - \lambda I$ being bounded above by $1/\delta$). Thus, by part (a), the sequence $(A_n - \lambda I)$ itself is stable. Hence, $\lambda \notin \sigma(\mathbf{A} + \mathcal{G}) = \limsup \sigma(A_n)$ by Corollary 6.3.12, which is a contradiction. □

Let us mention a partial converse to Theorem 6.4.8. Indeed, a *self-adjoint* sequence $(A_n) \in \mathcal{F}$ is fractal if and only if

$$\limsup \sigma(A_n) = \liminf \sigma(A_n).$$

A proof of this fact is in [73], Theorem 7.3.

Pseudospectra. Let $\varepsilon > 0$. An element b of a Banach algebra with identity e is said to be *ε-invertible* if it is invertible and if $\|b^{-1}\| < 1/\varepsilon$. Thus, the ε-pseudospectrum of b consists of all $\lambda \in \mathbb{C}$ for which $b - \lambda e$ is not ε-invertible. Accordingly, we call a sequence $(A_n) \in \mathcal{F}$ *ε-stable* if the coset $(A_n) + \mathcal{G}$ is ε-invertible in $\mathcal{F}/\mathcal{G}$.

Theorem 6.4.9 *Let $\mathcal{A}$ be a fractal C^*-subalgebra of $\mathcal{F}$ which contains the identity sequence, let $\varepsilon > 0$, and let $\mathbf{A} = (A_n) \in \mathcal{A}$. Then*

(a) *the sequence $\mathbf{A}$ is ε-stable if and only if it possesses an ε-stable subsequence.*

(b) *the sequence $(\sigma^\varepsilon(A_n))$ of pseudospectra converges with respect to the Hausdorff metric.*

Proof. (a) Let $\eta : \mathbb{N} \to \mathbb{N}$ be a strongly monotonically increasing sequence such that $R_\eta(\mathbf{A})$ is an ε-stable subsequence of $\mathbf{A}$. Then the sequence $\mathbf{A}$ is stable by Theorem 6.4.8 (a), and Theorem 6.4.6 (a) implies that

$$1/\varepsilon > \|((a_{\eta(n)}) + \mathcal{G}_\eta)^{-1}\|_{\mathcal{F}_\eta/\mathcal{G}_\eta} = \|((A_n) + \mathcal{G})^{-1}\|_{\mathcal{F}/\mathcal{G}}.$$

Hence, $\mathbf{A}$ is ε-stable.

(b) Assume that λ belongs to $\limsup \sigma^\varepsilon(A_n) \setminus \liminf \sigma^\varepsilon(A_n)$. Then there is a monotonically increasing sequence $\eta : \mathbb{N} \to \mathbb{N}$ such that $\lambda \notin \limsup \sigma^\varepsilon(A_{\eta(n)})$. By Theorem 6.3.14, the sequence $(A_{\eta(n)} - \lambda I)$ is ε-stable, which implies via assertion (a) the ε-stability of the complete sequence $(A_n - \lambda I)$. Again by Theorem 6.3.14, this yields $\lambda \notin \limsup \sigma^\varepsilon(A_n)$, which contradicts the assumption. □

Numerical ranges. Here is the corresponding result for numerical ranges.

Theorem 6.4.10 *Let $\mathcal{A}$ be a fractal C^*-subalgebra of $\mathcal{F}$ which contains the identity sequence, and let $\mathbf{A} = (A_n) \in \mathcal{A}$. Then the sequence $(N(A_n))$ of numerical ranges converges with respect to the Hausdorff metric.*

Proof. The inclusion $\liminf N(A_n) \subseteq \limsup N(A_n)$ is obvious. For the reverse inclusion recall that numerical ranges are convex, and that the limes inferior of a sequence of convex sets is convex again. Thus, the inclusion $\limsup N(A_n) \subseteq \liminf N(A_n)$ holds if and only if $\operatorname{conv} \limsup N(A_n) \subseteq \liminf N(A_n)$. So, by Theorem 6.3.16, what we have to prove is that

$$N_{\mathcal{F}/\mathcal{G}}(\mathbf{A} + \mathcal{G}) \subseteq \liminf N(A_n).$$

From the inverse closedness of C^*-algebras with respect to numerical ranges and from the third isomorphy theorem we conclude that

$$N_{\mathcal{F}/\mathcal{G}}(\mathbf{A} + \mathcal{G}) = N_{(\mathcal{A}+\mathcal{G})/\mathcal{G}}(\mathbf{A} + \mathcal{G}) = N_{\mathcal{A}/(\mathcal{A}\cap\mathcal{G})}(\mathbf{A} + \mathcal{A} \cap \mathcal{G}).$$

Thus, let $\lambda \in N_{\mathcal{A}/(\mathcal{A}\cap\mathcal{G})}(\mathbf{A} + \mathcal{A} \cap \mathcal{G})$, and let ϕ be a state of $\mathcal{A}/(\mathcal{A} \cap \mathcal{G})$ such that $\lambda = \phi(\mathbf{A} + \mathcal{A} \cap \mathcal{G}) = \phi(\pi(\mathbf{A}))$ where again π refers to the canonical homomorphism from $\mathcal{A}$ onto $\mathcal{A}/(\mathcal{A} \cap \mathcal{G})$. Since π is fractal, one has $\lambda = \phi(\pi_\eta R_\eta(\mathbf{A}))$ for every monotonically increasing sequence $\eta : \mathbb{N} \to \mathbb{N}$.

The functional $\phi \circ \pi_\eta$ is a state on $\mathcal{A}_\eta := R_\eta \mathcal{A}$. Hence,

$$\lambda \in N_{\mathcal{A}_\eta}(R_\eta(\mathbf{A})) = N_{\mathcal{F}}(R_\eta(\mathbf{A})),$$

whence, by Theorem 6.3.6,

$$\lambda \in \operatorname{conv} \sup N(A_{\eta(n)}) \quad \text{for every } \eta. \tag{6.25}$$

Now assume there exists an $\lambda \in N_{\mathcal{F}/\mathcal{G}}(\mathbf{A} + \mathcal{G})$ which does not belong to the limes inferior of the sequence $(N(A_n))_{n\in\mathbb{N}}$. Then there is a strongly monotonically increasing sequence $\eta^* : \mathbb{N} \to \mathbb{N}$ such that $\operatorname{dist}(\lambda, N(A_{\eta^*(n)})) \geq d > 0$ for all n. Due to the compactness and the convexity of numerical ranges, there exist points $\lambda_{\eta^*(n)} \in N(A_{\eta^*(n)})$ such that

$$|\lambda - \lambda_{\eta^*(n)}| = \operatorname{dist}(\lambda, N(A_{\eta^*(n)})),$$

and the points $\lambda_{\eta^*(n)}$ are unique for every n. Since $|\lambda_{\eta^*(n)}| \leq \|A_{\eta^*(n)}\| \leq \|\mathbf{A}\|_{\mathcal{F}}$, there is at least one cluster point m^* of the sequence $(\lambda_{\eta^*(n)})_{n\in\mathbb{N}}$.

Given $\varepsilon > 0$, consider those $\eta^*(n)$ for which the point $m_{\eta^*(n)}$ belongs to the ε-neighborhood of m^*. The values $\eta^*(n)$ single out a subsequence η_ε of η^*. A little thought reveals that, for this subsequence,

$$\text{dist}\,(\lambda, \text{conv} \sup_{\eta_\varepsilon(n)} N(A_{\eta_\varepsilon(n)}) \geq d/2 \tag{6.26}$$

if only ε is small enough. Since (6.25) holds for every sequence (in particular, for η_ε), (6.26) contradicts (6.25). □

The limes inferior of a sequence of convex sets is convex again. Hence, from Theorem 6.3.16 and the theorem just proved we get:

Corollary 6.4.11 *Let $\mathcal{A}$ be a fractal C^*-subalgebra of $\mathcal{F}$ which contains the identity sequence. Then, for every sequence $(A_n) \in \mathcal{A}$,*

$$\limsup N(A_n) = \liminf N(A_n) = N_{\mathcal{F}/\mathcal{G}}((A_n) + \mathcal{G}).$$

6.4.4 Fractality of the finite section method for a class of band-dominated operators

In general the finite section method for a band-dominated operator fails to be fractal. An archetypal example is the block diagonal operator

$$A = \text{diag}\,(\ldots, B, B, B, \ldots) \quad \text{where} \quad B = \begin{pmatrix} 0 & 1 \\ 1 & 0 \end{pmatrix},$$

thought of as acting on $l^2(\mathbb{Z})$, in which case

$$0 \in \limsup \sigma(P_n A P_n) \setminus \liminf \sigma(P_n A P_n)$$

where P_n is as in Section 6.2. Thus, the sequence $(P_n A P_n)$ cannot be fractal by Theorem 6.4.8 (b). It is the goal of this section to single out a class of band-dominated operators for which the finite section method is fractal.

For simplicity, we let $E = l^2(\mathbb{Z}^2, H)$ with a Hilbert space H, and we let the algebra $\mathcal{A}_E(C_{L(H)})$ be as in Section 2.3.6. Let further Ω be a compact and convex polygon in $\mathbb{R}^2$ with vertices in $\mathbb{Z}^2$ and with the point (0, 0) in its interior. The boundary of Ω will be denoted by $\partial\Omega$. Further, for $m \geq 1$, let $P_m \in L(E)$ stand for the operator of multiplication by the characteristic function of $m\Omega \cap \mathbb{Z}^2$, and set $Q_m := I - P_m$. For every operator $A \in \mathcal{A}_E(C_{L(H)})$, we consider its finite sections

$$A_m := P_m A P_m \tag{6.27}$$

thought of as acting on the space $\text{Im}\, P_m = l^2(m\Omega \cap \mathbb{Z}^2, H)$. Equivalently, we can consider the sequence $(P_n A P_n + Q_n)$ of operators acting on $L(E)$. Let $\mathcal{C}$ refer to the smallest closed subalgebra of $\mathcal{F}$ which contains all sequences of the form $(P_n A P_n + Q_n)$ with $A \in \mathcal{A}_E(C_{L(H)})$. We are interested in a stability result for the sequences in $\mathcal{C}$, and this result should be in a form which allows us to deduce the fractality of this algebra. The crucial idea to reach this goal is to define a family $(W_t)_{t\in T}$ of *fractal* homomorphisms from $\mathcal{C}$ into $L(E)$ such that the set of all operators $W_t((P_nAP_n+Q_n)_{n\in\mathbb{N}})$ with $t \in T$ essentially coincides with $\sigma_{stab}(A)$. Then, as in Section 2.2.2, a similar result can be derived for arbitrary sequences in $\mathcal{C}$. With this result, the fractality of $\mathcal{C}$ follows from Theorem 6.4.4.

Let the notations u_j, H_j, K_j and H_j^0, K_j^0 be as in Section 6.2.3, and let again Π_j stand for the projection $\hat{\chi}_{H_j^0}I$. Further, for $x \in \partial\Omega$, μ_x refers to the point at which the ray $\{tx : t \geq 0\} \subseteq \mathbb{R}^2$ intersects the sphere S^1. From Proposition 2.3.23 (c) we know that the local operator spectrum $\sigma_{\mu_x}(A)$ is a singleton for every operator $A \in \mathcal{A}_E(C_{L(H)})$. We denote the (unique) element of this spectrum by A_{μ_x}. With these remarks, we get the following result as a special case of Theorem 6.2.7.

Theorem 6.4.12 *Let $A \in \mathcal{A}_E(C_{L(H)})$. The sequence $(P_mAP_m + Q_m)$ is stable if and only if*

(a) *the operator A is invertible on E.*

(b) *for every $x \in (u_j, u_{j+1})$, the operator $\Pi_j A_{\mu_x} \Pi_j + (I - \Pi_j)$ is invertible.*

(c) *for every $x \in \{u_1, \ldots, u_k\}$, the operator $\Pi_j\Pi_{j-1}A_{\mu_x}\Pi_{j-1}\Pi_j + (I - \Pi_j\Pi_{j-1})$ is invertible.*

In order to define the announced homomorphisms, we associate with every point $x \in \partial\Omega$ and with every positive integer m a point $x_m \in \mathbb{Z}^2$ as follows: If x is the vertex u_j of Ω, then $x_m := mu_j$. In case x lies on some open interval (u_j, u_{j+1}), we choose $\lambda_0 \in (0, 1)$ such that $x = \lambda_0 u_j + (1 - \lambda_0)u_{j+1}$, and then define $x_m := [m\lambda_0]u_j + (m - [m\lambda_0])u_{j+1}$ where $[y]$ refers to the integer part of y. In any case, the point x_m belongs to the boundary of mH_j, and the sequence $(x_m)_{m\geq 1}$ converges to $\mu_x \in S^1$ with respect to the Gelfand topology. With these remarks, the proof of the following proposition, establishing the existence of certain $\mathcal{P}$-strong limits, is straightforward.

Proposition 6.4.13

(a) *If $\mathbf{A} = (A_m) \in \mathcal{C}$, then the $\mathcal{P}$-strong limit $W(\mathbf{A}) := \mathcal{P}\text{-lim}A_m$ exists. In particular, if $\mathbf{B} = (P_mAP_m + Q_m)$ is one of the generating sequences of $\mathcal{C}$, then $W(\mathbf{B}) = A$.*

(b) *Let $x \in \partial\Omega$ and $\mathbf{A} = (A_m) \in \mathcal{C}$. Then the $\mathcal{P}$-strong limit*

$$W_x(\mathbf{A}) := \mathcal{P}\text{-lim}V_{-x_m}A_mV_{x_m}$$

exists. If, in particular, $\mathbf{B} = (P_mAP_m + Q_m)$ and $x \in (u_j, u_{j+1})$, then

$$W_x(\mathbf{B}) = \Pi_j A_{\mu_x}\Pi_j + (I - \Pi_j),$$

whereas in case $x = u_j$,

$$W_x(\mathbf{B}) = \Pi_j\Pi_{j-1}A_{\mu_x}\Pi_{j-1}\Pi_j + (I - \Pi_j\Pi_{j-1}).$$

With these observations, we can reformulate Theorem 6.4.12 as follows: The sequence $\mathbf{A} := (P_mAP_m + Q_m)$ with $A \in \mathcal{A}_E(C_{L(H)})$ is stable if and only if the operator $W(\mathbf{A}) = A$ as well as all operators $W_x(\mathbf{A})$ with $x \in \partial\Omega$ are invertible.

Obviously, the mappings W and W_x are *-homomorphisms from $\mathcal{C}$ into $L(E)$. So it is easy to extend this result to arbitrary sequences in $\mathcal{C}$ (compare Section 2.2.2).

Theorem 6.4.14 *A sequence $\mathbf{A} \in \mathcal{C}$ is stable if and only if the operator $W(\mathbf{A})$ and all operators $W_x(\mathbf{A})$ with $x \in \partial\Omega$ are invertible.*

Now the fractality of the algebra $\mathcal{C}$ follows almost at once: the point is that the homomorphisms W and W_x are fractal. For example, if $\mathbf{A} = (A_m) \in \mathcal{C}$ and if $\eta : \mathbb{N} \to \mathbb{N}$ is strongly monotonically increasing, then

$$W_x(\mathbf{A}) = \mathcal{P}\text{-}\lim_{n\to\infty} V_{-x_{\eta(n)}} A_{\eta(n)} V_{x_{\eta(n)}}.$$

Hence, the operator $W_x(\mathbf{A})$ is uniquely determined by each of the subsequences of $\mathbf{A}$. Now Theorem 6.4.4 applies to get the following.

Corollary 6.4.15 *The algebra $\mathcal{C}$ is fractal.*

We conclude by specifying the Theorems 6.4.6, 6.4.8, 6.4.9 and 6.4.10 to the present context. Recall that the mappings W and W_x with $x \in \partial\Omega$ (more correctly: their quotients by $\mathcal{C} \cap \mathcal{G}$) form not only a weakly sufficient, but a *sufficient* family of homomorphisms for the algebra $\mathcal{C}/(\mathcal{C} \cap \mathcal{G})$. Thus, the suprema in Theorems 2.2.7, 6.3.2 and 6.3.3 are actually maxima.

Corollary 6.4.16

(a) *Let $\mathbf{A} = (A_n) \in \mathcal{C}$. Then the sequence of the norms $\|A_n\|$ converges, and*

$$\lim \|A_n\| = \|\mathbf{A} + \mathcal{G}\| = \max\{\|W(A)\|,\ \|W_x(\mathbf{A})\| : x \in \partial\Omega\}.$$

(b) *Let $\mathbf{A} = (A_n) \in \mathcal{C}$ be a stable sequence. Then the sequence of the condition numbers* cond A_n *converges, and*

$$\begin{aligned}\lim \operatorname{cond} A_n &= \max\{\|W(A)\|,\ \|W_x(\mathbf{A})\| : x \in \partial\Omega\} \\ &\quad \cdot \max\{\|W(A)^{-1}\|,\ \|W_x(\mathbf{A})^{-1}\| : x \in \partial\Omega\}.\end{aligned}$$

Corollary 6.4.17 *Let $\mathbf{A} = (A_n) \in \mathcal{C}$ be a sequence of normal operators. Then the sequence of the spectra $\sigma(A_n)$ converges with respect to the Hausdorff metric, and*

$$\text{h-}\lim \sigma(A_n) = \sigma(W(\mathbf{A})) \cup \bigcup_{x\in\partial\Omega} \sigma(W_x(\mathbf{A})).$$

Corollary 6.4.18 *Let $\mathbf{A} = (A_n) \in \mathcal{C}$ and $\varepsilon > 0$. Then the sequence of the ε-pseudospectra $\sigma^\varepsilon(A_n)$ converges with respect to the Hausdorff metric, and*

$$\text{h-}\lim \sigma^\varepsilon(A_n) = \sigma^\varepsilon(W(\mathbf{A})) \cup \bigcup_{x\in\partial\Omega} \sigma^\varepsilon(W_x(\mathbf{A})).$$

Corollary 6.4.19 *Let $\mathbf{A} = (A_n) \in \mathcal{C}$. Then the sequence of the numerical ranges $N(A_n)$ converges with respect to the Hausdorff metric, and*

$$\text{h-}\lim N(A_n) = \operatorname{conv}\left(N(W(\mathbf{A})) \cup \sup_{x\in\partial\Omega} N(W_x(\mathbf{A}))\right).$$

6.5 Comments and references

There is a vast literature dealing with the finite section method for various concrete classes of band-dominated operators, see [30, 32, 59, 86, 87, 137]. The pioneers in developing the Banach algebra language of numerical analysis which allowed them to apply local principles to the study of the applicability of an approximation method were Kozak and Simonenko [86, 87].

The only paper (we know) which deals with the finite section method of general band-dominated operators is Gohberg/Kaashoek [62] where the case $N = 1$ is considered. In this paper, the problem of the stability of the sequence $(P_n A P_n)$ is reduced to the stability problem for two related sequences. In this connection, one should also consult [15] where it is shown (in our notations) that, for *each* five-diagonal self-adjoint operator A on $l^2(\mathbb{N})$,

$$\sigma(A) = \liminf_{n \to \infty} \sigma(P_n A P_n).$$

In a special case (for so-called homogeneous symbols), the close relationship between Fredolmness in two dimensions and the finite section method in one dimension was discovered by Douglas and Howe [49]. Subsequently, Gorodetsky [67, 68] became to understand that also in the general case the problem of the applicability of the finite section method over homothetic polyhedra and similar sets in N dimensions is equivalent to the Fredholmness of a convolution operator over an appropriate cone in $N + 1$ dimensions. The idea was developed further in [137, 138, 139]. The remarkable simple stability theorems presented in Section 6.2.2 have been proved in the very recent paper [100].

The results of Sections 6.3 and 6.4 can be found also in [151, 73] in a slightly different form. The point is that [151, 73] exclusively deal with algebras of approximation sequences for which *sufficient* families of homomorphisms are known, whereas the appearance of *weakly sufficient* families is typical for approximations of general band-dominated operators. We also refer to [73] for a more detailed study of spectral approximation, and also for some historical comments. We here confine ourselves to mentioning the papers [20, 33, 144, 147, 180] for approximations of ε-pseudospectra and [145] for approximations of numerical ranges. An extension of the convergence theorems for pseudospectra to the finite section method for operators on L^p-spaces with $1 < p < \infty$ was achieved in [21].

The notion of fractality of an approximation sequence has been introduced in [151] and is studied in detail in [73, 146]. It should be mentioned that fractality of an algebra does not only make approximation processes more uniform; it turns out that it also determines to a large extent the structure of the algebra (see [148] for some results in this direction).

The finite section method for general band-dominated operators seems to be the first instance of an non-fractal approximation method where complete necessary and sufficient stability criteria are available. In general, the non-fractality of an approximation method raises serious difficulties. Here are two examples of stability problems for approximation sequences which seem to be not tractable at the present moment.

Example 1: Spline Galerkin methods for singular integral equations on intervals. Consider a singular integral equation $(aI + bS)u = f$ on the Hilbert space $L^2(\mathbb{R})$ with piecewise continuous coefficients a and b and with the operator S of singular integration along the real line defined by

$$(Su)(t) := \frac{1}{\pi i} \int_{-\infty}^{\infty} \frac{u(s)}{s-t}\, ds \quad \text{where} \quad t \in \mathbb{R}.$$

For the approximate solution of this equation let S_n stand for the smallest closed subspace of $L^2(\mathbb{R})$ which contains all functions which are constant on each of the intervals $(k/n,\ (k+1)/n)$ with $k \in \mathbb{Z}$, and let L_n stand for the orthogonal projection from $L^2(\mathbb{R})$ onto the spline space S_n. It is not too hard to analyze the stability of the spline Galerkin sequence $(L_n(aI + bS)|_{S_n})$ in case a and b are piecewise continuous functions which are continuous on $\mathbb{R} \setminus \mathbb{Z}$ (see, e.g., [72] for an approach using Banach algebra techniques and for a stability criterion). But, as far as we know, nobody succeeded in deriving a general necessary and sufficient criterion if the coefficients a and b are merely supposed to be piecewise continuous on $\mathbb{R}$ (without further assumptions on the location of their discontinuities). We believe that the non-fractality of the sequence $(L_n(aI + bS)|_{S_n})$ is a main obstacle for these difficulties.

The non-fractality of spline Galerkin methods for singular integrals with piecewise continuous coefficients can be seen most easily for the special self-adjoint sequence $(L_n \chi I|_{S_n})$ where χ is the characteristic function of the interval $(\sqrt{2},\ \infty)$. Indeed, the functions φ_{kn}, $k \in \mathbb{Z}$, which are defined as $\sqrt{n}$ times the characteristic function of the interval $[k/n,\ (k+1)/n)$, form an orthonormal basis of S_n, and the matrix representation of $L_n \chi I|_{S_n}$ with respect to this basis is the diagonal matrix $\operatorname{diag}_k(\alpha_k)$ where

$$\alpha_k := \begin{cases} 0 & \text{if} \quad k < \{\sqrt{2}n\} - 1 \\ \{\sqrt{2}n\} - \sqrt{2}n & \text{if} \quad k = \{\sqrt{2}n\} - 1 \\ 1 & \text{if} \quad k > \{\sqrt{2}n\} - 1, \end{cases}$$

with $\{x\}$ referring to the smallest integer which is not smaller than $x \in \mathbb{R}$. Since the numbers $\{\sqrt{2}n\} - \sqrt{2}n$ form a dense subset of $[0,\ 1]$ (the Jacobi-Kronecker theorem), we conclude that

$$\limsup \sigma(L_n \chi I|_{S_n}) = \limsup\{0,\ 1,\ \{\sqrt{2}n\} - \sqrt{2}n\} = [0,\ 1],$$

whereas

$$\liminf \sigma(L_n \chi I|_{S_n}) = \liminf\{0,\ 1\, \{\sqrt{2}n\} - \sqrt{2}n\} = \{0,\ 1\}.$$

Due to Theorem 6.4.8, the sequence $(L_n \chi I|_{S_n})$ is not fractal. It is interesting to observe that certain collocation methods for singular integral equations with arbitrary piecewise continuous coefficients are fractal (see again [72]). □

Example 2: Spline Galerkin methods for singular integral equations on domains. The same difficulties as in Example 1 arise for spline Galerkin methods for equations on bounded domains. A typical example is the spline Galerkin method for the singular integral equation $(aI + bS_\Omega)u = f$ on $L^2(\Omega)$ with Ω a convex domain in $\mathbb{R}^2$ with C^2-boundary (as considered in [55]). Here, S_Ω is the operator $\chi_\Omega S_{\mathbb{R}^2}\chi_\Omega I$, and $S_{\mathbb{R}^2}$ is the singular integral operator

$$(S_{\mathbb{R}^2}u)(x) := -\frac{1}{\pi}\int_{\mathbb{R}^2}\frac{u(y)}{(x-y)^2}\,dy.$$

A direct application of a spline Galerkin projection (based on a spline space over a suitable lattice of $\mathbb{R}^2$) to this equation leads necessarily to a non-fractal behavior in the neighborhood of the boundary of Ω. In [55] it is tried to manage these effects by modifying the spline space by an appropriate cutting off technique. □

Of course, the successful approach to the stability of the finite section method for band-dominated operators by means of the limit operators method gives some hope that these techniques might also prove to be useful for analyzing other non-fractal and still intractable approximation sequences.

Chapter 7

Axiomatization of the Limit Operators Approach

The goal of this final chapter is to develop an axiomatic approach to the limit operators method which contains many of the applications mentioned before as special cases. Thereby, we restrict ourselves to the Hilbert space case. As an application of this approach, we examine the Fredholm property of operators in C^*-subalgebras of $L(L^2(\mathbb{X}))$ which are generated by operators of multiplication, operators of convolution and operators of shift, and where $\mathbb{X}$ is a homogeneous non-commutative group.

7.1 An axiomatic approach to the limit operators method

Let H be a complex Hilbert space and $L(H)$ the C^*-algebra of all bounded linear operators acting on H. Assume further that we are given three families of operators in $L(H)$:

- a sequence $(P_n)_{n\in\mathbb{N}}$ of self-adjoint operators which converge strongly to the identity operator I on H,
- a sequence $(Q_m)_{m\in\mathbb{N}}$, and
- a countable family $(V_y)_{y\in Y}$ of unitary operators.

Of course, one can take every countable set in place of Y; in particular one can choose $Y = \mathbb{N}$. But for several applications, it will prove more advantageous to identify Y with $\mathbb{Z}^N$ or with a countable discrete subgroup of a locally compact group, for example.

Here are the axioms which relate the families (P_n), (Q_m) and (V_y) with each other. For each $n \in \mathbb{N}$ and $y \in Y$, write $P_{n,y} := V_y P_n V_y^*$.

Axiom 1

(a) *For fixed $m, n \in \mathbb{N}$, there are only finitely many elements $y \in Y$ such that $Q_m P_{n,y} \neq 0$ or $P_{n,y} Q_m \neq 0$.*

(b) *For fixed $n \in \mathbb{N}$ and $y \in Y$, there are only finitely many $m \in \mathbb{N}$ such that $(I - Q_m)P_{n,y} \neq 0$ or $P_{n,y}(I - Q_m) \neq 0$.*

Axiom 2

(a) *For each $n \in \mathbb{N}$, there is an M such that $P_n P_m = P_m P_n = P_n$ for all $m \geq M$.*

(b) *For each $n \in \mathbb{N}$, there is an infinite subset Y_n of Y such that*

$$\sum_{y \in Y_n} \|P_{n,y} u\|^2 = \|u\|^2 \quad \textit{for all } u \in H.$$

(c) *For each $n \in \mathbb{N}$ and $l > n$, there is a constant $C_{n,l}$ such that*

$$\sum_{y \in Y_n} \|P_{l,y} u\|^2 \leq C_{n,l} \|u\|^2 \quad \textit{for all } u \in H.$$

Thus, the family (P_n) forms an increasing approximate identity in the sense of Sections 1.1.2 and 1.1.4. Let $\mathcal{H}$ denote the set of all injective sequences from $\mathbb{N}$ into Y.

Definition 7.1.1 *The operator $B \in L(H)$ is called a* limit operator *of the operator $A \in L(H)$ with respect to the sequence $h \in \mathcal{H}$ if, for every $n \in \mathbb{N}$,*

$$\|(V_{h(k)}^* A V_{h(k)} - B) P_n\| \to 0 \quad \textit{and} \quad \|P_n (V_{h(k)}^* A V_{h(k)} - B)\| \to 0$$

as $k \to \infty$. The set of all limit operators of A will be denoted by $\sigma_{op}(A)$.

Lemma 7.1.2 *Every operator $A \in L(H)$ has at most one limit operator with respect to a given sequence $h \in \mathcal{H}$.*

Proof. Let both B_1 and B_2 be limit operators of A with respect to the sequence h. Then $(B_1 - B_2) P_n = 0$ for all n. Taking the strong limit as $n \to \infty$ we get $B_1 - B_2 = 0$. □

This lemma justifies to denote the limit operator of A with respect to the sequence h by A_h. The following proposition describes some elementary properties of limit operators.

Proposition 7.1.3 *Let $h \in \mathcal{H}$, and let $A, B, C_1, C_2, \ldots \in L(H)$ be operators for which the limit operator with respect to h exists. Then*

(a) $\|A_h\| \leq \|A\|$.

(b) *$(A + B)_h$ exists, and $(A + B)_h = A_h + B_h$.*

(c) *$(A^*)_h$ exists, and $(A^*)_h = (A_h)^*$.*

(d) *if $C \in L(H)$ and $\|C_n - C\| \to 0$, then C_h exists and $\|(C_n)_h - C_h\| \to 0$.*

Proof. We will prove assertion (a) only. The other ones follow as in Proposition 1.2.2. Let $\varepsilon > 0$. For every $n \in \mathbb{N}$, choose $k_n \in \mathbb{N}$ such that the sequence $k = (k_n)_{n \in \mathbb{N}}$ is strongly increasing and that

$$\|(V_{h(k_n)}^* A V_{h(k_n)} - A_h) P_n\| < \varepsilon \quad \text{for all } n \in \mathbb{N}.$$

Clearly, the sequence $h \circ k$ belongs to $\mathcal{H}$, and the limit operator $A_{h \circ k}$ exists and coincides with A_h. Now, for every $u \in H$ and $n \in \mathbb{N}$,

$$\|A_h P_n u\| \leq \|V_{h(k_n)}^* A V_{h(k_n)} P_n u\| + \varepsilon \|u\| \leq \|A\| \, \|P_n u\| + \varepsilon \|u\|.$$

Letting n go to infinity (recall that $P_n \to I$ strongly), we get

$$\|A_h u\| \leq (\|A\| + \varepsilon)\|u\| \quad \text{for all } u \in H.$$

Thus, $\|A_h\| \leq \|A\| + \varepsilon$, which holds for every $\varepsilon > 0$. □

Definition 7.1.4 *Let $\mathcal{A}_0$ denote the set of all operators $A \in L(H)$ enjoying the following properties*

(a) $\lim_{n\to\infty} \sup_{y \in Y} \|P_{n,y}A - AP_{n,y}\| = 0.$

(b) *every sequence in $\mathcal{H}$ possesses a subsequence h for which the limit operator A_h exists.*

(c) *for every $n \in \mathbb{N}$, there is an $l \in \mathbb{N}$ such that*

$$P_{n,y}A = P_{n,y}AP_{l,y} \quad \textit{and} \quad AP_{n,y} = P_{l,y}AP_{n,y} \quad \textit{for all} \quad y \in Y.$$

Let $\mathcal{A}$ denote the closure of $\mathcal{A}_0$ in $L(H)$.

An operator $A \in L(H)$ which satisfies condition (b) of this definition, will be called *rich*. Note also that if an operator A is subject to condition (c), then this condition holds for all sufficiently large l. Indeed, let $P_{n,y}A = P_{n,y}AP_{l,y}$ and choose L such that $P_{l,y} = P_{l,y}P_{m,y} = P_{m,y}P_{l,y}$ for all $m \geq L$, which can be done by Axiom 2 (a). Then

$$P_{n,y}AP_{m,y} = P_{n,y}AP_{l,y}P_{m,y} = P_{n,y}AP_{l,y}P_{n,y}A$$

for all $m \geq L$.

Theorem 7.1.5 *$\mathcal{A}$ is a C^*-subalgebra of $L(H)$ which contains the identity operator.*

Proof. We have to show that $\mathcal{A}_0$ is a *-algebra which contains the identity operator. Obviously, the operator I satisfies (a) and (b), and it satisfies (c) due to Axiom 2 (a). Further, if $A \in \mathcal{A}_0$, then $A^* \in \mathcal{A}_0$, too. Indeed, A^* satisfies (a) and (c) since the P_n are self-adjoint, and it satisfies (b) due to Proposition 7.1.3 (c).

Let now $A, B \in \mathcal{A}_0$. Then the sum $A + B$ satisfies (a) obviously, and it satisfies (b) due to Proposition 7.1.3 (b). So, let $n \in \mathbb{N}$, and let $l_1, l_2 \in \mathbb{N}$ be such that

$$P_{n,y}A = P_{n,y}AP_{l_1,y} \quad \text{and} \quad P_{n,y}B = P_{n,y}BP_{l_2,y} \tag{7.1}$$

for all $y \in Y$. Choose l such that $P_{l_1,y} = P_{l_1,y}P_{l,y}$ and $P_{l_2,y} = P_{l_2,y}P_{l,y}$. Then, multiplying the identities (7.1) by $P_{l,y}$ from the right-hand side, we get

$$P_{n,y}AP_{l,y} = P_{n,y}AP_{l_1,y} \quad \text{and} \quad P_{n,y}BP_{l,y} = P_{n,y}BP_{l_2,y}$$

for all $y \in Y$. Together with (7.1), this yields

$$P_{n,y}A = P_{n,y}AP_{l,y} \quad \text{and} \quad P_{n,y}B = P_{n,y}BP_{l,y}$$

for all $y \in Y$. Adding these equalities, we get (c) for the sum $A + B$. Thus, $A + B \in \mathcal{A}_0$, and it remains to show that the product AB also belongs to $\mathcal{A}_0$. Clearly,

$$\begin{aligned} \|P_{n,y}AB - ABP_{n,y}\| &\leq \|P_{n,y}AB - AP_{n,y}B\| + \|AP_{n,y}B - ABP_{n,y}\| \\ &\leq \|P_{n,y}A - AP_{n,y}\| \, \|B\| + \|A\| \, \|P_{n,y}B - BP_{n,y}\|. \end{aligned}$$

Thus, AB is subject to condition (a).

Let now $h \in \mathcal{H}$. Then there is a subsequence $\tilde{h}$ of H such that $A_{\tilde{h}}$ exists, and there is a subsequence g of $\tilde{h}$ such that B_g exists. Of course, then also A_g exists and is equal to $A_{\tilde{h}}$. We claim that

If $A, B \in \mathcal{A}_0$ and if A_g, B_g exist, then $(AB)_g$ exists, and $(AB)_g = A_g B_g$.

Indeed, for every $n \in \mathbb{N}$,

$$\begin{aligned} &\|(V^*_{g(k)} ABV_{g(k)} - A_g B_g)P_n\| \\ &\quad = \|(V^*_{g(k)} AV_{g(k)} V^*_{g(k)} BV_{g(k)} - A_g B_g)P_n\| \\ &\quad \leq \|(V^*_{g(k)} AV_{g(k)} - A_g)V^*_{g(k)} BV_{g(k)} P_n\| + \|A_g(V^*_{g(k)} BV_{g(k)} - B_g)P_n\|. \end{aligned} \tag{7.2}$$

Since $B \in \mathcal{A}_0$, there is an $l \in \mathbb{N}$ such that

$$P_{l,y}BP_{n,y} = BP_{n,y} \quad \text{for all } y \in Y.$$

In particular,

$$P_l V^*_{g(k)} BV_{g(k)} P_n = V^*_{g(k)} BV_{g(k)} P_n \quad \text{for all } k \in \mathbb{N}.$$

Thus, we can estimate (7.2) by

$$\|(V^*_{g(k)} AV_{g(k)} - A_g)P_l\| \, \|V^*_{g(k)} BV_{g(k)} P_n\| + \|A_g\| \, \|(V^*_{g(k)} BV_{g(k)} - B_g)P_n\|$$

which goes to 0 as $k \to \infty$. Consequently,

$$\|(V^*_{g(k)} ABV_{g(k)} - A_g B_g)P_n\| \to 0 \quad \text{as } k \to \infty.$$

The dual condition follows similarly. This proves our claim which, on its hand, shows that AB satisfies condition (b) of Definition 7.1.4. For condition (c), let $n \in \mathbb{N}$. Then there is an $\tilde{n} \in \mathbb{N}$ such that

$$P_{n,y}AB = P_{n,y}AP_{\tilde{n},y}B \quad \text{for all } y \in Y.$$

Let further $l \in \mathbb{N}$ be such that $P_{\tilde{n},y}B = P_{\tilde{n},y}BP_{l,y}$ for all $y \in Y$. Then

$$P_{n,y}AB = P_{n,y}AP_{\tilde{n},y}BP_{l,y} = P_{n,y}ABP_{l,y} \quad \text{for all } y \in Y.$$

Similarly, one checks the dual condition of (c) for AB. Thus, $AB \in \mathcal{A}_0$. □

One easily checks that every operator $A \in \mathcal{A}$ satisfies the conditions (a) and (b) of Definition 7.1.4 (use Proposition 7.1.3 (d) to get that A satisfies (b)). But, in general, condition (c) is not satisfied for all operators in $\mathcal{A}$.

Definition 7.1.6 *Let A, $B \in L(H)$ and $B \neq 0$. The* lower norm of A relative to B *is the number*

$$\nu(A|B) := \inf\{\|ABu\| : u \in H, \|Bu\| = 1\},$$

and $\nu(A) := \nu(A|I)$ is called the lower norm of A.

It is well known that the operator A is invertible from the left if and only if $\nu(A) > 0$, and it is invertible from the right if and only if $\nu(A^*) > 0$. Thus, A is invertible if and only if both $\nu(A) > 0$ and $\nu(A^*) > 0$. Moreover, A is a Φ_+-operator if and only if there exists an operator $P \in L(H)$ such that $I - P$ is compact and $\nu(A|P) > 0$ ([108], Ch. I, Lemma 2.1). Recall that an operator $A \in L(H)$ is a Φ_+-*operator* if A has a closed range and a finite-dimensional kernel, whereas A is called a Φ_--*operator* if its adjoint A^* is a Φ_+-operator. Thus, if A is both a Φ_+ and a Φ_--operator, then A is Fredholm. $\Phi_\pm$-operators are also called semi-Fredholm operators.

The following three propositions prepare the proof of the main result of this section.

Proposition 7.1.7 *Let $h \in \mathcal{H}$, and let $A \in L(H)$ be an operator, for which the limit operator A_h exists. Then*

$$\nu(A|I - Q_m) \leq \nu(A_h) \quad \textit{for all } m \in \mathbb{N}.$$

Since $P_n \to I$ strongly, there is a P_n which is non-zero. Hence, Axiom 1 (a) implies that $Q_m \neq I$ for all m, and the relative lower norms $\nu(A|I - Q_m)$ are well defined.

Proof of Proposition 7.1.7. By the definition of the relative lower norm,

$$\|A(I - Q_m)u\| \geq \nu(A|I - Q_m)\, \|(I - Q_m)u\| \quad \text{for all } u \in H,$$

whence

$$\|V^*_{h(k)} A V_{h(k)} V^*_{h(k)} (I - Q_m)u\| \geq \nu(A|I - Q_m)\, \|V^*_{h(k)}(I - Q_m)u\|$$

for all k. We choose $u := V_{h(k)} P_n v$ with $n \in \mathbb{N}$ and $v \in H$. Then

$$\|V^*_{h(k)} A V_{h(k)} V^*_{h(k)} (I - Q_m) V_{h(k)} P_n v\| \geq \nu(A|I - Q_m)\, \|V^*_{h(k)}(I - Q_m) V_{h(k)} P_n v\|.$$

By the triangle inequality,

$$\begin{aligned} &\|V^*_{h(k)} A V_{h(k)} P_n v\| + \|V^*_{h(k)} A V_{h(k)} V^*_{h(k)} Q_m V_{h(k)} P_n v\| \\ &\qquad \geq \nu(A|I - Q_m)\, (\|P_n v\| - \|V^*_{h(k)} Q_m V_{h(k)} P_n v\|). \end{aligned} \tag{7.3}$$

From Axiom 1 (a) we infer that $Q_m V_{h(k)} P_n = Q_m P_{n,h(k)} V_{h(k)} = 0$ for all sufficiently large k. Thus, if k is large enough, then (7.3) gives

$$\|V_{h(k)}^* A V_{h(k)} P_n v\| \geq \nu(A|I - Q_m)\, \|P_n v\|.$$

Letting k go to infinity, we obtain

$$\|A_h P_n v\| \geq \nu(A|I - Q_m)\, \|P_n v\| \quad \text{for all } n \in \mathbb{N} \text{ and } v \in H,$$

and letting now n go to infinity, we finally arrive at

$$\|A_h v\| \geq \nu(A|I - Q_m)\, \|v\| \quad \text{for all } v \in H.$$

This is the assertion. □

Proposition 7.1.8 *Let $A \in L(H)$ be rich, and let $h \in \mathcal{H}$ and $n \in \mathbb{N}$. Then,*

$$\inf_{A_g \in \sigma_{op}(A)} \nu(A_g) \leq \liminf_{j \to \infty} \nu(A|P_{n,h(j)}).$$

Proof. Abbreviate $\nu_A := \liminf_{j\to\infty} \nu(A|P_{n,h(j)})$ and let $\varepsilon > 0$. We choose a subsequence g of h such that

$$\nu_A = \lim_{j\to\infty} \nu(A|P_{n,h(j)}) \tag{7.4}$$

and that the limit operator A_g exists. Further, we choose vectors $u_{g(j)} \in H$ with $\|P_{n,g(j)} u_{g(j)}\| = 1$ for all j such that

$$\nu_A = \lim_{j\to\infty} \|AP_{n,g(j)} u_{g(j)}\|,$$

and we choose $l \in \mathbb{N}$ such that $P_n = P_l P_n$. Then $\|AP_{n,g(j)} u_{g(j)}\| < \nu_A + \varepsilon$ as well as $\|(V_{g(j)}^* A U_{g(j)} - A_g) P_l\| < \varepsilon$ for all sufficiently large j, whence

$$\begin{aligned}
&\|A_g P_n V_{g(j)}^* u_{g(j)}\| \\
&\quad \leq \|(V_{g(j)}^* A U_{g(j)} - A_g) P_n V_{g(j)}^* u_{g(j)}\| + \|V_{g(j)}^* A V_{g(j)} P_n V_{g(j)}^* u_{g(j)}\| \\
&\quad = \|(V_{g(j)}^* A U_{g(j)} - A_g) P_l P_n V_{g(j)}^* u_{g(j)}\| + \|AP_{n,g(j)} u_{g(j)}\| \\
&\quad \leq \|(V_{g(j)}^* A U_{g(j)} - A_g) P_l\|\, \|V_{g(j)} P_n V_{g(j)}^* u_{g(j)}\| + \|AP_{n,g(j)} u_{g(j)}\| \\
&\quad = \|(V_{g(j)}^* A U_{g(j)} - A_g) P_l\| + \|AP_{n,g(j)} u_{g(j)}\| \nu_A + 2\varepsilon
\end{aligned}$$

for sufficiently large j. Hence,

$$\frac{\|A_g P_n V_{g(j)}^* u_{g(j)}\|}{\|P_n V_{g(j)}^* u_{g(j)}\|} = \frac{\|A_g P_n V_{g(j)}^* u_{g(j)}\|}{\|P_{n,g(j)} u_{g(j)}\|} \leq \nu_A + 2\varepsilon,$$

which gives $\nu(A_g) \leq \nu_A + 2\varepsilon$. Since $\varepsilon > 0$ is arbitrary, this implies the assertion. □

Proposition 7.1.9 *Let $A \in \mathcal{A}_0(H)$. Then*

$$\inf_{n\in\mathbb{N}} \inf_{h\in\mathcal{H}} \liminf_{j\to\infty} \nu(A|P_{n,h(j)}) \le \liminf_{m\to\infty} \nu(A|I - Q_m).$$

Proof. Abbreviate $\mu_A := \liminf_{m\to\infty} \nu(A|I - Q_m)$, and let $\varepsilon > 0$. Choose $n \in \mathbb{N}$ such that

$$\sup_{y\in Y} \|P_{n,y}A - AP_{n,y}\| \le \varepsilon \tag{7.5}$$

which is possible by Definition 7.1.4 (a). Further, let $m : \mathbb{N} \to \mathbb{N}$ be a strongly monotonically increasing sequence such that

$$\nu(A|I - Q_{m(r)}) \le \mu_A + \varepsilon \quad \text{for every } r \in \mathbb{N},$$

and choose vectors $w_{m(r)} = (I - Q_{m(r)})v_{m(r)}$ with $\|w_{m(r)}\| = 1$ and

$$\|Aw_{m(r)}\| \le \mu_A + 2\varepsilon \quad \text{for every } r \in \mathbb{N}. \tag{7.6}$$

Finally, choose $l > n$ such that

$$P_n = P_nP_l \quad \text{and} \quad P_{n,y}A = P_{n,y}AP_{l,y} \quad \text{for all } y \in Y. \tag{7.7}$$

Then, estimating by the Cauchy-Schwarz inequality, we get for all $r \in \mathbb{N}$,

$$\begin{aligned}
&\sum_{y\in Y_n} \|AP_{n,y}w_{m(r)}\|^2 \\
&\quad= \sum_{y\in Y_n} \|AP_{n,y}P_{l,y}w_{m(r)}\|^2 \\
&\quad\le \sum_{y\in Y_n} \left(\|P_{n,y}AP_{l,y}w_{m(r)}\| + \|(P_{n,y}A - AP_{n,y})P_{l,y}w_{m(r)}\|\right)^2 \\
&\quad\le \sum_{y\in Y_n} \left(\|P_{n,y}Aw_{m(r)}\| + \varepsilon\|P_{l,y}w_{m(r)}\|\right)^2 \\
&\quad= \sum_{y\in Y_n} \left(\|P_{n,y}Aw_{m(r)}\|^2 + 2\varepsilon\|P_{n,y}Aw_{m(r)}\|^2\,\|P_{l,y}w_{m(r)}\| + \varepsilon^2\|P_{l,y}w_{m(r)}\|^2\right) \\
&\quad\le \sum_{y\in Y_n} \|P_{n,y}Aw_{m(r)}\|^2 + \varepsilon^2 \sum_{y\in Y_n} \|P_{l,y}w_{m(r)}\|^2 \\
&\qquad\quad + 2\varepsilon \left(\sum_{y\in Y_n} \|P_{n,y}Aw_{m(r)}\|^2\right)^{1/2} \left(\sum_{y\in Y_n} \|P_{l,y}w_{m(r)}\|^2\right)^{1/2}.
\end{aligned}$$

Conditions (b) and (c) of Axiom 2 yield

$$\begin{aligned}
&\sum_{y\in Y_n} \|AP_{n,y}w_{m(r)}\|^2 \\
&\quad\le \|Aw_{m(r)}\|^2 + \varepsilon^2C_{n,l}^2\|w_{m(r)}\|^2 + 2\varepsilon\|Aw_{m(r)}\|\,C_{n,l}\|w_{m(r)}\| \\
&\quad\le (\mu_A + 2\varepsilon)^2 + \varepsilon^2C_{n,l}^2 + 2\varepsilon(\mu_A + 2\varepsilon)C_{n,l} \\
&\quad= (\mu_A + 2\varepsilon + \varepsilon C_{n,l})^2.
\end{aligned}$$

Taking into account the identity

$$1 = \|w_{m(r)}\|^2 = \sum_{y \in Y_n} \|P_{n,y} w_{m(r)}\|^2,$$

we conclude that

$$\frac{\sum_{y \in Y_n} \|AP_{n,y} w_{m(r)}\|^2}{\sum_{y \in Y_n} \|P_{n,y} w_{m(r)}\|^2} \le (\mu_A + (2 + C_{n,l})\varepsilon)^2.$$

In particular, for every $r \in \mathbb{N}$, there is an $y(r) \in Y_n$ such that $P_{n,y(r)} w_{m(r)} \neq 0$ and

$$\frac{\|AP_{n,y(r)} w_{m(r)}\|}{\|P_{n,y(r)} w_{m(r)}\|} \le \mu_A + (2 + C_{n,l})\varepsilon, \tag{7.8}$$

and we have

$$P_{n,y(r)}(I - Q_{m(r)}) \neq 0 \quad \text{for all } r \in \mathbb{N}.$$

Consider the sequence $y : \mathbb{N} \to Y$, $r \mapsto y(r)$. Suppose for a moment that this sequence takes only finitely many values. Then there is one value, y^* say, such that

$$P_{n,y^*}(I - Q_{m(r)}) \neq 0$$

for infinitely many $r \in \mathbb{N}$. This contradicts Axiom 1 (b). Thus, the sequence y has an infinite range. In this case, we can choose a sequence $h \in \mathcal{H}$ (i.e., an injective mapping from $\mathbb{N}$ into Y) as well as a sequence $j : \mathbb{N} \to \mathbb{N}$ such that

$$\frac{\|AP_{n,h(k)} w_{j(k)}\|}{\|P_{n,h(k)} w_{j(k)}\|} \le \mu_A + (2 + C_{n,l})\varepsilon \quad \text{for all } k \in \mathbb{N}.$$

(To be precise: we choose a strongly monotonically increasing sequence $r : \mathbb{N} \to \mathbb{N}$ such that $y \circ r$ is an injective mapping from $\mathbb{N}$ onto the range of y and set $h := y \circ r$ and $j := m \circ r$.) Consequently,

$$\nu(A|P_{n,h(k)}) \le \mu_A + (2 + C_{n,l})\varepsilon \quad \text{for all } k \in \mathbb{N}.$$

This finally shows that

$$\inf_{n \in \mathbb{N}} \inf_{h \in \mathcal{H}} \liminf_{k \to \infty} \nu(A|P_{n,h(k)}) \le \mu_A + (2 + C_{n,l})\varepsilon.$$

Since $\varepsilon > 0$ is arbitrary, this is the assertion. □

Theorem 7.1.10 *Let Axioms* 1 *and* 2 *be satisfied. Then, for every* $A \in \mathcal{A}$,

$$\liminf_{m \to \infty} \nu(A|I - Q_m) = \inf_{A_g \in \sigma_{op}(A)} \nu(A_g). \tag{7.9}$$

Proof. First let $A \in \mathcal{A}_0$. From Proposition 7.1.7 we infer that

$$\liminf_{m\to\infty} \nu(A|I-Q_m) \le \inf_{A_g \in \sigma_{op}(A)} \nu(A_g),$$

and in Proposition 7.1.9 we have just proved that

$$\inf_{n\in\mathbb{N}} \inf_{h\in\mathcal{H}} \liminf_{j\to\infty} \nu(A|P_{n,h(j)}) \le \liminf_{m\to\infty} \nu(A|I-Q_m).$$

Finally, we conclude from Proposition 7.1.8 that

$$\inf_{A_g \in \sigma_{op}(A)} \nu(A_g) \le \inf_{n\in\mathbb{N}} \inf_{h\in\mathcal{H}} \liminf_{j\to\infty} \nu(A|P_{n,h(j)}).$$

These inequalities yield the assertion (7.9) in case $A \in \mathcal{A}_0$. Let now $A \in \mathcal{A}$, and abbreviate the left-hand side of (7.9) by S_l and the right-hand side by S_r. Let $\varepsilon > 0$ be arbitrarily given. Then there is an m_0 such that

$$\nu(A|I-Q_m) \ge S_l - \varepsilon \quad \text{for all } m \ge m_0.$$

Thus,

$$\|A(I-Q_m)u\| \ge (S_l-\varepsilon)\,\|(I-Q_m)u\| \quad \text{for all } m \ge m_0 \text{ and all } u \in H.$$

Choose $A' \in \mathcal{A}_0$ such that $\|A - A'\| < \varepsilon$. Then

$$\begin{aligned}&\|A'(I-Q_m)u\| \\ &\qquad \ge \|A(I-Q_m)u\| - \|A-A'\|\,\|(I-Q_m)u\| \ge (S_l-2\varepsilon)\,\|(I-Q_m)u\|\end{aligned}$$

for all $m \ge m_0$ and $u \in H$. Consequently, $\liminf_{m\to\infty} \nu(A'|I-Q_m) \ge S_l - 2\varepsilon$. By what has already been proved, this implies that

$$\inf_{A'_g \in \sigma_{op}(A')} \nu(A'_g) \ge S_l - 2\varepsilon$$

or, equivalently,

$$\|A'_g u\| \ge (S_l - 2\varepsilon)\|u\| \tag{7.10}$$

for all limit operators A'_g of A' and for all $u \in H$. If the limit operator of A with respect to the sequence h exists, then there is a subsequence g of h such that the limit operator of A' with respect to g exists, and by Proposition 7.1.3 (a) one has

$$\|A'_g - A_h\| = \|A'_g - A_g\| \le \|A' - A\| \le \varepsilon.$$

Thus, (7.10) implies

$$\|A_h u\| \ge \|A'_g u\| - \|A'_g - A_h\|\,\|u\| \ge (S_l - 3\varepsilon)\|u\|$$

for all $u \in H$, whence $\nu(A_h) \ge S_l - 3\varepsilon$. Consequently, $S_r \ge S_l$, and the reverse inequality can be proved in the same way. □

We will see now how Theorem 7.1.10 applies to the study of Fredholmness of operators in the algebra $\mathcal{A}$. For this, we have to assume the following.

Axiom 3 *The operators Q_m are self-adjoint projections which converge strongly to the identity operator on H, and $Q_m Q_{m+1} = Q_{m+1} Q_m = Q_m$ for all $m \in \mathbb{N}$.*

Under this assumption, the sequence of the ranges of $I - Q_m$ is monotonically decreasing,

$$(I - Q_m)(I - Q_{m+1}) = I - Q_m - Q_{m+1} + Q_m Q_{m+1} = I - Q_{m+1}.$$

Thus, the sequence of the relative lower norms $\nu(A|I - Q_m)$ is monotonically increasing, and the limes inferior on the left-hand side of (7.9) is actually a limes (and a supremum).

Definition 7.1.11 *An operator $A \in \mathcal{A}$ is* locally compact *if the operators AQ_m and $Q_m A$ are compact for every $m \in \mathbb{N}$. The set of all locally compact operators in $\mathcal{A}$ will be denoted by $\mathcal{K}$.*

It is easy to see that $\mathcal{K}$ is a C^*-subalgebra of $\mathcal{A}$.

Now we will have to distinguish between two cases. In the first case we assume that the identity operator is locally compact. In this case, every projection Q_m is compact, and every operator in $\mathcal{A}$ is locally compact.

Theorem 7.1.12 *Let Axioms 1–3 be satisfied, and assume that the identity operator is locally compact. Then an operator $A \in \mathcal{A}$ is*

(a) *a Φ_+-operator if and only if*

$$\inf_{A_g \in \sigma_{op}(A)} \nu(A_g) > 0. \tag{7.11}$$

(b) *a Φ_--operator if and only if*

$$\inf_{A_g \in \sigma_{op}(A)} \nu(A_g^*) > 0. \tag{7.12}$$

(c) *a Fredholm operator if and only if all limit operators of A are invertible and if the norms of their inverses are uniformly bounded.*

Proof. We will prove assertion (a) only. Assertion (b) follows by passing to the adjoint operator, and (c) is a consequence of (a) and (b).

Let condition (7.11) be satisfied. Then, by Theorem 7.1.10, there are an $m \in \mathbb{N}$ and a constant $C > 0$ such that $\nu(A|I - Q_m) \geq C$. Consequently,

$$\|A(I - Q_m)u\| \geq C\|(I - Q_m)u\| \quad \text{for all } u \in H$$

or, equivalently,

$$\langle (I - Q_m)A^*A(I - Q_m)u, (I - Q_m)u\rangle \geq C^2\|(I - Q_m)u\|^2 \quad \text{for all } u \in H.$$

Thus, the self-adjoint operator

$$(I - Q_m)A^*A(I - Q_m) : \operatorname{Im}(I - Q_m) \to \operatorname{Im}(I - Q_m) \tag{7.13}$$

has a trivial kernel and a closed range and is, consequently, invertible. We let $B \in L(\operatorname{Im}(I - Q_m))$ be the inverse of the operator (7.13) and extend B to the operator $C := B(I - Q_m) + Q_m$ acting on all of H. Then,

$$\begin{aligned} CA^*A &= B(I - Q_m)A^*A + Q_mA^*A \\ &= B(I - Q_m)A^*A(I - Q_m) + B(I - Q_m)A^*AQ_m + Q_mA^*A \\ &= I - Q_m + B(I - Q_m)A^*AQ_m + Q_mA^*A = I + K \end{aligned}$$

with a compact operator K. Hence, A is a Φ_+-operator.

Let, conversely, A be a Φ_+-operator. Then there are a compact operator K and a positive constant C such that

$$C\|Au\| \geq \|u\| - \|Ku\| \quad \text{for all } u \in H.$$

In particular,

$$C\|A(I - Q_m)u\| \geq \|(I - Q_m)u\| - \|K(I - Q_m)u\|$$

for all $u \in H$ and $m \in \mathbb{N}$. Due to the strong convergence of the $I - Q_m$ to zero, there is an m_0 such that $\|K(I - Q_m)\| \leq 1/2$ for all $m \geq m_0$. Thus,

$$C\|A(I - Q_m)u\| \geq \|(I - Q_m)u\| - \frac{1}{2}\|(I - Q_m)u\|,$$

whence

$$\|A(I - Q_m)u\| \geq \frac{1}{2C}\|(I - Q_m)u\|$$

for all $u \in H$ and $m \geq m_0$. This implies that

$$\liminf_{m\to\infty} \nu(A|I - Q_m) \geq 1/(2C),$$

and the assertion follows via Theorem 7.1.10. □

Now we assume that the identity operator is not locally compact. In this case, the set

$$\mathcal{B} := \{\gamma I + K : \gamma \in \mathbb{C},\, K \in \mathcal{K}\}$$

is a unital C^*-subalgebra of $\mathcal{A}$ which has $\mathcal{K}$ as its proper ideal. Observe also that the representation of an operator $A \in \mathcal{B}$ in the form $\gamma I + K$ with $\gamma \in \mathbb{C}$ and $K \in \mathcal{K}$ is unique. Moreover, only a finite number of projections Q_m can be compact. Indeed, the existence of an infinite sequence of compact projections Q_m would imply that all projections Q_m are compact and, hence, the identity I would be locally compact.

Theorem 7.1.13 *Let Axioms 1–3 be satisfied and assume that the identity operator is not locally compact, and let* $A = \gamma I + K \in \mathcal{B}$ *with* $\gamma \in \mathbb{C} \setminus \{0\}$ *and* $K \in \mathcal{K}$. *Then* A *is*

(a) *a* Φ_+*-operator if and only if*

$$\inf_{A_g \in \sigma_{op}(A)} \nu(A_g) > 0. \tag{7.14}$$

(b) *a* Φ_-*-operator if and only if*

$$\inf_{A_g \in \sigma_{op}(A)} \nu(A_g^*) > 0. \tag{7.15}$$

(c) *a Fredholm operator if and only if all limit operators of* A *are invertible and if the norms of their inverses are uniformly bounded.*

Proof. Again, we will prove assertion (a) only. It follows as in the proof of Theorem 7.1.12 that if A is a Φ_+-operator, then (7.14) holds. Let, conversely, (7.14) be satisfied. Then we get, as in the proof of Theorem 7.1.12, that the operator

$$(I - Q_m)A^*A(I - Q_m) : \text{Im}\,(I - Q_m) \to \text{Im}\,(I - Q_m)$$

is invertible for all sufficiently large m. We choose m large enough such that the operator Q_m is not compact (recall that only a finite number of the Q_m can be compact). Clearly, the operator

$$(I - Q_m)A^*A(I - Q_m) + Q_m : H \to H$$

is invertible, too, and its inverse C belongs to the smallest C^*-subalgebra $\mathcal{C}$ of $L(H)$ which contains the algebra $\mathcal{B}$ and the operator Q_m (due to the inverse closedness of C^*-algebras). We set $R' := CA^*$ and obtain

$$\begin{aligned} R'A - I &= CA^*A - I \\ &= C(I - Q_m)A^*A(I - Q_m) + CQ_m \\ &\quad + C(I - Q_m)A^*AQ_m + CQ_mA^*A - CQ_m - I \\ &= C(I - Q_m)A^*AQ_m + CQ_mA^*A - CQ_m. \end{aligned}$$

Thus, the operator $R'A - I$ belongs to the smallest closed ideal $\mathcal{J}$ of $\mathcal{C}$ which contains the operator Q_m. We further set $R := \gamma R' - AR' + I$ and find

$$RA - \gamma I = \gamma R'A - AR'A + A - \gamma I = (\gamma I - A)\,(R'A - I) \in \mathcal{K}\mathcal{J}.$$

A little thought shows that

$$\mathcal{C} \subseteq \mathbb{C}I + \mathbb{C}Q_m + \mathcal{K} + K(H) \quad \text{and} \quad \mathcal{J} \subseteq \mathbb{C}Q_m + K(H)$$

and, thus, $\mathcal{K}\mathcal{J} \subseteq K(H)$. Hence, $RA - \gamma I$ is compact. Since $\gamma \neq 0$ by hypothesis, A is a Φ_+-operator. □

The unpleasant hypothesis $\gamma \neq 0$ in Theorem 7.1.13 can be avoided if also the differences of the projections Q_m are non-compact.

Corollary 7.1.14 *Let Axioms* 1–3 *be satisfied and assume that the identity operator is not locally compact. Furthermore, assume that, for every $m \in \mathbb{N}$, there is an $l > m$ such that $Q_l - Q_m$ is not compact. Then the assertion of Theorem* 7.1.13 *holds for every operator $A \in \mathcal{B}$.*

Proof. In view of the proof of Theorem 7.1.13, we only have to check that $\gamma \neq 0$ if $A = \gamma I + K \in \mathcal{B}$ is subject to condition (7.14). Contrary to what we want, assume that $\gamma = 0$, whence $A \in \mathcal{K}$. The operator R' introduced in the proof of Theorem 7.1.13 belongs to $\mathcal{C} \subseteq \mathbb{C}I + \mathbb{C}Q_m + \mathcal{K} + K(H)$, which implies that $R'A \in \mathcal{C}\mathcal{K} \subseteq \mathcal{K} + K(H)$. Further, as we have seen above, $R'A - I \in \mathcal{J} \subseteq \mathbb{C}Q_m + K(H)$. Consequently, $I \in \mathbb{C}Q_m + \mathcal{K} + K(H)$.

Let $\alpha \in \mathbb{C}$, $K \in \mathcal{K}$ and $L \in K(H)$ be such that $I = \alpha Q_m + K + L$. Multiplying this equality by $Q_l - Q_m$ we get

$$\begin{aligned} Q_l - Q_m &= \alpha(Q_l - Q_m)Q_m + (Q_l - Q_m)K + (Q_l - Q_m)L \\ &= (Q_l - Q_m)K + (Q_l - Q_m)L \in K(H) \end{aligned}$$

for all $l > m$. This contradicts the hypothesis. □

We will finally extend these results to a more general class of operators of the form $M + K$ where $K \in \mathcal{K}$ and where M is a "completely not locally compact" operator. To do this, we have slightly to restrict the class of operators under consideration.

Definition 7.1.15 *Let $\mathcal{A}_{00}$ denote the class of all operators $A \in \mathcal{A}_0$ with the following property: for every $m \in \mathbb{N}$, there is an $l > m$ such that*

$$Q_m A = Q_m A Q_l \quad \textit{and} \quad A Q_m = Q_l A Q_m.$$

Let further $\mathcal{A}'$ stand for the smallest closed subalgebra of $\mathcal{A}$ which contains the set $\mathcal{A}_{00}$ and the algebra $\mathcal{K}$.

Proposition 7.1.16 *$\mathcal{A}'$ is a C^*-subalgebra of $L(H)$ which contains the identity operator, and $\mathcal{K}$ is a proper closed ideal of $\mathcal{A}'$.*

Proof. Let $K \in \mathcal{K}$ and $A \in \mathcal{A}_{00}$. We show that then AK and KA belong to $\mathcal{K}$. Indeed, $Q_m KA$ is compact for every m. If we choose $l > m$ such that $Q_m A = Q_m A Q_l$, then we also get the compactness of $Q_m AK = Q_m A Q_l K$. Thus, $AK \in \mathcal{K}$, and the inclusion $KA \in \mathcal{K}$ follows similarly. Now it is clear that $\mathcal{A}_{00} + \mathcal{K}$ is a symmetric algebra, which has $\mathcal{K}$ as its proper ideal. □

Let now $\mathcal{M}$ be a C^*-subalgebra of $\mathcal{A}'$ which contains the identity operator as well as all projections Q_m, and which satisfies the condition

$$\mathcal{M} \cap (\mathcal{K} \cap K(H)) = \{0\}. \tag{7.16}$$

Since $\mathcal{K}$ is a closed ideal of $\mathcal{A}'$, the set $\mathcal{C} := \mathcal{M} + \mathcal{K}$ is a C^*-subalgebra of $\mathcal{A}'$, and condition (7.16) guarantees that every operator $A \in \mathcal{C}$ can be uniquely written as $A = M + K$ with $M \in \mathcal{M}$ and $K \in \mathcal{K}$. Indeed, if $M \in \mathcal{M}$ belongs to $\mathcal{K}$, then $MQ_m \in \mathcal{K}Q_m \subseteq \mathcal{K} \cap K(H)$. Hence, $MQ_m = 0$. Letting m go to infinity, we get $M = 0$.

Theorem 7.1.17 *Let the Axioms 1–3 be satisfied, let $\mathcal{M}$ and $\mathcal{K}$ be as above, and let $A = M + K \in \mathcal{C}$ with $M \in \mathcal{M}$ and $K \in \mathcal{K}$. Then A is a Fredholm operator if and only if the operator M is invertible and if all limit operators of A are uniformly invertible.*

Proof. Let A be a Fredholm operator. Then, by inverse closedness of C^*-algebras, A has a regularizer R which belongs to $\mathcal{C} + K(H)$. We write $A = M + K$ and $R = N + K' + L$ where $M, N \in \mathcal{M}$, $K, K' \in \mathcal{K}$ and $L \in K(H)$. From $RA - I \in K(H)$ we conclude that $NM - I \in \mathcal{K} + K(H)$. Hence, for every $m \in \mathbb{N}$, we have $NMQ_m - Q_m \in \mathcal{K} \cap K(H)$. Since $NMQ_m - Q_m \in \mathcal{M}$, this implies via (7.16) that $NMQ_m - Q_m = 0$. Letting m go to infinity, we obtain $NM = I$. Analogously, $MN = I$, whence the invertibility of M. Further, it can be shown as in the proof of Theorem 7.1.12, that the Fredholm property of A implies the uniform invertibility of the set of the limit operators of A.

For the converse direction, we are going to show that if M is invertible from the left-hand side and if

$$\inf_{A_g \in \sigma_{op}(A)} \nu(A_g) > 0, \tag{7.17}$$

then A is a Φ_+-operator. The corresponding result with respect to invertibility from the right-hand side and to the Φ_--property follows by taking adjoints, and these two results combine to give the assertion.

Let (7.17) be satisfied. As in the proof of Theorem 7.1.13, we conclude that the operator

$$(I - Q_m)A^*A(I - Q_m) + Q_m : H \to H$$

is invertible and that its inverse C belongs to the algebra $\mathcal{C}$. We set $R' := CA^*$ and obtain that the operator $R'A - I$ belongs to the smallest closed ideal $\mathcal{J}$ of $\mathcal{C}$ which contains all operators Q_m. Further, for $A = M + K$ with $M \in \mathcal{M}$ and $K \in \mathcal{K}$, we set $R := KR' + I$ and find

$$RA - M = KR'A + A - M = K(R'A - I) \in \mathcal{K}\mathcal{J}.$$

Since $\mathcal{K}$ is an ideal in $\mathcal{A}$, we have $\mathcal{K}\mathcal{J} \subseteq K(H)$. Thus, $RA - M \in K(H)$. Multiplication from the left-hand side by a left inverse M^{-1} of M yields

$$M^{-1}RA - M^{-1}M = M^{-1}RA - I \in K(H).$$

Thus, A is a Φ_+-operator. □

We are going to illustrate these results by a few simple examples where $H = L^2(\mathbb{R}^N)$. Let $f : \mathbb{R}^N \to [0, 1]$ be a continuous function with

$$f(x) \begin{cases} = 1 & \text{if} \quad |x|_\infty \le 1/3 \\ > 0 & \text{if} \quad |x|_\infty < 2/3 \\ = 0 & \text{if} \quad |x|_\infty \ge 2/3, \end{cases}$$

and let $\varphi(x) \geq 0$ be defined by

$$\varphi(x)^2 := \frac{f(x)}{\sum_{k\in\mathbb{Z}^N} f(x-k)}.$$

Further, let P_n denote the operator of multiplication by the function $\varphi_n : x \mapsto \varphi(x/n)$, write Q_m for the operator of multiplication by the characteristic function of the cube $[-m, m]^N$, and let V_y refer to the shift operator on H given by $(V_y f)(x) = f(x-y)$. Since φ_n is identically 1 on the cube $[-n/3, n/3]^N$, Axioms 1, 2 (a) and 3 are satisfied with $Y := \mathbb{Z}^N$.

We will check now that also Axiom 2 (b) and (c) is satisfied with $Y_n := n\mathbb{Z}^N$. Let $u \in L^2(\mathbb{R}^N)$ have a compact support. Then

$$\begin{aligned}
\sum_{y\in Y_n} \|P_{n,y}a\|_2^2 &= \sum_{y\in Y_n} \int_{\mathbb{R}^N} \left|\varphi\left(\frac{x-y}{n}\right) u(x)\right|^2 dx \\
&= \sum_{y\in\mathbb{Z}^N} \int_{\mathbb{R}^N} \left|\varphi\left(\frac{x-nz}{n}\right) u(x)\right|^2 dx \\
&= \int_{\mathbb{R}^N} \sum_{y\in\mathbb{Z}^N} \left|\varphi\left(\frac{x}{n} - z\right)\right|^2 |u(x)|^2 dx \\
&= \int_{\mathbb{R}^N} |u(x)|^2 dx \;=\; \|u\|_2^2
\end{aligned}$$

where the change of summation and integration is justified by the compactness of the support of u, which makes the sum finite. This proves Axiom 2 (b). Let now $l > n$. We choose a finite subset L of Y such that

$$\sum_{z\in L} P_{l,0} P_{n,z} = P_{l,0}.$$

Then we also have

$$\sum_{z\in L+y} P_{l,y} P_{n,z} = P_{l,y} \quad \text{for every } y \in Y.$$

Using the Cauchy-Schwarz inequality, we get for $u \in L^2(\mathbb{R}^N)$

$$\begin{aligned}
\sum_{y\in Y_n} \|P_{l,y}u\|_2^2 &= \sum_{y\in Y_n} \Big\| \sum_{z\in L+y} P_{l,y}P_{n,z}u\Big\|_2^2 \\
&\leq |L| \sum_{y\in Y_n} \sum_{z\in L+y} \|P_{l,y}P_{n,z}u\|_2^2 \\
&\leq |L| \sum_{y\in Y_n} \sum_{z\in L+y} \|P_{n,z}u\|_2^2 \\
&\leq |L|^2 \sum_{y\in Y_n} \|P_{n,y}u\|_2^2 \;=\; |L|^2 \|u\|_2^2
\end{aligned}$$

since each item in the last double sum occurs $|L|$ times. This settles the second axiom.

Next we are going to describe some classes of operators which belong to the algebra $\mathcal{A}_0$ corresponding to the chosen specifications of H, P_n and V_y. If $A = aI$ is the operator of multiplication by a BUC function on $\mathbb{R}^N$, then A evidently satisfies conditions (a) and (c) of Definition 7.1.4 (since A commutes with the operators P_n). By Proposition 3.3.6, A also satisfies condition (b).

Let now k be a function in $L^1(\mathbb{R}^N)$ with compact support and consider the operator $A = C(k)$ of convolution by k. Then condition (a) of Definition 7.1.4 follows from Proposition 3.3.2, and condition (b) is evident since the operator A is shift invariant. For condition (c), it is sufficient to show that, for every $n \in \mathbb{N}$, there is an $l \in \mathbb{N}$ such that $AP_n = P_l AP_n$ (again due to the shift invariance of A and by the self-adjointness of the P_n). Let, for definiteness, the supports of φ_n and k be contained in the balls with center 0 and with radii r and R, respectively. Then, for every function $u \in L^2(\mathbb{R}^N)$, the support of $AP_n u$ is contained in the ball with center 0 and radius $r + R$. Now choose l such that φ_l is identically 1 on that ball. Then $P_l AP_n = AP_n$ as desired.

A third class of operators in $\mathcal{A}_0$ is provided by the shift operators $A = V_\alpha$ with $\alpha \in \mathbb{Z}^N$. Here, the conditions (b) and (c) of Definition 7.1.4 are evident, and condition (a) follows easily from the shift invariance of V_α and the uniform continuity of φ. Moreover, all three classes of operators belong to the subalgebra $\mathcal{A}_{00}$ of $\mathcal{A}_0$ introduced in Definition 7.1.15. This can be checked in a similar way as condition (c) of Definition 7.1.4.

In accordance with the above notations, we introduce the algebras $\mathcal{A}$ and $\mathcal{A}'$ as well as the ideal $\mathcal{K}$ of $\mathcal{A}'$. Note that the algebra $\mathcal{A}'$ contains all convolutions by functions in $L^1(\mathbb{R}^N)$ (since the compactly supported functions are dense in $L^1(\mathbb{R}^N)$), whereas $\mathcal{K}$ contains, for example, all products of operators in the above-mentioned three classes provided that at least one of the factors is a convolution operator by an $L^1(\mathbb{R}^N)$-function.

We let further $\mathcal{M}$ stand for the smallest closed subalgebra of $L(H)$ which contains all multiplication operators aI with $a \in BUC(\mathbb{R}^N)$, all projections Q_n with $n \in \mathbb{N}$ and all shifts V_α with $\alpha \in \mathbb{Z}^N$.

Proposition 7.1.18 *The only compact operator in $\mathcal{M}$ is the zero operator. In particular, the algebra $\mathcal{M}$ satisfies condition* (7.16).

Proof. Let Γ be the discretization operator introduced in Section 3.1.3. Thus, Γ is a *-isomorphism from $L(L^2(\mathbb{R}^N))$ onto $L(l^2(\mathbb{Z}^N, L^2(I_0)))$ where $I_0 = [0, 1]^N$, and we can think of $\Gamma(A)$ for an operator A on $L^2(\mathbb{R}^N)$ as the infinite matrix $(A_{ij})_{i,j\in\mathbb{Z}^N}$ where

$$A_{ij} = \chi_{I_0} V_{-i} A V_j \chi_{I_0} I \quad \text{for } i, j \in \mathbb{Z}^N.$$

If, in particular, A belongs to $\mathcal{M}$, then one easily checks that each of the operators A_{ij} is the operator of multiplication by a function a_{ij} in $C(I_0)$. Hence, if $A \in \mathcal{M}$

is compact, then each of the operators A_{ij} is a compact multiplication operator, whence $a_{ij} = 0$. Since Γ is an isomorphism, this implies that $A = 0$. □

Thus, the operators in $\mathcal{M} + \mathcal{K}$ are subject to Theorem 7.1.17. That is, an operator $A = M + K$ with $M \in \mathcal{M}$ and $K \in \mathcal{K}$ is a Fredholm operator if and only if the operator M is invertible and if all limit operators of A are uniformly invertible.

The invertibility of an operator $A \in \mathcal{M}$ can be checked by means of the following observation, where we let the functions a_{ij} be as in the proof of Proposition 7.1.18. For each $t \in I_0$, we consider the operator $\gamma_A(t)$ on $l^2(\mathbb{Z}^N)$ which is given by its matrix representation $(a_{ij}(t))_{i,j \in \mathbb{Z}^N}$. It turns out that the functions a_{ij}, $i, j \in \mathbb{Z}^N$, are equicontinuous, which is mainly a consequence of the uniform continuity of the functions in BUC. Hence, $\gamma_A : I_0 \to L(l^2(\mathbb{Z}^N))$ is a continuous function. This continuity implies that the operator $A \in \mathcal{M}$ is invertible if and only if its discretization $\Gamma(A)$ is invertible if and only if the operators $\gamma_A(t)$ are invertible for every $t \in I_0$.

7.2 Operators on homogeneous groups

Here we collect some basic facts on homogeneous groups, and we introduce the basic classes of operators which will be considered in the forthcoming section: multiplication, convolution, and shift operators on a homogeneous group.

7.2.1 Homogeneous groups

A homogeneous group $\mathbb{X}$ arises by equipping $\mathbb{R}^N$ with a Lie group structure and with a family of dilations that act as group automorphisms on $\mathbb{X}$. More precise, we are given smooth operations of multiplication and inversion

$$\mathbb{R}^N \times \mathbb{R}^N \to \mathbb{R}^N : (x, y) \mapsto x \cdot y \quad \text{and} \quad \mathbb{R}^N \to \mathbb{R}^N : x \mapsto x^{-1}$$

which provide $\mathbb{R}^N$ with a Lie group structure and which are such that the origin of $\mathbb{R}^N$ is the identity element of the associated Lie group. Furthermore, there are positive integers $a_1 \leq \cdots \leq a_N$ (with the monotonicity being no essential restriction) such that the dilations

$$D_\delta : \mathbb{R}^N \to \mathbb{R}^N, \quad (x_1, \ldots, x_N) \mapsto (\delta^{a_1} x_1, \ldots, \delta^{a_N} x_N)$$

provide group automorphisms for every $\delta > 0$. In particular,

$$D_\delta(x \cdot y) = D_\delta x \cdot D_\delta y \quad \text{for all } x, y \in \mathbb{X}.$$

These assumptions imply that the multiplication on $\mathbb{X}$ is necessarily of the form

$$x \cdot y = x + y + Q(x, y), \quad x, y \in \mathbb{X},$$

where $Q : \mathbb{R}^N \times \mathbb{R}^N \to \mathbb{R}^N$ satisfies

$$Q(0, 0) = Q(x, 0) = Q(0, x) = 0.$$

Moreover, if one writes $Q = (Q_1, \ldots, Q_N)$, then each Q_k is a polynomial in $2N$ real variables which is homogeneous of degree a_k. Thus, Q contains no pure monomials in x or y.

Since the Euklidean measure dx on $\mathbb{R}^N$ is both left and right invariant with respect to the group multiplication, it is the Haar measure on $\mathbb{X}$. Note also that $d(D_\delta x) = \delta^a dx$, where $a := a_1 + \cdots + a_N$.

There is a natural norm function ρ on $\mathbb{X}$ defined by

$$\rho(x) := \max\{|x_j|^{1/a_j} : 1 \leq j \leq N\}.$$

Note that $\rho(x) \geq 0$ and $\rho(x) = 0$ if and only if $x = 0$. Also, $\rho(D_\delta x) = \delta\rho(x)$ for $\delta > 0$, and there is a constant $C > 0$ such that

$$\rho(x \cdot y) \leq C\,(\rho(x) + \rho(y)) \quad \text{and} \quad \rho(x^{-1}) \leq C\,\rho(x).$$

Set $\rho(x, y) := \rho(x^{-1} \cdot y)$. Then the collection of all balls

$$B(x, \varepsilon) := \{y \in \mathbb{X} : \rho(x, y) < \varepsilon\}, \quad \varepsilon > 0,$$

forms an open neighborhood base of the point $x \in \mathbb{X}$. Since $\rho : \mathbb{X} \times \mathbb{X} \to \mathbb{R}$ is invariant with respect to multiplication from the left-hand side, one also has $B(x, \varepsilon) = x \cdot B(0, \varepsilon)$, and since the Haar measure is left invariant,

$$|B(x, \varepsilon)| = |B(0, \varepsilon)| = \varepsilon^a |B(0, 1)|.$$

An archetypal example of a non-commutative homogeneous group is the Heisenberg group $\mathbb{H}^{2n+1}$ which can be identified with $\mathbb{C}^n \times \mathbb{R}$, provided with the group operation

$$(w, s) \cdot (z, t) = (w + z, s + t + 2\,\mathrm{Im}\,\langle w, z\rangle)$$

where $\langle w, z\rangle := \sum_{j=1}^n z_j \overline{w}_j$. In this case, $(a_1, \ldots, a_{2n+1}) \in \mathbb{N}^{2n+1}$ is given by $(1, 1, \ldots, 1, 2)$.

7.2.2 Multiplication operators

Throughout what follows, let $\mathbb{X}$ be a homogeneous group. By $C_b(\mathbb{X})$ we denote the C^*-algebra of all bounded continuous functions on $\mathbb{X}$ with norm $\|f\|_\infty := \sup_{x\in\mathbb{X}} |f(x)|$, and we let $BUC(\mathbb{X})$ stand for the C^*-subalgebra of $C_b(\mathbb{X})$ which consists of the uniformly continuous functions, i.e., $a \in C_b(\mathbb{X})$ belongs to $BUC(\mathbb{X})$ if, for each $\varepsilon > 0$, there is a $\delta > 0$ such that $|a(x_1) - a(x_2)| < \varepsilon$ whenever $\rho(x_1, x_2) < \delta$. Further, we let $Q_{SC}(\mathbb{X})$ refer to set of all functions $a \in L^\infty(\mathbb{X})$ such that

$$\limsup_{y\to\infty} \int_M |a(y^{-1} \cdot x)|\, dx = 0$$

for every compact $M \subset \mathbb{X}$, and we write $Q_C(\mathbb{X})$ for the set of all functions $a \in L^\infty(\mathbb{X})$ such that, for each open and pre-compact subset D of $\mathbb{X}$, the function

$$t \mapsto \int_D (a(t) - a(t \cdot s))\, ds$$

belongs to $Q_{SC}(\mathbb{X})$. Finally, a function $a \in C_b(\mathbb{X})$ belongs to the class $SO(\mathbb{X})$ of the *slowly oscillating* functions if

$$\lim_{t\to\infty} \sup_{h\in M} |a(t) - a(t \cdot h)| = 0$$

for every compact subset M of $\mathbb{X}$.

One can check as in Section 3.2 that $BUC(\mathbb{X})$ and $Q_C(\mathbb{X})$ are C^*-subalgebras of $L^\infty(\mathbb{X})$, that $SO(\mathbb{X})$ is a C^*-subalgebra of $BUC(\mathbb{X})$, and that $Q_{SC}(\mathbb{X})$ is a closed ideal in $L^\infty(\mathbb{X})$. Furthermore, $Q_C(\mathbb{X}) = Q_{SC}(\mathbb{X}) + SO(\mathbb{X})$.

7.2.3 Partition of unity

Let $\mathbb{Y}$ be a discrete subgroup of the group $\mathbb{X}$ which acts freely on $\mathbb{X}$ such that $\mathbb{X}/\mathbb{Y}$ is a compact manifold. Let M be a fundamental domain of $\mathbb{X}$ with respect to the action of $\mathbb{Y}$ on $\mathbb{X}$ by left multiplication, i.e., M is a bounded domain in $\mathbb{X}$ such that

$$\mathbb{X} = \bigcup_{\alpha\in\mathbb{Y}} \alpha \cdot \overline{M}.$$

Let M' be an open neighborhood of $\overline{M}$ such that the family $\{\alpha \cdot M'\}_{\alpha\in\mathbb{Y}}$ provides a covering of $\mathbb{X}$ of finite multiplicity, i.e., there is a number l such that each point of $\mathbb{X}$ belongs to at most l of the sets $\alpha \cdot M'$. Let $f : \mathbb{X} \to [0, 1]$ be a continuous function with $f(x) = 1$ on $\overline{M}$ and $f(x) = 0$ outside M', and let φ be the non-negative function on $\mathbb{X}$ which satisfies

$$\varphi^2(x) := \frac{f(x)}{\sum_{\beta\in\mathbb{Y}} f(\beta \cdot x)}. \tag{7.18}$$

Evidently, $0 \le \varphi(x) \le 1$, and if M' is chosen sufficiently small, then φ is identically 1 on a certain open subset of $\mathbb{X}$.

For every positive integer n and every $y \in \mathbb{Y}$, we set $\varphi_n(x) := \varphi(D_{1/n}x)$ and $\varphi_{n,y}(x) := \varphi_n(y \cdot x)$. If we further let $\mathbb{Y}_n := D_n\mathbb{Y}$, then

$$\sum_{y\in\mathbb{Y}_n} \varphi^2_{n,y}(x) = 1 \quad \text{for every } x \in \mathbb{X}. \tag{7.19}$$

Indeed, for $x \in \mathbb{X}$,

$$\begin{aligned}\sum_{y\in\mathbb{Y}_n} \varphi^2_{n,y}(x) &= \sum_{y\in\mathbb{Y}_n} \varphi^2(D_{1/n}(y \cdot x)) \\ &= \sum_{z\in\mathbb{Y}} \varphi^2(D_{1/n}(D_n z \cdot x)) = \sum_{z\in\mathbb{Y}} \varphi^2(z \cdot D_{1/n}x),\end{aligned}$$

and the assertion follows from the definition of φ (the denominator on the right-hand side of (7.18) is shift invariant). In that sense, the family $\{\varphi_{n,y}\}_{y\in\mathbb{Y}}$ forms a partition of unity on $\mathbb{X}$.

7.2.4 Convolution operators

Let $k \in L^1(\mathbb{X})$. The operator $C(k) = C_r(k)$ of right convolution by k,

$$(C(k)u)(x) := \int_{\mathbb{R}^m} k(x^{-1} \cdot y)u(y)dy = \int_{\mathbb{R}^m} k(z)u(x \cdot z)dz, \quad x \in \mathbb{X},$$

is bounded on $L^2(\mathbb{X})$ and invariant with respect to the left shift, i.e.,

$$U_{l,g}C(k) = C(k)U_{l,g} \quad \text{where} \quad (U_{l,g}f)(x) := f(g \cdot x) \quad \text{for } g \in \mathbb{X}.$$

We denote by $\mathcal{C}_r(\mathbb{X})$ the set of all operators of right convolution by a function in $L^1(\mathbb{X})$. The following results go back to Steinberg [158]. They can be proved as their counterparts, the Theorems 3.2.2 and 3.2.10.

Theorem 7.2.1 *The following conditions are equivalent for $a \in L^\infty(\mathbb{X})$:*

(a) *the operators BaI and aB are compact on $L^2(\mathbb{X})$ for every $B \in \mathcal{C}_r(\mathbb{X})$,*

(b) *$a \in Q_{SC}(\mathbb{X})$,*

(c) *There is an open pre-compact subset D of $\mathbb{X}$ such that $\lim_{t\to\infty} \int_D |a(t\cdot s)|\, ds = 0$.*

Theorem 7.2.2 *The following assertions are equivalent for $a \in L^\infty(\mathbb{X})$:*

(a) *the operators $BaI - aB$ are compact on $L^2(\mathbb{X})$ for every $B \in \mathcal{C}_r(\mathbb{X})$,*

(b) *$a \in Q_C(\mathbb{R}^N)$,*

(c) *$a \in Q_{SC}(\mathbb{R}^N) + SO(\mathbb{R}^N)$.*

The following property of convolution operators is basic for the applicability of the axiomatic scheme.

Proposition 7.2.3 *Let $A \in \mathcal{C}_r(\mathbb{X})$. Then $\lim_{n\to\infty} \|[\varphi_{n,\alpha}I, A]\| = 0$ uniformly with respect to $\alpha \in \mathbb{Y}$.*

Proof. Let $A = C_r(k)$ with $k \in L^1(\mathbb{X})$, and set

$$\gamma_1(n, \alpha) := \sup_{x\in\mathbb{X}} \int_{\mathbb{X}} |k(x^{-1} \cdot y)|\, |\varphi_{n,\alpha}(x) - \varphi_{n,\alpha}(y)|dy,$$

$$\gamma_2(n, \alpha) := \sup_{y\in\mathbb{X}} \int_{\mathbb{X}} |k(x^{-1} \cdot y)|\, |\varphi_{n,\alpha}(x) - \varphi_{n,\alpha}(y)|dx.$$

Further, we suppose for a moment that there is an $R > 0$ such that $k(x) = 0$ if $\rho(x, 0) \geq R$. Then, for $j = 1, 2$,

$$\begin{aligned} \gamma_j(n, \alpha) &\leq \sup_{\rho(x,\, y)\leq R} |\varphi_{n,\alpha}(x) - \varphi_{n,\alpha}(y)| \int_{\mathbb{X}} |k(x)|dx \\ &\leq \sup_{\rho(x,\, y)\leq R/n} |\varphi(x) - \varphi(y)| \int_{\mathbb{X}} |k(x)|dx \end{aligned}$$

where we used that $\rho(D_\delta x, D_\delta y) = \delta\rho(x, y)$. The function φ is compactly supported and, hence, uniformly continuous on $\mathbb{X}$. Thus, for each $\varepsilon > 0$, we find an $n_0 \in \mathbb{N}$ such that $\gamma_j(n, \alpha) \leq \varepsilon$ for all $j = 1, 2, n \geq n_0$, and $\alpha \in \mathbb{Y}$. This shows that

$$\|[\varphi_{n,\alpha}I, K]\| \leq \max\{\sup_{\alpha\in\mathbb{Y}} \gamma_1(n, \alpha), \sup_{\alpha\in\mathbb{Y}} \gamma_2(n, \alpha)\} \to 0$$

as $n \to \infty$. Since the set of all functions with compact support is dense in $L^1(\mathbb{X})$, we can use a standard approximation argument to get the assertion also for arbitrary kernels $k \in L^1(\mathbb{X})$. □

7.2.5 Shift operators

Let $C_b^1(\mathbb{X})$ stand for the set of all bounded real-valued functions on $\mathbb{X}$ which possess a bounded derivative in the common sense, i.e., when $\mathbb{X}$ is realized as $\mathbb{R}^N$, and let $g =: (g_1, \ldots, g_N) : \mathbb{X} \to \mathbb{R}^N$ be a function which satisfies

(α) $g_j \in C_b^1(\mathbb{X})$ for all $j = 1, \ldots, N$,

(β) the mapping $F_g : \mathbb{X} \to \mathbb{X}$, $x \mapsto x \cdot g(x)$ is bijective,

(γ) $\liminf_{x\to\infty} |\det(dF_g(x))| > 0$. (Here, as usual, $df(x)$ refers to the derivative of the function $f : \mathbb{R}^N \to \mathbb{R}^N$ at x.)

We consider shift operators of the form

$$(T_g u)(x) := u(x \cdot g(x)).$$

Proposition 7.2.4 *If g satisfies the conditions (α)–(γ), then the operator T_g is bounded on $L^2(\mathbb{X})$.*

Proof. For $u \in L^2(\mathbb{X})$, we have

$$\|T_g u\|^2 = \int_{\mathbb{X}} |u(F_g(x))|^2\, dx = \int_{\mathbb{X}} |u(y)|^2 |\det dF_g^{-1}(y)|\, dy \leq C\|u\|^2$$

where $C := \sup_{y\in\mathbb{X}} |\det dF_g^{-1}(y)| < \infty$ due to conditions (β) and (γ). □

We call the function g *slowly oscillating* if, in addition to conditions (α)–(γ),

(δ) $\lim_{x\to\infty} \|dg(x)\| = 0$,

and we denote the class of all shifts T_g with a slowly oscillating function g by $\mathcal{R}(\mathbb{X})$.

Proposition 7.2.5 *Let $T_g \in \mathcal{R}(\mathbb{X})$. Then $\lim_{n\to\infty} \|[\varphi_{n,\alpha}I, T_g]\| = 0$ uniformly with respect to $\alpha \in \mathbb{Y}$.*

Proof. Let $u \in L^2(\mathbb{X})$ with norm 1. Then

$$\begin{aligned} \|[\varphi_{n,\alpha}I,\, T_g]\, u\| &\le \sup_{x\in\mathbb{X}} |\varphi_{n,\alpha}(x) - \varphi_{n,\alpha}(x \cdot g(x))|\, \|T_g u\| \\ &\le C \sup_{x\in\mathbb{X}} |\varphi_{n,\alpha}(x) - \varphi_{n,\alpha}(x \cdot g(x))| \\ &\le C \sup_{x\in\mathbb{X}} |\varphi(D_{1/n}(\alpha \cdot x)) - \varphi(D_{1/n}(\alpha \cdot x \cdot g(x)))|. \end{aligned}$$

Let now $\varepsilon > 0$ be arbitrarily given. Since φ is uniformly continuous on $\mathbb{X}$, there is a $\delta > 0$ such that, for all $x_1,\, x_2 \in \mathbb{X}$ with $\rho(x_1^{-1} \cdot x_2) < \delta$,

$$|\varphi(x_1) - \varphi(x_2)| < \varepsilon.$$

Thus, if $\rho(D_{1/n}(g(x))) < \delta$, then

$$|\varphi(D_{1/n}(\alpha \cdot x)) - \varphi(D_{1/n}(\alpha \cdot x \cdot g(x)))| < \varepsilon.$$

Since the function g is bounded by assumption,

$$\rho(D_{1/n}(g(x))) = \max_{1\le j\le N} |n^{-a_j} g_j(x)|^{1/a_j} \le n^{-1} \max_{1\le j\le N} \|g_j\|_\infty^{1/a_j}.$$

Thus, if we choose n_0 such that

$$n_0^{-1} \max_{1\le j\le N} \|g_j\|_\infty^{1/a_j} < \delta,$$

then $\|[\varphi_{n,\alpha}I,\, T_g]\| \le C\,\varepsilon$ for all $n \ge n_0$ with a constant C independent of α. □

Here are a few instances where the requirements (α)–(δ) are satisfied.

Example A. If g is a constant function then T_g belongs to $\mathcal{R}(\mathbb{X})$. □

Example B. Let $\mathbb{X}$ be the commutative group $\mathbb{R}^m$ where

$$(T_g u)(x) = u(x + g(x)),$$

and let the conditions (α) and (δ) be fulfilled. If one of the conditions

$$\max_{1\le j\le m} \sum_{k=1}^{m} \sup_x \left| \frac{\partial g_j(x)}{\partial x_k} \right| < 1, \quad \max_{1\le k\le m} \sum_{j=1}^{m} \sup_x \left| \frac{\partial g_j(x)}{\partial x_k} \right| < 1 \tag{7.20}$$

holds, then $T_g \in \mathcal{R}(\mathbb{X})$. Indeed, each of the conditions (7.20) implies that g is a contraction. Thus, by the Banach fixed point theorem,

$$F_g : \mathbb{R}^m \to \mathbb{R}^m, \quad x \mapsto x + g(x)$$

is invertible, and it follows from condition (δ) that

$$\lim_{x\to\infty} \det(dF_g(x)) = 1$$

which gives condition (γ). □

Example C. Let $\mathbb{H}^{2n+1}$ be the Heisenberg group with coordinates $(x, y, t) \in \mathbb{R}^n \times \mathbb{R}^n \times \mathbb{R}$. Consider the function $g(x, y, t) := (p(x, y), q(x, y), \tau(x, y, t))$ where we assume that the mapping

$$(x, y) \mapsto (p(x, y), q(x, y)) : \mathbb{R}^{2n} \to \mathbb{R}^{2n}$$

satisfies a condition analogous to (7.20) and where, consequently, the mapping

$$(x, y) \mapsto \Phi(x, y) := (x + p(x, y), y + q(x, y))$$

is invertible. Thus, for arbitrary $x', y' \in \mathbb{R}^n$, the system

$$x' = x + p(x, y), \quad y' = y + q(x, y)$$

possesses a unique solution

$$x = \phi(x', y'), \quad y = \psi(x', y').$$

Moreover we suppose that g is slowly oscillating in the sense that

$$\lim_{(x,y)\to\infty} d_x p(x, y) = \lim_{(x,y)\to\infty} d_y p(x, y) = 0, \tag{7.21}$$

$$\lim_{(x,y)\to\infty} d_x q(x, y) = \lim_{(x,y)\to\infty} d_y q(x, y) = 0, \tag{7.22}$$

$$\lim_{(x,y,t)\to\infty} d_x \tau(x, y, t) = \lim_{(x,y,t)\to\infty} d_y \tau(x, y, t) = \lim_{(x,y,t)\to\infty} d_t \tau(x, y, t) = 0. \tag{7.23}$$

Let, finally,

$$\sup_{(x,y,t)\in\mathbb{R}^{2n+1}} |d_t \tau(x, y, t)| < 1. \tag{7.24}$$

Then the mapping $F_g : \mathbb{H}^{2n+1} \to \mathbb{H}^{2n+1}$ sending (x, y, t) to

$$\begin{aligned}&(x, y, t) \cdot g(x, y, t)\\ &\quad = (x + p(x, y), y + q(x, y), t + \tau(x, y, t) + 2(\langle(x, q(x, y)\rangle - \langle y, p(x, y)\rangle))\end{aligned}$$

is invertible. Indeed, for arbitrary $(x', y', t') \in \mathbb{H}^{2n+1}$, the equation

$$t' = t + \tau(\phi(x', y'), \psi(x', y'), t) + 2\,\Psi(x', y')$$

with $\Psi(x', y')$ given by

$$\langle \phi(x', y'), q(\phi(x', y'), \psi(x', y'))\rangle - \langle \psi(x', y'), p(\phi(x', y'), \psi(x', y'))\rangle$$

has a unique solution t due to (7.24). This proves condition (β), and condition (γ) follows since (7.21)–(7.23) imply that

$$\lim_{x\to\infty} \det(d_x F_g(x)) = 1.$$

Consequently, $T_g \in \mathcal{R}(\mathbb{H}^{2n+1})$. □

7.3 Fredholm criteria for convolution type operators with shift

We derive Fredholm criteria for convolution type operators with shift in two settings: for operators acting on the homogeneous group $\mathbb{X}$ itself, and for operators acting on a discrete subgroup $\mathbb{Y}$ of $\mathbb{X}$.

7.3.1 Operators on homogeneous groups

To study the Fredholmness of convolution type operators on $L^2(\mathbb{X})$, we specify the operator families from the axiomatic approach in Section 7.1 as follows.

- For $n \in \mathbb{N}$, let P_n stand for the operator of multiplication by φ_n (recall the definitions of M, M', φ and φ_n from Section 7.2.3).
- For $m \in \mathbb{N}$, let Q_m be the operator of multiplication by the characteristic function of $\{x \in \mathbb{X} : \rho(x, 0) > m\}$.
- Choose $Y := \mathbb{Y}$, and for $y \in Y$, let V_y be the operator of left shift by y,

$$(V_y u)(x) := (U_{l,y} u)(x) = u(y \cdot x).$$

Observe that $P_{n,y}$ is then the operator of multiplication by $\varphi_{n,y}$. It is elementary to check that the families (P_n) and (Q_m) consist of self-adjoint operators which converge strongly to the identity operator, that the Q_m are projections and the V_y are unitary operators, and that the first and third axiom from Section 7.1 are satisfied. Condition (a) of the second axiom is a consequence of the fact that the functions φ are identically 1 on a certain open subset of $\mathbb{X}$. If we further choose $Y_n := D_n\mathbb{Y}$, then the remainder of the second axiom follows from (7.19) and from the finite multiplicity of the covering of $\mathbb{X}$ by the family $\{y \cdot M'\}_{y\in\mathbb{Y}}$ as in the example discussed at the end of Section 7.1. So we can form the algebras $\mathcal{A}_{00}$, $\mathcal{A}_0$, $\mathcal{A}$, $\mathcal{A}'$ as well as $\mathcal{K}$ with respect to this choice of the operators P_n, Q_m and V_y.

Let us further mention that, due to the discreteness of $\mathbb{Y}$, the set $\mathcal{H}$ of all injective sequences from $\mathbb{N}$ into $\mathbb{Y}$ coincides with the set of all sequences $h : \mathbb{N} \to \mathbb{Y}$ which tend to infinity in the sense that $\rho(0, h(n)) \to \infty$ as $n \to \infty$.

Next we will single out classes of operators of convolution type which belong to the algebras $\mathcal{A}$ and $\mathcal{K}$. We denote by $\mathcal{K}_0(\mathbb{X})$ the set of all operators in $L(L^2(\mathbb{X}))$ of the form

$$\sum_{j=1}^{N} \prod_{k=1}^{L} S_{jk} a_{jk} K_{jk} b_{jk} T_{jk} \tag{7.25}$$

with positive integers L and N, where a_{jk}, $b_{jk} \in BUC(\mathbb{X})$, S_{jk}, $T_{jk} \in \mathcal{R}(\mathbb{X})$, and where $K_{jk} \in \mathcal{C}_r(\mathbb{X})$ are convolution operators with compactly supported kernel functions. We let further $\mathcal{K}(\mathbb{X})$ stand for the closure of $\mathcal{K}_0(\mathbb{X})$ in $L(L^2(\mathbb{X}))$.

Proposition 7.3.1 *The set $\mathcal{K}(\mathbb{X})$ is a closed subalgebra of the algebra $\mathcal{A}$. Moreover, $\mathcal{K}(\mathbb{X})$ is contained in the set $\mathcal{K}$ of the locally compact operators, and it contains all compact operators on $L^2(\mathbb{X})$.*

Proof. It is evident that $\mathcal{K}_0(\mathbb{X})$ is an algebra and that $\mathcal{K}(\mathbb{X})$ is a closed subalgebra of $L(L^2(\mathbb{X}))$. Next we show that if A is one of the operators aI with $a \in BUC(\mathbb{X})$, $C(k)$ with $k \in L^1(\mathbb{X})$ compactly supported, or $T_g \in \mathcal{R}(\mathbb{X})$, then A belongs to the algebra $\mathcal{A}_0$.

Let us start with condition (a) of Definition 7.1.4, i.e., with

$$\lim_{n\to\infty} \|[A, P_{n,y}]\| = 0 \quad \text{uniformly with respect to } y \in \mathbb{Y}. \tag{7.26}$$

For A a multiplication operator, this is evident. If A is a convolution operator, then (7.26) follows from Proposition 7.2.3, and if A is a shift operator, the assertion has been shown in Proposition 7.2.5.

Let now $h \in \mathcal{H}$. For condition (b) of Definition 7.1.4, we have to check that, for each choice of A, there is a subsequence g of h such that the limit operator A_g exists.

If $A = aI$ is the operator of multiplication by the function $a \in BUC(\mathbb{X})$, then $V_{h(k)}^{-1} a V_{h(k)}$ is the operator of multiplication by the function $x \mapsto a(h(k) \cdot x)$. The functions in $BUC(\mathbb{X})$ are bounded and uniformly continuous by definition. Hence, by the Arzelà-Ascoli theorem, the sequence h possesses a subsequence g such that the functions $x \mapsto a(g(k) \cdot x)$ tend uniformly on compact subsets of $\mathbb{X}$ to a certain bounded function a_g as $k \to \infty$. Consequently, the operators $V_{h(k)}^{-1} a V_{h(k)}$ converge strongly to $a_g I$, and the strong convergence of the adjoint sequence follows analogously.

If A is the convolution operator $C(k)$, then A commutes with each of the operators V_y, whence the strong convergence of $V_{h(k)}^{-1} A V_{h(k)}$ to A for every sequence h. To get the strong convergence of the adjoint sequence, note that the adjoint of $C(k)$ is the operator of right convolution by the function $x \mapsto \overline{k(x^{-1})}$, which belongs to $L^1(\mathbb{X})$ and which has a compact support whenever k has these properties.

Let, finally, $A = T_r \in \mathcal{R}(\mathbb{X})$ be the operator of shift by the function r. Then,

$$(V_{h(k)}^{-1} T_r V_{h(k)} u)(x) = u(x \cdot r(h(k) \cdot x)).$$

Since the functions $x \mapsto r(h(k) \cdot x)$ are uniformly bounded with respect to $k \in \mathbb{N}$ and equicontinuous on compact subsets of $\mathbb{X}$, the Arzelà-Ascoli theorem implies the existence of a subsequence g of h such that the functions $x \mapsto r(g(k) \cdot x)$ converge uniformly on compact subsets of $\mathbb{X}$ to a certain bounded function r_g. Since r is slowly oscillating, the function r_g is constant. We show that then the strong limit of the operators $V_{g(k)}^{-1} T_r V_{g(k)}$ as $k \to \infty$ exists and that

$$\text{s-lim}_{k\to\infty} V_{g(k)}^{-1} T_r V_{g(k)} = T_{r_g}. \tag{7.27}$$

Let u be a compactly supported continuous function on $\mathbb{X}$. Then u is uniformly continuous on $\mathbb{X}$, and there exists a compact subset Ω of $\mathbb{X}$ such that

$$u(x \cdot r(g(k) \cdot x)) - u(x \cdot r_g) = 0 \quad \text{whenever} \quad x \notin \Omega$$

(recall that r is bounded). It is further evident from the definition of r_g that, for arbitrary $\delta > 0$, there exists a $k_0 \in \mathbb{N}$ such that, for all $k \geq k_0$ and all $x \in \Omega$, $\rho(r(g(k) \cdot x), r_g) < \delta$. Since u is uniformly continuous, this implies that for each $\varepsilon > 0$, there exists a $k_0 \in \mathbb{N}$ such that

$$\sup_{x \in \Omega} |u(x \cdot r(g(k) \cdot x)) - u(x \cdot r_g)| < \varepsilon \quad \text{for all } k \geq k_0.$$

Thus,

$$\lim_{k \to \infty} V_{g(k)}^{-1} T_r V_{g(k)} u = T_{r_g} u$$

for every continuous and compactly supported function u on $\mathbb{X}$. Since these functions form a dense subset of $L^2(\mathbb{X})$, this implies (7.27). For the strong convergence of the adjoint sequence, note that, by the substitution rule, T_r^* is the operator $T_s bI$ where

$$s(x) = r(F_r^{-1}(x))^{-1} \quad \text{and} \quad b(x) = |\det dF_r^{-1}(x)|.$$

Both functions s and b are bounded and uniformly continuous. Thus, given $h \in \mathcal{H}$, we get as above the existence of a subsequence h' of h such that the strong limit of the sequence $V_{h'(k)}^{-1} T_s V_{h'(k)}$ exists, and then we can choose a subsequence g of h' such that the strong limit of the sequence $V_{g(k)}^{-1} b V_{g(k)}$ exists. Thus, the strong limit of $V_{g(k)}^{-1} T_r^* V_{g(k)}$ exists, too (and, of course, one can choose the same sequences g here and in (7.27)).

Our next goal is condition (c) of Definition 7.1.4, i.e., we want to show that, for every $n \in \mathbb{N}$, there is an $l \in \mathbb{N}$ such that

$$P_{n,y} A = P_{n,y} A P_{l,y} \quad \text{and} \quad A P_{n,y} = P_{l,y} A P_{n,y} \quad \text{for all } y \in \mathbb{Y}.$$

This condition can be shown in the same way as in the example at the end of Section 7.1 (recall once more that the function φ is identically equal to 1 on a certain open subset of $\mathbb{X}$). Since $\mathcal{A}_0$ is an algebra, it follows that $\mathcal{K}_0(\mathbb{X}) \subseteq \mathcal{A}_0$, whence $\mathcal{K}(\mathbb{X}) \subseteq \mathcal{A}$.

In order to show that $\mathcal{K}_0(\mathbb{X})$ is contained in $\mathcal{K}$, the ideal of the locally compact operators, consider the operator $Q_m T_g a C(k)$ where $T_g \in \mathcal{R}(\mathbb{X})$, $a \in BUC(\mathbb{X})$ and $k \in L^1(\mathbb{X})$. It is easy to see that there is a compactly supported and bounded function b such that $Q_m T_g a C(k) = Q_m T_g a b C(k)$. Since the operator $bC(k)$ is compact by Theorem 7.2.1, we conclude that $Q_m A$ is compact for every operator $A \in \mathcal{K}_0(\mathbb{X})$. Analogously, $A Q_m$ is compact. Thus, $\mathcal{K}_0(\mathbb{X}) \subseteq \mathcal{K}$, whence $\mathcal{K}(\mathbb{X}) \subseteq \mathcal{K}$.

Finally, the inclusion $K(L^2(\mathbb{X})) \subseteq \mathcal{K}(\mathbb{X})$ can be shows as Proposition 3.3.1. □

Specifying Theorem 7.1.13 and its Corollary 7.1.14 to the present context, we obtain the following.

Theorem 7.3.2 *Let $K \in \mathcal{K}(\mathbb{X})$ and $\gamma \in \mathbb{C}$. Then the operator $A := \gamma I + K$ is a*

(a) *Φ_+-operator if and only if*

$$\inf_{A_\beta \in \sigma_{\mathfrak{B}}(A)} \nu(A_\beta) > 0. \tag{7.28}$$

(b) *a Φ_--operator if and only if*

$$\inf_{A_\beta \in \sigma_{\mathfrak{B}}(A)} \nu(A_\beta^*) > 0. \tag{7.29}$$

(c) *Fredholm operator if and only if all limit operators of A are invertible and if the norms of their inverses are uniformly bounded.*

Let $\mathcal{M}$ denote the smallest closed subalgebra of $L(L^2(\mathbb{X}))$ which contains all operators of multiplication by functions in $BUC(\mathbb{X})$ as well as all operators Q_m. Then condition (7.16) is satisfied, and Theorem 7.1.17 has the following consequence.

Theorem 7.3.3 *Let $K \in \mathcal{K}(\mathbb{X})$ and $a \in BUC(\mathbb{X})$. Then the operator $A := aI + K$ is Fredholm if and only if the function a is invertible in $L^\infty(\mathbb{X})$, if all limit operators of A are invertible, and if the norms of their inverses are uniformly bounded.*

Finally, we would like to mention a special class of operators in $\mathbb{C}I + \mathcal{K}(\mathbb{X})$ for which the invertibility of their limit operators can be efficiently checked.

For, let $\mathcal{K}(Q_C(\mathbb{X}))$ be the closure of the algebra $\mathcal{K}_0(Q_C(\mathbb{X}))$ of all operators of the form (7.25), but now with a_{jk}, $b_{jk} \in Q_C(\mathbb{X})$. We claim that all limit operators of operators belonging to $\mathbb{C}I + \mathcal{K}(Q_C(\mathbb{X}))$ are invariant with respect to left shifts.

If $a \in Q_{SC}(\mathbb{X})$ and $K \in \mathcal{C}_r(\mathbb{X})$, then the operators aK and KaI are compact (Theorem 7.2.1). Hence, the limit operators of these operators exist with respect to every sequence $h \in \mathcal{H}$, and they are equal to zero. Further, if a is slowly oscillating, then every limit operator of aI is a scalar operator, as we have already observed in Proposition 3.3.9. The same holds for the limit operators of the shifts $T_g \in \mathcal{R}(\mathbb{X})$ as we observed in the proof of Proposition 7.3.1.

Thus, every limit operator of an operator in $\mathbb{C}I + \mathcal{K}(Q_C(\mathbb{X}))$ belongs to the smallest C^*-subalgebra $\mathcal{B}(\mathcal{C}_r(\mathbb{X}), \mathcal{R}^c(\mathbb{X}))$ of $L(L^2(\mathbb{X}))$ which contains all convolution operators in $\mathcal{C}_r(\mathbb{X})$ and all shift operators $\mathcal{R}(\mathbb{X})$ by a constant function (i.e., by an element of the group $\mathbb{X}$). So, in this special setting, Theorem 7.3.2 reduces the problem of (semi-) Fredholmness for operators in $\mathbb{C}I + \mathcal{K}(Q_C(\mathbb{X}))$ to the problem of invertibility of operators in the algebra $\mathcal{B}(\mathcal{C}_r(\mathbb{X}), \mathcal{R}^c(\mathbb{X}))$ which are invariant with respect to left shifts by elements in $\mathbb{X}$. To study this invertibility problem, methods of (noncommutative) harmonic analysis are available (cf. [177]). For example, in case of the usual commutative group $\mathbb{R}^N$, the operator

$$A := \gamma I + \sum_{j=1}^{n} K_j T_j$$

where $\gamma \in \mathbb{C}$, K_j is a convolution with kernel $k_j \in L^1(\mathbb{R}^N)$ and T_j is the shift by $g_j \in \mathbb{R}^N$, is invertible on $L^2(\mathbb{R}^N)$ if and only if

$$\inf_{\xi \in \mathbb{R}^N} |\gamma + \sum_{j=1}^{n} \hat{k}_j(\xi) e^{i\langle \xi, g_j \rangle}| > 0$$

where $\hat{k}_j$ refers to the Fourier transform of k_j.

7.3.2 Operators on discrete subgroups

Let $l^2(\mathbb{Y})$ be the space of all complex-valued functions u on the discrete group $\mathbb{Y}$ for which

$$\|u\|^2_{l^2(\mathbb{Y})} := \sum_{x \in \mathbb{Y}} |u(x)|^2 < \infty,$$

and write $l^\infty(\mathbb{Y})$ for the space of all bounded complex-valued functions on $\mathbb{Y}$, provided with the norm

$$\|a\|_{l^\infty(\mathbb{Y})} := \sup_{x \in \mathbb{Y}} |a(x)|.$$

By aI we denote the operator of multiplication by $a \in l^\infty(\mathbb{Y})$ thought of as acting on $l^2(\mathbb{Y})$. Further, given $z \in \mathbb{Y}$, we let $U_{z,l}$ and $U_{z,r}$ stand for the unitary operators of left and right shift acting at $u \in l^2(\mathbb{Y})$ by

$$(U_{z,l}u)(y) := u(z \cdot y) \quad \text{and} \quad (U_{z,r}u)(y) := u(y \cdot z), \quad y \in \mathbb{Y}.$$

Finally, for every function ψ on $\mathbb{X}$, we denote its restriction onto $\mathbb{Y}$ by $\hat{\psi}$.

Definition 7.3.4 *Let $\mathcal{B}(l^\infty(\mathbb{Y}), \{U_{z,r}\}_{z \in \mathbb{Y}})$ denote the closure in $L(l^2(\mathbb{Y}))$ of the set of all operators of the form*

$$\sum_{\gamma \in \Gamma} a_\gamma U_{\gamma,r} \quad \text{with } a_\gamma \in l^\infty(\mathbb{Y}) \tag{7.30}$$

where Γ runs through the finite subsets of $\mathbb{Y}$.

It turns out that $\mathcal{B}(l^\infty(\mathbb{Y}), \{U_{z,r}\}_{z \in \mathbb{Y}})$ is a C^*-subalgebra of $L(l^2(\mathbb{Y}))$. The Fredholm properties of operators in this algebra can be studied by the axiomatic approach from Section 7.1 again. The needed families of operators are specified as follows.

- For $n \in \mathbb{N}$, let P_n stand for the operator of multiplication by $\hat{\varphi}_n$.
- For $m \in \mathbb{N}$, let Q_m be the operator of multiplication by the restriction of the characteristic function of $\{x \in \mathbb{X} : \rho(x, 0) > m\}$ onto $\mathbb{Y}$.
- Choose $Y := \mathbb{Y}$, and for $z \in Y$, let V_z be the operator of left shift by z,

$$(V_z u)(y) := (U_{l,z}u)(y) = u(z \cdot y) \quad \text{for } y \in \mathbb{Y}.$$

One can check (essentially in the same way as in the previous section) that the Axioms 1–3 are satisfied and that all operators of the form (7.30) belong to the algebra $\mathcal{A}_0$ (defined with respect to the above-specified operators). For the existence of the limit operators, one has to employ a Cantor diagonalization argument again. Thus, the algebra $\mathcal{B}(l^\infty(\mathbb{Y}), \{U_{z,r}\}_{z\in\mathbb{Y}})$ proves to be a C^*-subalgebra of the corresponding algebra $\mathcal{A}$.

In the discrete setting at hand, the identity operator is locally compact, and all operators $I - Q_m$ are compact. So Theorem 7.1.12 implies the following.

Theorem 7.3.5 *Let* $A \in \mathcal{B}(l^\infty(\mathbb{Y}), \{U_{g,r}\}_{g\in\mathbb{Y}})$. *Then* A *is a*

(a) Φ_+*-operator if and only if* $\inf\{\nu(A_g) : A_g \in \sigma_{op}(A)\} > 0$.

(b) Φ_-*-operator if and only if* $\inf\{\nu(A_g^*) : A_g \in \sigma_{op}(A)\} > 0$.

(c) *Fredholm operator if and only if all limit operators of* A *are invertible and if the norms of their inverses are uniformly bounded.*

7.4 Comments and references

The axiomatic approach to the limit operators method was developed in [130], where one also finds the Fredholm criteria for convolution operators on homogeneous groups considered in Section 7.2. The extension of these results to convolution operators with variable shifts can be found in [132]. We owe the suggestion to look for an axiomatic approach to the referee of our paper [137], who also proposed a first version of this approach in her/his report.

In Section 7.2.1, where we have collected some basic facts on homogeneous groups, we follow [174], Chapter XIII, Section 5. Nice introductions to the Heisenberg group can be found in [77, 155, 177]. The results cited in Section 7.2.2 as well as Theorems 7.2.1 and 7.2.2 are due to Shteinberg [158].

Singular integral operators and pseudodifferential operators on the Heisenberg group have been intensively studied by many authors (compare the monographs [174, 112, 176, 177], which contain extensive bibliographies). Let us also mention the papers [183, 184, 84, 85], which are devoted to the analysis of double convolutions on a class of step two nilpotent Lie groups. The Fredholm property of operators in certain algebras generated by convolution and multiplication operators on non-commutative locally compact groups was considered in [159, 160] by means of Simonenko's local principle (see [166]). A comprehensive study of integral operators with constant coefficients and constant shifts on the real half-line is undertaken in [59]. Note also that there is an extensive bibliography devoted to singular integral and convolution type operators with shifts (see, for instance, [7, 9, 88, 91, 81] and the references therein).

Bibliography

[1] S. AGMON, Lectures on Exponential Decay of Solutions of Second Order Elliptic Equations. – Princeton University Press, Princeton 1982.

[2] M.S. AGRANOVICH, Spectral properties of elliptic pseudodifferential operators on a closed curve. – Funkts. Anal. Prilozh. **13**(1979), 4, 54–56 (Russian, Engl. transl.: Funct. Anal. Appl. **13**(1979), 4, 279–281).

[3] M.S. AGRANOVICH, On elliptic pseudodifferential operators on a closed curve. – Trudi Mosk. mat. ob-va **47**(1984), 22–67 (Russian).

[4] M.S. AGRANOVICH, Elliptic Operators on Closed Manifolds. – Sovr. Probl. Mat., Fund. Napravl., **63**(1990), 5–130 (Russian, Engl. transl.: Encyclopedia Math. Sci. **63**, Springer-Verlag, New York, Berlin, Heidelberg 1994).

[5] M.S. AGRANOVICH, B.A. AMOSOV, On matrix elliptic pseudodifferential operators on a closed curve. – Funkts. Anal. Prilozh. **15**(1981), 3, 79–81 (Russian, Engl. transl.: Funct. Anal. Appl. **15**(1981), 3, 217–219).

[6] G.R. ALLAN, Ideals of vector valued functions. – Proc. Lond. Math. Soc. **18**(1968), 3, 193–216.

[7] A.B. ANTONEVICH, Linear Functional Equations. Operator Approach. – Birkhäuser Verlag, Basel, Boston, Berlin 1995.

[8] A. ANTONEVICH, M. BELOUSOV, A. LEBEDEV, Functional Differential Equations II: C^*-applications. – Longman, Harlow 1998 (two volumes).

[9] A. ANTONEVICH, A. LEBEDEV, Functional Differential Equations I: C^*-theory. – Longman, Harlow 1994.

[10] W. ARVESON, C^*-algebras and numerical linear algebra. – J. Funct. Anal. **122**(1994), 2, 333–360.

[11] W. ARVESON, The role of C^*-algebras in infinite dimensional numerical linear algebra. – Contemp. Math. **167**(1994), 115–129.

[12] J. AVRON, P.H.M. MOUCHE, B. SIMON, On the measure of the spectrum for the almost Mathieu operator. – Comm. Math. Phys. **132**(1990), 1, 103–118.

[13] J. AVRON, B. SIMON, Almost periodic Schrödinger operators, II. The integrated density of states. – Duke Math. J. **50**(1983), 1, 369–391.

[14] R. BEALS, Characterization of pseudodifferential operators and applications. – Duke Math. J. **44**(1977), 1, 45–57.

[15] A. BEN-ARTZI, On approximation spectrum of bounded self-adjoint operators. – Integral Equations Oper. Theory **9**(1986), 2, 266–274.

[16] A. BEN-ARTZI, I. GOHBERG, Fredholm properties of band matrices and dichotomy. – *In:* Topics in operator theory. The Constantin Apostol memorial issue (Ed.

I. GOHBERG), Operator Theory: Advances and Appl. **32**, Birkhäuser Verlag, Basel 1988, 37–52.

[17] C.J. BISHOP, A. BÖTTCHER, YU.I. KARLOVICH, I. SPITKOVSKY, Local spectra and index of singular integral operators with piecewise continuous coefficients on composed curves. – Math. Nachr. **206**(1999), 5–83.

[18] B. BLACKADAR, K-Theory for Operator Algebras. – M. S. R. I. Monographs, Vol. 5, Springer-Verlag, Berlin, New York 1986 (Second ext. ed.: Cambridge Univ. Press, Cambridge, New York, Melbourne 1998).

[19] F.-P. BOCA, Rotation C^*-Algebras and Almost Mathieu Operators. – Theta, Bucharest 2001.

[20] A. BÖTTCHER, Pseudospectra and singular values of large convolution operators. – J. Integral Equations Appl. **6**(1994), 3, 267–301.

[21] A. BÖTTCHER, S. GRUDSKY, B. SILBERMANN, Norms of inverses, spectra, and pseudospectra of large truncated Wiener-Hopf operators and Toeplitz matrices. – New York J. Math. **3**(1997), 1–31.

[22] A. BÖTTCHER, YU.I. KARLOVICH, Toeplitz and singular integral operators on general Carleson Jordan curves. – *In*: Singular Integral Operators and Related Topics (Eds. A. BÖTTCHER, I. GOHBERG), Operator Theory: Advances and Appl. **90**, Birkhäuser Verlag, Basel, Boston, Berlin 1996, 119–152.

[23] A. BÖTTCHER, YU.I. KARLOVICH, Carleson Curves, Muckenhoupt Weights, and Toeplitz Operators. – Progr. Math. **154**, Birkhäuser Verlag, Basel, Boston, Berlin 1997.

[24] A. BÖTTCHER, YU.I. KARLOVICH, Toeplitz operators with PC symbols on general Carleson Jordan curves with arbitrary Muckenhoupt weights. – Trans. Am. Math. Soc. **351**(1999), 8, 3143–3196.

[25] A. BÖTTCHER, YU.I. KARLOVICH, V.S. RABINOVICH, Emergence, persistence, and disappearance of logarithmic spirals in the spectra of singular integral operators. – Integral Equations Oper. Theory **25**(1996), 4, 406–444.

[26] A. BÖTTCHER, YU.I. KARLOVICH, V.S. RABINOVICH, Mellin pseudodifferential operators with slowly varying symbols and singular integrals on Carleson curves with Muckenhoupt weights. – manuscripta math. **95**(1998), 363–376.

[27] A. BÖTTCHER, YU.I. KARLOVICH, V.S. RABINOVICH, The method of limit operators for one-dimensional singular integral operators with slowly oscillating data. – J. Oper. Theory **43**(2000), 1, 171–198.

[28] A. BÖTTCHER, A.E. PASENCHUK, On the invertibility of Wiener-Hopf operators on the quarter plane with kernels whose support is located in a half-plane. – *In*: Differ. Integr. Uravn. Prilozh., Izd. Kalmyk. and Rostov. Univ., Elista 1982, 9–19 (Russian).

[29] A. BÖTTCHER, M. SEYBOLD, Discrete Wiener-Hopf equations on spaces with Muckenhoupt weight. – Studia Math. **143**(2000), 2, 121–144.

[30] A. BÖTTCHER, B. SILBERMANN, Analysis of Toeplitz Operators. – Akademie-Verlag, Berlin 1989 and Springer-Verlag, Berlin, Heidelberg, New York 1990.

[31] A. BÖTTCHER, B. SILBERMANN, Infinite Toeplitz and Hankel matrices with operator valued entries. – SIAM J. Math. Anal. **27**(1996), 3, 805–822.

[32] A. BÖTTCHER, B. SILBERMANN, Introduction to Large Truncated Toeplitz Matrices. – Springer-Verlag, New York 1999.

[33] A. BÖTTCHER, H. WOLF, Spectral approximation for Segal-Bargmann space Toeplitz operators. – *In*: Linear Operators, Banach Center Publ. **38**, PAN, Warsaw 1997, 25–48.

[34] F.F. BONSALL, J. DUNCAN, Numerical Ranges of Operators on Normed Spaces and of Elements of Normed Algebras. – Cambridge University Press, Cambridge 1971.

[35] F.F. BONSALL, J. DUNCAN, Numerical Ranges II. – Cambridge University Press, Cambridge 1973.

[36] A. BOULKHEMAIR, Remarks on a Wiener type pseudodifferential operators and Fourier integral operators. – Math. Res. Lett. **4**(1997), 53–67.

[37] H. BOURBAKI, Éléments de Mathématique. Intégration. Chapitres 7, 8. – Hermann, Paris 1963.

[38] L.M. BREKHOVSKIKH, O.A. GODIN, Acoustics in stratified media. – Nauka, Moskva 1989 (Russian).

[39] L.A. COBURN, R.D. MOYER, I.M. SINGER, C^*-Algebras of almost periodic pseudodifferential operators. – Acta Math. **130**(1973), 3–4, 279–307.

[40] H.O. CORDES, On compactness of commutators of multiplications and convolutions and boundedness of pseudodifferential operators. – J. Funct. Anal. **18**(1975), 2, 115–131.

[41] H.L. CYCON, R.G. FROESE, W. KIRSH, B. SIMON, Schrödinger Operators with Applications to Quantum Mechanics and Global Geometry. – Springer-Verlag, Berlin, Heidelberg, New York 1987.

[42] D. DAVID, Opérateurs intégraux singuliers sur certaines courbes du plan complexe. – Ann. Sci. École Norm. Sup. **17**(1984), 157–189.

[43] K.R. DAVIDSON, C^*-Algebras by Example. – Fields Institute Monographs, Vol. 6, Am. Math. Soc., Providence, Rhode Island, 1996.

[44] E.B. DAVIES, Spectral Theory and Differential Operators. – Cambridge Univ. Press, Cambridge, New York 1996.

[45] P. DEIFT, Orthogonal Polynomials and Random Matrices: A Riemann-Hilbert Approach. – Courant Lectures Notes in Mathematics **3**, Am. Math. Soc., Providence, Rhode Island, 2000.

[46] V.M. DEUNDYAK, B.YA. SHTEINBERG, Index of convolution operators with slowly varying coefficients. – Funkts. Anal. Prilozh. **19**(1985), 4, 84–85 (Russian, Engl. transl.: Funct. Anal. Appl. **19**(1985), 4, 321–323.

[47] J. DIXMIER, C^*-Algebras. – North Holland Publ. Comp., Amsterdam, New York, Oxford 1982.

[48] R.G. DOUGLAS, On the invertibility of a class of Toeplitz operators on the quarter plane. – Indiana Univ. Math. J. **21**(1972), 11, 1031–1036.

[49] R.G. DOUGLAS, R. HOWE, On the C^*-algebra of Toeplitz operators on the quarter plane. – Trans. Am. Math. Soc. **158**(1971), 1, 203–217.

[50] R.V. DUDUCHAVA, Integral Equations with Fixed Singularities. – Teubner, Leipzig 1979.

[51] E.M. DYNKIN, B.P. OSILENKER, Weighted norm estimates for singular integrals and their applications. – J. Sov. Math. **30**(1985), 2094–2154.

[52] T. EHRHARDT, S. ROCH, B. SILBERMANN, Symbol calculus for singular integrals with operator-valued coefficients. – *In:* Singular Integral Operators and Related Topics (Eds. A. BÖTTCHER, I. GOHBERG), Operator Theory: Advances and Appl. **90**, Birkhäuser Verlag, Basel, Boston, Berlin 1996.

[53] J. FAVARD, Sur les équations différentielles à coefficients presque periodiques. – Acta Math. **51**(1927), 31–81.

[54] B.V. FEDOSOV, M.A. SHUBIN, Index of random elliptic operators, I. – Mat. Sbornik **106(148)**(1978), 1, 108–140 (Russian, Engl. transl.: Math. USSR Sbornik **34**(1978), 5, 671–699).

[55] T. FINCK, Splinaapproximationsverfahren für singuläre Integralgleichungen auf ebenen, glattberandeten Gebieten. – PhD thesis, TU Chemnitz 1998.

[56] F.D. GAKHOV, YU.I. CHERSKI, Equations of Convolution Type. – Mir, Moskva 1978 (Russian).

[57] T.W. GAMELIN, Uniform Algebras. – Prentice-Hall, Inc., Englewood Cliffs, N.J., 1969.

[58] I.M. GELFAND, D.A. RAIKOV, G.E. SHILOV, Commutative Normed Rings. – Goz. Izd. Fiz.-Mat. Lit., Moskva 1960 (Russian, Engl. transl.: Chelsea Publ. Comp., New York 1964).

[59] I. GOHBERG, I. FELDMAN, Convolution Equations and Projection Methods for Their Solution. – Nauka, Moskva 1971 (Russian, Engl. transl.: Am. Math. Soc. Transl. of Math. Monographs **41**, Providence, R.I., 1974).

[60] I. GOHBERG, S. GOLDBERG, M.A. KAASHOEK, Classes of Linear Operators I. – Operator Theory: Advances and Appl. **49**, Birkhäuser Verlag, Basel, Boston, Berlin 1990.

[61] I. GOHBERG, S. GOLDBERG, M.A. KAASHOEK, Classes of Linear Operators II. – Operator Theory: Advances and Appl. **63**, Birkhäuser Verlag, Basel, Boston, Berlin 1993.

[62] I. GOHBERG, M.A. KAASHOEK, Projection method for Block Toeplitz operators with operator-valued symbols. – *In:* Toeplitz Operators and Related Topics (Eds. E.L. BASOR, I. GOHBERG), Operator Theory: Advances and Appl. **71**, Birkhäuser Verlag, Basel, Boston, Berlin 1994, 79–104.

[63] I.C. GOHBERG, M.G. KREIN, Integral equations on the semi-axis with kernels depending on the difference of the arguments. – Usp. Mat. Nauk **13**(1958), 2, 3–120 (Russian).

[64] I.C. GOHBERG, M.G. KREIN, Systems of integral equations on the semi-axis with kernels depending on the difference of the arguments. – Usp. Mat. Nauk **13**(1958), 5, 3–72 (Russian).

[65] I. GOHBERG, N. KRUPNIK, One-dimensional Linear Singular Integral Operators, Vol. I and II. – Birkhäuser Verlag, Basel, Boston, Stuttgart 1992.

[66] L.S. GOLDENSTEIN, I.C. GOHBERG, On a multidimensional integral equation on a half-space whose kernel is a function of the difference of the arguments, and on a discrete analogue of this equation. – Dokl. Akad. Nauk SSSR **131**(1960), 1, 4–12 (Russian, Engl. transl.: Sov. Math. **1**(1960), 1, 173–176).

[67] M.B. GORODETSKI, On discrete convolutions in the quarter-plane with infinitely differentiable symbol. – Mat. Zametki **27**(1980), 2, 217–224 (Russian, Engl. transl.: Math Notes **27**(1980), 104–108).

[68] M.B. GORODETSKI, Two-dimensional Toeplitz operators with analytic symbols and their applications. – Cand. Dissert., Rostov-on-Don State Univ. 1980 (Russian).

[69] J.M. GRACIA-BONDÍA, J.C. VÁRILLY, H. FIGUEROA, Elements of Non-commutative Geometry. – Birkhäuser Verlag, Boston, Basel, Berlin 2001.

[70] I.S. GRADSTEIN, I.M. RYSHIK, Tables of Integrals, Sums, Series and Products. – Nauka, Moskva 1971 (Russian, Engl. transl.: Academic Press, New York, London 2000).

[71] V.V. GRUSHIN, Pseudodifferential operators in $\mathbb{R}^n$ with bounded symbols. – Funkts. Anal. Prilozh. **4**(1970), 3, 37–50 (Russian, Engl. transl.: Funct. Anal. Appl. **4**(1971), 202–212).

[72] R. HAGEN, S. ROCH, B. SILBERMANN, Spectral Theory of Approximation Methods for Convolution Equations. – Operator Theory: Advances and Appl. **74**, Birkhäuser Verlag, Basel, Boston, Berlin 1995.

[73] R. HAGEN, S. ROCH, B. SILBERMANN, C^*-Algebras and Numerical Analysis. – Marcel Dekker, Inc., New York, Basel 2001.

[74] F. HAUSDORFF, Set theory. – Chelsea Publ. Comp., New York 1957.

[75] N. HIGSON, J. ROE, G. YU, A coarse Mayer-Vietoris principle. – Math. Proc. Cambridge Philos. Soc. **114**(1993), 85–97.

[76] L. HÖRMANDER, The Analysis of Linear Partial Differential Operators, Vol. 1–4. – Springer-Verlag, Berlin, Heidelberg, New York, Tokio 1983/1985.

[77] R. HOWE, On the role of the Heisenberg group in harmonic analysis. – Bull. Am. Math. Soc. **3**(1980), 2, 821–843.

[78] R. HUNT, B. MUCKENHOUPT, R. WHEEDEN, Weigthed norm inequalities for the conjugate function and Hilbert transform. – Trans. Am. Math. Soc. **176**(1973), 227–251.

[79] V. IVRII, Microlocal Analysis and Precise Spectral Asymptotics. – Springer-Verlag, Berlin, Heidelberg, New York 1998.

[80] A. JIRARI, Second Order Sturm-Liouville Difference Equations and Orthogonal Polynomials. – Memoirs Am. Math. Soc. **542**, Providence, Rhode Island, 1995.

[81] N. KARAPETIANTS, S. SAMKO, Equations with Involutive Operators. – Birkhäuser Verlag, Boston, Basel, Berlin, 2001.

[82] T. KATO, Perturbation Theory for Linear Operators. – Springer-Verlag, Berlin, Heidelberg, New York 1976 (second ed.).

[83] Y. KATZNELSON, An Introduction to Harmonic Analysis. – John Wiley & Sons, Inc., New York, London, Sydney, Toronto 1968.

[84] V.V. KISIL', The spectrum of the algebra generated by two-sided convolutions on the Heisenberg group and by operators of multiplication by continuous functions. – Dokl. Akad. Nauk **337**(1994), 4, 439–441 (Russian, Engl. Transl.: Russ. Acad. Sci. Dokl. Math. **50**(1995), 1, 92–96).

[85] V.V. KISIL', On an algebra of pseudo-differential operators which is generated by convolutions on the Heisenberg group. – Sib. Mat. Zh. **34**(1993), 6, 75–85 (Russian, Engl. transl.: Sib. Math. J. **34**(1993), 6, 1066–1075).

[86] A.V. KOZAK, A local principle in the theory of projection methods. – Dokl. Akad. Nauk SSSR **212**(1973), 6, 1287–1289 (Russian, Engl. transl.: Sov. Math. Dokl. **14**(1973), 5, 1580–1583).

[87] A.V. KOZAK, I.B. SIMONENKO, Projection methods for solving multidimensional discrete convolution equations. – Sib. Mat. Zh. **21**(1980), 2, 119–127 (Russian).

[88] V.G. KRAVCHENKO, G.S. LITVINCHUK, Introduction to the Theory of Singular Integral Operators with Shifts. – Kluwer Academic Publisher, Dordrecht, Boston, London 1994.

[89] P. KUCHMENT, Floquet Theory for Partial Differential Equations. – Operator Theory: Advances and Appl. **60**, Birkhäuser Verlag, Basel, Boston, Berlin 1993.

[90] H. KUMANO-GO, Pseudodifferential Operators. – MIT Press, Cambridge, Mass., 1981.

[91] V.G. KURBATOV, Linear differential-difference operators. – Voronezh. Univ. Press, Voronezh 1990 (Russian).

[92] V.G. KURBATOV, Functional Differential Operators and Equations. – Kluwer Academic Publ., Dordrecht, Boston, London 1999.

[93] B.V. LANGE, V.S. RABINOVICH, On the Noether property of multidimensional discrete convolution operators. – Mat. Zametki **37**(1985), 3, 407–421 (Russian).

[94] B.V. LANGE, V.S. RABINOVICH, On the Noether property of multi-dimensional operators of convolution type with measurable bounded coefficients. – Izv. Vyssh. Ucheb. Zaved., Mat. **6**(1985), 22–30 (Russian, Engl. transl.: Transl. Izv. Vuz. Mat. **29**(1985), 6, 26–35).

[95] B.V. LANGE, V.S. RABINOVICH, Pseudo-differential operators on $\mathbb{R}^n$ and limit operators. – Mat. Sbornik **129**(1986), 2, 175–185 (Russian, Engl. transl.: Math. USSR Sbornik **577**(1987), 1, 183–194).

[96] P.D. LAX, Functional Analysis. – Wiley-Interscience, 2002.

[97] B.M. LEVITAN, V.V. ZIKOV, Almost Periodic Functions and Differential Equations. – Cambridge Univ. Press, Cambridge, New York 1982.

[98] M. LINDNER, The finite section method in the space of essentially bounded functions: an approach using limit operators. – Numer. Funct. Anal. Optimization **24**(2003), 7 & 8, 863–893.

[99] M. LINDNER, Limit Operators and Applications on the Space of Essentially Bounded Functions. – PhD thesis, TU Chemnitz 2003.

[100] M. LINDNER, V.S. RABINOVICH, S. ROCH, Finite sections of band operators with slowly oscillating coefficients. – Submitted to Lin. Alg. Appl.

[101] M. LINDNER, B. SILBERMANN, Invertibility at infinity of band-dominated operators on the space of essentially bounded functions. – *In*: Erhard Meister Memorial Volume, Birkhäuser Verlag (to appear).

[102] L.H. LOOMIS, An Introduction to Abstract Harmonic Analysis. – D. Van Nostrand Comp., Inc., Princeton, New Jersey, 1953.

[103] V.A. MALYSHEV, On the solution of discrete Wiener-Hopf equations in a quarter plane. – Dokl. Akad. Nauk SSSR **187**(1969), 6, 1243–1246 (Russian, Engl. transl.: Sov. Math. Dokl. **10**(1969), 4, 1032–1036).

[104] V.A. MALYSHEV, Wiener-Hopf equations in a quadrant of the plane, discrete groups, and automorphic functions. – Mat. Sb. **84**(1971), 4, 499–525 (Russian, Engl. transl.: Math. USSR Sbornik **13**(1971), 4, 491–516).

[105] E. MEISTER, F.-O. SPECK, Some multidimensional Wiener-Hopf equations with applications. – *In:* Trends in Applications of Pure Mathematics to Mechanics, Vol. II (Ed. H. ZORSKI), Pitman, London et al. 1979, 217–262.

[106] E. MEISTER, F.-O. SPECK, Wiener-Hopf operators on three-dimensional wedge-shaped regions. – Appl. Anal. **10**(1980), 31–45.

[107] E. MEISTER, F.-O. SPECK, A contribution to the quarter-plane problem in diffraction theory. – J. Math. Anal. Appl. **130**(1988), 223–236.

[108] S.G. MIKHLIN, S. PROESSDORF, Singular Integral Operators. – Springer-Verlag, Berlin, Heidelberg, New York, Tokio 1986.

[109] E.M. MUKHAMADIJEV, On the invertibility of differential operators in partial derivatives of elliptic type. – Dokl. Akad. Nauk SSSR **205**(1972), 6, 1292–1295 (Russian).

[110] E.M. MUKHAMADIJEV, On the normal solvability and the Noether property of elliptic operators on functions on $\mathbb{R}^n$. – Zap. nauchn. sem. LOMI **110**(1981), 120–140 (Russian).

[111] G.J. MURPHY, C^*-Algebras and Operator Theory. – Academic Press, Inc., Boston, San Diego, New York, London, Sydney, Tokyo, Toronto 1990.

[112] A. NAGEL, E.M. STEIN, Lectures on Pseudo-Differential Operators: Regularity Theorems and Applications to Non-Elliptic Problems. – Math. Notes, Princeton Univ. Press, Princeton, New Jersey, 1979.

[113] L. PASTUR, A. FIGOTIN, Spectra of random and almost-periodic operators. – Springer-Verlag, Berlin, Heidelberg, New York 1992.

[114] G.K. PEDERSEN, C^*-Algebras and Their Automorphism Groups. – Academic Press, New York 1979.

[115] M. PIMSNER, D. VOICULESCU, Exact sequences for K-groups and Ext-groups of certain crossed products. – J. Operator Theory **4**(1980), 1, 93–118.

[116] S. PRÖSSDORF, Einige Klassen singulärer Gleichungen. – Akademie-Verlag, Berlin 1974 (German, Engl. transl.: Some Classes of Singular Equations, North Holland Publ. Comp., Amsterdam, New York, Oxford 1978, Russian transl.: Mir, Moskva 1979).

[117] S. PRÖSSDORF, B. SILBERMANN, Numerical Analysis for Integral and Related Operator Equations. – Akademie-Verlag, Berlin 1991, and Birkhäuser Verlag, Basel, Boston, Stuttgart 1991.

[118] V.S. RABINOVICH, On the Fredholmness of pseudodifferential operators in the scale of $L_{2,p}$ spaces. – Sib. Mat. Zh. **29**(1988), 4, 149–161 (Russian, Engl. transl.: Sib. Math. J. **29**(1988), 4, 635–646).

[119] V.S. RABINOVICH, Singular integral operators on a composed contour with oscillating tangent and pseudodifferential Mellin operators. – Soviet Math. Dokl. **44**(1992), 791–796.

[120] V.S. Rabinovich, Pseudodifferential operators with operator symbols: local invertibility and limit operators. – Linear Topological Spaces and Complex Analysis **1**(1994), 58–73.

[121] V.S. Rabinovich, Mellin pseudo-differential operators with operator symbols and its applications. – *In*: Partial Differential Operators and Mathematical Physics (Eds. M. Demuth, B. W. Schulze), Operator Theory: Advances and Appl. **78**, Birkhäuser Verlag, Basel, Boston, Berlin 1995, 271–279.

[122] V.S. Rabinovich, Pseudodifferential operators with analytic symbols and some of their applications. – *In*: Proceedings of the Seminar on Linear Topological Spaces and Complex Analysis, Metu-Tubitak, Ankara 1995, 79–98.

[123] V.S. Rabinovich, Singular integral operators on composed contours and pseudodifferential operators. – Mat. Zametki **58**(1995), 1, 65–85 (Russian, Engl. transl.: Math. Notes **58**(1995), 1, 772–734).

[124] V.S. Rabinovich, Algebras of singular integral operators on composed contours with nodes that are logarithmic whirl points. – Izv. Ross. Akad. Nauk, Ser. Mat., **60**(1996), 169–200 (Russian, Engl. transl.: Izv. Math. **60**(1996), 6, 1261–1292).

[125] V.S. Rabinovich, Mellin pseudodifferential operators techniques in the theory of singular integral operators on some Carleson curves. – *In*: Differential and Integral Operators (Eds. I. Gohberg, R. Mennicken, C. Tretter), Operator Theory: Advances and Appl. **102**, Birkhäuser Verlag, Basel, Boston, Berlin 1998, 201–218.

[126] V.S. Rabinovich, Criterion for local invertibility of pseudodifferential operators with operator symbols and some applications. – Tr. St. Peterbg. Mat. Obshch. **5**(1998), 349–379 (Russian, Engl. transl.: Am. Math. Soc. Transl. Ser. 2, **193**(1999), 239–259).

[127] V.S. Rabinovich, An Introductory Course on Pseudodifferential Operators. – Textos de Matemática **1**, Instituto Superior Técnico, Lisboa 1998.

[128] V.S. Rabinovich, Pseudodifferential operators with analytic symbols and estimates for eigenfunctions of Schrödinger operators. – Z. Anal. Anwend. **21**(2002), 2, 351–370.

[129] V.S. Rabinovich, S. Roch, Local theory of the Fredholmness of band-dominated operators with slowly oscillating coefficients. – *In*: Toeplitz Matrices, Convolution Operators, and Integral Equations (Eds. A. Böttcher, I. Gohberg, P. Junghanns), Operator Theory: Advances and Appl. **135**, Birkhäuser Verlag, Basel, Boston, Berlin 2002, 267–291.

[130] V.S. Rabinovich, S. Roch, An axiomatic approach to the limit operators method. – *In*: Singular Integral Operators, Factorization and Applications (Eds. A. Böttcher, M. A. Kaashoek, A. B. Lebre, A. F. dos Santos, F.-O. Speck), Operator Theory: Advances and Appl. **142**, Birkhäuser Verlag, Basel 2003, 263–285.

[131] V.S. Rabinovich, S. Roch, Algebras of approximation sequences: Spectral and pseudospectral approximation of band-dominated operators. – Preprint 2186 Technical University Darmstadt 2001, 22 p.

[132] V.S. Rabinovich, S. Roch, Integral operators with shifts on homogeneous groups. – Preprint 2213 Technical University Darmstadt 2002, 18 p.

[133] V.S. Rabinovich, S. Roch, Fredholmness of convolution type operators. – Preprint 2264 Technical University Darmstadt 2003, 28 p.

[134] V.S. RABINOVICH, S. ROCH, Wiener algebras of operators, and applications to pseudodifferential operators. – Preprint 2302 Technical University Darmstadt 2003, 46 p.

[135] V.S. RABINOVICH, S. ROCH, Pseudodifference operators on weighted spaces, and applications to discrete Schrödinger operators. – Preprint 2306 Technical University Darmstadt 2003, 44 p.

[136] V.S. RABINOVICH, S. ROCH, J. ROE, Fredholm indices of band-dominated operators. – Preprint 2285 Technical University Darmstadt 2003, 18 p, to appear in Integral Equations Oper. Theory.

[137] V.S. RABINOVICH, S. ROCH, B. SILBERMANN, Fredholm theory and finite section method for band-dominated operators. – Integral Equations Oper. Theory **30**(1998), 4, 452–495.

[138] V.S. RABINOVICH, S. ROCH, B. SILBERMANN, Band-dominated operators with operator-valued coefficients, their Fredholm properties and finite sections. – Integral Equations Oper. Theory **40**(2001), 3, 342–381.

[139] V.S. RABINOVICH, S. ROCH, B. SILBERMANN, Algebras of approximation sequences: Finite sections of band-dominated operators. – Acta Appl. Math. **65**(2001), 315–332.

[140] M. REED, B. SIMON, Methods of Modern Mathematical Physics, I. Functional Analysis. – Academic Press, New York, London 1972.

[141] M. REED, B. SIMON, Methods of Modern Mathematical Physics, II. Fourier Analysis, Self-Adjointness. – Academic Press, New York, San Francisco, London 1975.

[142] M. REED, B. SIMON, Methods of Modern Mathematical Physics, III. Scattering Theory. – Academic Press, New York, San Francisco, London 1979.

[143] M. REED, B. SIMON, Methods of Modern Mathematical Physics, IV. Analysis of Operators. – Academic Press, New York, San Francisco, London 1978.

[144] L. REICHEL, L.N. TREFETHEN, Eigenvalues and pseudo-eigenvalues of Toeplitz matrices. – Linear Algebra Appl. **162**(1992), 153–185.

[145] S. ROCH, Numerical ranges of large Toeplitz matrices. – Linear Algebra Appl. **282**(1998), 185–198.

[146] S. ROCH, Algebras of approximation sequences: Fractality. *In*: Problems and Methods in Mathematical Physics (Eds. J. ELSCHNER, I. GOHBERG, B. SILBERMANN), Operator Theory: Advances and Appl. **121** Birkhäuser Verlag, Basel, Boston, Berlin 2001, 471–497.

[147] S. ROCH, Pseudospectra of operator polynomials. – *In*: Recent Advances in Operator Theory (Eds. A. DIJKSMA, M.A. KAASHOEK, A.C.M. RAN), Operator Theory: Advances and Appl. **124**, Birkhäuser Verlag, Basel, Boston, Berlin 2001, 545–558.

[148] S. ROCH, Algebras of approximation sequences: Fredholm theory in fractal algebras. – Studia Math. **150**(2002), 1, 53–77.

[149] S. ROCH, B. SILBERMANN, Non-strongly converging approximation methods. – Demonstr. Math. **22**(1989), 3, 651–676.

[150] S. ROCH, B. SILBERMANN, Algebras of convolution operators and their image in the Calkin algebra. – Report R-MATH-05/90 des Karl-Weierstraß-Instituts für Mathematik, Berlin 1990, 157 S.

[151] S. ROCH, B. SILBERMANN, C^*-algebra techniques in numerical analysis. – J. Oper. Theory **35**(1996), 2, 241–280.

[152] J. ROE, Index Theory, Coarse Geometry and Topology of Manifolds. – CBMS Lecture Notes Vol. **90**, Amer. Math. Soc., Providence, Rhode Island, 1996.

[153] M. RØRDAM, F. LARSEN, N.J. LAUSTSEN, An Introduction to K-Theory for C^*-Algebras. – London Math Soc. Student Texts 49, Cambridge University Press, Cambridge, New York, Melbourne 2000.

[154] J. SARANEN, W.L. WENDLAND, The Fourier series representation of pseudodifferential operators on closed curves. – Complex Variables **8**(1987), 55–64.

[155] S. SEMMES, An introduction to Heisenberg groups in analysis and geometry. – Notices Am. Math. Soc. **50**(2003), 6, 640–646.

[156] V.N. SEMENYUTA, I.B. SIMONENKO, On the index of multidimensional discrete convolutions. – Mat. Issled. **4**(1969), 2, 88–94 (Russian).

[157] V.N. SEMENYUTA, I.B. SIMONENKO, Computation of the index of multidimensional discrete convolutions. – Mat. Issled. **4**(1969), 4, 134–141 (Russian).

[158] B.YA. SHTEINBERG, Operators of convolution type on locally compact groups. – Rostov-na-Donu 1979, Dep. at VINITI 715-80, 65 p. (Russian).

[159] B.YA. SHTEINBERG, On convolution operators on locally compact groups. – Funkts. Anal. prilozh. **15**(1981), 95–96 (Russian).

[160] B.YA. SHTEINBERG, Boundedness and compactness of convolution operators on locally compact groups. – Mat. Zametki **38**(1985), 2, 278–292 (Russian).

[161] B.YA. SHTEINBERG, Compactification of locally compact groups and Fredholmness of convolution operators with coefficients in factor groups. – Tr. St-Peterbg. Mat. Obshch. **6**(1998), 242–260 (Russian).

[162] M.A. SHUBIN, Almost-periodic functions and partial differential operators. – Uspekhi Mat. Nauk **33**(1978), 2, 3–135 (Russian).

[163] M.A. SHUBIN, Spectral theory and index of elliptic partial differential operators with almost-periodic coefficients. – Uspekhi Mat. Nauk **34**(1979), 2, 96–135 (Russian).

[164] M.A. SHUBIN, Pseudodifferential Operators and Spectral theory. – Springer-Verlag, Berlin, Heidelberg, New York, Tokio 2001 (second ed.).

[165] B. SIMON Almost periodic Schrödinger operators: a review. – Adv. Appl. Math. **3**(1982), 463–490.

[166] I.B. SIMONENKO, A new general method of investigating linear operator equations of singular integral equation type, I. – Izv. Akad. Nauk SSSR, Ser. Mat., **29**(1965), 3, 567–586 (Russian).

[167] I.B. SIMONENKO, Operators of convolution type in cones. – Mat. Sb. **74 (116)**(1967), 2, 567–586 (Russian, Engl. transl.: Math. USSR Sbornik **3**(1967), 2, 279–293).

[168] I.B. SIMONENKO, On multidimensional discrete convolutions. – Mat. Issled. **3**(1968), 1, 108–127 (Russian).

[169] I.B. SIMONENKO, CHIN NGOK MINH, A Local Method in the Theory of One-Dimensional Singular Integral Equations with Piecewise Continuous Coefficients. Fredholmness. – Izd. Rostov. Univ., Rostov-na-Donu 1986 (Russian).

[170] J. SJÖSTRAND, Wiener type algebras of pseudodifferential operators. – Preprint Unité de Recherche Associée D 0169, Centre de Mathématiques, Ecole Politechnique, Séminaire 1994–1995, 20 décembre 1994.

[171] J. SJÖSTRAND, An algebra of pseudodifferential operators. – Math. Res. Lett. **1**(1994), 185–192.

[172] F.-O. SPECK, Eine Erweiterung des Satzes von Rakovsik und ihre Anwendung in der Simonenko-Theorie. – Math. Ann. **228**(1977), 2, 93–100.

[173] I. SPITKOVSKY, Singular integral operators with *PC* symbols on the spaces with general weights. – J. Funct. Anal. **105**(1992), 1, 129–143.

[174] E.M. STEIN, Harmonic Analysis: Real-Variable Methods, Orthogonality, and Oscillatory Integrals. – Princeton Univ. Press, Princeton, New Jersey, 1993.

[175] M.E. TAYLOR, Pseudodifferential Operators. – Princeton University Press, Princeton, New Jersey, 1981.

[176] M.E. TAYLOR, Noncommutative Microlocal Analysis, I. – Memoirs Am. Math. Soc. **313**(1984).

[177] M.E. TAYLOR, Noncommutative Harmonic Analysis. – Math. Surveys and Monographs **22**, Am. Math. Soc., Providence, Rhode Island, 1986.

[178] M.E. TAYLOR, Partial Differential Equations I. – Springer-Verlag, New York. Berlin, Heidelberg 1996.

[179] G. TESCHL, Jacobi Operators and Completely Integrable Nonlinear Lattices. – Mathematical Surveys and Monographs **72**, Am. Math. Soc., Providence, Rhode Island, 1999.

[180] L.N. TREFETHEN, Pseudospectra of matrices. – *In*: Numerical Analysis 1991 (Eds. D. F. GRIFFITHS, G. A. WATSON), Longman 1992, 234–266.

[181] F. TREVES, Introduction to Pseudodifferential and Fourier Integral Operators. – Plenum Press, New York 1980 (two volumes).

[182] H. TRIEBEL, Interpolation Theory, Function Spaces, Differential Operators. – North-Holland Publishing Company, Amsterdam, New York, Oxford 1978.

[183] N.L. VASILEVSKI, Convolution operators on standard CR-manifolds, II. Algebras of convolution operators on the Heisenberg group. – Integral Equations Oper. Theory **19**(1994), 3, 327–358.

[184] N.L. VASILEVSKI, R. TRUJILLO, Group convolution operators on standard CR-manifolds, I. Structural properties. – Integral Equations Oper. Theory **19**(1994), 1, 66–104.

[185] V.S. VLADIMIROV, Methods of the Theory of Functions of Many Complex Variables. – Nauka, Moskva 1964 (Russian, Engl. transl.: MIT Press, Cambridge, Mass., London 1966).

[186] H. WIDOM, Singular integral equations on L^p. – Trans. Am. Math. Soc. **97**(1960), 1, 131–160.

[187] C.H. WILCOX, Sound Propagation in Stratified Fluids. – Applied Mathematical Science vol. 50, Springer-Verlag, New York, Berlin, Heidelberg, Tokio 1984.

[188] G. YU, K-theoretic indices of Dirac type operators and the Roe algebra. – PhD thesis, SUNY at Stony Brook 1991.

[189] V.JU. ZAVADSKII, Grid methods for waveguides. – Nauka, Moskva 1986 (Russian).

Index

MIX
Papier aus verantwortungsvollen Quellen
Paper from responsible sources
FSC® C105338

If you have any concerns about our products,
you can contact us on
ProductSafety@springernature.com

In case Publisher is established outside the EU,
the EU authorized representative is:
Springer Nature Customer Service Center GmbH
Europaplatz 3, 69115 Heidelberg, Germany

Printed by Libri Plureos GmbH
in Hamburg, Germany

BIN TRAVERLER FORM

Cut By: JOHANNA A **Qty** 45 **Date** 10.12.16

Scanned By:________ **Qty**____ **Date**____

Scanned Batch ID's

________ ________ ________

Notes / Exceptions

____